Latest Research on Eye Tracking Applications

Latest Research on Eye Tracking Applications

Guest Editors

Vassilios Krassanakis
Erwan David
Olivier Le Meur

MDPI

Basel • Beijing • Wuhan • Barcelona • Belgrade • Novi Sad • Cluj • Manchester

Guest Editors
Vassilios Krassanakis
Department of Surveying and Geoinformatics Engineering
University of West Attica
Egaleo
Greece

Erwan David
LIUM
Le Mans University
Le Mans
France

Olivier Le Meur
InterDigital R&D
Cesson-Sévigné
France

Editorial Office
MDPI AG
Grosspeteranlage 5
4052 Basel, Switzerland

This is a reprint of the Special Issue, published open access by the journal *Applied Sciences* (ISSN 2076-3417), freely accessible at: https://www.mdpi.com/journal/applsci/special_issues/99E2K700SK.

For citation purposes, cite each article independently as indicated on the article page online and as indicated below:

Lastname, A.A.; Lastname, B.B. Article Title. *Journal Name* **Year**, *Volume Number*, Page Range.

ISBN 978-3-7258-8233-5 (Hbk)
ISBN 978-3-7258-8234-2 (PDF)
https://doi.org/10.3390/books978-3-7258-8234-2

Contents

About the Editors

Vassilios Krassanakis

Vassilios Krassanakis, Diploma of Rural and Surveying Engineering, Doctor of Engineering, National Technical University of Athens (NTUA), is an Associate Professor at the Department of Surveying and Geoinformatics Engineering, Faculty of Engineering, University of West Attica, specialized in Cartography and Visual Perception of Maps. Currently, he teaches courses mainly in the areas of cartography, visual perception of maps, web mapping, and programming in geoinformatics. He has also taught as a Visiting Professor, Invited Speaker or Instructor, at the National and Kapodistrian University of Athens, the Harokopio University, the Hellenic Military Geographical Service, and the Technical Chamber of Greece. Previously, he worked as a Postdoctoral Researcher at the Université de Nantes and the NTUA, as a Visiting Researcher at Ghent University, as an External Research Collaborator at the National Observatory of Athens and the Hellenic Centre for Marine, and as a Technical Researcher at the NTUA. He is a member of the Editorial Board of the *Journal of Eye Movement Research* and Guest Editor of Special Issues in *Geographies*, *ISPRS International Journal of Geo-Information*, and *Applied Sciences* as well as one Collection in *Discover Computing* and *Discover Imaging*. He has participated in several research projects funded by European, French, or Greek grants. He has authored numerous scientific publications, including journal papers, conference proceedings, book chapters, and an academic book. His current research activities are mainly focused on the areas of Map and Landscape Perception, Eye Movement Analysis and Gaze Behavior Modeling, Geovisualization and Spatial Data Infrastructures (SDIs).

Erwan David

Erwan David is an Associate Professor of Computer Science at Le Mans University, France, affiliated with the Computer Science Lab and the Technology Enhanced Learning team in Laval. His research sits at the intersection of computer science and cognitive sciences, focusing on gaze movement dynamics and modeling. He specializes in leveraging Virtual Reality (VR) to study visual behavior in complex, naturalistic situations, with a particular emphasis on eye tracking methods and visual attention in 3D environments. Dr. David holds master's degrees in Psychology (2014) and Cognitive Sciences (2016), and earned his PhD in Computer Science in 2019. He conducted postdoctoral research at Goethe University Frankfurt (Germany) from 2019 to 2023, where he expanded his expertise in VR-based eye tracking and cognitive modeling. He is active in the scientific community, regularly organizing workshops and delivering invited talks on eye tracking in extended reality. A central contribution to his work is the development of computational methods for processing gaze data in extended reality (3D), notably the Salient360! toolbox, which provides researchers with efficient tools for handling, visualizing, and analyzing gaze data. His publications address topics such as visuo-motor tendencies, visual search, saliency modeling, and peripheral vision, appearing in journals such as *Journal of Vision*, *Scientific Reports*, and *Computers & Graphics*.

Olivier Le Meur

Olivier Le Meur received a Ph.D. degree in computer vision from the University of Nantes, France, in 2005. He received accreditation to supervise research from the University of Rennes 1, France, in 2014. He has more than ten years of industrial experience in image and video processing, coding, and analysis with Thomson Video Network, Technicolor, and Xiaomi Research and Development. From 2009 to 2021, he was an Associate Professor with the University of Rennes 1, where he supervised the research team Percept, IRISA. In 2021, he went on leave from the University of Rennes 1 and is currently a Principal Scientist with InterDigital, leading the "Energy-Aware Media" research project and "Learned Image Compression" project. His main research interest deals with visual perception, visual attention modeling, image analysis, quality metrics, applied machine learning, and energy-aware image processing. He has authored or co-authored many papers on the aforementioned topics. He was an Editorial Board Member of the *Journal of Imaging* and an Associate Editor of *IET Image Processing*.

applied sciences MDPI

Editorial

Latest Research on Eye Tracking Applications

Vassilios Krassanakis [1,*], Erwan David [2] and Olivier Le Meur [3]

1 Department of Surveying & Geoinformatics Engineering, University of West Attica, Egaleo Park Campus, Ag. Spyridonos Str., Egaleo, 12243 Athens, Greece
2 LIUM, Le Mans Université, F-72000 Le Mans, France; erwan.david@univ-lemans.fr
3 Interdigital, 35510 Cesson-Sévigné, France; olivier.lemeur@interdigital.com
* Correspondence: krasvas@uniwa.gr

1. Introduction

Undoubtedly, eye tracking has undergone a fundamental transformation over the past decade. Previously, it functioned as a laboratory instrument requiring specialist hardware, controlled illumination, and chin rests. Today, it is a technology deployed across neuroscience, education, architecture, consumer research, clinical diagnostics, and human–computer interaction. This expansion has been driven by concurrent advances on two fronts. First, algorithmic methods for extracting reliable signals from raw gaze data have matured. Second, domain-specific knowledge about gaze behaviour in applied contexts has accumulated. The collected papers in this Special Issue reflect both trajectories. Together, they illustrate a field navigating the productive tension between methodological rigour and applied ambition. The Special Issue curated a wide array of eye tracking studies, which demonstrate the field's diversity. This breadth resolves into two explicit, coherent organising principles that are worth emphasising because they carry implications for how eye tracking research is evaluated and how it should be designed.

The first organising principle concerns the direction of inference. Eye tracking studies can be oriented "inward" by using gaze, pupil response, and oculomotor events as proxies for internal states. These include cognitive load, affective arousal, attentional strategy, or linguistic processing difficulty. In this case, the dependent variable of interest is a process within the observer. The visual stimulus serves as the instrument of elicitation. Alternatively, eye tracking studies can be oriented "outward". This involves deploying the same measurements to evaluate external objects, stimuli, environments, or communication designs. In this case, the observer's gaze is the instrument. The object of investigation is the stimulus itself. This distinction contrasts eye tracking as a window into cognition with eye tracking as a tool for design evaluation.

The second organising principle concerns methodological infrastructure. Gaze data must be correctly acquired, segmented, and represented before it can serve the inferential direction. Several contributions of the Special Issue address that foundational layer directly, such as the classification of eye movement events from time-series signals, the generation and augmentation of eye-image data for gaze estimation, and the characterisation of how low-level oculomotor parameters respond to stimulus properties. These works do not belong to a single applied domain, but instead form the substrate on which the applied work depends.

A third, less-structured but equally valid reading of the Issue worth mentioning is the reading organised by the domain of deployment. Across fifteen studies, eye tracking is applied to translation cognition, reading development, music and text interaction, urban landscape perception, architectural colour, public health communication, financial marketing, and food retail. This breadth is not incidental—it reflects the progressive normalisation of eye tracking as a general-purpose behavioural measurement tool, analogous in scope to questionnaire-based methods, but with access to temporal and spatial dimensions of attention that self-reporting cannot provide. The significance of that normalisation, and its methodological obligations, is taken up in the concluding remarks.

2. Methodological Foundations: Signal, Estimation, and Oculomotor Characterisation

The value of any eye tracking study depends on decisions made prior to domain interpretation. These include how raw positional data are segmented into meaningful events and how gaze direction is estimated from image input, as well as how oculomotor behaviour is characterised under controlled conditions. The accurate classification of eye movement events (fixations, saccades, smooth pursuits, etc.) has long been one of the field's core technical challenges. Although threshold-based algorithms are still widely used, their susceptibility to noise, sampling rate, and individual differences remains a significant concern that may compromise a study's conclusions.

Birawo and Kasprowski (contribution 1) address this problem with a hybrid deep-learning architecture combining two-dimensional convolutional neural networks and long short-term memory layers. Their key contribution is systematic: rather than proposing a single feature set, they evaluate all combinations of velocity, acceleration, jerk, and direction as inputs, identifying which combinations drive classification performance for each event type. The finding that velocity combined with direction (augmented by acceleration or jerk) substantially improves post-saccadic oscillation detection is important because PSOs are frequently conflated with short saccades or fixation onsets by classical algorithms. The residual difficulty with PSO detection simultaneously marks an open problem for future work. This paper establishes a benchmark and a feature selection rationale that will be important for future research.

While Birawo and Kasprowski (contribution 1) focus on temporal signal classification, Katrychuk and Komogortsev (contribution 2) examine gaze estimation from images. Appearance-based gaze estimation (i.e., inferring gaze direction from the visual appearance of the eye region rather than from corneal reflection geometry) requires large, diverse training sets that are expensive to collect. Thus, their proposed solution improves gaze estimation; the release of the accompanying dataset extends the article's value beyond its immediate experimental findings. This work is representative of a broader move in the field toward data-efficient methods that reduce the cost of entry for gaze estimation research, with implications for low-cost and webcam-based eye tracking implementations.

3. Eye Tracking as a Window into Cognition

The majority of the presented works in this Special Issue employ eye tracking, particularly pupillometry, to access internal cognitive and affective states that are not directly observable. These states include cognitive load, translation effort, reading difficulty, emotional reactivity, and attentional strategy during complex tasks. The corresponding works are characterised by an inferential approach: gaze is the measuring instrument, while the phenomenon of interest lies within the observer.

3.1. Cognitive State and Attentional Processing

Genova et al. (contribution 3) investigate how the statistical learning of visual regularities shapes oculomotor behaviour across repeated exposure. Participants differentiated visual pattern pairs whose order was repeated six times (over three sessions), while fix-

 https://doi.org/10.3390/app16104915

ation duration and saccade amplitude values, as well as saccade number and fixation count, were collected across sequential epochs. The stimulus manipulation varied the frequency of occurrence and the presence of common chunks, with bottom-up salience held approximately constant, so that observed shifts in eye movement parameters could be attributed to top-down attentional processes driven by learning. Results show a progressive increase in fixation duration and a decrease in saccade amplitude across experimental blocks. This reflects the canonical transition from ambient to focal processing as knowledge of stimulus structure accumulates. More unexpectedly, saccade number increased while fixation count decreased, a dissociation the authors interpret as narrowing attentional focus combined with reduced memory load once the relevant stimulus characteristics have been determined. This work tracks the temporal dynamics of statistical learning as it unfolds within the experiment rather than measuring its outcome in a subsequent test phase, using oculomotor parameters as continuous indices of an ongoing cognitive transition from uncertainty-driven exploration to knowledge-guided focal analysis.

Wu et al. (contribution 4) establish the empirical relationship between pupil dilation and subjective workload ratings (NASA-TLX) across two task types, including a visual tracking and a laparoscopic visuomotor task. Their results indicate a moderate positive correlation applicable to both types. The higher cognitive demands of the laparoscopic task, which requires coordinated visuomotor processing, produced larger pupil responses and higher NASA-TLX scores, while task repetition systematically reduced both. The moderate, rather than strong, correlation is informative. It quantifies the gap between conscious participant reports and autonomic system registrations, which is important for applied contexts. In such settings, workload monitoring must be both sensitive and interpretable. This work serves as a methodological anchor for pupillometry work, confirming the measure's validity while specifying its precision limits.

Gugerell et al. (contribution 5) extend the pupillometry framework to a more complex inferential territory with a simulated clinical interview. In this interview, cognitive load (single vs. dual task) and emotional content (neutral vs. negative interviewee responses) are manipulated independently. Their central question focuses on whether automatic cognitive load tracking via pupil size remains reliable when emotional processing varies. This has direct implications for any real-world deployment of pupillometry outside the laboratory. Their results confirm that pupil size reliably discriminates high from low cognitive load. However, this discrimination is attenuated under negative emotional content. The implication is that pupillometry-based load monitoring requires emotional context to be either controlled or co-measured, a constraint that significantly affects study design in clinical, educational, and operational settings. The clinical interview framework enables deployment in a professional environment, while the implemented approach prioritises immediate practical application over future research directions.

Sáiz-Manzanares et al. (contribution 6) extend the multimodal approach furthest. They integrate eye tracking with skin conductance and psychogalvanic response in a virtual laboratory setting. Then, they apply supervised and unsupervised machine learning to identify cognitive load components. Their sample includes health sciences and biomedical engineering students. This population and educational context place the paper at the intersection of cognitive measurement and learning environment design. The finding that viewing a task preview before execution increases the relative load while decreasing intrinsic load has direct instructional design implications. The proposal of an explanatory model with acceptable fit indices moves the contribution from descriptive toward structural. This ambition is not common in eye tracking studies. Its success here invites replication across different task domains. This work also functions as a methodological bridge since it

 https://doi.org/10.3390/app16104915

uses cognitive state measurement to derive conclusions about instructional design. Hence, this study participates in both the "inward" and "outward" frames simultaneously.

Liu et al. (contribution 7) deploy pupillometry in Rhesus macaques performing a face-fixation task. By recording pupil size responses to faces varying in species, emotional expression, viewing angle, and orientation, they demonstrate that macaques show differential autonomic responses. Specifically, these responses occur during exposure to carnivore faces and negative conspecific expressions. These findings align with threat detection predictions from comparative cognition. The negative correlation between pupil size and gaze duration on particular stimulus categories indicates that longer visual engagement does not directly reflect greater interest. Instead, approach and avoidance motivations produce distinct oculomotor signatures. Beyond its comparative findings, this work validates pupillometry as a tool for non-verbal and non-cooperative subjects. This methodological contribution is relevant to developmental, clinical, and animal cognition research, all of which require measures that circumvent verbal report.

3.2. Reading, Language Processing, and Educational Contexts

Eye tracking has a long history in reading research, where the temporal and spatial precision of eye tracking data provides access to incremental language processing in ways that reaction time or comprehension tests cannot.

Andreou and Gkantaki (contribution 8) examine whether background music modulates text comprehension and visual processing patterns in undergraduate readers. Specifically, they compare the effects of preferred music, non-preferred music, and café noise. Their finding that fixation measures remain stable across auditory conditions while reading performance declines under non-preferred music dissociates early visual processing from downstream comprehension. This is theoretically significant; it suggests that auditory distraction does not disrupt the oculomotor mechanics of reading but does impair the cognitive processing that reading supports. The implication for study design in reading research is that fixation-based measures alone may miss comprehension effects that are mediated at a post-perceptual stage. This work provides a clean template for follow-up studies with larger and more diverse populations, utilising a design that systematically varies auditory conditions while holding text difficulty constant.

Wang et al. (contribution 9) use eye tracking and keystroke logging to examine how novice Chinese–English translators distribute cognitive resources depending on translation direction. Their study of 24 postgraduate translation students finds that the two directions impose structurally different cognitive demands. In English-to-Chinese translation, source text reading is relatively light, while target text production, particularly under prolonged fixation, increases cognitive load. The pattern reverses in Chinese-to-English translation, where producing the English target is less demanding than simultaneous source-target processing. One of the study's findings is a systematic perception gap: most participants believed comprehending the English source was their primary cognitive burden during L2-to-L1 translation, while objective data showed they were expending far greater resources on Chinese target production than they recognised. The authors attribute this to the automaticity of native-language output, which conceals its actual processing cost. These findings have direct implications for translator training, particularly regarding how self-report methods are used to assess cognitive load.

Ceple et al. (contribution 10) examine whether oculomotor performance in non-reading tasks, such as saccades, smooth pursuit, and fixation stability, differs between children with and without reading difficulties. Their large-scale study involved 377 children between the ages of 6 and 13. Their results indicate no statistically significant difference in oculomotor performance between the groups. This challenges the body of dyslexia

 https://doi.org/10.3390/app16104915

literature that attributes reading difficulties to oculomotor deficits. The authors argue that previous positive findings may reflect participant selection artefacts rather than genuine group differences. The use of a validated reading assessment tool (Acadience Reading) and a school-based recruitment strategy strengthens the ecological validity of these findings, reporting no significant oculomotor differences. The contribution of this work does not lie in detecting an effect, but rather in carefully delineating its absence.

4. Eye Tracking as a Tool for Design Evaluation

The contributions reported in this section reverse the inferential direction. Rather than using gaze to diagnose cognitive states, they use it to evaluate the perceptual effectiveness of designed artefacts and environments, including buildings, urban landscapes, advertisements, packaging, and financial communications. The observer's attention is not the object of study; it is the measuring instrument applied to external objects. This reorientation has practical consequences: the relevant metrics shift from pupil diameter and cognitive load indices toward area-of-interest fixation proportions, time-to-first-fixation, and total dwell time on specific design elements.

4.1. Built Environment and Urban Landscape Perception

Wang and Fu (contribution 11) investigate how residents and tourists allocate visual attention to elements of dense urban landscapes viewed from elevated observation decks, combining eye tracking with psychometric scales measuring aesthetic response and place identity. Their findings indicate that residents and tourists show similar oculomotor patterns when viewing the same scenes, but divergent perceptual responses to specific elements. This separates the bottom-up attentional pull of salient features from the top-down interpretive processes shaped by familiarity and belonging. The relationship between the aesthetic evaluation of specific elements—such as the colour vividness of water, mountains, and landmarks—and place identity scores has direct implications for heritage preservation and urban planning. It identifies which visual attributes carry the heaviest psychological weight for different user groups, providing information that cannot be recovered through behavioural observation or questionnaires alone.

Wang et al. (contribution 12) apply eye tracking towards the evaluation of colour combinations on building facades. Using digitally modelled residential buildings in Shanghai's Yangpu District with systematically varied hue, brightness, and saturation combinations, they cross-validate physiological indicators (fixation duration, count, and area-of-interest metrics) against subjective quality ratings. Their principal findings provide actionable guidance for architectural colour specification. Specifically, they demonstrate that brightness differences drive visual quality more strongly than hue differences. However, they also find that large hue variations can achieve high evaluations when paired with appropriate brightness and saturation contrasts. The mathematical support for colour combination outcomes represents a level of quantitative rigour unusual in architectural perception research and exemplifies how eye tracking can move design evaluation from preference-based intuition toward empirically constrained decision-making.

4.2. Visual Communication, Advertising, and Consumer Behaviour

Milošević et al. (contribution 13) examine whether monochromatic illustrations increase visual attention to public health advertisements targeting young people, comparing text-only and illustrated versions of alcohol- and tobacco-prevention messages. Their consistent and direct results show that illustrated advertisements attract more fixations, longer dwell times, and higher noticeability than text-only equivalents, regardless of the topic. The use of eye tracking in this study functions as a validity check on design assumptions

https://doi.org/10.3390/app16104915

that practitioners often make without empirical grounding. Furthermore, this finding adds to the growing body of evidence that visual embellishment in health communication serves a purpose beyond mere aesthetics, as it fundamentally determines whether a message is processed at all.

Pluskota et al. (contribution 14) deploy eye tracking, individual in-depth interviews, and text mining towards the examination of how young consumers perceive promotional and informational messages from cooperative banks. The study's primary contribution lies in its multi-method design. By triangulating oculomotor data with verbal and textual analysis, the authors can distinguish what attracts attention from what is consciously processed and retained, and further separate both from what participants are able or willing to articulate. Their results indicate that excessive text and visual complexity provoke negative reactions and disengagement, whereas infographics and colour are strongly linked to brand recognition.

Malheiros et al. (contribution 15) shift the evaluative context from communication to retail product choice, examining how visual attention to quality cues in fresh and cooked beef influences purchase probability among Brazilian consumers. The study's integration of eye tracking with a discrete choice model, in which nine hypothetical products vary across multiple attributes simultaneously, allows the authors to connect fixation metrics directly to choice outcomes instead of treating attention as a proxy for preference. For fresh beef, colour dominates both attention and choice; for cooked beef, tenderness, price, and flavour cues drive fixation patterns and purchase likelihood. The differential attention profiles across product types demonstrate that visual salience is category-specific and cannot be generalised from one food context to another, a finding with implications for product labelling strategy and in-store communication design.

5. Concluding Remarks and Future Perspectives

The contributions of the Special Issue collectively show eye tracking as a genuinely interdisciplinary technology, not in the sense that the same tool is applied uncritically across different fields, but in the more demanding sense that its deployment now requires field-specific knowledge of what gaze data can and cannot reveal in a given context.

Several cross-cutting observations warrant emphasis, starting with the contributions in Section 2, which make clear that the methodological substrate of the field remains an active research front. Event detection is not a solved problem, as evidenced by a proposed CNN-LSTM approach that improves on classical algorithms but leaves PSO classification with limited accuracy. However, as the field moves toward image-based estimation and complex machine learning models, the need for "explainable AI" in eye tracking becomes paramount to ensure that automated insights remain theoretically grounded. While specific augmentation strategies address data scarcity, the gap between laboratory-grade and head-mounted or webcam-grade data remains a barrier to broad ecological deployment in naturalistic settings.

In Section 3, the contributions reveal a consistent and underappreciated finding that self-reports diverge from physiological measurement in systematic, context-dependent ways. Researchers show that translators' introspective accounts of their cognitive effort are directionally biassed by task structure, while another study only finds a moderate correlation between NASA-TLX scores and pupil dilation. Equally significant is the role of eye tracking as a vital bridge to understanding non-verbal or non-cooperative subjects, such as infants, macaques, or clinical populations with reading difficulties. By eliminating the requirement for verbal report, these studies demonstrate that oculomotor data can provide access to cognitive processes that would otherwise remain hidden.

 https://doi.org/10.3390/app16104915

Finally, the contributions in Section 4 make an implicit case for eye tracking as a standard tool in design evaluation, analogous to the role that usability testing plays in interface design. Studies of building colour, urban landscape, health advertising, financial communication, and food retail all demonstrate that eye tracking data can discriminate between design alternatives in ways that preference ratings alone cannot. What they also demonstrate is that the inferential chain from gaze to preference to behaviour is not automatic, as attention is a necessary but not sufficient condition for positive evaluation or purchase, and the contributions that integrate eye tracking with choice models or psychometric scales extract, correspondingly, richer conclusions than those relying on fixation metrics alone.

The domain-level reading of the Special Issue, covering eye tracking applied across translation cognition, dyslexia, music and reading, urban perception, architectural colour, public health advertising, financial communication, food retail, and comparative animal cognition, carries a final implication. The normalisation of eye tracking as a general-purpose behavioural measure creates methodological obligations that the field has not yet fully codified. Reporting standards, minimum sample sizes, and the conditions under which particular metrics are valid proxies for particular constructs all require domain-specific guidance that does not yet exist in consolidated form. Several works in this Special Issue contribute to that codification implicitly through their methodological transparency, yet explicit meta-scientific effort in this direction would accelerate the field's capacity to accumulate knowledge across studies. That remains, perhaps, the most productive frontier that this Special Issue leaves open.

Conflicts of Interest: The authors declare no conflict of interest.

List of Contributions:

1. Birawo, B.A.; Kasprowski, P. Performance Analysis of Eye Movement Event Detection Neural Network Models with Different Feature Combinations. *Appl. Sci.* **2025**, *15*, 6087. https://doi.org/10.3390/app15116087.
2. Katrychuk, D.; Komogortsev, O.V. Appearance-Based Gaze Estimation as a Benchmark for Eye Image Data Generation Methods. *Appl. Sci.* **2024**, *14*, 9586. https://doi.org/10.3390/app14209586.
3. Genova, B.; Bocheva, N.; Hristov, I. Effect of Stimulus Regularities on Eye Movement Characteristics. *Appl. Sci.* **2024**, *14*, 10055. https://doi.org/10.3390/app142110055.
4. Wu, Y.; Zhang, Y.; Zheng, B. Workload Assessment of Operators: Correlation Between NASA-TLX and Pupillary Responses. *Appl. Sci.* **2024**, *14*, 11975. https://doi.org/10.3390/app142411975.
5. Gugerell, D.; Gollan, B.; Stolte, M.; Ansorge, U. Studying Pupil-Size Changes as a Function of Task Demands and Emotional Content in a Clinical Interview Situation. *Appl. Sci.* **2024**, *14*, 11714. https://doi.org/10.3390/app142411714.
6. Sáiz-Manzanares, M.C.; Marticorena-Sánchez, R.; Sáez-García, J.; González-Díez, I. Analysing Virtual Labs Through Integrated Multi-Channel Eye-Tracking Technology: A Proposal for an Explanatory Fit Model. *Appl. Sci.* **2024**, *14*, 9831. https://doi.org/10.3390/app14219831.
7. Liu, X.; Zhang, Z.; Dai, J. Evaluating Pupillometry as a Tool for Assessing Facial and Emotional Processing in Nonhuman Primates. *Appl. Sci.* **2025**, *15*, 3022. https://doi.org/10.3390/app15063022.
8. Andreou, G.; Gkantaki, M. An Eye-Tracking Study on Text Comprehension While Listening to Music: Preliminary Results. *Appl. Sci.* **2025**, *15*, 3939. https://doi.org/10.3390/app15073939.
9. Wang, Y.; Li, S.; Rasmussen, Y.Z. Translators' Allocation of Cognitive Resources in Two Translation Directions: A Study Using Eye-Tracking and Keystroke Logging. *Appl. Sci.* **2025**, *15*, 4401. https://doi.org/10.3390/app15084401.
10. Ceple, I.; Krauze, L.; Serpa, E.; Svede, A.; Goliskina, V.; Vasiljeva, S.; Kassaliete, E.; Ganebnaya, A.; Volberga, L.; Truksa, R.; et al. Eye Movement Parameters in Children with Reading Difficulties. *Appl. Sci.* **2025**, *15*, 954. https://doi.org/10.3390/app15020954.

 https://doi.org/10.3390/app16104915

11. Wang, P.; Fu, H. The Influence of Different Visual Elements of High-Density Urban Observation Decks on the Visual Behavior and Place Identity of Tourists and Residents. *Appl. Sci.* **2025**, *15*, 3875. https://doi.org/10.3390/app15073875.
12. Wang, Z.; Shen, M.; Huang, Y. Combining Eye-Tracking Technology and Subjective Evaluation to Determine Building Facade Color Combinations and Visual Quality. *Appl. Sci.* **2024**, *14*, 8227. https://doi.org/10.3390/app14188227.
13. Milošević, M.; Kovačević, D.; Brozović, M. The Influence of Monochromatic Illustrations on the Attention to Public Health Messages: An Eye-Tracking Study. *Appl. Sci.* **2024**, *14*, 6003. https://doi.org/10.3390/app14146003.
14. Pluskota, P.; Słupińska, K.; Wawrzyniak, A.; Wąsikowska, B. The Design of Informational and Promotional Messages by Cooperative Banks and Their Perception Among Young Consumers—An Eye-Tracking Analysis Versus Conscious Identification Based on Empirical Research. *Appl. Sci.* **2025**, *15*, 9635. https://doi.org/10.3390/app15179635.
15. Malheiros, B.A.; Spers, E.E.; Contreras Castillo, C.J.; Aroeira, C.N.; De Lima, L.M. The Role of Visual Attention and Quality Cues in Consumer Purchase Decisions for Fresh and Cooked Beef: An Eye-Tracking Study. *Appl. Sci.* **2025**, *15*, 7360. https://doi.org/10.3390/app15137360.

applied sciences

MDPI

Article

Performance Analysis of Eye Movement Event Detection Neural Network Models with Different Feature Combinations

Birtukan Adamu Birawo and Pawel Kasprowski *

Department of Applied Informatics, Silesian University of Technology, Akademicka 16, 44-100 Gliwice, Poland; birtukan.birawo@polsl.pl
* Correspondence: pawel.kasprowski@polsl.pl

Abstract: Event detection is the most important element of eye movement analysis. Deep learning approaches have recently demonstrated superior performance across various fields, so researchers have also used them to identify eye movement events. In this study, a combination of two-dimensional convolutional neural networks (2D-CNN) and long short-term memory (LSTM) layers is proposed to simultaneously classify input data into fixations, saccades, post-saccadic oscillations (PSOs), and smooth pursuits (SPs). The first step involves calculating features (i.e., velocity, acceleration, jerk, and direction) from positional points. Various combinations of these features have been used as input to the networks. The performance of the proposed method was evaluated across all feature combinations and compared to state-of-the-art feature sets. Combining velocity and direction with acceleration and/or jerk demonstrated significant performance improvement compared to other feature combinations. The results show that the proposed method, using a combination of velocity and direction with acceleration and/or jerk, improves PSO identification performance, which has been difficult to distinguish from short saccades, fixations, and SPs using classic algorithms. Finally, heuristic event measures were applied, and performance was compared across different feature combinations. The results indicate that the model combining velocity, acceleration, jerk, and direction achieved the highest accuracy and most closely matched the ground truth. It correctly classified 82% of fixations, 90% of saccades, and 88% of smooth pursuits. However, the PSO detection rate was only 73%, highlighting the need for further research.

Keywords: event detection; eye movement types; deep learning; feature extraction

1. Introduction

Eye tracking is highly interdisciplinary and can be used in numerous fields, including cognitive science, psychology, neurology, engineering, medicine, and marketing [1,2]. The application of eye tracking in human–computer interaction allows disabled people to interact with a computer through gaze [3]. Eye monitoring can also be used to track car drivers [4] or pilots [5].

Extracting events from eye movement data is the most important step for automated visual behavior analysis and for many applications [6]. The movement of the eye is typically divided into several types of events. The most important is fixation, a low-speed movement that occurs when the person focuses their eyes on an object. The purpose of fixation is to stabilize the object in the fovea for clear vision [7]. Saccade is a high-speed eye movement from one fixation to another [8]. A smooth pursuit (SP) event is performed only when the eyes follow a slowly moving object [9,10]. Post-saccadic oscillation (PSO) events are rapid oscillatory movements that may occur immediately after a saccade. They can be described as instabilities that appear at the end of a saccade and are characterized by a slight wobbling movement that leads to fixation after a saccade [10,11].

The problem of automatic detection of eye movement events has been studied for many years. However, it is associated with many challenges. For instance, raw eye-tracking data frequently consist of disturbances and noise, which may make it difficult for detection methods to identify events correctly. Moreover, differentiating between PSOs, SPs, fixations, and saccades, as well as finding precise onsets and offsets of events, may be difficult even for manual coders, and there is evidence that there are differences in event classification even between experienced coders [12,13].

Today, event detection is performed by processing raw eye-tracking data with specialized detection algorithms. Threshold-based and fully automated event detection algorithms have been used in previous works. For a long time, two broad classes of threshold-based algorithms have been used to detect eye movement events. The first class includes dispersion threshold-based algorithms that detect fixations [14]. The most popular dispersion-based algorithm is the I-DT algorithm by Salvucci and Goldberg [15]. These algorithms detect events, assuming that, for the fixation, the raw recorded data must fit in a spatial box for a specified minimum time. The second category is the velocity-based algorithms that focus on detecting saccades. The most popular velocity-based algorithm is the I-VT algorithm [16,17]. These algorithms classify eye movement data by calculating velocities and comparing them with a predefined threshold. Although the algorithms above differentiate between fixations and saccades only, Komogortsev et al. [18] proposed the first automated event detection method that may also be used to classify smooth pursuits. Larsson et al. [19] proposed another threshold-based approach for identifying fixations and smooth pursuits during intersaccadic intervals. The approach performed the classifications in three stages, as follows: a preliminary segmentation evaluating the characteristics of each such segment and reorganizing the preliminary segments into fixations and smooth pursuit events. This approach was proposed to address the misclassification between fixations and smooth pursuits. However, like all other threshold-based approaches, this approach also requires the end users to set threshold parameters manually, and the three-stage identification process makes the approach complex and time-consuming.

As event detection involves classifying each raw gaze sample into a separate class (event), applying machine learning algorithms to this task seems straightforward. Several attempts have been published using machine learning and, specifically, deep learning techniques to perform event detection [20–22]. Details are provided in the Section 2.

However, currently available event detection methods are unable to detect fixations, saccades, PSOs, and smooth pursuits simultaneously. In this study, we proposed a combination of 2D-CNN with an LSTM deep learning model that simultaneously identifies fixations, saccades, PSOs, and smooth pursuits. We aim to find the combination of features that gives the best results for different events. We compared the proposed model for different sets of features extracted from the raw data and with the baseline method, which is the most similar to our approach, presented in [21]. We made this comparison on two levels, i.e., a sample level and an event level. Finally, we evaluated event measures for the proposed methods and compared them to the results of manual classification. The event-measuring metrics were the number of events and the mean duration of events. We tested the performance of our methods on the publicly available dataset described in the Section 3.

This work has significant practical implications for eye-tracking applications. For example, in human–computer interaction (HCI), accurate eye movement classification can improve gaze-based interfaces, enabling smoother and more intuitive user experiences. In medical diagnosis, the precise identification of eye movement events can help in the early detection of neurological disorders such as Parkinson's disease or multiple sclerosis, where abnormalities in eye movements serve as critical diagnostic markers. By improving the classification performance for multiple eye movement events simultaneously, this model could provide more reliable and detailed eye-tracking analytics, facilitating advancements in both HCI systems and clinical diagnostics.

The main contributions of the paper are as follows:

- Testing the performance of the models for different feature sets consisting of velocity, acceleration, jerk, and direction.
- Creating a model that simultaneously classifies four types of events, i.e., fixations, saccades, smooth pursuits, and post-saccadic oscillations.
- Comparison of the usability of different features in classifying specific events.
- Proposing the use of event statistics as a simple tool to evaluate the quality of a model.

The remainder of this paper is structured as follows: The related work is reviewed in Section 2. Section 3 describes the dataset used in this study, including its characteristics and potential biases. The proposed method is discussed in Section 4. Results and discussion are presented in Section 5. Section 6 evaluates the proposed methods using event measures. Finally, the paper concludes in Section 7, where future directions are outlined.

2. Related Works

Event detection can be performed manually by human coders or algorithms. In manual event detection, one or more human experts classify raw eye-tracking data into different event types based on subjective threshold values. Manual event classification is still treated as a "golden standard" [12]. However, manual classification is ineffective at classifying eye movement events. Firstly, it is time-consuming, and secondly, different coders might also use distinctive subjective selection regulations that give different results [23].

Threshold-based methods were the first automated eye movement event classification algorithms and are still frequently used [15]. For example, Salvucci and Goldberg evaluated five different threshold-based event detection algorithms that were introduced by different authors. The evaluation was based on spatial and temporal criteria. The algorithms were namely, I-VT, I-DT [15], I-HMM [24], I-AOI, and I-MST [15]. The I-DT is a simple eye movement event detection algorithm that distinguishes fixation and saccade points based on the dispersion of subsequent sample coordinates. The algorithm classifies gaze samples as elements of fixation when they are located within a spatially limited area (for example, 0.5°) for the minimum allowed fixation duration [25]. It follows the assumption that fixation points generally occur close to each other. The saccades are then implicitly classified as everything else [15]. However, dispersion threshold methods poorly detect fixations and saccades when the signal is noisy [26]. Therefore, choosing the optimum threshold greatly impacts the classification performance and can lead to biased results and misclassifications.

The I-VT event detection algorithm is another threshold-based algorithm and the foundation for automated, objective standard event detection. Many studies have adopted this approach [27,28]. It assumes that saccadic eye movements are characterized by high velocity, and fixational movements by low velocity [15]. However, just like I-DT, the classic I-VT method is designed to classify eye-tracking input data into fixations and saccades only. The other event types, like smooth pursuits, post-saccadic oscillations, and noises, are not considered. Finding the optimum threshold value is also challenging.

One of the main problems with threshold-based event detection methods discussed in this section is that they can only identify fixation and saccade movements. In order to address this problem, Komogortsev and Karpov [18] proposed the first ternary automated event detection methods. The methods classify raw eye-tracking data into fixations, saccades, and smooth pursuits. The methods are I-VVT, I-VDT, and I-VMP. The I-VVT algorithm identifies fixations, saccades, and smooth pursuits (SPs) by first applying the I-VT algorithm to classify fixations and saccades using the existing threshold, then using one more threshold to distinguish SPs from fixations. The I-VMP algorithm distinguishes smooth pursuits from fixations using the movement direction. The I-VDT algorithm integrates both I-VT and I-DT to classify fixations, saccades, and smooth pursuits. As in I-VVT, I-VDT first applies the velocity threshold to classify saccades and fixations, and then the dispersion threshold is applied to distinguish fixations and smooth pursuits. The drawback of these methods is that they are still threshold-based, so the user must manually set the optimum threshold values. In addition, the identification process is done in two steps and requires two optimum threshold values.

Finding the optimum threshold value in threshold-based event detection methods is challenging, as threshold values vary depending on the type of task, data quality, and the user. In order to address these issues, Ref. [29] proposed an automated velocity threshold data-driven event classification method. The algorithm can find the threshold adaptively and avoid the influence of noise by using adaptive VT values, considering different levels of noise occurrence. It also detects glissades as separate event types. However, glissades are detected based on duration only, which may lead to short saccades being classified as glissades or long glissades being classified as saccades. The algorithm can identify fixations, saccades, and glissades only. It does not consider smooth pursuits.

Threshold-based event detection methods require several threshold parameters, and finding the optimal values can be challenging. Machine learning and deep learning approaches can address these problems by applying end-to-end, fully automated event detection without human intervention or manually setting parameters like thresholds. These approaches are becoming popular in eye movement event detection. For instance, Ref. [30] proposed a fully automated eye movement event classification using a random forest classifier. The method classifies the input data into fixations, saccades, and PSOs. According to the authors, the machine learning approach outperforms other event detection methods. However, this method does not consider the important event type (smooth pursuit). The other fully automated event detection method was proposed in [22]. It is based on a deep convolutional neural network that predicts probabilities for each sample to belong to a fixation, saccade, or smooth pursuit based on a sequence of gaze samples. The approach uses the signal's frequency to classify the data into event types. The method classifies fixations, saccades, and smooth pursuits only. The comparison is done with velocity and dispersion threshold algorithms only.

Ref. [21] showed another example of applying deep learning to event detection. The proposed network combines a 1D convolutional network with a BLSTM layer. It classifies raw eye movement data into fixations, saccades, and smooth pursuits. Individual feature sets used for the model are raw xy coordinates, speed, acceleration, and direction. According to the authors, the combination of direction and speed provided good results. Acceleration as an additional feature did not improve the average event classification, probably because it cannot be used to distinguish smooth pursuits from fixations. The algorithm identifies fixations, saccades, and smooth pursuits. It does not consider PSO event types. Another deep learning approach in [31] is proposed for online event classification for fixations, saccades, and smooth pursuits.

In summary, there have been many attempts to utilize machine learning and deep learning for eye movement event detection. However, to the authors' knowledge, none of them attempted to combine the four eye movement features used in this work or to develop a model that simultaneously classifies eye movement data into four event types, i.e., fixations, saccades, smooth pursuits, and post-saccadic oscillations.

3. Dataset

To train and test the model, a manually annotated eye-tracking dataset was used, which had also been used in many previous studies [9,21,22,30]. The dataset was recorded with a Hi-Speed 1250 eye tracker from SensoMotoric Instruments at 500 Hz. The data were labeled by two manual raters, Marcus Nyström (MN) and Richard Andersson (RA), and annotated into six event types, i.e., fixations, saccades, PSOs, smooth pursuits, blinks, and undefined.

The subjects were presented with static images, video clips, and simple moving dot stimuli. The evaluation focused on the video domain with event classes including fixations, saccades, PSOs, and smooth pursuits, as the study was centered on these four types. Image viewing and moving dot data were not included because image viewing datasets lacked smooth pursuit events, and the moving dot domain did not contain a time-stamp feature, which was essential to calculate features such as velocity, acceleration, and jerk, which are key features in building the model. Six files with video stimuli were selected for the experiment, containing a total of 16,952 gaze samples. Among these, 36.4% of the samples were labeled as fixations, 6.8% as saccades, 3.1% as PSOs, and 53.6% as smooth pursuits. The statistics of the files are shown in Table 1.

While this dataset has provided a robust foundation for evaluating the model, potential biases are acknowledged due to the focus on video stimuli and the exclusion of other stimulus types, such as static images and text reading. Eye movement patterns can vary significantly across different tasks and stimulus domains. This limitation may affect the generalizability of our approach to tasks outside the video domain. Additionally, although the dataset was annotated by expert raters, potential inter-rater variability could introduce annotation bias. Future work will explore cross-domain evaluations to further assess the robustness of the proposed method.

Table 1. The description of files used to evaluate the models.

Files	Fixations Points	Saccades Points	PSOs Points	SPs Points
File 1	625	65	101	819
File 2	962	339	134	1611
File 3	932	59	99	520
File 4	1401	390	107	2748
File 5	1158	156	38	1468
File 6	1100	147	50	1523

4. Proposed Method

In this study, a combination of 2D-CNN and LSTM was proposed to classify raw eye movement data into eye movement events. The CNN part filters the input data, and the LSTM part classifies the pre-processed data into event types. A similar approach to Startsev et al. [21] was used. Before feeding the input to the network, velocity, acceleration, jerk, and direction were calculated from the raw positional coordinate points (x, y). The model classified each input data sample into one of four classes: fixations, saccades, PSOs, and smooth pursuits. Heuristic measures were then applied to the classifier's output, such as merging nearby fixation points and determining the duration and number of events to produce the final events.

Unlike most event detection methods, the proposed approach simultaneously classifies fixation, saccade, PSO, and SP. Identifying PSOs in the presence of fixations, saccades, and smooth pursuits is particularly challenging due to their similar movement behaviors. Because of this, PSOs are frequently misclassified when only velocity is used. For example, PSOs may resemble fixations because of the small oscillatory movements in fixations. They can also be misclassified as saccades due to their similar velocity behaviors. In order to address these issues, the model incorporates not only velocity but also higher-order derivatives of velocity as features. In addition, the impacts of different feature combinations on event detection accuracy were evaluated.

4.1. Network Architecture

The proposed model is a combination of CNN and LSTM approaches. Such an architecture is frequently referred to as a long-term recurrent convolutional network (LRCN) [32]. The model takes a sequence of samples, and then subsequent convolutional layers are used to extract features that are important for further classification. While the original work [32] worked with images, our model works with eye movement data samples containing preprocessed attributes such as velocity or acceleration. The features extracted from the input data by the convolutional layers are then passed as a sequence to the recurrent layers. Recurrent layers function as a memory-based network. Their subsequent outputs are generated based on the current input, and all inputs proceed in a sequence. In contrast to [21], only the last output of the network is used. Thanks to the combination of convolutional and recurrent layers, the network is able to utilize both spatial and temporal features of the signal.

Therefore, the network takes a stream of samples as input. Each sample is a set of different eye movement features, such as direction, velocity, and their higher-order derivatives. The stream of samples is analyzed in windows of 50 samples (which gives a time span of 100 ms in our dataset). To obtain a prediction for each sample, the window moves over the sequence one by one, and the sample in the middle of the window is classified according to the neighboring samples. The network comprises different layers, precisely three convolutional layers with 32, 16, and 16 filters and a kernel size of 5. After the convolutional layers, a permute function reshapes the CONV2D output into an LSTM input shape, two LSTM layers with 32 and 16 units, batch normalization layers, and a dense output layer. A sequence of samples with a shape of $50xN$, where N is the number of features for each feature's combination, serves as the input to the network. The network architecture is shown in Figure 1.

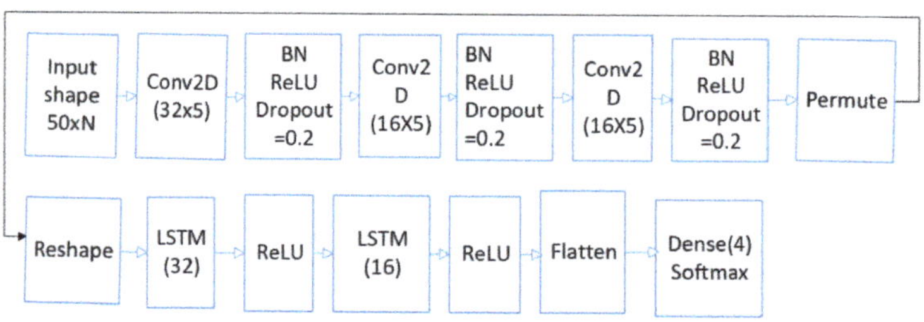

Figure 1. Architecture of the proposed network.

The model was evaluated using a leave-one-file-out cross-validation approach. In each fold, one file was reserved for testing, while the remaining files were used for training. Sequences of 50 samples were created from the raw data to form input windows for the model. The model was compiled using the RMSprop optimizer with the default learning rate of 0.001, categorical cross-entropy as the loss function, and categorical accuracy as the evaluation metric. During training, the model was trained for 25 epochs with a batch size of 50. For each training iteration, the labels were binarized using LabelBinarizer to encode the categorical labels into a one-hot format.

4.2. Feature Extraction

The main goal was to determine which features and feature combinations would correctly classify fixations, saccades, PSOs, and smooth pursuit events simultaneously.

The features used for classification were velocity, acceleration, direction, and jerk. It was hypothesized that velocity and direction alone were insufficient to distinguish PSOs due to the overlapping nature of different event types. For example, velocity and direction were used in [21], and it worked well, but only for fixation, saccade, and smooth pursuit detection. The velocity feature can distinguish saccades from fixations and smooth pursuits because fixations and smooth pursuits are low-velocity, and saccades are high-velocity movement types. Additionally, direction can distinguish smooth pursuits from fixations due to the uniform distribution of direction in fixations but not in smooth pursuits. However, using only velocity parameters to distinguish fixations from smooth pursuits will not work correctly, as both fixations and SPs are low-velocity movement types. Saccades and PSOs will also be misclassified, as both are high-speed movement types.

Because different features were important in classifying different events, tests were performed to verify which combination of features gave the best event detection results. Velocity, acceleration, and jerk were calculated from the raw x, y, coordinates, as shown in Figure 2, while the direction was calculated as shown in Equation (1).

$$Direction_i = atan2(x_{i+1} - x_i, y_{i+1} - y_i) \tag{1}$$

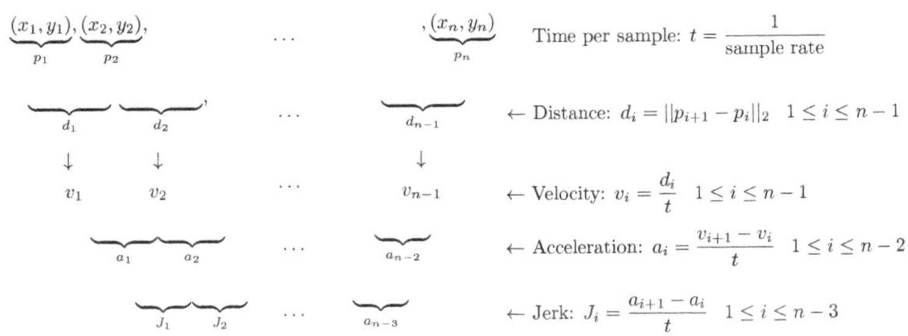

Figure 2. Calculation of features.

4.3. Performance Evaluation

Currently used fully automated event detection algorithms cannot simultaneously detect all four events (i.e., fixations, saccades, PSOs, and smooth pursuits). Additionally, there is no standard performance metric for eye movement event detectors [33]. So, for a fair comparison, it is recommended that they be compared using the same metrics (either against other algorithms or against the same algorithm under different conditions), using an applicable evaluation method [33].

Threshold-based methods were not considered for comparison, as they are statically designed for specific event types based on the threshold value. For example, I-DT and I-VT are designed to detect fixation and saccades only. However, the dataset used for evaluation is annotated with four different event types. Therefore, comparing our method with threshold-based methods would be unfair.

The method was compared with the one most similar to ours, as presented in [21]. This method also used a neural network built from a combination of CNN and LSTM layers, but they used 1D-CNN and BLSTM layers. Another difference was that their model output a sequence of event points, while ours classified only one point at a time. In addition, their method was designed to identify only fixations, saccades, and smooth pursuits. It did not consider PSOs, which were tricky to distinguish from fixations, saccades, and smooth pursuits due to their behavioral overlap with these event types.

To compare their model with ours, the model was reimplemented, and the final layer was changed to return only one value, as in ours. In addition, a 50-sample window was used as input to the model. The model was then trained on the same data as used in our experiments.

However, the primary purpose was to compare the same model for different feature sets. This comparison was made on the following two levels: sample-level and event-level. For the sample level, the output of the trained model was compared with the ground truth, and classic measures such as the F1 score and Cohen's kappa were calculated for each sample. Furthermore, the corresponding confusion matrices for each combination of features—i.e., velocity (V), acceleration (A), direction (D), and jerk (J)—were analyzed.

Classification of single-gaze data points is just the first step in event detection [9]. The next step is to join identically classified neighboring points to form eye movement events. The process is visualized in Figure 3. There are many possible measures to compare such results. However, there is no commonly accepted measure for this event-level comparison [33]. Therefore, we compared general event statistics, i.e., the number of events and their mean duration. We believe that such information gives a reliable way to compare different results.

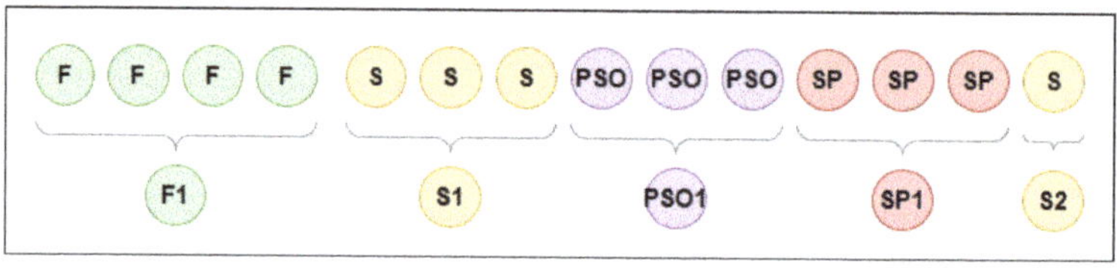

Figure 3. Merging neighboring events. F, S, PSO, and SP stand for fixation, saccade, post-saccadic oscillation, and smooth pursuit, respectively.

The train and test dataset split was done using the leave-one-file-out (LOFO) cross-validation method to avoid biased and overestimated model performance using a single test set, because we repeatedly fit a model to a dataset containing n-1 files. In our case, six files from the dataset described in Section 3 were used.

5. Results and Discussion

In this section, the results and discussions on the impacts of various feature combinations on the classification performances of different eye movement events are presented. Direction plays a crucial role in identifying smooth pursuit (SP), as it captures the change in the direction of eye movement, which is key to distinguishing SP from other event types. Velocity is essential for distinguishing saccades from fixations and smooth pursuits, as it provides a clear differentiation between high-speed and low-speed movements. Additionally, acceleration and jerk contribute significantly to detecting post-saccadic oscillations (PSOs), as they help to separate PSOs from other movement types based on their rapid, oscillatory nature. Our analysis shows that combinations of velocity, acceleration, jerk, and direction (such as VAJD, VAD, and VJD) improve classification performance, particularly for PSO detection. We also found that while the absence of jerk in the VAD combination did not significantly affect performance, the lack of velocity in the AJD combination severely impacted the model's ability to correctly classify events.

The performances of different models are presented in Tables 2 and 3. The models using different feature combinations were compared with the ground truth (human coders) as a reference. The baseline model with velocity and direction features, originally used in [21], was also evaluated for fixation, saccades, and smooth pursuit classification. Precision and F1 score were calculated for the proposed models. In order to measure the overall agreement between the manual coders and the proposed methods, Cohen's kappa was calculated between each manual coder and the proposed methods. Moreover, confusion matrix analysis was used to analyze the performance of the models.

The experiment started with the combinations of VD and AD feature combinations, which were used in the baseline method [21]. It was discovered that the AD feature combination exhibited very low event detection performance due to its inability to distinguish fixations from smooth pursuits and PSOs from all other event types. A combination of velocity and direction (VD) can identify saccades and PSOs, as velocity can identify saccades from fixations and smooth pursuits. However, the VD combination performs poorly in identifying PSOs due to their speed-related behavior being similar to that of saccades.

The results show that using acceleration or jerk in addition to velocity and direction helps the model to distinguish PSOs from other event types. This implies that the model's classification performance with the VAJD, VAD, and VJD feature combinations is superior to that of the AJD, VD, and AD combinations. Not surprisingly, the AJD feature combination cannot classify all event types, especially PSOs. This implies that combining velocity and direction with acceleration and/or jerk shows significant performance improvement for all event types, including PSO.

Table 2. Comparison of F1 score (F1) for each event type, mean F1 score, and mean Cohen's kappa (K).

	F1 Score					
Algorithms	**Fixations**	**Saccades**	**PSOs**	**SPs**	**Mean F1**	**Mean K**
VAJD	0.792	0.918	0.815	0.80	0.83	0.733
VAD	0.803	0.914	0.818	0.803	0.834	0.735
VJD	0.798	0.908	0.795	0.805	0.826	0.723
VD	0.772	0.914	0.797	0.794	0.819	0.708
1D-CNN-BLSTM [21] *	0.667	0.72	-	0.663	0.7	0.50
1D-CNN-BLSTM(VD) **	0.52	0.79	0.61	0.61	0.63	0.48
AJD	0.644	0.679	0.468	0.641	0.608	0.526
AD	0.37	0.543	0.31	0.464	0.42	0.235

* The 1D-CNN-BLSTM row shows the results reported in [21], classifying the three event types—fixation, saccade, and SP—using the VD feature set on the dataset used in this paper. ** The 1D-CNN-BLSTM (VD) refers to our changed and re-implemented version of their model (classifying four events).

Table 3. Precision of each event type and feature combination. We compare the performance of the proposed model with different feature combinations.

	Precision Score				
Algorithms	**Fixations**	**Saccades**	**PSOs**	**SPs**	**Mean Precision**
VAJD	0.766	0.937	0.935	0.733	0.842
VAD	0.788	0.919	0.935	0.739	0.845
VJD	0.778	0.917	0.921	0.735	0.837
VD	0.7642	0.919	0.932	0.717	0.833
1D-CNN-BLSTM(VD)	0.64	0.78	0.79	0.56	0.69
AJD	0.607	0.642	0.804	0.559	0.765
AD	0.517	0.458	0.617	0.371	0.490

As the results show, feature selection significantly impacts the event classification results. On the one hand, velocity is crucial for correctly detecting saccades from fixations and smooth pursuits. Hence, these events have low scores for AD and AJD sets. On the other hand, velocity is not sufficient to correctly distinguish between fixations and SPs, as both are low-speed movement types. Therefore, the direction is used to detect fixations from SPs. A classic VD set performs better than the former sets for SPs, but its performance is still low for PSOs. Adding acceleration and jerk improves the general performance of the models, especially for PSO detection. A comparison of VAD, VJD, and VAJD sets shows that both acceleration and jerk have a similar impact on the results, and their combination is only marginally better than using only one of them (see Tables 4 and 5).

A comparison with the results of the 1D-CNN-BLSTM model on the same dataset, as reported in [21], shows that our model performs better and can also distinguish the PSO events. Our modified implementation of the same model does not perform well. However, one of the reasons may be that we changed the model to return only one value instead of the sequence, and skipped the model optimization step.

Table 4. Accuracy for each file and feature combination for each file in the LOFO cross-validation. We compare the performance of the proposed model with different feature combinations and the state-of-the-art approach.

	Accuracy Score					
Features	**File1**	**File2**	**File3**	**File4**	**File5**	**File6**
VAJD	0.75	0.85	0.78	0.85	0.92	0.92
VAD	0.76	0.85	0.78	0.85	0.92	0.92
VJD	0.75	0.84	0.77	0.85	0.92	0.92
VD	0.72	0.83	0.762	0.84	0.92	0.92
1D-CNN-BLSTM(VD)	0.59	0.57	0.55	0.62	0.61	0.66
AJD	0.59	0.74	0.74	0.54	0.79	0.89
AD	0.69	0.47	0.48	0.58	0.18	0.62

Table 5. Cohen's Kappa for each file and feature combination for each file in the LOFO cross-validation. We compare the performance of the proposed model with different feature combinations and the state-of-the-art approach.

	Cohen's Kappa					
Features	**File1**	**File2**	**File3**	**File4**	**File5**	**File6**
VAJD	0.60	0.73	0.64	0.73	0.86	0.85
VAD	0.61	0.73	0.63	0.73	0.85	0.85
VJD	0.58	0.72	0.62	0.73	0.86	0.86
VD	0.51	0.70	0.61	0.72	0.85	0.80
1D-CNN-BLSTM(VD)	0.31	0.35	0.25	0.33	0.08	0.40
AJD	0.34	0.54	0.57	0.28	0.64	0.80
AD	0.48	0.21	0.19	0.30	0.04	0.19

Misclassifications Analysis

In this section, the misclassifications between events for each feature set are discussed using the confusion matrices, which help to analyze which events were mistaken for others in each feature set.

The confusion matrix analysis for the combinations of AD and VD features is shown in Figures 4a and 4b, respectively. Each row in the matrices shows the percentage of samples for a given event that were classified as the event shown in the corresponding column. The results for the AD combination clearly show that fixations and smooth pursuits are often confused (45% of fixations are classified as smooth pursuits). Moreover, the recall of PSOs is very low, and most of the actual PSOs are misclassified (only 21% are classified correctly).

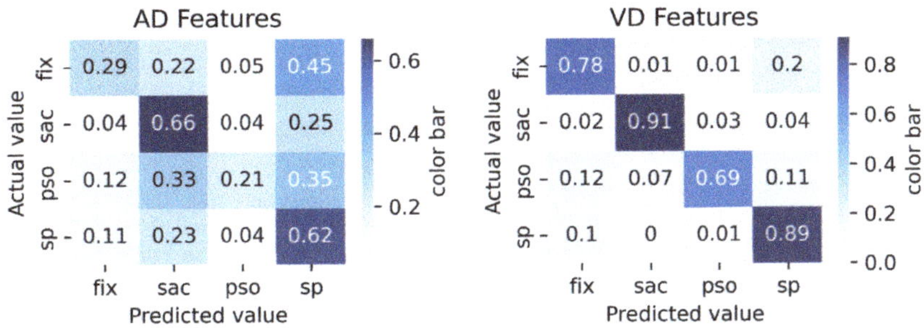

(**a**) confusion matrix for AD (**b**) confusion matrix for VD

Figure 4. The confusion matrix for AD and VD feature combinations.

The velocity and direction (VD) combination shows a noticeable improvement over the AD combination. The combination significantly improves the performance of fixations, saccades, and smooth pursuits. However, its performance in correctly identifying PSOs from fixations and smooth pursuits is lower. So, we propose using the VAJD, VAD, AJD, and VJD combinations to improve PSO identification performance and reduce misclassifications between PSOs and other event types.

Combining velocity and direction with acceleration and/or jerk improved the overall identification performance for all event types compared to other feature combinations, especially for PSO detection. However, the performance of PSO identification is still low, and many PSO points are misclassified as fixations and smooth pursuits. The results for these feature combinations are shown in Figure 5. For example, the results of the VAJD feature combination in Figure 5b show that 11% of PSOs are misclassified as smooth pursuits and 11% as fixations. The VAD and VJD feature combinations perform approximately the same as VAJD.

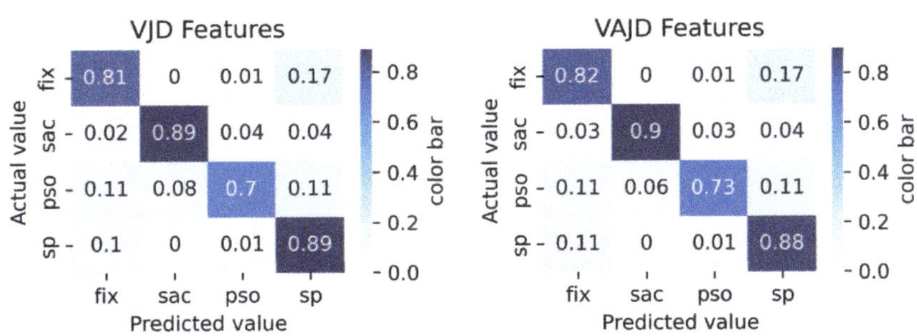

(**a**) confusion matrix for VJD (**b**) confusion matrix for VAJD

Figure 5. The confusion matrix for VJD and VAJD feature combinations.

In addition, experiments were conducted with combinations without velocity or jerk. A combination of VAD and AJD features was used for the experiment. The results in Figure 6a show that the absence of jerk in the VAD combination does not decrease performance. Regarding the effect of velocity, the AJD in Figure 6b shows that the absence of velocity significantly affects the overall performance. The results imply that combining acceleration and jerk with direction (and with the absence of velocity) does not improve overall performance for all event types.

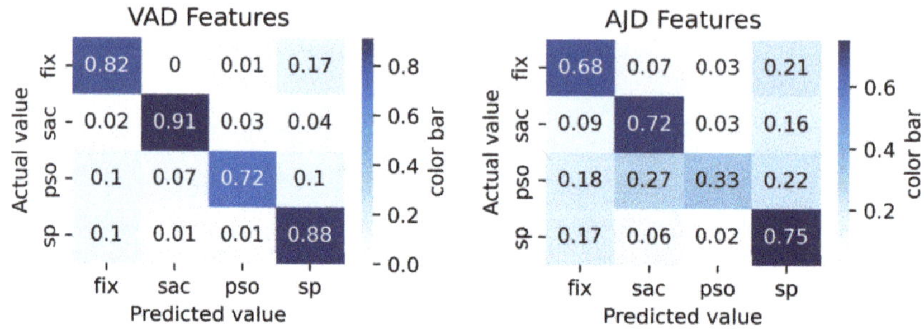

(**a**) confusion matrix for VAD (**b**) confusion matrix for AJD

Figure 6. The confusion matrix for VAD and AJD feature combinations.

6. Event Measures

As stated in Section 4.3, classifying each gaze point as a specific event was just the first step in event detection. The next step was to merge the neighboring points with the same classification into one event, as shown in Figure 3. In this way, events with their own properties were created. For instance, fixations were characterized by a location (calculated as the mean or median of gaze points), as well as an onset and offset, while saccades and smooth pursuits were defined by their start and end points.

The main problem with all threshold-based methods is their high susceptibility to outliers. When one gaze point in the middle of a fixation is classified as any other event, it automatically divides the fixation into two separate fixations. The popular solution is to merge subsequent fixations based on their low spatial and temporal distance. However, this introduces two more thresholds that must be manually assigned.

Machine learning-based algorithms are more robust to this problem as they classify gaze points, taking into account not only the sample's properties but also its neighborhood. However, the problem still exists. To analyze this phenomenon for our results, we merged the neighboring event points to form events. Then, for each event type, we calculated two metrics, that is, the mean number of events and the mean event duration.

The results in Table 6 show the number of events and mean duration of events for different combinations of features and ground truth (manual coding).

Table 6. The event measures detected in the test data for different feature combinations by both manual annotation and the proposed model. The column event measure shows the list of event-measuring metrics, and the columns VAJD, VJD, VAD, AJD, VD, and AD are the results for feature sets.

Metrics	VAJD	VAD	VJD	AJD	VD	AD	GT
Mean No. of Fix	12.5	13.5	14.5	48	18.3	40	10
Mean No of Sacc	14.6	15	15	37	15.3	47.5	15.5
Mean No. of PSOs	11.3	11.3	11.1	19.6	11	26.1	11
Mean No. of SP	9.8	10.8	11.5	58.5	15.3	71	7
Mean FIX-dur	217.2	209.1	203.1	49.3	158.2	23.9	226
Mean Sac-dur	25.1	25.3	25.0	17.4	24.9	25.5	24
Mean PSO-dur	16.8	16.3	16.8	11.0	17.0	11.7	16
Mean SP-dur	403.8	366.2	365.2	60.5	280.3	73.0	508

Generally, our methods find more fixations than GT, which probably means that some actual fixations are classified as two or more separate fixations by the model. This is especially visible for AD and AJD combinations, where the number of detected fixations is about five times larger than it should be. A similar situation can be observed for smooth pursuits. Not surprisingly, when more separate events were detected, the mean durations of events were shorter.

The results show that the calculations of event-level statistics, such as the number and duration of events, may reveal additional information about the quality of the model. For instance, the confusion matrix of the VD model in Figure 4b suggests that its performance is similar to (or even better than) the VAD, VJD, and VAJD models for smooth pursuit detection. However, event-measuring metrics in Table 6 show that using the VD model resulted in the creation of much smoother pursuit events than for the VAD, VJD, and VAJD models and a shorter duration of these events. So, it may be assumed that the smooth pursuit detection performance of the VD model is lower than that of the VAD, VJD, and VAJD models, even when the sample-level accuracy is better.

As a conclusion from the results, the proposed models with feature combinations VAJD, VAD, and VJD perform approximately the same, outperforming AJD, VD, and AD feature combinations.

7. Conclusions and Future Works

A deep learning method was proposed to extract eye movement events from raw eye-tracking data. The main objective of this study was to identify the best feature sets for building a model that simultaneously classifies four event types. The velocity, acceleration, jerk, and direction characteristics were used. Combinations of these different features were used to detect events.

The classification performance of the proposed model across different feature combinations was compared with that of the state-of-the-art model. The results revealed that combining velocity and acceleration with either direction, jerk, or both jerk and acceleration features significantly improved performance compared to other feature sets. This indicated that the VAJD, VAD, and VJD feature combinations outperformed the AJD, VD, and AD feature combinations.

Furthermore, we analyzed the impact of each feature on the detection of specific events, i.e., fixations, saccades, PSOs, and smooth pursuits. To the best of our knowledge, this is the first fully automated event detection method that simultaneously classifies fixations, saccades, PSOs, and smooth pursuits. The event measures for the proposed methods were also evaluated and compared with the manual classification results. The event-measuring metrics included the number of events and the mean duration of events.

Obviously, the work carries certain limitations. Our results, even for the best-performing models, are far from perfect, and there is room for improvement. In particular, the accuracy of the PSO classification is relatively low. In this paper, we focus on the analysis of the best combinations of different features. The next step should be to use the best-performing feature combinations to build more robust models. Future work will focus on extending the research through an extensive analysis of various network architectures and hyperparameters. Furthermore, more data will be used for training to further improve the model's performance.

As stated in the performance evaluation section, comparing different algorithms with different conditions and objectives is difficult [33]. Therefore, we were only able to compare the proposed models with the one that was the most similar to ours. However, this model originally did not consider PSO events, so the comparison was not entirely fair. No model was found that automatically classifies all four events.

In addition, it would be valuable to test the performance of the algorithms using other datasets. Specifically, this work focuses on video stimulus datasets labeled with fixations, saccades, PSOs, and smooth pursuits. Therefore, testing the models on other types of data could tell more about the generality of our methods. Since PSOs are the most problematic, future work will analyze how they differ across different types of stimuli and attempt to develop models tailored to specific types.

The other line of enhancement is based on improving the accuracy of the PSO classification in the presence of fixations, saccades, and smooth pursuits. The proposed method improves existing methods by simultaneously classifying fixations, saccades, PSOs, and SPs. However, we believe that the proposed approach can be further optimized in such a way that the accuracy of the PSO is taken into account.

Author Contributions: conceptualization, P.K.; software, B.A.B.; calculations, B.A.B., data curation, B.A.B., P.K.; writing—original draft preparation, B.A.B.; writing—review and editing, P.K.; supervision, P.K. All authors have read and agreed to the published version of the manuscript.

Funding: This research was funded by the Silesian University of Technology, Poland grant number BKM/572/RAU7/2024.

Institutional Review Board Statement: Ethical review and approval were waived for this study because no experiments on humans or animals were done.

Informed Consent Statement: Not applicable.

Data Availability Statement: The Python code for data preparation and all classifications performed is available in the GitHub repository at https://github.com/mebirtukan/Notebook_2D_CNN_LSTM-for-Eye-Movement-Event-Detections (accessed on 22 May 2025).

Conflicts of Interest: The authors declare no conflicts of interest.

References

1. Punde, P.A.; Jadhav, M.E.; Manza, R.R. A study of eye tracking technology and its applications. In Proceedings of the 2017 1st International Conference on Intelligent Systems and Information Management (ICISIM), Aurangabad, India, 5–6 October 2017; IEEE: Piscataway, NJ, USA, 2017; pp. 86–90.
2. Braunagel, C.; Geisler, D.; Stolzmann, W.; Rosenstiel, W.; Kasneci, E. On the necessity of adaptive eye movement classification in conditionally automated driving scenarios. In Proceedings of the Ninth Biennial ACM Symposium on Eye Tracking Research & Applications, Charleston, SC, USA, 14–17 March 2016; pp. 19–26.
3. Naqvi, R.A.; Arsalan, M.; Park, K.R. Fuzzy system-based target selection for a NIR camera-based gaze tracker. *Sensors* **2017**, *17*, 862. [CrossRef] [PubMed]
4. Naqvi, R.A.; Arsalan, M.; Batchuluun, G.; Yoon, H.S.; Park, K.R. Deep learning-based gaze detection system for automobile drivers using a NIR camera sensor. *Sensors* **2018**, *18*, 456. [CrossRef] [PubMed]
5. Nurhasan, A.A.; Kasprowski, P.; Harezlak, K.; Birawo, B.A. Object Detection in Movies—Case Study. In Proceedings of the Asian Conference on Intelligent Information and Database Systems, Ras Al Khaimah, United Arab Emirates, 15–18 April 2024; Springer: Singapore, 2024; pp. 3–12.
6. Salvucci, D.D.; Anderson, J.R. Automated eye-movement protocol analysis. *Hum.-Comput. Interact.* **2001**, *16*, 39–86. [CrossRef]
7. Larsson, L. Event Detection in Eye-Tracking Data for Use in Applications with Dynamic Stimuli. Ph.D. Thesis, Lund University, Lund, Sweden, 2016.
8. Holmqvist, K.; Nyström, M.; Andersson, R.; Dewhurst, R.; Jarodzka, H.; Van de Weijer, J. *Eye Tracking: A Comprehensive Guide to Methods and Measures*; OUP: Oxford, UK, 2011.
9. Birawo, B.; Kasprowski, P. Review and evaluation of eye movement event detection algorithms. *Sensors* **2022**, *22*, 8810. [CrossRef] [PubMed]
10. Wyatt, H.J.; Pola, J. Smooth pursuit eye movements under open-loop and closed-loop conditions. *Vis. Res.* **1983**, *23*, 1121–1131. [CrossRef] [PubMed]
11. Nyström, M.; Hooge, I.; Holmqvist, K. Post-saccadic oscillations in eye movement data recorded with pupil-based eye trackers reflect motion of the pupil inside the iris. *Vis. Res.* **2013**, *92*, 59–66. [CrossRef] [PubMed]
12. Hooge, I.T.; Niehorster, D.C.; Nyström, M.; Andersson, R.; Hessels, R.S. Is human classification by experienced untrained observers a gold standard in fixation detection? *Behav. Res. Methods* **2018**, *50*, 1864–1881. [CrossRef] [PubMed]
13. Andersson, R.; Larsson, L.; Holmqvist, K.; Stridh, M.; Nyström, M. One algorithm to rule them all? An evaluation and discussion of ten eye movement event-detection algorithms. *Behav. Res. Methods* **2017**, *49*, 616–637. [CrossRef] [PubMed]

14. Zemblys, R.; Diederick, N.; Holmqvist, K. End-to-end eye-movement event detection using deep neural networks. In Proceedings of the Journal of Eye Movement Research: Vol. 10, iss. 6: Abstracts of the 19th European Conference on Eye Movements, Wuppertal, Germany, 20–24 August 2017; International Group for Eye Movement Research: Heidelberg, Germany, 2017; Volume 10.
15. Salvucci, D.D.; Goldberg, J.H. Identifying fixations and saccades in eye-tracking protocols. In Proceedings of the 2000 Symposium on Eye Tracking Research & Applications, Palm Beach Gardens, FL, USA, 6–8 November 2000; pp. 71–78.
16. Tafaj, E.; Kübler, T.C.; Kasneci, G.; Rosenstiel, W.; Bogdan, M. Online classification of eye tracking data for automated analysis of traffic hazard perception. In Proceedings of the Artificial Neural Networks and Machine Learning–ICANN 2013: 23rd International Conference on Artificial Neural Networks, Sofia, Bulgaria, 10–13 September 2013; Proceedings 23; Springer: Berlin/Heidelberg, Germany, 2013; pp. 442–450.
17. Hartridge, H.; Thomson, L. Methods of investigating eye movements. *Br. J. Ophthalmol.* **1948**, *32*, 581. [CrossRef] [PubMed]
18. Komogortsev, O.V.; Karpov, A. Automated classification and scoring of smooth pursuit eye movements in the presence of fixations and saccades. *Behav. Res. Methods* **2013**, *45*, 203–215. [CrossRef] [PubMed]
19. Larsson, L.; Nyström, M.; Andersson, R.; Stridh, M. Detection of fixations and smooth pursuit movements in high-speed eye-tracking data. *Biomed. Signal Process. Control* **2015**, *18*, 145–152. [CrossRef]
20. Zemblys, R.; Niehorster, D.C.; Holmqvist, K. gazeNet: End-to-end eye-movement event detection with deep neural networks. *Behav. Res. Methods* **2019**, *51*, 840–864. [CrossRef] [PubMed]
21. Startsev, M.; Agtzidis, I.; Dorr, M. 1D CNN with BLSTM for automated classification of fixations, saccades, and smooth pursuits. *Behav. Res. Methods* **2019**, *51*, 556–572. [CrossRef] [PubMed]
22. Hoppe, S.; Bulling, A. End-to-end eye movement detection using convolutional neural networks. *arXiv* **2016**, arXiv:1609.02452.
23. Friedman, L.; Prokopenko, V.; Djanian, S.; Katrychuk, D.; Komogortsev, O.V. Factors affecting inter-rater agreement in human classification of eye movements: A comparison of three datasets. *Behav. Res. Methods* **2023**, *55*, 417–427. [CrossRef] [PubMed]
24. Salvucci, D.D.; R Anderson, J. Tracing eye movement protocols with cognitive process models. In *Proceedings of the Twentieth Annual Conference of the Cognitive Science Society*; Routledge: London, UK, 1998.
25. Stark, L. Scanpaths revisited: Cognitive models, direct active looking. In *Eye Movements: Cognition and Visual Perception*; Routledge: London, UK, 1981; pp. 193–226.
26. Harężlak, K.; Kasprowski, P. Evaluating quality of dispersion based fixation detection algorithm. In Proceedings of the Information Sciences and Systems 2014: Proceedings of the 29th International Symposium on Computer and Information Sciences, Krakow, Poland, 27–28 October 2014; Springer: Cham, Switzerland, 2014; pp. 97–104.
27. Sen, T.; Megaw, T. The effects of task variables and prolonged performance on saccadic eye movement parameters. In *Advances in Psychology*; Elsevier: Amsterdam, The Netherlands, 1984; Volume 22, pp. 103–111.
28. Erkelens, C.J.; Vogels, I.M. The initial direction and landing position of saccades. In *Studies in Visual Information Processing*; Elsevier: Amsterdam, The Netherlands, 1995; Volume 6, pp. 133–144.
29. Nyström, M.; Holmqvist, K. An adaptive algorithm for fixation, saccade, and glissade detection in eyetracking data. *Behav. Res. Methods* **2010**, *42*, 188–204. [CrossRef] [PubMed]
30. Zemblys, R.; Niehorster, D.C.; Komogortsev, O.; Holmqvist, K. Using machine learning to detect events in eye-tracking data. *Behav. Res. Methods* **2018**, *50*, 160–181. [CrossRef] [PubMed]
31. Elmadjian, C.; Gonzales, C.; Costa, R.L.d.; Morimoto, C.H. Online eye-movement classification with temporal convolutional networks. *Behav. Res. Methods* **2022**, *55*, 3602–3620. [CrossRef] [PubMed]
32. Donahue, J.; Anne Hendricks, L.; Guadarrama, S.; Rohrbach, M.; Venugopalan, S.; Saenko, K.; Darrell, T. Long-term recurrent convolutional networks for visual recognition and description. In Proceedings of the IEEE Conference on Computer Vision and Pattern Recognition, Boston, MA, USA, 7–12 June 2015; pp. 2625–2634.
33. Startsev, M.; Zemblys, R. Evaluating eye movement event detection: A review of the state of the art. *Behav. Res. Methods* **2022**, *55*, 1653–1714. [CrossRef] [PubMed]

applied sciences

MDPI

Article

Appearance-Based Gaze Estimation as a Benchmark for Eye Image Data Generation Methods

Dmytro Katrychuk * and Oleg V. Komogortsev

Department of Computer Science, Texas State University, 601 University Drive, San Marcos, TX 78666, USA; ok@txstate.edu
* Correspondence: d_k139@txstate.edu

Abstract: Data augmentation is commonly utilized to increase the size and diversity of training sets for deep learning tasks. In this study, we propose a novel application of an existing image generation approach in the domain of realistic eye images that leverages data collected from 40 subjects. This hybrid method combines the benefits of precise control over the image content provided by 3D rendering, while introducing the previously lacking photorealism and diversity into synthetic images through neural style transfer. We prove its general efficacy as a data augmentation tool for appearance-based gaze estimation when generated data are mixed with a sparse train set of real images. It improved the results for 39 out of 40 subjects, with an 11.22% mean and a 19.75% maximum decrease in gaze estimation error, achieving similar metrics for train and held-out subjects. We release our data repository of eye images with gaze labels used in this work for public access.

Keywords: eye-tracking; gaze estimation; machine learning; generative adversarial networks; data augmentation; style transfer

1. Introduction

Eye-tracking (ET) is a technology that at its core enables the estimation of the eyeball's orientation with respect to a chosen frame of reference. In the most general case, it means solving this estimation problem relative to the subject's head. Most modern human–computer interaction applications that ET enables involve finding the point of gaze in the device coordinates, such as the pixel of interest on the computer screen. ET has enabled many important applications since desktop computer workstations have become available. They range from basic gaze-guided human–computer interfaces [1–3] to more demanding approaches that depend on complex features extracted from eye movement patterns. The visual system is one of our main tools for interacting with the surrounding environment, so ET empowers a natural dimension of human–computer interaction. This capability is crucial for providing accessibility for disabled individuals who are unable to use traditional input modalities. Real-time ET may also be used to deploy foveated rendering that produces high-quality content only near the point of gaze [4,5]. This strategy is essential in delivering improved rendering quality in extended reality devices that operate in a constrained power envelope. Specific traits of each subject's oculomotor system, eye shape, periocular features, etc., were shown to have great potential for biometrics [6] and health assessment. With access to data on the order of ten thousand subjects, it is possible to build an eye movement biometrics system for user authentication with a false rejection rate of 2.4% at the false acceptance rate of 1 in 50,000 [7]. The main advantage of such a system is that there is no known way to spoof it, which is not the case for a fingerprint sensor or a facial recognition system. In regards to the health assessment, the detection of various diseases and states with performance above random chance, including schizophrenia, mild traumatic brain injury, intoxication, and fatigue, has been demonstrated using eye movement signals [8–10].

The first key component of any ET pipeline is the hardware equipment that enables it. We limit the scope of interest to solutions based on one or more cameras, as they are the most relevant in modern applications [11]. The next key component of the pipeline is the method used to estimate the gaze from camera image(s). These methods are commonly categorized as model-based, appearance-based, and hybrid between the two [12]. Overall, every method has a step of extracting the relevant features from the eye image and the main difference is in the way these are used for the gaze estimation. The first category usually involves a parametric eye model that is designed with a certain level of realism. In the purest model-based case, the extracted features are used to estimate the set of model parameters that is sufficient to make a direct gaze inference [13]. A fully appearance-based method is typically not backed by an explicit eye model and aims to produce the gaze directly from an image. With the emergence of desktop and embedded hardware for high-resolution image capture and powerful deep learning techniques for image processing [14–16], the appearance-based gaze estimation became the most common in modern ET solutions [17]. These solutions are mostly data-driven, and the representative feature distributions and large quantity of their training data are essential in achieving good performance.

Data augmentation is a valuable tool in general machine learning tasks for increasing the size and diversity of the collected data set. It is commonly seen in studies that work with visual or audio information, where cropping, affine transformations, or adding random noise are applied to artificially increase the number of data samples [18–20]. These methods can usually prevent overfitting of deep learning models to the training set, but they are unlikely to help in achieving better robustness in real-world deployment of the model where it can experience completely unseen samples yet from the same domain. For complex supervised learning tasks, it is unfeasible to collect the labeled training set that completely represents the distribution of the data used for model inference. As random sub-sampling of the existing data would not help to close the gap, it may be possible to introduce the lacking variability by the synthetic generation of the new data. It is challenging not only to produce a realistic extension of the data yet also to preserve the label consistency in the generated output; otherwise the learned model has a high chance of showing poor generalization in testing scenarios.

Synthetic data generation is especially appealing in ET research. The reason for this is two-fold. First, it is expensive time- and resource-wise to collect labeled data that exhibit all potential real-world variability, such as sensor positions, lighting conditions, and subject appearance. Second, it provides great research flexibility for novel ET sensors that have not yet been built in consumer-grade hardware [21,22]. In this work, we focused on creating a generative framework for the domain of near-eye monocular images. We propose a hybrid solution that fuses two most common approaches from the relevant literature: synthetic rendering and end-to-end learning from a large pool of data. This solution enables us to map the real subject identities from pre-collected video-oculography data onto a Blender artificial model [23]. It combines the best of both worlds, as it allows precise control over the rendering scene while introducing the lacking diversity and photorealism in the synthetic renders. After posing this problem as a style transfer, we investigated the applicability of a StarGANv2 [24] architecture, a proven solution in that field. We applied our own modification of this method and discussed the potential further improvements specific to the ET domain. The generated images were evaluated in terms of their photorealism, visual and content consistency. The results were compared with the Blender rendering tool [23] and EyeGAN architecture [25]. The proposed method was finally evaluated as a data augmentation tool for increasing the training set size of an appearance-based gaze estimation neural network.

The main contributions of our work can be summarized as follows:

- We present an expandable benchmark for synthetic eye image generation methods that measures their utility for data augmentation in appearance-based gaze estimation tasks. We prove its effectiveness by showing the unsatisfactory performance of existing generative methods in the eye-tracking domain achieved under its testing conditions.
- We modify a method for general image domain transfer and adapt it for real and synthetic eye images. The resulting hybrid approach combines the powerful control

over the ground-truth eye-tracking features afforded by 3D rendering with the photorealism and subject diversity of pre-collected image data. It improves the results for 39 out of 40 subjects with a 11.22% mean and an 19.75% maximum decrease in gaze estimation error. The proposed method outperforms all prior approaches under our benchmark, and we show the difference to be statistically significant.

- We release for public access our real data set of 1.24 M eye images and their corresponding gaze labels collected from 40 human subjects in a random saccades task using the custom video-oculography desktop setup: https://doi.org/10.5281/zenodo.13852050 (accessed on 9 October 2024).

2. Prior Work

2.1. Image Rendering Tools

A number of general-purpose 3D rendering tools have been adopted for the ET domain, the most notable examples being Unity and Blender. A Blender head model with a highly detailed eye region was presented in [23]. The public availability of the source code allows for fine-grain control over most parameters of the rendered scene through the Python API. The most significant downside of the model is its use of only one artificial subject, which makes it impossible to introduce subject variability in the rendered images. Moreover, the visual appearance of the rendered images is substantially different from what can be seen in real eye image data sets, such as MPIIGaze [14]. This is problematic, as a lack of realism in synthetic images has been shown to reduce gaze estimation performance [26].

Improving the realism of eye image rendering is an active research area. A Unity-based solution was introduced in [27]. It offered substantial variability in subject appearance and ambient conditions with extremely fast rendering speed. Nair et al. [28] presented RIT-Eyes. Implemented in Blender, this rendering pipeline generates synthetic images and 2D/3D segmentation masks using eye attributes, gaze position, and environmental/sensor parameters (camera pose, etc.) as input. One key advancement of this pipeline versus prior work is the ability to generate temporally coherent eye videos that mimic the blink, head, and eye movements found in publicly available data sets [29]. Kim et al. in [22] introduced novel large-scale data for near-eye gaze estimation. They presented a synthetic data set based on anatomically informed eye and face models with variations in face shape, gaze direction, pupil and iris, skin tone, and external conditions (2 M images with the 1280×960 resolution), and a real-world data set collected from 35 subjects (2.5 M images with the 640×480 resolution). Unfortunately, there is a common drawback in all these methods. The authors did not make the source code publicly available, at best only released either the already generated data (NvGaze) or the compiled executable with very little control over the generated sample domain (UnityEyes). These complex 3D environments are also challenging to reproduce from scratch, which greatly reduces the applicability of their solutions to novel settings.

2.2. Learning-Based Data Generation Methods

While synthetic data sets generated using 3D creation software have been utilized for multiple applications in the ET domain, many modern solutions for data generation rely on a framework of generative adversarial networks (GANs) introduced by Goodfellow et al. in [30]. Since its introduction, GAN-based methods have enabled breakthroughs in the generation of high-resolution photorealistic images [31] (BigGAN), [32] (StyleGAN3); high-fidelity speech audio synthesis [33] (Hifi-GAN); unsupervised image [34] (pix2pix), [35] (CycleGAN), [24] (StarGAN-v2); and audio [36] (Mic2Mic) domain translation. The original GAN framework training process is notoriously unstable. This issue inspired many follow-up works proposing alternative adversarial loss functions and optimization problem statements [37] (LS-GAN), [38] (WGAN-GP), [39] (DRAGAN). Unfortunately, none has been shown to be superior to others, and adjusting a GAN to a specific task usually requires an extensive hyperparameter search [40]. Another common challenge is to pick a set of metrics that allow for a fair and thorough evaluation of the generated output. One of the most comprehensive surveys on the pros and cons of various metrics available in the literature can be found in [41]. We are interested in applying the GAN framework in an attempt to combine the desirable characteristics of synthetic and real eye images, namely, to benefit from the efficiency of the synthetic rendering process (i.e., ability to rapidly vary a camera pose, a gaze vector, environmental lighting conditions, etc.) while capturing the inherent realism encountered in real-world data capture.

In the more recent literature, diffusion probabilistic models [42] were shown to outperform GANs in generating realistic image data [43]. They represent a class of likelihood-based models that learn the underlying parametric distribution of given data samples. The training of a diffusion model typically involves a two-stage process with the forward and backward parts. During the forward diffusion an input sample is corrupted with a varying level of Gaussian noise. This effect is reversed in the backward diffusion with the denoising neural network, which learns to predict the noise added to the sample during the whole forward process. At inference time, the denoising network is utilized to progressively transform the initial random noise into an output sample from the learned domain [42]. The diffusion-based generative models achieve state-of-the-art results of increasingly impressive visual fidelity, but require thousands of hours of GPU train time and hundreds of millions of training samples to achieve that [44–46]. While there is an active work in making the various parts of the diffusion method more efficient [47,48], the computational and data requirements still prohibit their use by smaller research groups with limited resources. We suggest these methods be considered in future work and decided to limit our scope to GAN-based approaches for image generation as a more attainable solution for the broader research community.

GANs have been utilized to augment image-based training data in the literature throughout various studies. Examples include the generation of new images from existing training sets to improve classification accuracy [49], along with the transformation of contrast computer tomography (CT) images to improve segmentation in non-contrast CT images [50]. Within the eye image domain, the most notable application of data augmentation using GANs was an attempt to address the aforementioned lack of realism within synthetically generated images. In this section, we describe three of these networks—SimGAN [26], GazeGAN [51], and EyeGAN [25]—and discuss their applicability in our problem setting.

2.2.1. SimGAN

SimGAN [26] was developed to improve the realism of synthetically generated labeled images by leveraging the large-scale availability of unlabeled real images. The main idea of this method was to use the generator network as the "refiner". It did not generate an output from any kind of low-dimensional representation, but rather learned the pixel-wise transformations of the input image. In other words, it "refined" the original synthetic image to make it look like it was sampled from the distribution of real data. The original study used 1.2 M synthetic images generated with UnityEyes and real eye images from the MPIIGaze data set.

2.2.2. GazeGAN

The GazeGAN [51] significantly improves upon SimGAN in output image resolution and photorealism by leveraging the large and diverse set of unlabeled real facial images. Their joint use with the labeled UnityEyes renders without a structured connection between the two was enabled by supplementing the pure adversarial training with the cycle consistency loss. The fundamental work first to apply this principle in training an image translation network with no paired train samples in different domains was published in [35]. The idea of cycle consistency is based on the observation that if the generated sample can be translated back to its original domain, the result of such inverse transformation should match the original input sample. Most importantly, it relaxes the requirement for having paired ground-truth translation samples if the inverse computation can be done within the same generative framework. In this study, we further explore the effectiveness of cycle consistency in the unsupervised domain translation between real and synthetic eye images.

2.2.3. EyeGAN

The EyeGAN study [25] improves on the SimGAN approach in a different manner. Instead of relying on the per-pixel refinement of the original image, the authors proposed using an eye segmentation mask as an input to the generator model. A ternary mask includes classes for the sclera, the pupil together with the iris, and the remaining periocular region. The labeled segmentation masks can be obtained as an additional output of the rendering tools, like UnityEyes that was used in the EyeGAN paper or Blender, as shown later in the "Synthetic Blender data" subsection. As it is very labor-intensive to manually label the real data, the segmentation network was initially trained only on the available synthetic data. To obtain the labeled masks for the real images, the generator facilitated the unsupervised domain transfer for the segmentation task. The segmentation network was updated as part of the GAN training process, and its training pool was periodically updated with the output of the generator network that was fed masks from the synthetic data set. The real eye images and their corresponding masks obtained from the segmentation network formed the training batch of the generator. Note that the image loss was no longer computed on the pair of images from different domains (synthetic and real) as in the case of SimGAN. This change, together with the guidance from domain-agnostic segmentation masks, resulted in a generator that much better preserves the eye gaze in the output images, as concluded by the EyeGAN authors. We picked EyeGAN as one of our baselines for realistic eye image generation, as the more recent example of purely GAN-based approaches. Our own schematic presentation of its framework is shown in Figure 1.

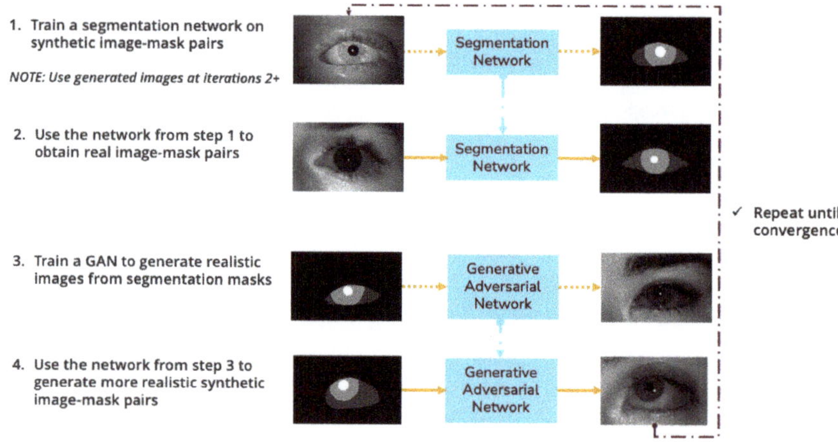

Figure 1. The outline of the original EyeGAN approach.

2.3. Data Augmentation via Style Transfer

While the referenced GAN studies showcased increasingly promising results, they had no mechanism to control the visual appearance of the subject in the generated image. In other words, they did not offer a way to impose the visual coherence of the network output when generating data for a given subject. It is reasonable to assume that this might limit the use of such a generative model as a data augmentation tool for appearance-based ET research. This statement is formulated as a formal hypothesis in the "Methodology" section to be tested on our data later. In order to mitigate this issue, we look into a related concept of style transfer [52]. It considers an image to be comprised of features related to two categories: style and content. The main problem posed by style transfer is to transform an input image to match the visual style of the target image while retaining its original content. The semantic segmentation mask can be used as a proxy for the content of a scene (e.g., gaze vector) [53], while the style can be represented via a set of high-level features, extracted from a pre-trained convolutional network [54].

2.3.1. Seg2Eye

The Seg2Eye method presented in [53] generated an eye image that is consistent in content with a given input segmentation mask and in style with a given real eye image. The content was preserved by using the SPADE normalization layer [55] throughout the generator network. The style was represented as a latent vector learned via the encoder network and fused into the generated image using the AdaIN layer [56] in conjunction with SPADE throughout the generator network. We found that the most challenging factor that limits the direct application of this method in novel scenarios is that an essential part of its training loss involves a ground-truth for the generated images that are rarely available. We arrived at that conclusion after trying to apply Seg2Eye using the original implementation (https://github.com/mcbuehler/Seg2Eye (accessed on 9 October 2024)) with the L1 loss dropped. We found that the generator network failed to start producing eye-like images from scratch with no per-pixel guidance provided.

2.3.2. Kaur et al. (2021) [57]

The authors of [57] proposed another method for generating a realistic eye image from a segmentation mask that follows the visual style of a provided image. It proved to be superior to Seg2Eye when tested in an unsupervised manner when no image ground-truth was available for training. The key difference compared to Seg2Eye that enabled the effectiveness in an unsupervised scenario was the cycle consistency, which reinforced the successful use of this loss in GazeGAN [51]. This time, its additional importance was shown in solving a task that only has indirect supervision for the components of the generated images, such as their content and style. However, the method in [57] also could not be applied directly to any novel task, as it was tailored to a specific target use case of gaze redirection. We think that a task-agnostic method would have more impact on the eye-tracking research community. As a result, we borrowed the SPADE+AdaIN layer from Seg2Eye [53], which proved to be successful in style transfer and propose incorporating it into another network designed for general unsupervised domain translation.

2.4. *StarGANv2*

StarGANv2 is a neural network designed for unpaired image-to-image domain translation [24]. It addressed a visual style transfer problem by considering the images to come from a discrete set of predefined domains, such as a human gender or an animal type. It built upon the first version of StarGAN [58] with the added ease of scalability to more than two domains, which was its main contribution. In addition to that, StarGANv2 also introduced updates to the underlying GAN framework that helped to achieve state-of-the-art results in generating photorealistic images even within just two domains. The problem we are solving can be easily reformulated as a domain transfer between two domains—synthetic renders and real eye images. Another point that makes this work directly relevant to our study is that it has been proven to produce great results in an adjacent domain of human faces. The image evidence provided in the original manuscript contains visually striking examples of changing the visual subject appearance while maintaining the important image content, such as a camera position, a head pose, an eye gaze, the placement of facial features, etc. We were interested in replicating a similar result in the more restricted case of eye images; therefore, we chose this method as one of the baselines to work with.

3. Materials

There are two data types utilized in this study: real and synthetic. In general, the real-world ET data are collected to represent a diverse subject population and to most closely resemble the images used in practical applications. However, the recording environments that sufficiently capture possible variability in gaze, camera position, lighting conditions, etc., are resource-prohibitive. Moreover, once the data are recorded, it is virtually impossible to make any post-hoc changes. On the contrary, the synthetic image renderings typically suffer from a severe visual domain mismatch compared to real data, but provide precise control over the modeled features that can be exercised as much as needed. In this section, we go over our real data collection setting, highlighting its feature set and the gaps in its coverage. Next, we present our synthetic pipeline that is designed to fill in the missing information by the means of image rendering. In the later sections, we explain how to fuse two sets, with the goal of bringing together the photorealism and subject diversity of real images and the ground-truth control over the scene rendering parameters of synthetic data.

3.1. Real VOG Data

The real data set is represented by the collection of real eye images previously reported in [59]. The data were collected from 40 different subjects and have a total of around 1.24 M samples. Every sample consists of a close-up eye image capture with the corresponding gaze information estimated via the means of video oculography (VOG). This custom VOG setup was used in place of standard commercial solutions, such as EyeLink 1000, due to the unique feature that allows saving raw eye images obtained by the ET camera [60]. The system was configured with an 850 nm IR illuminator, ThorLabs DCC1545M camera, and a hot mirror with a chin-rest from EyeLink 1000. The camera recorded monocular images with a resolution of 512 × 320 pixels at 120 Hz. The specific software and hardware details of the recording equipment are available in [60]. The schematic side view of the hardware setup can be found in Figure 2. For the purpose of the current study, it is important to note that this set was obtained using a fixed camera setup. Therefore, it captures a significant amount of subject and gaze variability but almost none in camera position and lighting condition domains. Producing synthetic data that cover the existing feature distribution gaps in real data would be a perfect use case for a data augmentation tool considered herein.

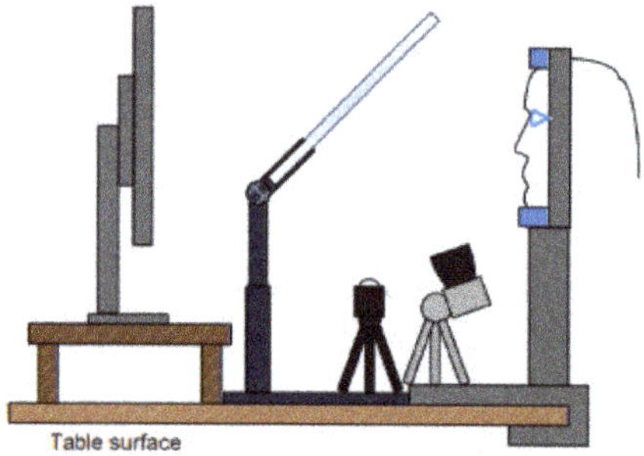

Figure 2. A schematic side view of the VOG data collection system (reproduced from [60] with the authors' permission).

The VOG data contain recordings of the random oblique saccades task that was originally designed to evaluate the effect of calibration grid density on the afforded gaze estimation error [59]. The task presents 174 on-screen fixation targets that densely cover the range of ±20.51° horizontally and ±16.7° vertically in degrees of visual angle. All targets were split into 3 groups, depending on their use case in the prior calibration study. The first 29 target points used to form various calibration grids were shown on the screen for a fixed time of 1.75 s each. The remaining points used for validation and testing purposes were shown for random time drawn from the uniform distribution within the [1 . . . 1.25] s range.

While the VOG data set was originally presented in [59], it was initially available only by private request. To facilitate future ET research, we make it publicly accessible at https://doi.org/10.5281/zenodo.13852050 (accessed on 9 October 2024). All recorded subjects gave informed consent under an experimental protocol approved by the Institutional Research Board of Texas State University (approval code 2018044), and their data were anonymized prior to public release.

Due to the nature of the employed stimulus, more than 90% of the data correspond to fixations on presented targets. It is unlikely that having a lot of similar images in the training set is beneficial for the performance of a generative network. If anything, when the coverage of the gaze feature is unevenly distributed, it creates a problem of imbalanced train data. This can directly hurt the network's generalization capabilities, as it is expected to have good performance across the whole gaze range. We solved this issue by resampling the original VOG data based on an eye movement classification label provided for each sample using the algorithm explained in [59]. Every image that corresponds to a saccade was included in the new set. For every consecutive range of fixation samples, only three images were included: the first, the last, and a randomly sampled one from within the range. This procedure reduced the total number of VOG samples from around 1.24 M to only 156 k, but resulted in a much more balanced set to be further used in training and evaluating the data generation methods considered herein.

3.2. Synthetic Blender Data

There were two different sets of Blender data generated for this study. The first one was used for training all image generation networks, described in the "Methodology" section later. One of the goals of the trained generative framework was to produce ET data that cover a wide range of features associated with it. Therefore, the required level of variability should be represented in its training set. The distribution of features already present in real VOG data together with a target scenario of simulating its missing camera position variability formed this requirement.

The second Blender data set was produced for the data augmentation study. The main idea behind that experiment was to benchmark various synthetic data sources as a substitute for real data. To achieve a fair assessment of employed data generation methods, we attempted to generate data that differ only in terms of the source they were produced from. This led to the task of producing a synthetic image set that mirrors the VOG data, where every image matches its real counterpart in everything but visual appearance. Note that every data augmentation method we tested directly relied on Blender renders. They were either used as is (Blender was tested as a separate method itself) or used as input to other tested methods (i.e., Blender segmentation masks in EyeGAN presented earlier; both images and masks in StarGANv2 described later). The Blender images were rendered in 320×200 resolution, which was chosen to maintain the original aspect ratio of 1.6 of VOG images while reducing the memory footprint of associated machine learning pipelines.

The task outlined above essentially requires recreating the data collection equipment setup in Blender, which is not completely feasible due to the following challenges. First, it is impossible to directly reproduce the exact lighting conditions of the real experiment in a synthetic setting. We simplified this problem by restricting the model space to just two lighting sources—one omni- and one single-directional (the same in both Blender sets). Both lights were placed at the same location in the rendering space as the camera, and their strength was randomly sampled from distributions manually tuned to mimic the range of lighting variability in real data. Second, while we could place a camera in the same spot employed in the real data collection, this does not actually achieve the image alignment we strive for. This is explained by the fact that subjects had their head in the chin-rest throughout the data collection. Even though the position of both the chin-rest and the camera were fixed, the relative positioning of eyes in the camera frame ended up being different for every subject due to the anatomical diversity. We solved this problem by manually adjusting the camera position for every subject. It was achieved by finding a camera placement, the render from which gives the best feasible alignment with the real eye image for a neutral gaze position of $(0.0°; 0.0°)$ used as a reference.

Several limitations of the original codebase were addressed in order to meet the specific needs of this experiment. Most importantly, the rendering pipeline was updated to have the segmentation masks computed as a part of its output, which was missing initially. In addition to that, the rendering script was modified to run in batches, which increased its efficiency. All specific details of our modifications together with the list of parameter values for generation of both Blender data sets are provided in the Appendix D.

It is worth re-emphasizing the set of pros and cons of utilized data sets, as it justifies the need for using two. On the one hand, the Blender model provides precise control over the image content (i.e., eye gaze, camera position, lighting setup, etc.), but it represents just a single artificial subject. On the other hand, the real VOG data offer the missing photorealism and subject appearance variability, while it is impossible to directly vary the content of already collected data. Our goal was to derive a method that would allow the mapping of real subject identities from VOG data onto synthetic renders from Blender data, while maintaining the content labels associated with the latter.

4. Methodology

4.1. RITNet Segmentation

The EyeGAN considered a segmentation mask to be a distilled version of the eye image that encodes information the most important for gaze tracking purposes. The SPADE applied the same general idea to any domain. It treated a segmentation mask as a convenient proxy for underlying image content that should remain consistent in the task of a visual style transfer. As many of our tested methods are SPADE-based, having an accurate segmentation mask for every eye image in our data was essential. The ground-truth masks for synthetic data were already available as a part of the Blender pipeline output. However, training a segmentation network for real VOG data would involve a process of image hand labeling that is extremely labor-intensive. Instead, we decided to employ an off-the-shelf network that provides a reasonable performance on our data without re-training. This was found to be the case with the RITNet network [61]—the winning model of the OpenEDS eye image segmentation contest held by Facebook, Inc. in 2019 [62]. We used the pre-trained model available in the official project repository at https://bitbucket.org/eye-ush/ritnet/src/master/ (accessed on 9 October 2024). All of the images we segmented were first resized from 320×200 to 640×400 to match the native resolution the network was trained on. The following list summarizes our use of RITNet:

- To compute the labels for the pupil class that were missing in the Blender synthetic masks.
- To segment the entire real VOG data (the examples are provided in the third column of Figure A2).
- To obtain segmentation masks of all the fake generated images in order to evaluate their content consistency with the input reference images.

4.2. EyeGAN and EyeGAN-RITNet

We chose EyeGAN as one of the baselines to evaluate, as it was shown to produce photorealistic eye images of good content consistency with input segmentation masks. We are thankful to Harsimran Kaur: https://scholar.google.com/citations?user=7KVRW0QAAAAJ (accessed on 9 October 2024), EyeGAN's main author, for providing us with the EyeGAN implementation to be used in our experiments. The provided codebase contains a few important updates to the architecture of networks comprising the GAN framework compared to the original publication. The generator is based on 7 ResNet building blocks that are updated to use the SPADE normalization layer. This architecture is largely based on the "SPADEGenerator" network comprised of "SPADEResNetBlock"s used in [55], the reference implementation of which is available on GitHub at https://github.com/NVlabs/SPADE/blob/master/models/networks/generator.py (accessed on 9 October 2024).

Our experiments with the EyeGAN (updated with SPADE) method revealed a severe domain-shift problem with the segmentation masks. In our case, the domain-shift is introduced the very first time the segmentation network was trained on Blender data and then evaluated on real data (step 2. from the first iteration as per Figure 1), as can be seen in the left column of Figure A2. These masks were used at the beginning of GAN training, and by the time the generator started to produce data of reasonable quality, the network already fit to incomplete masks. This becomes a problem when the valid segmentation masks are passed through this generator to form an updated training set for the segmentation network (step 4. in Figure 1). We found that the EyeGAN method never recovered from the domain shift introduced at the very beginning, as evidenced by the trajectory of training observed in the second column of Figure A2. We propose a minor modification to the EyeGAN based on the key observation that it struggled to achieve good segmentation of real eye images. We found that the RITNet segmentation of real images was much better than what could be achieved with the original EyeGAN pipeline described above, as can be seen by comparing the quality of masks depicted in Figure A2. Therefore, it was possible to get rid of the segmentation network in EyeGAN and avoid the complexity of its re-training altogether by simply switching to the use of RITNet masks. This change reduced the number of steps in the EyeGAN algorithm from four to just one that involved the training of the GAN itself (step 3. in Figure 1). The resulting method is dubbed EyeGAN-RITNet for the remainder of the manuscript.

4.3. StarGANv2 and StarGANv2-SPADE+AdaIN

We employed the official implementation of the base StarGANv2 from GitHub at https://github.com/clovaai/stargan-v2 (accessed on 9 October 2024). The minor changes were employed to make all but the mapping network architectures suitable for our image size of 320 × 200 that differed from the original size of 256 × 256. The networks were too deep otherwise and ran out of the smaller spatial dimensions to work with before the output was produced. The mapping network was the only one not suffering from this issue, and even though it was not used in our experiments, it was left unchanged for framework training consistency.

We proposed one key change to the base StarGANv2 that was inspired by the Seg2Eye's success in semantic-guided eye image synthesis. Our hypothesis was that the addition of SPADE would improve the content consistency of the generated images compared to the base method. We replaced every AdaIN layer in the decoding part of the base StarGANv2 generator with the combined SPADE+AdaIN layer. The new layer performs SPADE and AdaIN computations in parallel branches with their output combined by computing an average of two. The SPADE used input segmentation masks in a one-hot encoding that were downsized to match the layer's spatial dimensions using the 'nearest' down-sampling. The implementation of the combined layer was adapted from the original Seg2Eye [53] codebase at https://github.com/mcbuehler/Seg2Eye (accessed on 9 October 2024). The resulting StarGANv2-SPADE+AdaIN method is summarized in Figure 3.

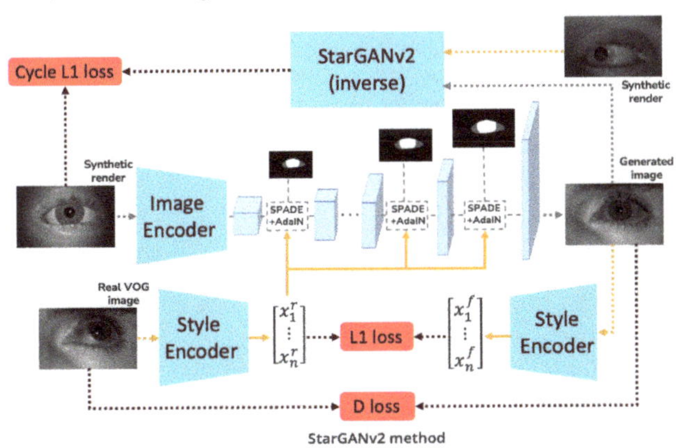

Figure 3. The outline of the proposed eye image generation method. The original StarGANv2 is modified with SPADE+AdaIN normalization layers in the image decoder and is adapted for domain transfer between real and synthetic eye images.

4.4. Train–Test Data Split

The first Blender set described earlier in the "Synthetic Blender data" subsection was used for training all data generation models. The whole set was split randomly into train/test subsets using a 90%/10% ratio. This resulted in 45,721 train and 5080 test Blender samples. The reduced real VOG set (mentioned in the text right before the Blender referenced above) was split on a per-subject basis using a 80%/20% ratio. This allowed us to test whether a given method could be an effective data augmentation tool for novel subjects that were not included in this study. The VOG data split resulted in 125,560 images from 32 train subjects and 30,515 images from 8 test subjects.

4.5. Evaluation Metrics

Using a synthetic framework as a data augmentation tool requires certain key aspects to be maintained in the generated images. First, they should be photorealistic and follow the visual style of a given reference image. Otherwise, their perceptual features may suffer a domain shift from a distribution mismatch with the real data. Second, the output of a generative method should be consistent in content with the input reference image and its segmentation mask. Failure to achieve good content consistency will introduce systematic bias in feature labels. For instance, the ground-truth gaze vector of a Blender render will no longer be an accurate label for an image obtained by passing that render through a StarGANv2-based method with poor content consistency. This will most likely prevent any improvement in the performance of end-to-end models when fake data are mixed with the real samples in the training set.

The Frechet Inception Distance (FID) is a metric proposed for assessing the perceptual quality of the generated images [63]. It is one of the most common metrics of photorealism used in the literature, as it was found to be consistent with human judgment [40]. The FID builds on the assumption that the perceptual features extracted from a set of images follow a Gaussian distribution. The Wasserstein-2 distance is then computed between the mean (μ) and covariance (cov) of features extracted from the real and generated image sets (P_r and P_g, respectively). The VGG-16 pre-trained on an ImageNet is typically used as a feature extractor ϕ. As a distance metric, the lower FID means better image quality, but the lowest achievable FID is data-dependent. The FID is then defined as follows, where $tr()$ is the matrix trace operator:

$$\text{FID}(\phi(P_r), \phi(P_g)) = ||\mu_r - \mu_g|| + tr(cov_r + cov_g - 2\sqrt{cov_r cov_g}) \quad (1)$$

The segmentation masks were used in this study as a proxy for image content. The most common metric of agreement between segmentation masks is an intersection-over-union (IoU), which is a ratio of the area of overlap between two masks and the area of their union. It was computed for every segmentation class (i.e., background, sclera, iris, and pupil) separately as a mean ratio across all masks. The single metric of content consistency for the whole data was computed as a mean over the per-class IoUs.

5. Experiments

5.1. Data Augmentation for Gaze Estimation

The data augmentation study was designed to test the real-world validity of the generated images. We chose the task of appearance-based gaze estimation to be our performance benchmark based on the following reasons. First, estimating gaze is the most fundamental task in image-based ET research. Second, modern appearance-based approaches learn a wide range of features that the underlying network finds to be useful for ET purposes. Making sure that the distribution of such features ends up being similar between the real and generated samples is an essential step in evaluating a synthetic data source. A severe distribution mismatch would cause the domain shift problem when the end-to-end solution is trained on some mix of real and fake data and then evaluated purely on real samples. Overall, a good generation method would be a tool for improving the performance of the end task when the data produced by it are mixed in with the real training set, which should be especially useful when the availability of real data is scarce. This exact use case was assessed with the benchmark we define as follows.

A subject-specific neural network was trained to regress eye images to gaze. The same reduced VOG real data set was used as in training all data generation methods. For every subject, the samples were randomly split into train and test sets using an 80%/20% ratio, and the following experiment was repeated 5 times in a cross-validation manner, each time with a different fold picked as a test set. The test set was set aside and used only at the evaluation step. The training set was further subdivided with the goal of simulating the scarce real data availability. There were 3 evaluation scenarios that each subject went through:

1. "The sparse real training set": the baseline where only 5% of the original training set was retained for training by random subsampling.
2. "Data generation method as a data augmentation tool": for that same 5% of real samples, the remaining 95% were added back from a data generation method that was being evaluated. The second Blender data set defined earlier in the "Synthetic Blender Data" subsection was used as input to all data generation methods. Note that as explained there, it was generated in such a way that the resulting fake images should follow the real ones they replace in the gaze position and camera location as closely as possible.
3. "The full real training set": the gaze estimation error on the full 100% training set was used to set the highest bar for a data generation method the output of which would be indistinguishable from real data.

The results were normalized to show the change in the gaze estimation error once the generated data were mixed in with the sparse training set of real images. The data point of 0% would represent no change in performance, and reaching 100% would mean that the performance on the full real training set was matched. This was done on a per-subject basis using the following formula (where $GE^{\%}_{rel}$ is the gaze error change as a percentage relative to the baseline performance on the sparse training set, and GE°_{i} is the gaze error in ° achieved in the i-th evaluation scenario from the list above):

$$GE^{\%}_{rel} = \frac{GE^{\circ}_{1} - GE^{\circ}_{2}}{GE^{\circ}_{1} - GE^{\circ}_{3}} * 100.0\ (\%) \tag{2}$$

5.2. Statistical Framework

The linear mixed-effects model was utilized to assess whether the observed differences were statistically significant. We used the implementation from the "lmerTest" package in R language [64]. The data generation method was considered a fixed effect. The subject and k-fold were initially modeled as random effects; however, the resulting model failed to converge. The k-fold random effect showed a zero variance estimate and per [65] was removed from the random effects, which led to a proper model fit. The Kenward–Roger approximation method [66] was used for estimating degrees of freedom. The multiple comparison of fixed effect factors was handled using Tukey's method with the implementation from the "emmeans" package in R language [67].

6. Results

6.1. Image Photorealism

The evaluation of all data generation methods on a test set is summarized in Table 1. The FID score was computed between 1000 fake and 1000 real images (both sets were randomly subsampled), and the mean of 5 repetitions is reported. All data have a different feasible FID score that can be estimated by computing it between two sets of real samples. In our case, the best practically achievable FID is 4.7. We can see that all methods that rely on learning from data had an order of magnitude improvement in the FID compared to the 262.95 score achieved by the Blender rendering tool. Using RITNet segmentation masks in EyeGAN training led to a slight decrease in the FID metric, which may be attributed to the trade-off between image quality and content consistency. Adding SPADE normalization to the StarGANv2 generator did not have a significant impact on the resulting image photorealism, but resulted in the method that achieved the best FID score of 51.08. However, there is still a lot of room for improvement in terms of closing the gap with real data that should be explored in future work.

Table 1. The comparison of all tested synthetic data generation methods using **a.** Frechet Inception Distance (FID) as a metric of image photorealism (lower is better, 4.7 is best); **b.** Mean Intersection-over-Union (mIoU) as a segmentation mask consistency metric, provided separately for sclera, iris and pupil segmentation classes (higher is better, 1.0 is best); **c.** whether it allows control over visual subject appearance in generated images.

Method	**a. FID** (4.7)	**b. mIoU** (1.0) All (Sclera/Iris/Pupil)	**c. Appearance Control**
Blender	262.95	≈1.0 (1.0/1.0/≈1.0)	−
EyeGAN	56.87	0.71 (0.48/0.66/0.78)	−
EyeGAN-RITNet	58.43	0.87 (0.69/0.88/0.93)	−
StarGANv2	51.55	0.50 (0.43/0.45/0.16)	+
StarGANv2-SPADE+AdaIN	51.08	0.57 (0.53/0.56/0.23)	+

6.2. Content Consistency

The content consistency was evaluated by computing the mIoU between the segmentation masks of the input and the output of every data generation method on a per-class basis and for all classes combined, as reported in Table 1. The background class typically occupies the majority of the eye image and is the easiest and at the same time the least important for us to get right, as it includes everything not related to the eyeball itself. Therefore, it is misleading to judge the eye image content consistency by a single combined mIoU, as it will be somewhat artificially inflated by the inclusion of the large mIoU for the background class. Moreover, the per-class scores may provide additional insight into the priorities of future work. Blender delivered the ideal content consistency, as the segmentation masks obtained via the rendering pipeline are pixel-perfect, and the "input" and the "output" are essentially the same mask. (Note that the original Blender masks lacked the pupil label that was added with the help of RITNet, which means that the mIoU for the pupil class, and, in turn, the combined mIoU, were very close but not technically equal to 1.0.) These Blender masks were used as input to all other data generation methods and were used as a ground-truth in computing their mIoU metrics. The missing output segmentation masks were computed by segmenting the generated images using RITNet. Our hypothesis that the use of RITNet masks to train EyeGAN would lead to improved content consistency was proved to be true. EyeGAN-RITNet significantly outperformed EyeGAN across all segmentation classes and achieved the best mIoU of 0.87 among the tested methods. We also hypothesized the same outcome from the use of SPADE in the StarGANv2 generator, and it turned out to be confirmed as well. Incorporating SPADE led to an increased mIoU of 0.57 with the base StarGANv2 achieving only 0.50. Nevertheless, the StarGANv2-SPADE+AdaIN method failed to match the mIoU scores of both EyeGAN-based methods. We propose two possible reasons to explain this result. First, EyeGAN was designed to have ground-truth mask-image pairs available for training, while StarGANv2 solved a much harder task of an unsupervised domain translation with no requirement of the paired training data. Second, StarGANv2 refined a synthetic render, rather than generating an image starting with its segmentation mask. The former approach was also utilized in SimGAN, and it was already shown to be inferior to the latter in terms of the resulting content consistency with the EyeGAN introduction. Therefore, we suggest that the StarGANv2 results may be further improved by borrowing the EyeGAN approach to image generation and exploring the way to leverage only a partial availability of ground-truth image-mask pairs in the training set.

6.3. Visual Inspection

The EyeGAN-RITNet generated images can be seen in Figure A3. They tend to have good perceptual quality, gaze and camera position consistency with the input segmentation mask of the depicted Blender render. The examples of image quality issues are provided in Figure A4. We can see that the images with certain edge gaze angles exhibit distorted iris shapes, the imperfectly segmented pupil leads to image artifacts, and the case of an almost closed eyelid is challenging. Moreover, we found that good consistency with the input segmentation mask does not always result in good gaze alignment between the synthetic render that corresponds to this mask and the generated image. This finding suggests the exploration of entities other than a segmentation mask that would be a better compact representation of ET image content. The major conceptual issue with the EyeGAN method is the complete lack of control over the output subject appearance, which makes it unusable as a subject-specific data generator. Note that this could be fixed by training multiple EyeGAN models—one per subject. Nevertheless, this solution carries a significant computational burden and has poor scalability.

The StarGANv2-based methods are the only ones that provide control over the visual subject appearance in their output. The StarGANv2-SPADE+AdaIN generated images shown in Figure A5 have improved perceptual quality, while it is visually noticeable that the overall content consistency is degraded when compared to EyeGAN-RITNet. The poor consistency is the most notable for the pupil, an observation that is supported by the lowest per-class mIoU in Table 1. The overall gaze consistency is manifested the most by the correctness of the iris placement, which is indeed mediocre, according to the corresponding per-class mIoU of 0.56. There is another major issue with the StarGANv2-SPADE+AdaIN results that can be seen in Figure A6. The resulting appearance can be controlled via a given VOG real image only to a certain extent, as the network suffers from mode collapse for certain subject identities. We manually went over all 40 subjects in the training set to assess the severity of the mode collapse issue of the StarGANv2 method. The 24 subjects had good visual consistency in the output images, and 5 of them can be seen in Figure A5. The remaining 16 were deemed to have bad consistency (with 5 of them shown in Figure A6). The common theme across all StarGANv2 issues seems to be some level of network overfitting. The observed results may suggest that the network had mostly learned to reproduce an image from the training set with the segmentation mask that is the closest to the input one. Therefore, we propose adding two following steps for future work on improving the StarGANv2 results. First, the overfitting should be addressed with the appropriate network regularization techniques. Second, other methods in addition to the AdaIN should be explored to improve the visual style transfer of the subject appearance. For instance, the features that are learned by the style encoder may not be as relevant to the subject appearance, which can be improved by training the encoder in a multi-task scenario with the added goal of the subject identity classification.

6.4. Data Augmentation Study

The results of the data augmentation study are presented in Figure 4 as boxplots of the gaze estimation improvement as a percentage. The improvement was computed as a change in performance relative to the baseline of the sparse real training set once the generated images were mixed into it. Every data point corresponds to a per-subject mean over 5 test folds. One subject turned out to be an outlier in the performance of the best method and was separated from this analysis, which is explained in more detail below. The mean gaze estimation error achieved on the sparse training set of real images was 1.01°. The use of the full real training set resulted in the gaze error of 0.45°. Note that the results were normalized to be in the percentage range on a per-subject basis. Therefore, any single data point at 0% does not correspond to the gaze estimation error in 1.01°, as this value depends on the baseline performance of the corresponding subject. Accordingly, if all data points of a given method are at 100%, only then their mean gaze estimation error is equal to 0.45°.

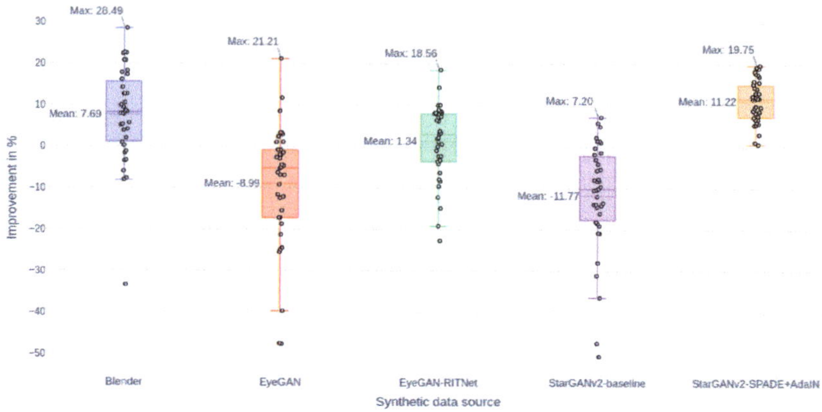

Figure 4. Comparison of synthetic data generation methods when used for data augmentation in appearance-based gaze estimation. The gaze estimation error improvement relative to the sparse training set of real images is summarized as a boxplot of a per-subject mean over 5 test folds.

We found that both methods that we applied directly as designed by their original author—EyeGAN and the base StarGANv2—are unsuitable for data augmentation purposes as they degraded the gaze estimation error on average by 8.99% and 11.77%, respectively. In both cases, our proposed changes led to a substantial improvement. Using RITNet masks for EyeGAN training allowed us to achieve the change in gaze estimation error that is positive on average (+1.34%). However, the use of EyeGAN-RITNet data in the training set was still detrimental to the gaze estimation performance for 15 out of 39 subjects. The addition of SPADE normalization into the base StarGANv2 turned out to be much more impactful. StarGANv2-SPADE+AdaIN is the only method that has proved to be a reliable tool for appearance-based gaze estimation. It degraded the performance for only one subject, which was removed from this analysis. For reporting completeness, the gaze estimation results for the excluded subject were the following: 0.84° on the sparse real set, 0.42° on the full real set; the use of Blender data resulted in the +0.28% gaze error improvement, EyeGAN - −31.4%, EyeGAN-RITNet - −38.87%, StarGANv2 - −0.08%, StarGANv2-SPADE+AdaIN - −31.32%. For the remaining 39 subjects, our best method achieved the mean gaze estimation increase of 11.22% with a maximum per-subject improvement of 19.75%. Moreover, it is the method with the smallest deviation from the mean performance. The performance difference between Blender, EyeGAN-RITNet, and StarGANv2-SPADE+AdaIN was statistically significant, as shown in Table A1. The other two methods were excluded from the statistical analysis, as they proved to be mostly harmful for data augmentation.

The most promising contribution of our StarGANv2-SPADE+AdaIN method is that it can be used to generate useful data even for novel subjects. Figure 5 shows that there was no significant degradation in performance for test subjects that were not included in the training set. However, this claim is supported by data from only 8 people, which may not be enough for far-reaching conclusions. Aside from collecting more data, the easiest way to improve the reliability of this study is to repeat it in a 5-fold cross-validation manner, splitting all 40 available subjects into disjoint test sets. Blender is another method that achieved similar performance for the train and test subjects, which was an expected outcome, as it always generates data from the same single identity.

There are two most surprising observations from the data augmentation study results. First, while the mean IoU value in Table 1 for the EyeGAN-based methods was the highest, that did not translate to good performance when assessed in an end-to-end learning scenario. Second, considering a dramatically low level of photorealism in Blender renders, it performed just slightly worse than StarGANv2 in this experiment. Nevertheless, most of the remaining absolute performance gap of 92.31% between the sparse and full training sets should be attributed to the complete lack of photorealism in this data set. We discuss the implications of these unexpected findings on future work in the next section.

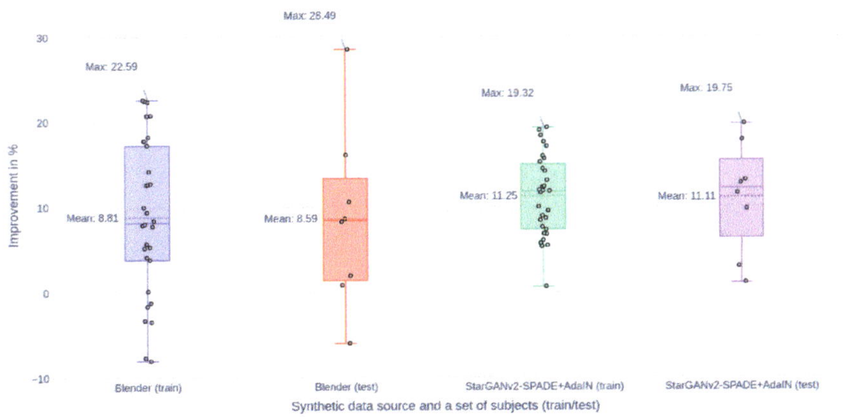

Figure 5. Comparison of data augmentation performance between train and test subjects for Blender and StarGANv2-SPADE+AdaIN methods. The plot format follows Figure 4.

7. Discussion and Future Work

To the best of our knowledge, the StarGANv2-SPADE+AdaIN method is the first method to generate realistic eye image data that does not require ground-truth supervision during training, as in [53]; a costly data collection setup akin to [68]; and can be used for any appearance-based research, unlike [57] designed purely for gaze redirection. The effectiveness of our method is proved by a novel benchmark that directly evaluates any synthetic data source as a tool for data augmentation.

We hope that the proposed data augmentation benchmark will be valuable to the eye-tracking community, as it is easily extensible to other evaluation scenarios. For instance, it is straightforward to adapt it for the task of zero- or few-shot gaze estimation. While StarGANv2-SPADE+AdaIN showed only a mild mean gaze estimation improvement of 11.22%, the average size of the sparse training set of real images used as a baseline was around 195 samples. It is reasonable to assume that the smaller this set is, the larger the improvement would be. Even if no subject-specific data are available for training, the task of zero-shot gaze estimation may be solved by using data from many synthetic subject identities. Any of these hypotheses can be formally tested within essentially the same evaluation framework, with the only minor change in the size of real data available for training. To generate data for a novel identity, we may need as few as just one image to compute the rough estimate of its style latent code. The sensitivity of data augmentation performance to the variance in the style code estimation is another question that we leave for future work.

The data augmentation study evaluates the mismatch between the real and generated data based on the image features that the neural network learned to be useful for gaze estimation. As in a typical fashion of end-to-end learning from data, the resulting features are not easily mapped to tangible domain-specific concepts. The fact that these features are not human explainable makes it hard to derive a plan for addressing the large remaining gap between real and fake images in appearance-based ET research. However, it is reasonable to assume that some linear combination of the general metrics in Table 1 may be a good predictor of the data augmentation performance of a given method. Thinking in this framework makes future work planning more feasible, as the goal of improving any objective well-defined metric is straightforward to set. Unfortunately, it is not clear from our study alone to what extent FID, mIoU, or style control influence the data augmentation performance, yet alone which metric is the most contributing factor in the performance increase. The Blender had an inferior FID score to any EyeGAN-based method, but it beat both of them in data augmentation. EyeGAN-RITNet had the best image content consistency among learning-from-data methods, yet it was easily beaten by StarGANv2-SPADE+AdaIN with one of the worst mIoUs. The only limited evidence we have is that increasing the mIoU within a similar range of FID and the same style-control ability leads to improved performance in data augmentation. This happened to be the case for our proposed modifications to both the original EyeGAN and StarGANv2 methods that increased their baseline mIoU level. Therefore, we suggest that further improvement in content consistency while maintaining or improving photorealism in StarGANv2-SPADE+AdaIN generated images is the most promising step for future work.

To sum up, this work presented a benchmark of eye image generation methods. We tested their applicability in improving appearance-based gaze estimation through data augmentation. In that scenario, the generated samples were mixed in with a sparse set of real data used to train a neural network inferring gaze from eye images in an end-to-end manner. We showed that existing generative methods common in the eye image domain deliver unsatisfactory results, often degrading the error in the target task of gaze estimation. This finding was obtained in a novel evaluation based on the pre-collected set of 1.24M gaze-labeled eye images from 40 subjects that we released for public access. To address the deficiencies of prior work, we derived a novel adaptation of the general domain transfer method to real and synthetic eye images. It resulted in a hybrid approach that combines the realism of recorded data with the flexibility of 3D rendering frameworks. This method outperformed all the others, improving the gaze estimation for 39 out of 40 subjects, with a 11.22% mean and a 19.75% maximum decrease in gaze error. The reported differences were shown to be statistically significant.

Author Contributions: Conceptualization, D.K. and O.V.K.; methodology, software, validation, formal analysis, investigation, data curation, writing—original draft preparation, visualization, D.K.; resources, writing—review and editing, supervision, project administration, O.V.K. All authors have read and agreed to the published version of the manuscript.

Funding: This research received no external funding.

Institutional Review Board Statement: The study was conducted in accordance with the Declaration of Helsinki, and approved by the Institutional Review Board of Texas State University (approval code 2018044).

Informed Consent Statement: Informed consent was obtained from all subjects involved in the study.

Data Availability Statement: The raw data supporting the conclusions of this article are publicly available at https://doi.org/10.5281/zenodo.13852050 (accessed on 9 October 2024).

Conflicts of Interest: The authors declare no conflicts of interest.

Appendix A. Visual Examples of Utilized Images as Segmentation Masks

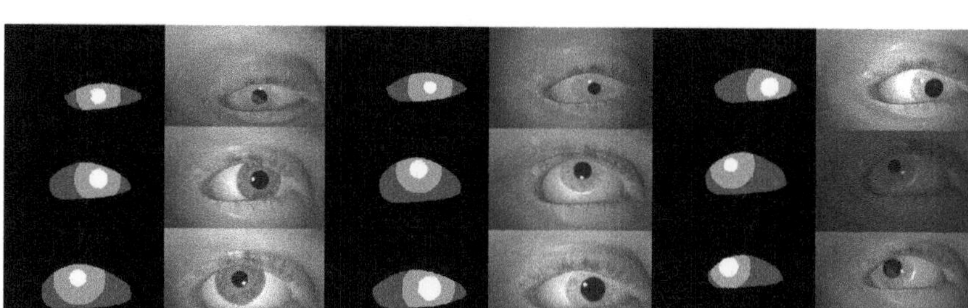

Figure A1. Image examples with corresponding segmentation masks that were rendered using the modified Blender model.

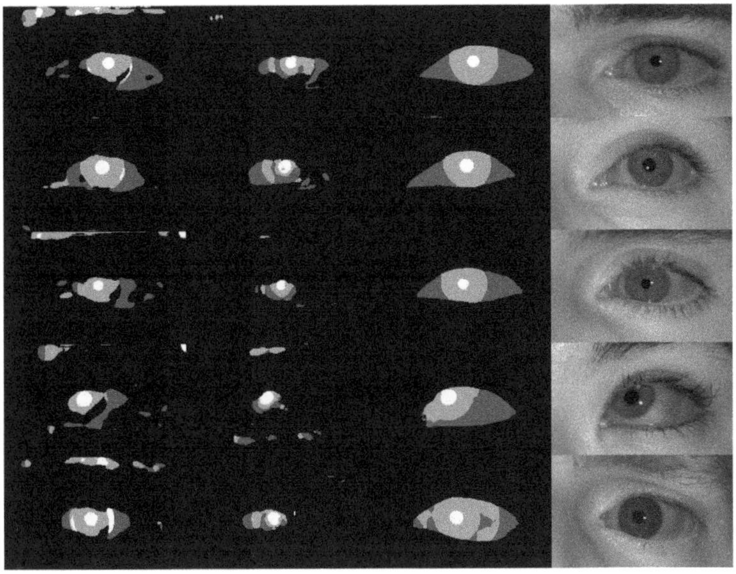

Figure A2. Real eye image examples (the fourth column) with corresponding segmentation masks that were computed at the first iteration of EyeGAN (the first column), at the second iteration of EyeGAN (the second column), and using the RITNet model (the third column).

Appendix B. Visual Examples of Generated Images

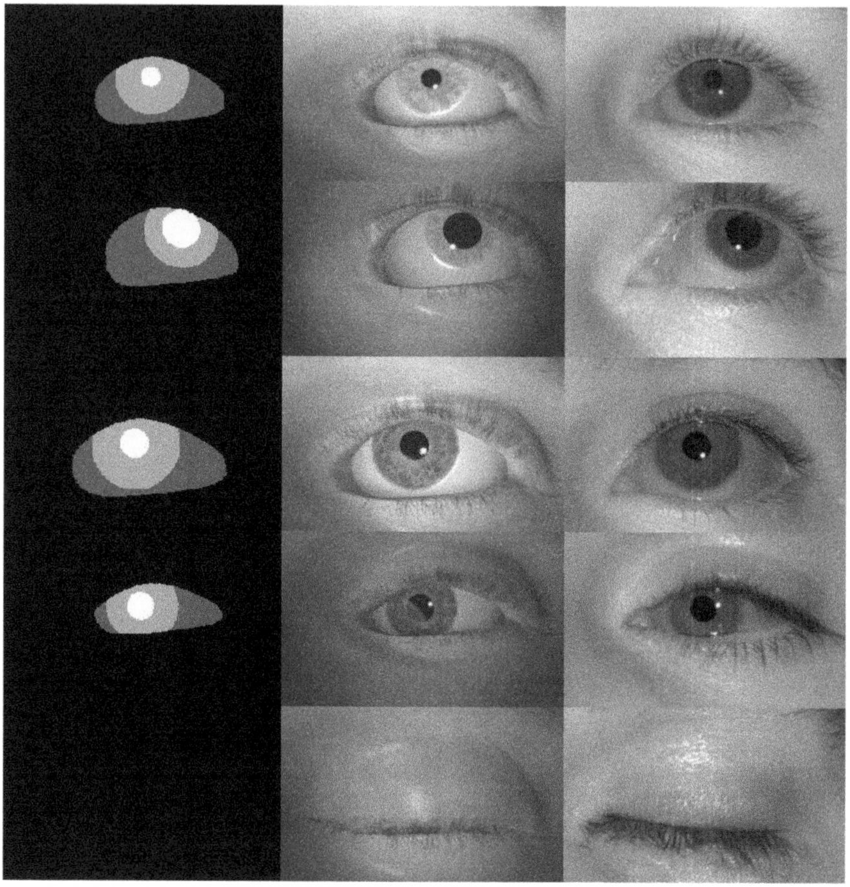

Figure A3. EyeGAN-RITNet generated images with good quality; each row is an input–output triplet (left—input segmentation mask; middle—a corresponding Blender render; right—EyeGAN-RITNet output).

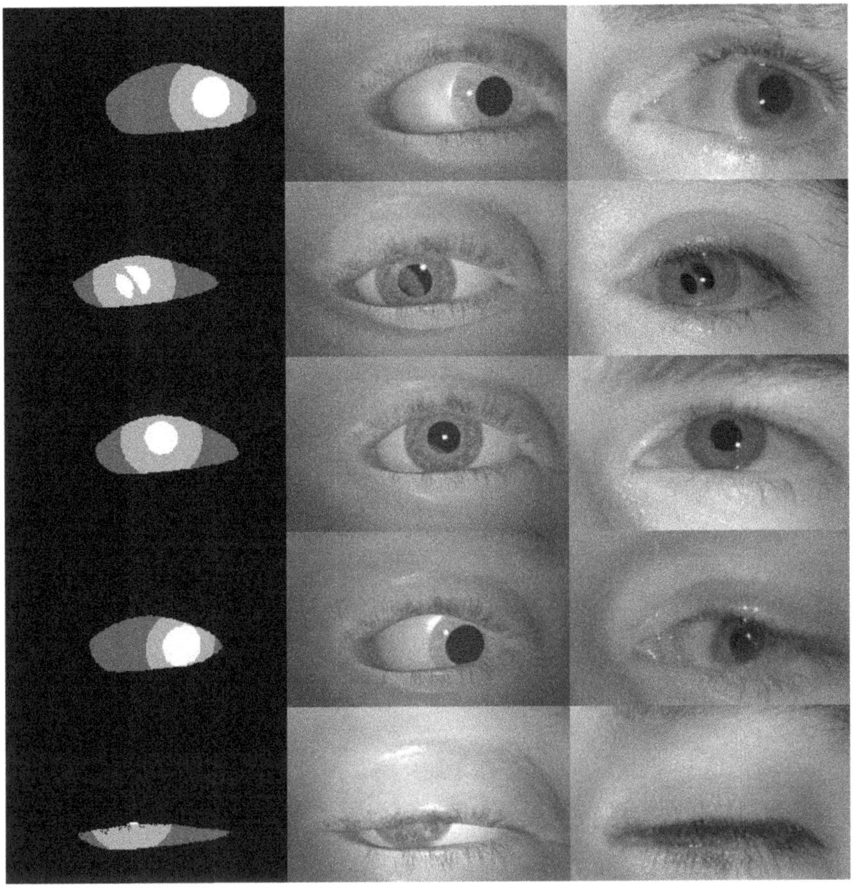

Figure A4. EyeGAN-RITNet generated images with quality issues; the image structure follows Figure A3.

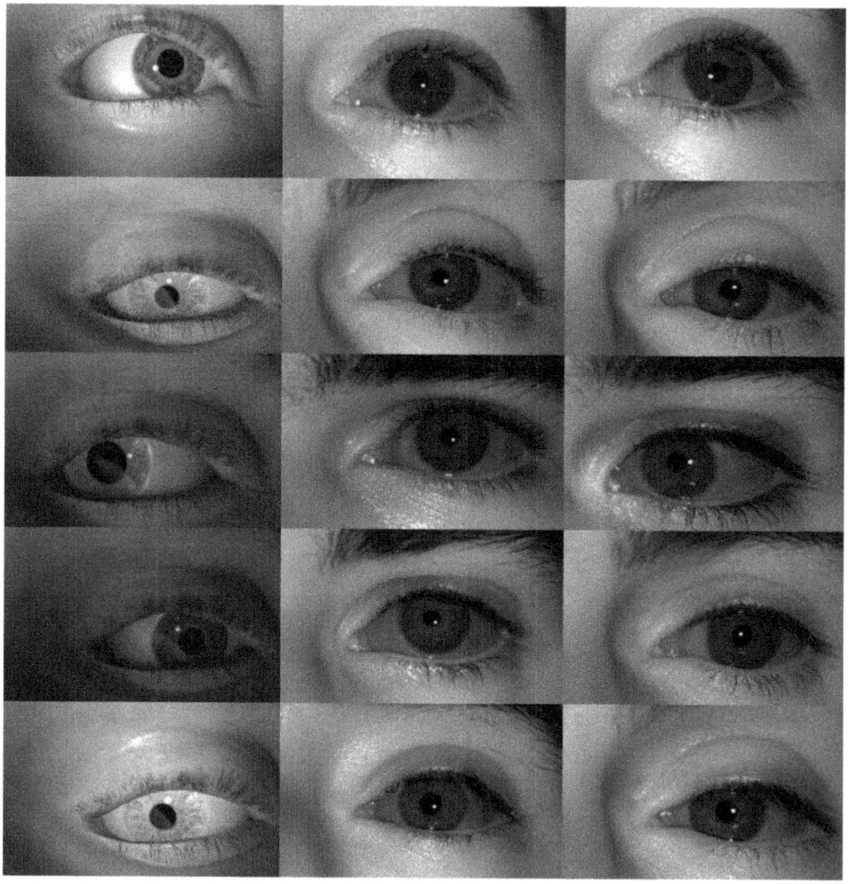

Figure A5. StarGANv2-SPADE+AdaIN generated images with good subject appearance consistency; each row is an input–output triplet (left—input Blender render, middle—input VOG image for target subject appearance, right—StarGANv2-SPADE+AdaIN output).

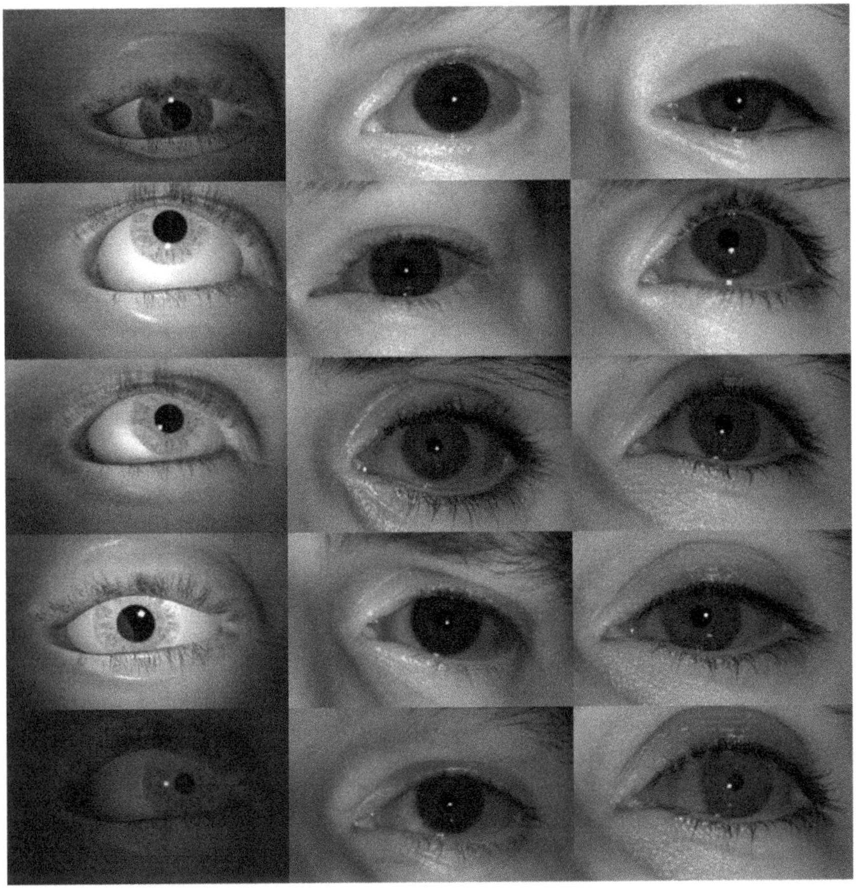

Figure A6. StarGANv2-SPADE+AdaIN generated images with bad subject appearance consistency; the image structure follows Figure A5.

Appendix C. Statistical Analysis of Difference in Data Augmentation Methods

Table A1. The statistical analysis of difference between data augmentation methods (the roman numerals follow the order of methods in Figure 4).

Comparison	Δ in Mean Error	DF	*t*-Value	*p*-Value
I vs. II	6.35	544	7.090	<0.0001
I vs. V	−3.54	544	−3.948	0.0003
II vs. V	−9.88	544	−11.038	<0.0001

Appendix D. Blender image rendering parameters

The first Blender set was used for training all image generation networks. One of the goals of the trained generative framework was to produce data covering a wide range of features, complementary to the variability present in real data. The following parameter values were utilized within the rendering pipeline (the values were sampled randomly, where U, N stand for uniform and normal distributions, respectively):

- Camera position: $U(\pm 2.5; \pm 2.5; \pm 5.0)$ mm—horizontal, vertical and depth-wise shift from the neutral position of $(0; 0; 0)$;
- Eye gaze: $U(\pm 20.51°; \pm 16.7°)$—covering the same range as in real data;

- Lighting conditions: a random omnidirectional point of light (i.e., light type "sun" in Blender 2.7) placed at $\{x = U(-90.0; -45.0), y = -45.0, z = U(-90.0; -45.0)\}$ location and strength property sampled from $U(1.0; 10.0)$; a random single-directional point of light (i.e., light type "led" in Blender 2.7) with location and strength sampled from the same distributions, intended to simulate varying lighting conditions;
- Dark/bright iris: 50%/50% split (as only two iris textures were available in the original Blender model);
- Pupil radius: $N(1.75; 0.33)$ mm limited to the range of $[1.0 \ldots 2.5]$;
- Blink: 5% chance with an eye "closedness" sampled from $U(0.0; 1.0)$.

The parameters for generating the second Blender set were manually adjusted, taken from the real data of a corresponding subject, or sampled randomly. The central camera position was aligned manually for each subject, as described in the main text. The gaze samples directly followed the real eye movement recordings, which also means that the number of generated samples both per-subject and overall was the same as in the reduced VOG set. The randomness was implemented both on across-subject and within-subjects levels. The remaining parameters, namely, lighting sources strength, pupil size, iris color, and eye "closedness" during a blink, were all non-trivial to adjust on a per-sample basis using corresponding real samples as reference points. Instead, they were randomly sampled. The overall scheme of rendering parameters is summarized below:

- Camera: manually adjusted per-subject base position + per-sample horizontal, vertical and depth shifts of $N(0; 0.1)$, $N(0; 0.05)$, $N(0; 0.25)$ (in mm), respectively;
- Eye gaze: taken from a corresponding sample of real VOG data for all subjects;
- Lighting conditions; Dark/bright iris: same parameters as in the first Blender set;
- Pupil radius: $N(1.5; 0.25)$ mm limited to the range of $[1.0 \ldots 2.0]$;
- Blink: 0% chance (no blinks, as in the reduced VOG data set).

References

1. Salvucci, D.D.; Anderson, J.R. Intelligent gaze-added interfaces. In Proceedings of the SIGCHI Conference on Human Factors in Computing Systems, The Hague, The Netherlands, 1–6 April 2000; pp. 273–280.
2. Møllenbach, E.; Hansen, J.P.; Lillholm, M. Eye movements in gaze interaction. *J. Eye Mov. Res.* **2013**, *6*. [CrossRef]
3. Fares, R.; Fang, S.; Komogortsev, O. Can we beat the mouse with MAGIC? In Proceedings of the SIGCHI Conference on Human Factors in Computing Systems, Paris, France, 27 April–2 May 2013; pp. 1387–1390.
4. Guenter, B.; Finch, M.; Drucker, S.; Tan, D.; Snyder, J. Foveated 3D graphics. *ACM Trans. Graph. (TOG)* **2012**, *31*, 164. [CrossRef]
5. Patney, A.; Salvi, M.; Kim, J.; Kaplanyan, A.; Wyman, C.; Benty, N.; Luebke, D.; Lefohn, A. Towards foveated rendering for gaze-tracked virtual reality. *ACM Trans. Graph. (TOG)* **2016**, *35*, 1–12. [CrossRef]
6. Rigas, I.; Friedman, L.; Komogortsev, O. Study of an extensive set of eye movement features: Extraction methods and statistical analysis. *J. Eye Mov. Res.* **2018**, *11*. [CrossRef]
7. Lohr, D.; Proulx, M.J.; Komogortsev, O. Establishing a Baseline for Gaze-driven Authentication Performance in VR: A Breadth-First Investigation on a Very Large Dataset. *arXiv* **2024**, arXiv:2404.11798.
8. Komogortsev, O.V.; Holland, C.D. The application of eye movement biometrics in the automated detection of mild traumatic brain injury. In *CHI'14 Extended Abstracts on Human Factors in Computing Systems*; Association for Computing Machinery: New York, NY, USA, 2014; pp. 1711–1716.
9. Keefe, R.S.; Silverman, J.M.; Mohs, R.C.; Siever, L.J.; Harvey, P.D.; Friedman, L.; Roitman, S.E.L.; DuPre, R.L.; Smith, C.J.; Schmeidler, J.; et al. Eye tracking, attention, and schizotypal symptoms in nonpsychotic relatives of patients with schizophrenia. *Arch. Gen. Psychiatry* **1997**, *54*, 169–176. [CrossRef]
10. Abel, L.A.; Friedman, L.; Jesberger, J.; Malki, A.; Meltzer, H. Quantitative assessment of smooth pursuit gain and catch-up saccades in schizophrenia and affective disorders. *Biol. Psychiatry* **1991**, *29*, 1063–1072. [CrossRef]
11. Klaib, A.F.; Alsrehin, N.O.; Melhem, W.Y.; Bashtawi, H.O.; Magableh, A.A. Eye tracking algorithms, techniques, tools, and applications with an emphasis on machine learning and Internet of Things technologies. *Expert Syst. Appl.* **2021**, *166*, 114037. [CrossRef]
12. Hansen, D.W.; Ji, Q. In the eye of the beholder: A survey of models for eyes and gaze. *IEEE Trans. Pattern Anal. Mach. Intell.* **2009**, *32*, 478–500. [CrossRef]
13. Guestrin, E.D.; Eizenman, M. General theory of remote gaze estimation using the pupil center and corneal reflections. *IEEE Trans. Biomed. Eng.* **2006**, *53*, 1124–1133. [CrossRef]
14. Zhang, X.; Sugano, Y.; Fritz, M.; Bulling, A. Mpiigaze: Real-world dataset and deep appearance-based gaze estimation. *IEEE Trans. Pattern Anal. Mach. Intell.* **2017**, *41*, 162–175. [CrossRef] [PubMed]

15. Kellnhofer, P.; Recasens, A.; Stent, S.; Matusik, W.; Torralba, A. Gaze360: Physically unconstrained gaze estimation in the wild. In Proceedings of the IEEE/CVF International Conference on Computer Vision, Seoul, Republic of Korea, 27 October–2 November 2019; pp. 6912–6921.
16. Cheng, Y.; Lu, F. Gaze estimation using transformer. In Proceedings of the 2022 26th International Conference on Pattern Recognition (ICPR), Montreal, QC, Canada, 21–25 August 2022; pp. 3341–3347.
17. Cheng, Y.; Wang, H.; Bao, Y.; Lu, F. Appearance-based gaze estimation with deep learning: A review and benchmark. *IEEE Trans. Pattern Anal. Mach. Intell.* **2024**, 1–20. [CrossRef] [PubMed]
18. Perez, L.; Wang, J. The effectiveness of data augmentation in image classification using deep learning. *arXiv* **2017**, arXiv:1712.04621.
19. Salamon, J.; Bello, J.P. Deep convolutional neural networks and data augmentation for environmental sound classification. *IEEE Signal Process. Lett.* **2017**, *24*, 279–283. [CrossRef]
20. Shorten, C.; Khoshgoftaar, T.M. A survey on Image Data Augmentation for Deep Learning. *J. Big Data* **2019**, *6*, 60. [CrossRef]
21. Katrychuk, D.; Griffith, H.K.; Komogortsev, O.V. Power-efficient and shift-robust eye-tracking sensor for portable VR headsets. In Proceedings of the 11th ACM Symposium on Eye Tracking Research & Applications, Denver, CO, USA, 25–28 June 2019; pp. 1–8.
22. Kim, J.; Stengel, M.; Majercik, A.; De Mello, S.; Dunn, D.; Laine, S.; McGuire, M.; Luebke, D. Nvgaze: An anatomically-informed dataset for low-latency, near-eye gaze estimation. In Proceedings of the 2019 CHI Conference on Human Factors in Computing Systems, Glasgow, UK, 4–9 May 2019; pp. 1–12.
23. Świrski, L.; Dodgson, N. Rendering synthetic ground truth images for eye tracker evaluation. In Proceedings of the Symposium on Eye Tracking Research and Applications, Safety Harbor, FL, USA, 26–28 March 2014; pp. 219–222.
24. Choi, Y.; Uh, Y.; Yoo, J.; Ha, J.W. Stargan v2: Diverse image synthesis for multiple domains. In Proceedings of the IEEE/CVF Conference on Computer Vision and Pattern Recognition, Seattle, WA, USA, 13–19 June 2020; pp. 8188–8197.
25. Kaur, H.; Manduchi, R. EyeGAN: Gaze–Preserving, Mask–Mediated Eye Image Synthesis. In Proceedings of the 2020 IEEE Winter Conference on Applications of Computer Vision (WACV), Snowmass, CO, USA, 1–5 March 2020.
26. Shrivastava, A.; Pfister, T.; Tuzel, O.; Susskind, J.; Wang, W.; Webb, R. Learning from simulated and unsupervised images through adversarial training. In Proceedings of the IEEE Conference on Computer Vision and Pattern Recognition, Honolulu, HI, USA, 21–26 July 2017; pp. 2107–2116.
27. Wood, E.; Baltrušaitis, T.; Morency, L.P.; Robinson, P.; Bulling, A. Learning an appearance-based gaze estimator from one million synthesised images. In Proceedings of the Ninth Biennial ACM Symposium on Eye Tracking Research & Applications, Charleston, SC, USA, 14–17 March 2016; pp. 131–138.
28. Nair, N.; Chaudhary, A.K.; Kothari, R.S.; Diaz, G.J.; Pelz, J.B.; Bailey, R. RIT-Eyes: Realistically rendered eye images for eye-tracking applications. In Proceedings of the Symposium on Eye Tracking Research and Applications, Stuttgart, Germany, 2–5 June 2020; pp. 1–3. [CrossRef]
29. Agtzidis, I.; Startsev, M.; Dorr, M. A Ground-Truth Data Set and a Classification Algorithm for Eye Movements in 360-degree Videos. *arXiv* **2019**, arXiv:1903.06474.
30. Goodfellow, I.; Pouget-Abadie, J.; Mirza, M.; Xu, B.; Warde-Farley, D.; Ozair, S.; Courville, A.; Bengio, Y. Generative adversarial nets. *Adv. Neural Inf. Process. Syst.* **2014**, *27*, 2672–2680.
31. Brock, A.; Donahue, J.; Simonyan, K. Large scale GAN training for high fidelity natural image synthesis. *arXiv* **2018**, arXiv:1809.11096.
32. Karras, T.; Aittala, M.; Laine, S.; Härkönen, E.; Hellsten, J.; Lehtinen, J.; Aila, T. Alias-free generative adversarial networks. *Adv. Neural Inf. Process. Syst.* **2021**, *34*, 852–863.
33. Kong, J.; Kim, J.; Bae, J. Hifi-gan: Generative adversarial networks for efficient and high fidelity speech synthesis. *Adv. Neural Inf. Process. Syst.* **2020**, *33*, 17022–17033.
34. Isola, P.; Zhu, J.Y.; Zhou, T.; Efros, A.A. Image-to-image translation with conditional adversarial networks. In Proceedings of the IEEE Conference on Computer Vision and Pattern Recognition, Honolulu, HI, USA, 21–27 July 2017; pp. 1125–1134.
35. Zhu, J.Y.; Park, T.; Isola, P.; Efros, A.A. Unpaired image-to-image translation using cycle-consistent adversarial networks. In Proceedings of the IEEE International Conference on Computer Vision, Venice, Italy, 22–29 October 2017; pp. 2223–2232.
36. Mathur, A.; Isopoussu, A.; Kawsar, F.; Berthouze, N.; Lane, N.D. Mic2mic: Using cycle-consistent generative adversarial networks to overcome microphone variability in speech systems. In Proceedings of the 18th International Conference on Information Processing in Sensor Networks, Montreal, QC, Canada, 16–18 April 2019; pp. 169–180.
37. Mao, X.; Li, Q.; Xie, H.; Lau, R.Y.; Wang, Z.; Paul Smolley, S. Least squares generative adversarial networks. In Proceedings of the IEEE International Conference on Computer Vision, Venice, Italy, 22–29 October 2017; pp. 2794–2802.
38. Gulrajani, I.; Ahmed, F.; Arjovsky, M.; Dumoulin, V.; Courville, A.C. Improved training of wasserstein gans. *Adv. Neural Inf. Process. Syst.* **2017**, *30* , 5769–5779.
39. Kodali, N.; Abernethy, J.; Hays, J.; Kira, Z. On convergence and stability of gans. *arXiv* **2017**, arXiv:1705.07215.
40. Lucic, M.; Kurach, K.; Michalski, M.; Gelly, S.; Bousquet, O. Are gans created equal? A large-scale study. *Adv. Neural Inf. Process. Syst.* **2018**, *31*, 700–709.
41. Borji, A. Pros and cons of GAN evaluation measures: New developments. *Comput. Vis. Image Underst.* **2022**, *215*, 103329. [CrossRef]
42. Ho, J.; Jain, A.; Abbeel, P. Denoising diffusion probabilistic models. *Adv. Neural Inf. Process. Syst.* **2020**, *33*, 6840–6851.
43. Dhariwal, P.; Nichol, A. Diffusion models beat gans on image synthesis. *Adv. Neural Inf. Process. Syst.* **2021**, *34*, 8780–8794.

44. Nichol, A.; Dhariwal, P.; Ramesh, A.; Shyam, P.; Mishkin, P.; McGrew, B.; Sutskever, I.; Chen, M. Glide: Towards photorealistic image generation and editing with text-guided diffusion models. *arXiv* **2021**, arXiv:2112.10741.
45. Podell, D.; English, Z.; Lacey, K.; Blattmann, A.; Dockhorn, T.; Müller, J.; Penna, J.; Rombach, R. Sdxl: Improving latent diffusion models for high-resolution image synthesis. *arXiv* **2023**, arXiv:2307.01952.
46. Betker, J.; Goh, G.; Jing, L.; Brooks, T.; Wang, J.; Li, L.; Ouyang, L.; Zhuang, J.; Lee, J.; Guo, Y.; et al. Improving image generation with better captions. *Comput. Sci.* **2023**, *2*, 8.
47. Rombach, R.; Blattmann, A.; Lorenz, D.; Esser, P.; Ommer, B. High-resolution image synthesis with latent diffusion models. In Proceedings of the IEEE/CVF Conference on Computer Vision and Pattern Recognition, New Orleans, LO, USA, 19–20 June 2022; pp. 10684–10695.
48. Wang, Z.; Jiang, Y.; Zheng, H.; Wang, P.; He, P.; Wang, Z.; Chen, W.; Zhou, M. Patch diffusion: Faster and more data-efficient training of diffusion models. *Adv. Neural Inf. Process. Syst.* **2024**, *36*, 72137–72154.
49. Antoniou, A.; Storkey, A.; Edwards, H. Data Augmentation Generative Adversarial Networks. *arXiv* **2017**, arXiv:1711.04340.
50. Sandfort, V.; Yan, K.; Pickhardt, P.J.; Summers, R.M. Data augmentation using generative adversarial networks (CycleGAN) to improve generalizability in CT segmentation tasks. *Sci. Rep.* **2019**, *9*, 16884. [CrossRef] [PubMed]
51. Sela, M.; Xu, P.; He, J.; Navalpakkam, V.; Lagun, D. Gazegan-unpaired adversarial image generation for gaze estimation. *arXiv* **2017**, arXiv:1711.09767.
52. Gatys, L.A.; Ecker, A.S.; Bethge, M. Image style transfer using convolutional neural networks. In Proceedings of the IEEE Conference on Computer Vision and Pattern Recognition, Las Vegas, NV, USA, 27–30 June 2016; pp. 2414–2423.
53. Buhler, M.; Park, S.; De Mello, S.; Zhang, X.; Hilliges, O. Content-consistent generation of realistic eyes with style. In Proceedings of the 2019 IEEE/CVF International Conference on Computer Vision Workshop (ICCVW), Seoul, Republic of Korea, 27–28 October 2019; pp. 1–5.
54. Johnson, J.; Alahi, A.; Fei-Fei, L. Perceptual losses for real-time style transfer and super-resolution. In Proceedings of the European Conference on Computer Vision, Amsterdam, The Netherlands, 11–14 October 2016; pp. 694–711.
55. Park, T.; Liu, M.Y.; Wang, T.C.; Zhu, J.Y. Semantic image synthesis with spatially-adaptive normalization. In Proceedings of the IEEE/CVF Conference on Computer Vision and Pattern Recognition, Long Beach, CA, USA, 15–20 June 2019; pp. 2337–2346.
56. Huang, X.; Belongie, S. Arbitrary style transfer in real-time with adaptive instance normalization. In Proceedings of the IEEE International Conference on Computer Vision, Venice, Italy, 22–29 October 2017; pp. 1501–1510.
57. Kaur, H.; Manduchi, R. Subject guided eye image synthesis with application to gaze redirection. In Proceedings of the IEEE/CVF Winter Conference on Applications of Computer Vision, Virtual, 5–9 January 2021; pp. 11–20.
58. Choi, Y.; Choi, M.; Kim, M.; Ha, J.W.; Kim, S.; Choo, J. Stargan: Unified generative adversarial networks for multi-domain image-to-image translation. In Proceedings of the IEEE Conference on Computer Vision and Pattern Recognition, Salt Lake City, UT, USA, 18–23 June 2018; pp. 8789–8797.
59. Katrychuk, D.; Griffith, H.; Komogortsev, O. A Calibration Framework for Photosensor-based Eye-Tracking System. In Proceedings of the Symposium on Eye Tracking Research and Applications, Stuttgart, Germany, 2–5 June 2020; pp. 1–5.
60. Abdulin, E.; Friedman, L.; Komogortsev, O. Custom video-oculography device and its application to fourth purkinje image detection during saccades. *arXiv* **2019**, arXiv:1904.07361.
61. Chaudhary, A.K.; Kothari, R.; Acharya, M.; Dangi, S.; Nair, N.; Bailey, R.; Kanan, C.; Diaz, G.; Pelz, J.B. Ritnet: Real-time semantic segmentation of the eye for gaze tracking. In Proceedings of the 2019 IEEE/CVF International Conference on Computer Vision Workshop (ICCVW), Seoul, Republic of Korea, 27–28 October 2019; pp. 3698–3702.
62. Facebook. 2019 OpenEDS Challenge for Eye Image Segmentation. Available online: https://research.facebook.com/openeds-challenge/ (accessed on 28 June 2023).
63. Heusel, M.; Ramsauer, H.; Unterthiner, T.; Nessler, B.; Hochreiter, S. Gans trained by a two time-scale update rule converge to a local nash equilibrium. *Adv. Neural Inf. Process. Syst.* **2017**, *30*, 6626–6637.
64. Kuznetsova, A.; Brockhoff, P.B.; Christensen, R.H. lmerTest package: Tests in linear mixed effects models. *J. Stat. Softw.* **2017**, *82*, 1–26. [CrossRef]
65. Barr, D.J.; Levy, R.; Scheepers, C.; Tily, H.J. Random effects structure for confirmatory hypothesis testing: Keep it maximal. *J. Mem. Lang.* **2013**, *68*, 255–278. [CrossRef]
66. Halekoh, U.; Højsgaard, S. A kenward-roger approximation and parametric bootstrap methods for tests in linear mixed models—The R package pbkrtest. *J. Stat. Softw.* **2014**, *59*, 1–32. [CrossRef]
67. Lenth, R.; Singmann, H.; Love, J.; Buerkner, P.; Herve, M. Package 'emmeans'. *Am. Stat.* **2018**, *24*, 216–221.
68. Li, G.; Meka, A.; Mueller, F.; Buehler, M.C.; Hilliges, O.; Beeler, T. EyeNeRF: A hybrid representation for photorealistic synthesis, animation and relighting of human eyes. *ACM Trans. Graph. (TOG)* **2022**, *41*, 1–16. [CrossRef]

applied sciences

Article

Effect of Stimulus Regularities on Eye Movement Characteristics

Bilyana Genova *, Nadejda Bocheva and Ivan Hristov

Institute of Neurobiology, Bulgarian Academy of Sciences, 1113 Sofia, Bulgaria; nadya@percept.bas.bg (N.B.)
* Correspondence: b.genova@abv.bg

Abstract: Humans have the unique ability to discern spatial and temporal regularities in their surroundings. However, the effect of learning these regularities on eye movement characteristics has not been studied enough. In the present study, we investigated the effect of the frequency of occurrence and the presence of common chunks in visual images on eye movement characteristics like the fixation duration, saccade amplitude and number, and gaze number across sequential experimental epochs. The participants had to discriminate the patterns presented in pairs as the same or different. The order of pairs was repeated six times. Our results show an increase in fixation duration and a decrease in saccade amplitude in the sequential epochs, suggesting a transition from ambient to focal information processing as participants acquire knowledge. This transition indicates deeper cognitive engagement and extended analysis of the stimulus information. Interestingly, contrary to our expectations, the saccade number increased, and the gaze number decreased. These unexpected results might imply a reduction in the memory load and a narrowing of attentional focus when the relevant stimulus characteristics are already determined.

Keywords: eye movements; fixation duration; saccade amplitude; saccade number; gaze number; learning; memory

1. Introduction

Humans detect and extract the temporal and spatial regularities in the environment through a process called statistical learning. While much research has been focused on distributional regularities (e.g., frequency of occurrence) and transitional probabilities (e.g., order of appearance), higher-order regularities remain underexplored [1]. Moreover, in typical studies on statistical learning, the acquired knowledge of the statistical regularities is not examined in its development but is evaluated in a subsequent test phase. One way to explore the learning process is by using non-observational tasks that require a response from the participants. This approach is used in studies applying either the Serial Reaction Time (SRT) task [2] or the Alternating Serial Reaction Time task (e.g., [2]). In both tasks, the stimulus location and order are manipulated, thus varying the predictability of the targets. The participants respond manually, and the response time change in the experimental trials is a measure of the learning success. In a recent study, [3] used eye movement records to evaluate the temporal dynamics of the learning process by examining the occurrence of anticipatory eye movements during the experiment. This study is one of the few studies that evaluate the changes in eye movement characteristics across trials to characterize the ongoing process of learning. However, the eye movement characteristics may provide richer information than those evaluated by the predictive gaze landings on targets with variable predictivity. Eye movements can provide insight into the unobservable processes of visual selection, decision making, and uncertainty reduction associated with learning.

It is generally believed that statistical learning occurs without awareness (e.g., [4]), i.e., it involves implicit memory. Ramey and co-authors [5] explored whether gaze behavior changes depending on the involvement of explicit or implicit memory in a visual search task. The study's results showed that the location of the first fixation depended on explicit memory, whereas the search efficiency evaluated by the eye scan path improved due to unconscious memory. The study also showed that the effects of the two memory types were independent.

Other authors (e.g., [6,7]) suggested that eye movements were not needed for implicit learning of spatial configurations. Arató and co-authors [8] provided evidence for the mutual connection between eye movement patterns and the acquired knowledge about the statistical structure of the scenes. Their study explored the interaction between eye movement characteristics and the learning process without a well-defined task. However, natural vision is an active process, and movements of the eyes allow continuous comparison of the current visual input with prior experience to guide our future behavior (e.g., make decisions, navigate [9], and use the acquired knowledge from a scene in future situations and conditions).

Eye movements are not random but reflect both bottom-up and top-down processes. The bottom-up processes are associated with the salience of the image features like edge distribution, contrast, color, size, and other low-level features. In the present study, we aimed to evaluate whether the eye movement behavior reflects the process of learning the high-order regularities in image characteristics when the salience of the stimuli does not differ significantly. Hence, the differences in eye movement characteristics primarily reflect top-down attentional control.

We manipulated the spatial regularities in visual displays, varying the repetition frequency, the co-occurrence of chunks, and the similarity of patterns, aiming to evaluate the interaction between the accumulated knowledge and the gaze behavior. Spatial regularities in a visual scene are related to the eye landings at relevant areas of interest in the visual scenes, whereas temporal regularities produce anticipatory eye movements reflecting the prediction of future stimulus positions based on the acquired knowledge from the stimulation.

We studied the potential changes in four characteristics of eye movements: fixation duration, saccade number, gaze number, and saccade amplitude. The fixation duration is related to the depth and speed of the information processing. Short fixations (50–150 msec) [10] are considered ambient, reflecting unconscious processing. Focal fixations [10] are supposed to reflect conscious processing, focusing on details and object identification.

The focal/ambient mode of image processing is also associated with changes in saccade amplitude. Short fixations are usually combined with large saccades, while the opposite is true for long fixations, reflecting the focal mode of processing. Unema and co-authors [11] showed a transition from an ambient mode of processing (short fixations and large saccade amplitudes) to a focal processing mode (long fixations and short saccades) during a single trial. In our study, conversely, we aim to explore the changes in the dynamics of eye movements during the learning process. We hypothesize that the participants will use the ambient mode of processing in the initial stages of the study to obtain maximal information about the spatial layout of the images. In contrast, when they acquire enough knowledge of the stimulus structure, they will switch to the focal mode with longer fixations, indicating deeper cognitive processes and a longer time for analyzing and interpreting the stimulus information. Hence, we expect the fixation duration to increase and the saccade amplitude to decrease in the learning and practice process.

Alternatively, the fixation duration might decrease. Short fixations are seen, for example, for expert readers (e.g., [12,13]). Practice is shown to reduce the location and identity uncertainty in visual search tasks, thereby reducing fixation duration [14,15]. Moreover, the participants were under time pressure (we limited the stimulus presentation to 3 s), a condition also known to reduce fixation duration [16]. Hence, in our experimental conditions, the repetition frequency of the patterns and the co-occurrence of visual chunks may allow participants to gain familiarity with the pattern combinations, adopting strategies that enable them to process relevant stimulus information more rapidly. Such an experimental outcome aligns with findings from previous studies, which suggest that expertise and time constraints can lead to shorter fixation durations as participants become more efficient in their visual search strategies.

If a transition from an ambient to a focal mode of processing occurs during the sequential trials, the saccade amplitude would be expected to decrease in the sequential blocks due to the correlation between the fixation duration and saccade amplitude. Moreover, it has been shown that the repetitive presentation of the patterns leads to a smaller attentional span and, thus, smaller saccade amplitudes [17,18].

The saccade number is tightly related to the number of fixations. It is supposed to reflect the efficiency of information search behavior, the complexity of the inference processes, and information acquisition. We expect that the saccade number will decrease in the sequential blocks of the trials due to the knowledge of the task and the improved discrimination of relevant from irrelevant stimulus information. However, in a study testing repetition effects in visual search [5], the authors related the improvement in search performance to unconscious memory. They showed that the unconscious memory did not affect the saccade number but made each saccade more efficient.

The gaze number reflects the number of regions of interest the participants select to perform the task. We expect that it will decrease due to learning the relevant stimulus characteristics for a task's performance and the increased efficiency in their selection.

In summary, we expect that acquiring knowledge about the image regularities would affect the gaze behavior, although the direction of this change is not well defined in advance. We hypothesize that it will also depend on the context, i.e., on the similarity of the patterns in the image and the difficulties of their discrimination.

2. Materials and Methods

2.1. Stimuli

The stimuli were twelve patterns of black squares of size 10×10 cm and 15 yellow squares (size 1×1 cm) inside them (Figure 1). The patterns were divided into three groups, and the members of each group were never presented with the members of another group. The patterns in Groups 1 and 2 always contained two chunks (groups of connected elements) of 5 squares. Two patterns (A and D) in these groups had the same 5-element chunks but in different positions. The other two patterns (B and C) had one common chunk with patterns A and D. The patterns in Group 3 had chunks of variable numbers.

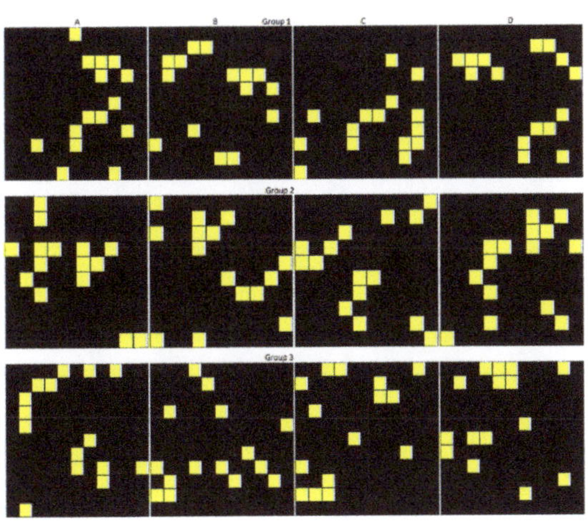

Figure 1. Pattern set used in the stimuli design. Each column contains patterns A, B, C, and D; each row—the patterns from different groups.

The patterns in each group were combined in pairs. The pattern on the left was presented in its original orientation, while the patterns on the right were rotated by 90°,

180°, or 270°. The patterns from each group were presented on both sides of a stimulus. The stimulus set contained 72 unique combinations of patterns. Only one of the patterns (A) in each group was combined with itself, although in different orientations.

We selected simple geometric shapes forming various chunks of different number and shape to avoid the effect of previous knowledge on the learning process. This choice is typical for many other studies of visual statistical learning (to name a few, [8,19–22]). Moreover, a recent study [23] showed that preexisting knowledge impacted statistical learning. Abstract shapes or geometric patterns have a better-defined structure that allows for a better identification of learning effects, without the confounding effect of previous experience or the insufficient control of other factors affecting the salience of the visual scenes or top-down factors.

2.2. Procedure

The patterns in a stimulus pair were presented on a gray background in the middle of a computer screen in a vertical direction and 2.5 cm to the left and right of the screen center in a horizontal direction so that the distance between their closest edges was 5 cm. The observers had to respond to whether the two patterns in a pair were the same or different. If the observers did not respond with a mouse click for 3 s, the stimulus disappeared and the screen turned gray. The stimulus duration was determined in a preliminary pilot study performed with experienced participants. This duration allows the task performance in the initial blocks to exceed slightly that of random guessing and avoids ceiling effects in the later experimental blocks. Other studies [24–26] also used a 3 s presentation time to explore the oculomotor anticipation in a serial reaction time task.

We included the patterns of Group 3 with a variable number of chunks to examine the effect of chunk co-occurrence on performance during the learning process. The choice of the patterns in the groups also allows for evaluation of the effect of perceptual load and task difficulty on the process of learning the regularities in the stimulus set.

The combination of the same patterns (combination AA) was presented nine times in the stimulus sequence, whereas the other combinations were presented only once. Thus, the stimulus sequence contained 90 stimulus pairs, presented in random order.

Each participant took part in three experimental sessions. In each session, the stimulus sequence was repeated twice; hence, six blocks of stimuli were presented in the same order. We will use the term epoch for the sequential blocks in the experiments.

Before each experimental session, the eye movements of the participants were calibrated using a 9-point calibration procedure. The eye movements were recorded with a head-mounted Jazz-Novo multisensor measurement system (Ober Consulting Sp. z o.o., Poznan, Poland) with a sampling frequency of 1000 Hz. It measures the eye movements of both eyes in horizontal and vertical directions by direct infrared oculography. High temporal precision is not essential for the studied eye movement characteristics, except for the fixation duration evaluation, but this system provides the spatial resolution needed for the present study. The participants sat in a dark room at a distance of 57 cm. Their heads were supported by a chin rest.

2.3. Participants

Ten naïve observers aged 26 to 59 (average age 41.7) participated in the study. They were recruited from the staff of the institutes of the Bulgarian Academy of Sciences and gave written informed consent to participate in the study. The sample size is relatively small, but we used a large set (72) of unique stimuli. Wilming and co-authors [27] showed that in the spatial analysis of eye movements, the sample size and the stimulus set can compensate for each other. The experiment was approved by the Ethical Board of the Institute of Neurobiology, Bulgarian Academy of Sciences (Protocol 48 from 6 June 2023).

2.4. Data Analyses and Measures

Eye tracking software (JazzManager 3.13 created by Ober Consulting Sp. z o.o., Poznan, Poland in cooperation with Nalecz Institute of Biocybernetics and Biomedical Engineering, Polish Academy of Sciences, Warsaw, Poland) was used to divide the eye movements into saccades and fixations using a threshold velocity of 30 deg/s and an acceleration threshold of 8000 deg/s^2. The potential calibration errors and eye drifts were corrected using the Iterative Closest Point algorithm [28].

We used the saccade number, the saccade amplitude, fixation duration, and gaze number to characterize the potential change in gaze behavior in the epochs. These characteristics were evaluated only during the stimulus presentation; thus, we excluded the memory-guided eye movements from the analyses.

The gaze number was estimated using the DBSCAN clustering algorithm [29] to aggregate the fixations into regions of interest. The clusters of fixation positions were required to have at least 200 data points (corresponding to a fixation duration of at least 200 msec for sequential eye records).

All statistical analyses were performed in an R environment [30]. We used generalized mixed models to evaluate the effect of stimulus similarity and the potential learning effects on the kinematics of eye movements. The models were fitted with the brms package [31]. This method is more robust for complex data structures and small sample sizes than frequentist methods. We needed to apply generalized mixed models to take into account the specifics of the data, the individual differences between the participants, and the non-normal distributions of the eye movement characteristics. Too often, the frequentist methods do not converge when applied to these types of models.

For all tested characteristics, we treated the pattern combinations not as random factors but as categorical variables, as the stimulus combinations were formed following specific rules. This choice allows us to evaluate the effect of pattern similarity on eye movement characteristics. In the modeling, we started with the complete set of 72 unique pattern pairs, i.e., assuming that the pattern position and orientation would change the gaze behavior. We also tested simpler models in which the pattern orientation or position was disregarded. Models, disregarding only the rotation angle of the patterns on the right, included 24 unique stimulus pairs, whereas when the relative position of the patterns in a pair was also disregarded, the unique stimuli were 15. We used cross-validation for model comparison. This approach depends less on the prior distribution than the model comparison based on the Bayesian factor (BF).

The fixation duration has a skewed distribution of positive values. We used a shifted lognormal distribution for the likelihood. Based on the earlier research data and the high variability in fixation duration, we used a normal distribution with a mean of 5 and a standard deviation of 0.8 as prior distribution for the pattern combinations. This choice implies that the expected saccade duration is about 245 msec and varies with the stimuli and the blocks in the 0 to 8000 msec range. The sigma was assumed to have a normal distribution with a zero mean and a standard deviation of 1.5. The standard deviation of the model parameters was modeled with normal distribution with a zero mean and a standard deviation of 0.5. We used the default values to model the shift in the distribution.

The data for the saccade amplitude have only positive values and could be considered as continuous variables. The exploratory analysis of the amplitudes showed that they had bimodal distributions. We modeled their likelihood function with a hurdle lognormal distribution. We used a normal distribution with a mean of 10 and a standard deviation of 3 for the prior distribution for the effect of the combinations and the epochs. This choice implies that the expected average number of saccades is about 15, and it could change for the different combinations or in the sequential blocks from 12 to 18.

The data for the saccade and gaze numbers are counts, i.e., positive integers. They were modeled first by a Poisson distribution. Both identity and log links were tested, but the post-predictive checks did not fit the experimental data well, so we used a negative binomial distribution for the modeling. This distribution relaxes the assumption of the equality of the mean and the variance characteristic of the Poisson distribution and allows for estimation of the dispersion directly from the data. The prior distribution for the main effects for the saccade number was set to a truncated normal (1, 1), implying the mean expected number of saccades of 10 and a range from 0 to 24. For the gaze number, we used a truncated normal distribution with a mean of three and a standard deviation of 3, meaning that we expected the range for the gaze number to vary from 0 to about 18 with a mean value of around 4.

The choice of prior distributions was governed by the recommendation of [32] to include a range of uncertainties larger than any plausible parameter value.

In modeling, we used 4 to 8 chains and 2000 to 30,000 iterations per chain, with the warm-ups being half of the iteration number. All model choices approximated the distribution of the response variables well, as shown by posterior predictive checks. Visual examination of the model trace plots did not indicate problems with chain mixing. Moreover, the R_{hat} values, i.e., the ratio of the variance across chains to the variance within chains [33], were less than 1.05, showing model conversion. The effective sample size measures exceed 5000, suggesting that the models provide stable parameter estimates.

We used the 95% credible intervals of the posterior distributions and their overlap to decide whether a difference in model parameters was significant. These intervals specify the uncertainty of the estimates and that the range of values of a parameter generated by the data lies within a 95% probability. Moreover, hypothesis testing on the effect of the experimental factors and the evidence ratio (ER) is provided to quantify the strength of evidence in favor of one hypothesis over another. We estimated the probability of direction (pd) of the model parameters. Its value varies between 50% and 100%, and it could indicate whether a parameter is strictly positive or negative. We assumed that a pd of 95% showed a notable change in a parameter.

As a measure of similarity between the patterns, we evaluated the cross-correlation between the patterns in a pair, the sum of the squared distances between the two patterns, the structure similarity index, the Jaccard similarity coefficient for binary images, and the Sørensen–Dice similarity coefficient. The Sørensen–Dice similarity coefficient gives the intersection between the point sets of the two patterns regarded as binary images divided by their sum.

3. Results

3.1. Fixation Duration

Typically, the fixation duration has skewed distributions to the right with median values around (200–250 msec) and larger mean values around 300–350 msec [10,11,34]. Our study has a few excessively long fixations (over 2 s) due to there being only one participant. These values did not exceed Tukey's fences and were not excluded from the data. Please note that we have restricted the eye movement analysis only to data obtained during the stimulus presentation. Hence, the largest possible fixation duration could equal the stimulus duration. The range of the fixation duration was 151–3000 msec with a median of 260 msec. The overall mean fixation duration was 314.19 msec. Figure 2 shows the distribution of the fixation duration in the various epochs of our study. It shows the skewed distribution of the fixation times and a decrease in the number of the shortest fixation durations in the sequential blocks. Overall, this trend shifts the median of the distributions to larger values consistent with our hypotheses of an increased fixation time in the sequential blocks.

We performed three models considering the effect of the pattern position and orientation in a stimulus pair (using all 72 pattern combinations), disregarding the rotation angle (using a stimulus set of 24 combinations), or disregarding the position swap and rotation angle (using a stimulus set of 15 combinations).

The model comparison shows that the simplest model, which disregards the position and rotation angle of the patterns in a stimulus, better describes the experimental data than the full model.

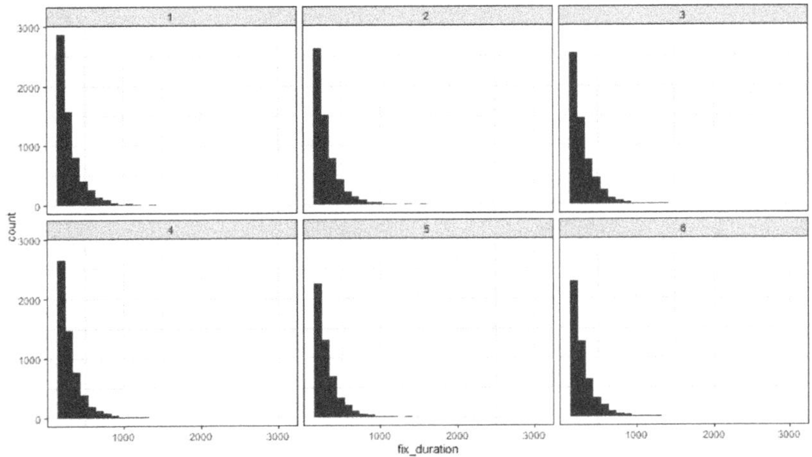

Figure 2. Distribution of the fixation duration in the sequential epochs (1–6) of the experiment.

After verifying the model's plausibility and the chosen prior values, we tested the hypothesis that the fixation duration shortens with practice. Unema and co-authors [11] suggested that the initial image viewing involved shorter fixations and larger saccade amplitudes, representing ambient processing. In contrast, smaller saccade amplitudes and longer fixation duration in the later viewing period represented focal processing. These results, however, concern the fixation duration in a single trial.

Figures 3 and 4 present the conditional effects obtained from the model and the changes in fixation duration for the different stimuli and epochs.

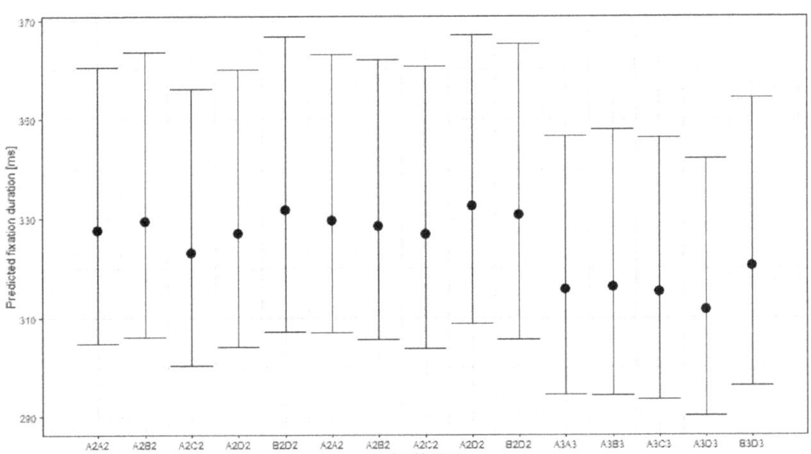

Figure 3. Predicted fixation duration for the different pattern combinations with 95% credible interval.

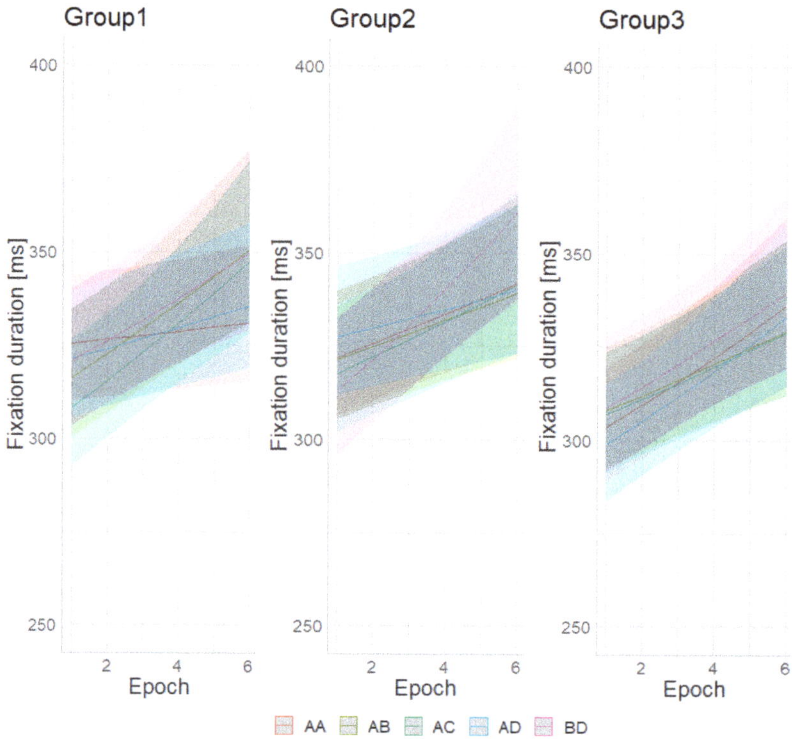

Figure 4. Predicted fixation duration for the different stimuli and epochs with 95% credible interval.

Figure 4 shows that the pattern combinations do not substantially change the fixation duration. The pairwise comparison of the fixation duration for the different pattern combinations and their 95% credible intervals (CIs) implies that the fixation duration for the pairs in each group and between Groups 1 and 2 did not differ significantly. Most of the stimuli from these groups induce significantly longer fixation durations than the combinations in Group 3. An exception is the combination BD from Group 3, which does not differ significantly in fixation duration from the combinations of Groups 1 and 2. It should be noted that the BD combination is the only one for which the two patterns in the pair did not swap positions.

The effect of learning, if any, is represented by the interaction between the sequential epochs and the combination pairs. Figure 4 shows that the change in fixation duration is largest for the BD pattern combination from Group 2. The epoch has the least effect for the same patterns from Group 1. Figure S1 (Supplementary File) shows the trend of change in the fixation duration in the sequential epochs. It is estimated by using the emtrends function from the package emmeans [35]. The 95% credible intervals of the estimated trend of change differ, implying an unequal effect of the epochs on discriminating the paired combinations. The pairwise comparisons of the trends (after correction for multiple comparisons) show significant differences in the increase in fixation duration for stimulus AA from Group 1, which is lower than that of all other combinations. For the rest of the combinations, the rate of change is similar.

3.2. Saccade Number

The saccade number is related to the number of fixations performed during the task. Figure 5 shows the distribution of the saccade number in each epoch. It shows that the saccade number has a skewed distribution.

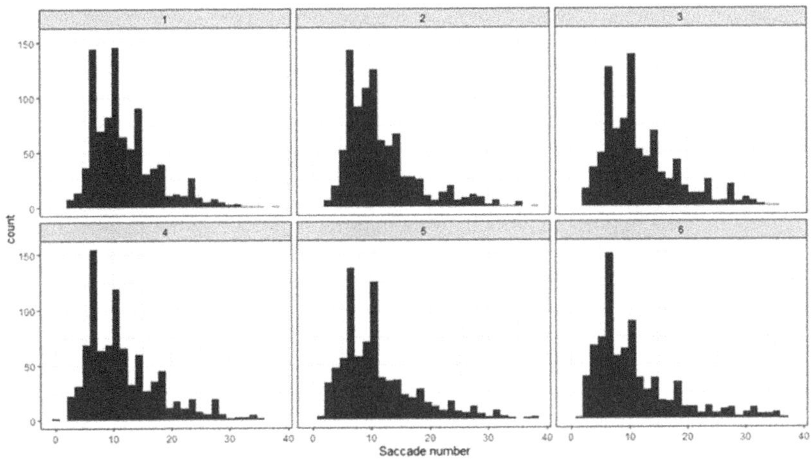

Figure 5. Distribution of the saccade number in the sequential epochs (1–6) of the experiment.

We tested three models for the effect of the stimuli and epochs on saccade number. The first uses all 72 stimuli, and thus, it assumes that the rotation angle and the position of the patterns affect the exploration activity of the participants. The second model assumes that the rotation angle is ineffective, but the position of the patterns (whether to the left or the right of the stimulus) changes the number of saccades. It includes 24 combinations of patterns. The third model assumes that neither the patterns' rotation nor position affects the saccade number. It includes 15 pairs of patterns.

The cross-validation for model comparison implied a better fit of the data by the model that ignores the rotation angle and the pattern positions and thus regards only 15 combinations of the patterns. In contrast to the fixation duration, the saccade number varies with the stimulus type. We compared the saccade number for the three groups of patterns and the combinations of the same and the different patterns. The evidence strongly supports the hypothesis that for each group, the same patterns require more saccades than the different ones (Figure 6; Group 1: estimate = 0.26; 95% CI [0.20–0.32]; evidence ratio = Inf; probability = 1.0; Group 2: estimate = 0.08; 95% CI [0.00–0.15]; evidence ratio = 24.13; probability = 0.96; Group 3: estimate = 0.26; 95% CI [0.19–0.34]; evidence ratio = Inf; probability = 1.0). Moreover, the same patterns from the different groups elicited an unequal number of saccades—the largest for Group 1 and about the same for Groups 2 and 3.

The analysis revealed an estimate of 0.15 and 95% credible intervals of 0.08–0.23 for the hypothesis A1A1 > A2A2 with an evidence ratio of 3332.33, and an estimate of 0.19 with 95% credible intervals for the hypothesis A1A1 > A3A3 and an evidence ratio of 9999. Based on the evidence ratio, the data provide strong evidence for more saccades elicited by the same patterns in Group 1 compared to the other groups. The evidence moderately supports the hypothesis that the saccade number for the same patterns in Group 2 exceeds those for the same patterns in Group 3 (hypothesis A2A2 > A3A3: estimate = 0.04; 95% CI [−0.05–0.13]; ER = 3.22; probability = 0.76).

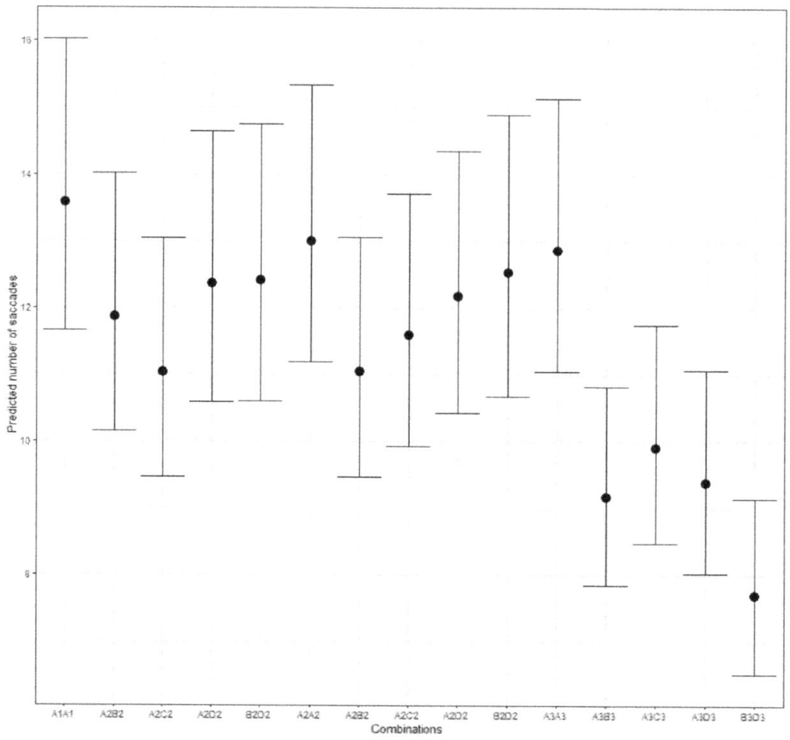

Figure 6. Predicted number of saccades for different pattern combinations.

Contrary to our expectations, the number of saccades increases in the different epochs (pd equals 100%; the evidence ratio for the hypothesis that the effect of epoch increases (epoch > 0) and equals Inf with an estimate 0.02; 95% CI [0.0–0.04]), providing strong support for the hypothesis of an increase in saccade number in the sequential blocks.

The saccade number in our study reflects the exploratory eye movements. Its increase in the sequential blocks might represent the gained knowledge of the spatial structure of the stimuli. For example, in [8], the authors showed that both the exploratory and the confirmatory eye movements increased in the sequential trials when the participants had an explicit knowledge about the underlying structure of the visual scenes or during long implicit learning. In our understanding, however, the increase in saccade number in our experimental conditions and task should be considered in combination with the change in the other eye movement characteristics. This result is further discussed in the Section 4.

We evaluated the trend of the epoch effect for the different combinations of patterns using the emtrends function [35]. For all combinations, the trend is positive. It is most prominent for the combination of the B and D patterns from Group 2, A and D from Group 1, and A and A for Group 3. The smallest increase in the saccade number was obtained for the AA pattern from Group 1 (Figure 7 and Figure S2 in the Supplementary File).

The insignificant effect of the rotation angle and pattern position implies that the dominant factor for searching for information and performing saccades is the number of chunks in the stimuli. The correlation between the estimated effect of pair combinations and the number of chunks is significant ($r = -0.59$, $p = 0.02$).

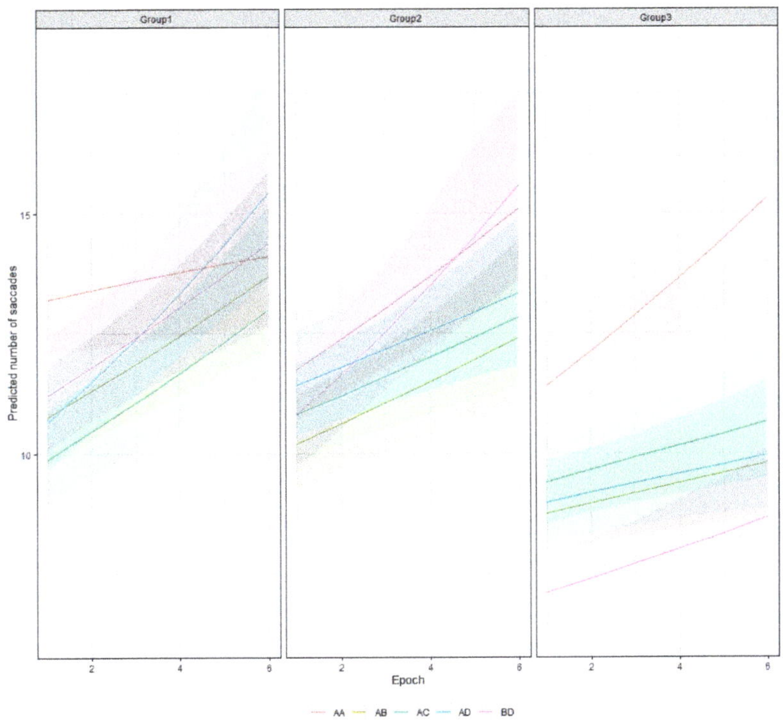

Figure 7. Predicted number of saccades for different stimuli and epochs with 95% credible interval.

3.3. Gaze Number

Whereas the saccade number in our study has a skewed distribution, the gaze number has a symmetric distribution (Figure 8). Again, we tested three different models for the effect of pattern combinations on gaze number: taking into account the position and rotation angle of the patterns in a pair, ignoring the position, or ignoring both the position and rotation of the patterns. The results suggest a better fit to the data by a model with 15 combinations, thus ignoring the relative position and rotation of the patterns.

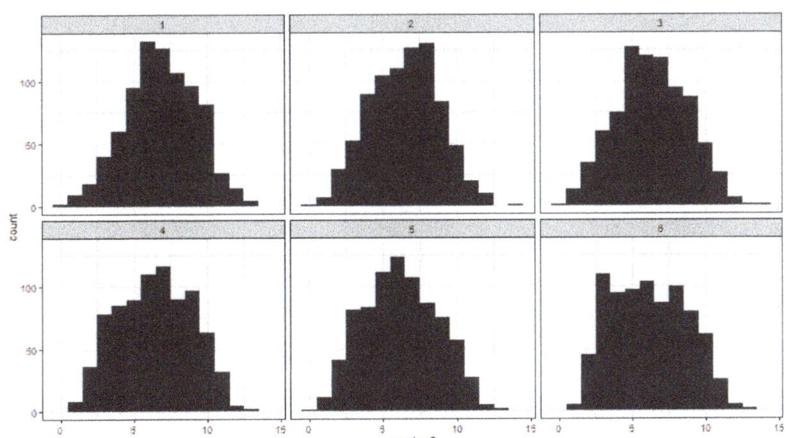

Figure 8. Distribution of the gaze number in the sequential epochs (1–6) of the experiment.

Figure 9 shows the predicted gaze number for the different pattern combinations. It demonstrates a larger gaze number for the stimuli with the same patterns compared to those with different ones for the patterns in Groups 1 and 3. This observation was confirmed by testing the hypothesis of a larger number of gazes for the same patterns in each group against the null hypothesis of no difference. The evidence strongly supports the alternative hypothesis for Groups 1 and 3, with an estimate of 0.13 and a 95% CI [0.05-−0.21] (ER = 332.33) for Group 1, and an estimate of 0.19 and a 95% CI [0.11-−0.27] (ER = 7999) for Group 3. There is no compelling evidence to suggest a larger gaze number for the same patterns than for the different patterns of Group 2 (estimate = 0, 95% CI [−0.08-−0.08]; ER = 0.91).

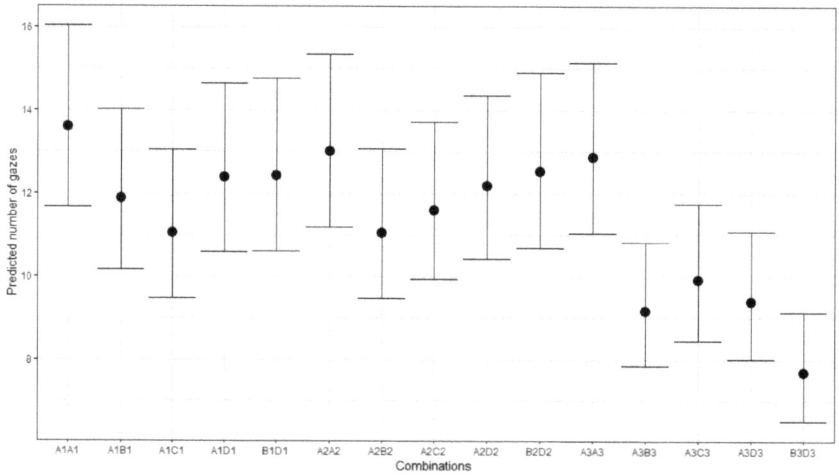

Figure 9. Predicted number of gazes for different pattern combinations.

The results also suggest that the number of gazes tends to decrease during the sequential epochs of trials. The estimated effect for the hypothesis for the negative effect of the epoch is −0.02 with 95% CI [−0.05, 0.01] and an evidence ratio of 20.24. Hence, whereas the saccade number tends to increase, the gazes tend to decrease, suggesting more localized eye landings in the sequential blocks. The change in the gaze number might reflect a better understanding of the stimulus structure and a better discrimination of the relevant and irrelevant parts of the stimuli. The reduction in gaze number may reflect a more local focus of attention typical for the focal mode of processing.

Figure 10 shows the interaction between the stimuli and the sequential blocks. It shows that the effects of learning varied with the different stimuli and that the reduction in the gaze number is more pronounced for the patterns of Group 3. The pairwise comparisons of the trend of the gaze number change (after correction for multiple comparisons) show significant differences between the stimuli from Groups 1 and 2 with at least two of the different pairs from Group 3 (AB and AD) and insignificant differences with patterns AA and BD from this group (Figure S3, Supplementary File).

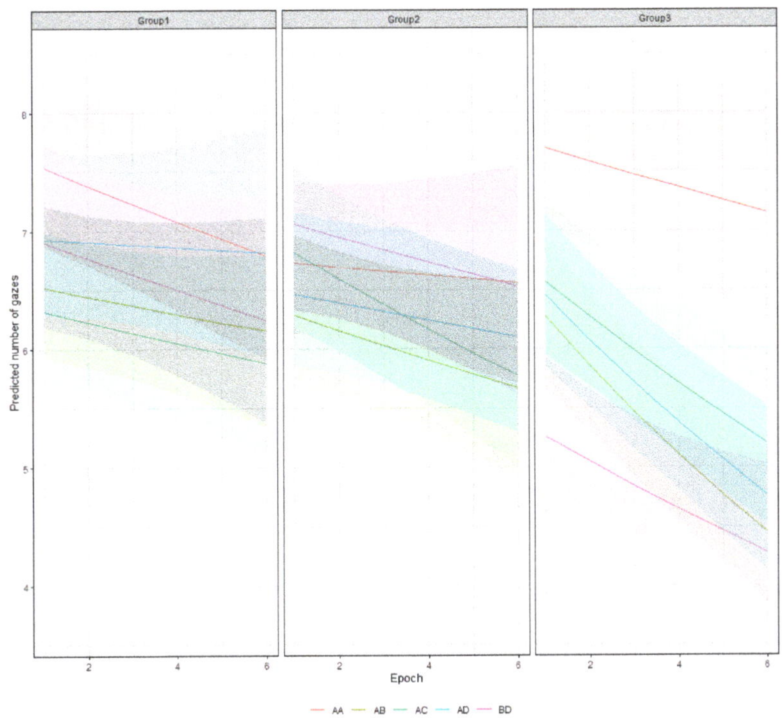

Figure 10. Predicted number of gazes for different pattern combinations and epochs.

3.4. Saccade Amplitude

The saccade amplitude would depend on the distinctive features of the patterns that allow for better discrimination. Thus, comparing the chunks in a pattern and between the patterns in the stimulus would be needed. Figure 11 shows the distributions of the saccade amplitudes in the sequential epochs. It clearly shows their bimodality. The two modes of these distributions occur because of the eye movements inside the patterns in the stimulus and between them.

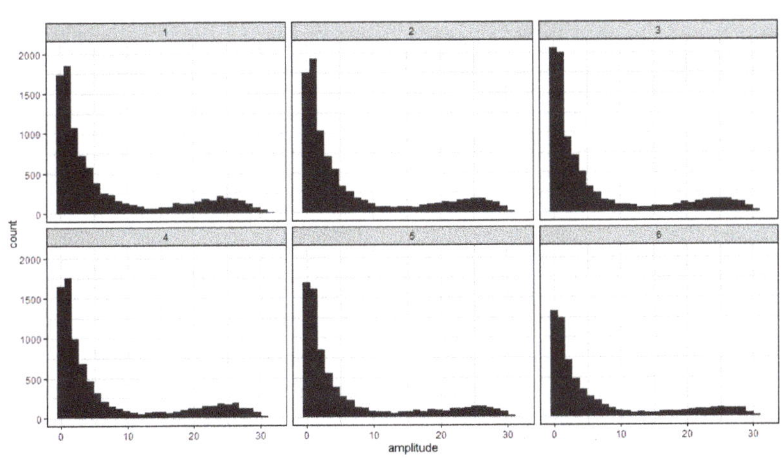

Figure 11. Distribution of the saccade amplitude in the sequential epochs (1–6) of the experiment.

We tested whether the saccade amplitude varies with the rotation angle of the patterns. Analyzing the overall effect of the rotation angle for the whole set of stimuli is not very meaningful since the rotation angle has different effects for the different stimuli, changing the relative distances between the chunks in the two patterns in a stimulus. The effect of the rotation angle might be shown as a difference between the three stimuli in a group that contains the same patterns in the same position, differing only by the rotation angle. Each group contains eight such triples, with three comparisons in each. The results show that the 90% credible intervals of the amplitudes do not overlap for several members of each group (Figure 12A–C).

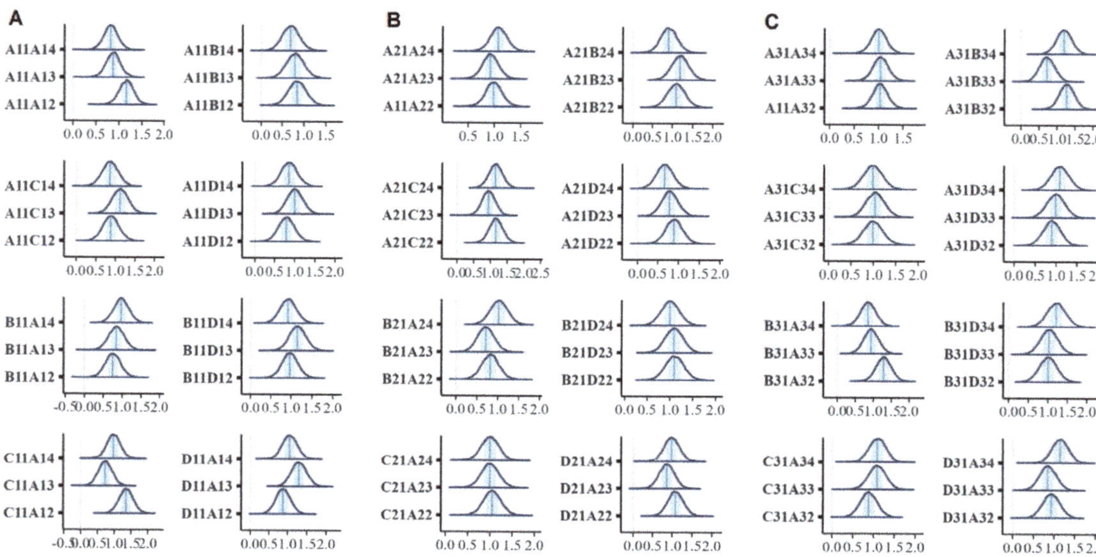

Figure 12. Posterior distributions for the saccade amplitudes for all stimulus patterns: (**A**) for the stimuli from Group 1; (**B**) for the stimuli from Group 2; (**C**) for the stimuli from Group 3. The shaded regions correspond to the 90% credible intervals of the median.

The comparison of the saccade amplitude shows that it is significantly larger for the same patterns than for the different patterns. The evidence strongly supports the alternative hypothesis of there being a larger saccade amplitude for the same patterns compared to the different patterns against the null hypothesis of no difference (estimate = 0.17; [0.10–0.25], evidence ratio (EF) = for Group 1, estimate = 0.10 [0.03–0.18] for Group 2, and 0.16 [0.08–0.24] for Group 3 (the values in brackets are the 95% credible intervals)).

The highest density interval (HDI) for the epochs is entirely negative, implying a decrease in saccade amplitude during the experiment (estimate = −0.06; 95% CI [−0.11, −0.02], EF = 91.59). For each combination of patterns, however, the HDI is strictly positive. The effect of the epoch is unequal for the different stimulus combinations. The trend in block effect is either negative or insignificant (estimated by the highest posterior density intervals of the emtrends function at the first and the last experimental blocks). No clear difference between the groups is evident.

The decrease in saccade amplitude may be interpreted as resulting from unconscious memory, which reduces the extent to which saccades are made to incorrect regions.

Contrary to the results of [36], we did not find a negative correlation between the fixation duration and saccade magnitude. Our data showed a lack of correlation between the two characteristics of the eye movements (r = 0.09).

3.5. Similarity Measures

We tried to understand whether the pattern similarity might predict the effect of the combinations on the different eye movement parameters. We used several different measures to describe the similarity of the patterns in a pair. Since the images could be regarded as matrices, we can use measures that are appropriate not only for comparing the images but also for matrix comparisons. Moreover, the patterns we used have only two colors. We calculated the mean differences between the values of the left and right patterns, the mean value of the coincidence of their values, the structure similarity index [37], the correlation, and the mean squared error between the left and right patterns. Converting the patterns to logical arrays, we also calculated the Sørensen–Dice similarity coefficient between binary images, the Jaccard similarity coefficient for binary images, and the BF (Boundary F1) contour matching score. All these similarity measures have higher values when the similarity of the patterns is stronger (Figure S4, Supplementary File). The results show that these measures are sensitive to pattern rotation, though typically, two rotation angles for the right pattern lead to equal values. Additionally, the average values of the similarity measures for pairs with swapped positions are equal. Moreover, most similarity measures give some of the lowest values for the pairs of the same but rotated patterns, except the mean squared error and the mean difference between the images (Figure S4, Supplementary File). The effects of the position and the rotation angle imply that the similarity of the patterns in a pair does not determine the performance.

4. Discussion

The study's results suggest that the attentional selection of relevant image characteristics varies with the experience obtained in experimental task performance. This change suggests an adaptive strategy of the observers that balances exploration and exploitation modes shown in the increased saccade number and the change in the saccade amplitude. They confirmed the usefulness of studying the eye movement characteristics to obtain information on the processes of attention, decision making, and memory.

Statistical learning is typically considered a process involving implicit memory. Studying the role of unconscious memory in a search task in real-world scenes, [5] showed no effect of unconscious memory on the number of saccades. Higuchi and Saiki [7] showed that eye movements were not necessary for implicit learning of spatial configurations, and keeping a stable fixation at the display center is beneficial for rapid learning of the spatial layout. Our results, however, show an increased number of eye movements in the sequential block of trials, but the gaze landings become more localized in the process of practice and learning. One distinctive feature of our study, compared to visual search studies, is that the relevant target or targets for successful task performance are not defined in advance. The participants have to determine which stimulus features to use for task performance. The large display size may facilitate an exploitative behavior with larger saccades in the initial stimulus blocks and more focused ones later on. The increase in fixation duration in the sequential blocks might be considered a switch from exploration to exploitation mode; however, the saccade number should decrease in this case.

Our findings indicate that learning spatial regularities in the stimulus set leads to a transition from ambient to focal processing, a phenomenon commonly observed in single trials [9,10]. Ambient processing is characterized by longer saccades and shorter fixations and is considered to be involved in processing spatial arrangements and extracting contextual information. In contrast, focal processing, which involves smaller saccades and longer fixations, reflects the processing of details and object identification. The increase in fixation duration and the decrease in saccade amplitude in the sequential blocks in our study may be considered as representing a more local focus of attention and a discrimination of the relevant from irrelevant stimulus features for task performance and their deeper exploration.

The study results show a marginal effect of the stimuli on the fixation duration. The more pronounced change in this gaze parameter suggests a transition from ambient to focal processing. The saccade amplitude change also suggests such a transition, as its magnitude decreases in the sequential epochs. Its independence from the rotation angle and the pattern position in a pair implies that the observers compare the relative positions of the chunks in the patterns. The unequal effects of the epoch on the different stimuli imply different learning rates. Based on the fixation duration and saccade amplitude outcomes, the data suggest more rapid learning for the different patterns.

In the present study, we varied the co-occurrence of chunks in the patterns that formed a stimulus pair. Contrary to typical studies manipulating the co-occurrence of objects, in our task, the presence of the same chunks in different patterns makes stimulus discriminability more difficult due to the increased pattern similarity. The large number of unique stimulus pairs would not allow the use of memory for the combinations to discriminate the patterns as the same or different. However, the stimulus set contains statistical regularities that could facilitate the performance in the sequential epochs.

The pattern A from each group appeared on the left side of the stimulus pairs in their original orientation. Moreover, out of the 90 trials in an epoch, each of these patterns appeared on the left 36 times, meaning that the three pattern As from the three groups were shown on the left on 60% of the trials in an epoch. If the participants memorize these patterns, they could use this knowledge to answer "different" whenever the pattern on the left side of the stimulus does not coincide with any of them. This strategy would affect both the saccade amplitude and the number of saccades and gazes—the saccades could explore predominantly the left pattern. This strategy would imply a decrease in saccade amplitude.

When pattern A is on the left side, the effective strategy for Groups 1 and 2 would be to seek the two 5-element chunks characteristic of pattern A in the right pattern. In this case, the number of fixations on pattern A might decrease due to greater exploration of the right pattern, leading to large saccade amplitude for pattern A on the left. This strategy would be less effective for the patterns from Group 3, as identifying pattern A on the left side of the pair would not benefit task performance. We tested the hypothesis that the saccade amplitude depends on the presence of pattern A on the left side of the pairs separately for each group. The results show that the saccade amplitudes for pattern A being on the left are less than the saccade amplitudes for the cases when any of the other patterns were on the left for the stimuli of Group 1 (estimate = −1.17, 95% CI [−2.34–−0.01], ER = 19.47, $p = 0.95$). No significant differences were observed depending on the position of pattern A in the pair for the stimuli of Groups 2 and 3 (estimate = 0.03; 95% CI [−1.13–−1.19], ER = 0.93; $p = 0.48$ for Group 2; estimate = −0.07, 95% CI [−1.31–1.17], ER = 1.16; $p = 0.54$ for Group 3).

These considerations imply changes in saccade amplitude and an effect of pattern position on it—an outcome observed in our results. They correspond to other studies showing that people learn not only the statistical regularities related to task-relevant targets, but also the regularities of the task-irrelevant chunks in the visual images [38].

One specific feature of our results is that only saccade amplitude depends on the rotation angle of the patterns. Interestingly, the saccade amplitude discriminates best between focal and ambient processing [39]. Most image similarity measures are insensitive to the relative positions of the patterns in the pair—being on the left or right—but are sensitive to rotation. Our data suggest that while the position of the patterns in a pair does not change the fixation duration and saccade number, it effectively changes the saccade amplitude. These results imply that similarity measures like cross-correlation or the Sørensen–Dice similarity coefficient could not adequately represent the effect of pattern characteristics on the saccade amplitude. Measures of similarity like Kullback–Liebler divergence might be better suited; however, they require probability distributions. The conversion of the images used in the study into probability distributions of grayscale values could not distinguish the patterns we used, as they contain an equal number of the same color elements. The effective similarity measure should consider both the brightness (color) of the patterns and their position. Strangely, measures of similarity related to image segmentation, like bfscore (Boundary F1), seem better suited to describing the effect of the image manipulations we used on the saccade amplitude size.

The observed change in the saccade number is unexpected as it reflects the efficiency of searching for information, whereas the number of gazes reflects the complexity of the interferential process. The greater number of fixations usually indicates insufficient information on the relevant stimulus or task. Thus, it is more reasonable to expect that the saccade number will decrease in the sequential epochs. Indeed, [9] demonstrated that the participants exhibited fewer saccades with practice with visual search tasks, indicating a shift towards more efficient information processing. Similarly, expert readers tend to make fewer, more targeted saccades, reflecting a refined strategy for extracting relevant information [10]. Wedel and co-authors [16] showed that short saccades with repeated fixations to small regions can be regarded as a way to reduce memory load and they could facilitate target identification. Thus, a potential effect of the increase in the saccade number and a decrease in gaze number in the sequential blocks is an efficient way of reducing the memory requirements of the task when the relevant stimulus characteristics are already determined. Eye movements reflect item memory [40], which may lead to decreased fixation duration and saccade number due to the repetition of patterns on the left stimulus side. The discrimination from the patterns on the right requires relational memory (e.g., [41]). The review on the interaction of eye movements and memory systems shows that relational memory requires at least 500–750 ms (about 2 s in total in a task with three faces, i.e., in a task that required studying face–scene pairs followed by a memory test with 3-face displays superimposed on the studied scenes). In our study, however, there are more objects (chunks) between which the relative positions need to be evaluated; thus, the time needed for relational memory should be longer. One way to reduce the memory load might be by increasing the number of saccades.

Another explanation for the increase in the saccade number might be a change in the number of strategies used by the participants. As suggested by [11], depending on the strategies of the observers and the various contexts, ambient and focal processing can change dynamically and this interplay would affect the eye movement characteristics. The shorter fixation duration in the first epochs of the study may also represent the use of partial information exploration due to making decisions under time pressure. With practice, the observers may switch to more complete information exploration due to a speeded-up allocation of attention to the different stimulus chunks. This change in strategy may allow them to explore more stimulus regions to reduce uncertainty and enhance decision-making accuracy.

5. Conclusions

The results of the study suggest various changes in eye movement characteristics that depend on the repetition frequency, practice, and stimulus characteristics of the patterns. These changes reflect attentional selection, memory involvement, and learning processes. Studying the modifications of the eye movement characteristics jointly provides a better understanding of the complex interaction between these cognitive processes. Learning the spatial regularities in a visual scene allows for the prediction of the future allocation of gaze and may be relevant in computer vision applications and the design of learning environments.

Supplementary Materials: The following supporting information can be downloaded at: https://www.mdpi.com/article/10.3390/app142110055/s1, Figure S1: Trend of change in fixation duration in the sequential epochs for different pattern combinations; Figure S2: Trend of change in saccadic number in the sequential epochs for different pattern combinations; Figure S3: Trend of change in gaze number in the sequential epochs for different pattern combinations; Figure S4: Similarity of the patterns in the pair combinations evaluated by different measures.

Author Contributions: Conceptualization, B.G., N.B. and I.H.; methodology B.G., N.B. and I.H.; software, N.B.; validation, B.G. and I.H.; formal analysis, N.B.; investigation, B.G. and I.H.; data curation, I.H.; writing—original draft preparation, B.G. and N.B.; writing—review and editing, B.G., N.B. and I.H.; visualization, B.G. and I.H.; supervision, B.G. and N.B.; funding acquisition, B.G., N.B. and I.H. Due to the unexpected death of I.H., only B.G. and N.B. approve the final version of the manuscript. All authors have read and agreed to the published version of the manuscript.

Funding: This research was funded by the Bulgarian National Science Fund, grant number KP-06-N52/6 from 12 November 2021. The APC was partially funded by the Bulgarian National Science Fund.

Institutional Review Board Statement: The study was approved with Protocol 48 by the Ethics Committee of the Institute of Neurobiology, Bulgarian Academy of Sciences on 6 June 2023 and was conducted in accordance with the Declaration of Helsinki.

Informed Consent Statement: Informed consent was obtained from all subjects involved in the study.

Data Availability Statement: The data presented in this study are available on request from the corresponding author due to ethical reasons.

Conflicts of Interest: The authors declare no conflicts of interest.

References

1. Fiser, J.; Lengyel, G. Statistical Learning in Vision. *Annu. Rev. Vis. Sci.* **2022**, *8*, 265–290. [CrossRef] [PubMed]
2. Nissen, M.J.; Bullemer, P. Attentional requirements of learning: Evidence from performance measures. *Cogn. Psychol.* **1987**, *19*, 1–32. [CrossRef]
3. Zolnai, T.; Dávid, D.R.; Pesthy, O.; Nemeth, M.; Kiss, M.; Nagy, M.; Nemeth, D. Measuring statistical learning by eye-tracking. *Exp. Results* **2022**, *3*, e10. [CrossRef]
4. Turk-Browne, N.B.; Jungé, J.; Scholl, B.J. The automaticity of visual statistical learning. *J. Exp. Psychol. Gen.* **2005**, *134*, 552–564. [CrossRef]
5. Ramey, M.M.; Yonelinas, A.P.; Henderson, J.M. Conscious and unconscious memory differentially impact attention: Eye movements, visual search, and recognition processes. *Cognition* **2019**, *185*, 71–82. [CrossRef] [PubMed]
6. Makovski, T.; Jiang, Y.V. Investigating the role of response in spatial context learning. *Q. J. Exp. Psychol.* **2011**, *64*, 1563–1579. [CrossRef] [PubMed]
7. Higuchi, Y.; Saiki, J. Implicit learning of spatial configuration occurs without eye movement. *Jpn. Psychol. Res.* **2017**, *59*, 122–132. [CrossRef]
8. Arató, J.; Rothkopf, C.A.; Fiser, J. Eye movements reflect active statistical learning. *J. Vis.* **2024**, *24*, 17. [CrossRef] [PubMed]
9. Ryan, J.D.; Cohen, N.J. The nature of change detection and online representations of scenes. *J. Exp. Psychol. Hum.* **2004**, *30*, 988–1015. [CrossRef]
10. Velichkovsky, B.B.; Khromov, N.; Korotin, A.; Burnaev, E.; Somov, A. Visual Fixations Duration as an Indicator of Skill Level in eSports. In *Human-Computer Interaction–INTERACT 2019*; Lamas, D., Loizides, F., Nacke, L., Petrie, H., Winckler, M., Zaphiris, P., Eds.; Lecture Notes in Computer Science; Springer: Cham, Switzerland, 2019; Volume 11746, pp. 397–405.
11. Unema, P.J.A.; Pannasch, S.; Joos, M.; Velichkovsky, B.M. Time course of information processing during scene perception: The relationship between saccade amplitude and fixation duration. *Vis. Cogn.* **2005**, *12*, 473–494. [CrossRef]
12. Gegenfurtner, A.; Lehtinen, E.; Säljö, R. Expertise differences in the comprehension of visualizations: A meta-analysis of eye-tracking research in professional domains. *Educ. Psychol. Rev.* **2011**, *23*, 523–552. [CrossRef]
13. Bertram, R.; Helle, L.; Kaakinen, J.K.; Svedström, E. The effect of expertise on eye movement behaviour in medical image perception. *PLoS ONE* **2013**, *8*, e66169. [CrossRef] [PubMed]
14. van der Lans, R.; Pieters, R.; Wedel, M. Online advertising suppresses visual competition during planned purchases. *J. Consum. Res.* **2021**, *48*, 374–393. [CrossRef]
15. Zelinsky, G.J.; Sheinberg, D.L. Eye movements during parallel-serial visual search. *J. Exp. Psychol. Hum.* **1997**, *23*, 244–262. [CrossRef]
16. Wedel, M.; Pieters, R.; van der Lans, R. Modeling Eye Movements During Decision Making: A Review. *Psychometrika* **2023**, *88*, 697–729. [CrossRef]
17. Kaspar, K.; König, P. Overt attention and context factors: The impact of repeated presentations, image type, and individual motivation. *PLoS ONE* **2011**, *6*, e21719. [CrossRef] [PubMed]
18. Kaspar, K.; König, P. Viewing behavior and the impact of low-level image properties across repeated presentations of complex scenes. *J. Vis.* **2011**, *11*, 26. [CrossRef]
19. Fiser, J.; Aslin, R.N. Unsupervised statistical learning of higher-order spatial structures from visual scenes. *Psychol. Sci.* **2001**, *12*, 499–504. [CrossRef]

20. Abla, D.; Okanoya, K. Visual statistical learning of shape sequences: An ERP study. *Neurosci. Res.* **2009**, *64*, 185–190. [CrossRef]
21. Siegelman, N.; Bogaerts, L.; Frost, R. Measuring individual differences in statistical learning: Current pitfalls and possible solutions. *Behav. Res. Methods* **2017**, *49*, 418–432. [CrossRef]
22. Growns, B.; Siegelman, N.; Martire, K.A. The multi-faceted nature of visual statistical learning: Individual differences in learning conditional and distributional regularities across time and space. *Psychon. Bull. Rev.* **2020**, *27*, 1291–1299. [CrossRef] [PubMed]
23. Rogers, L.; Park, S.; Vickery, T. Visual statistical learning is modulated by arbitrary and natural categories. *Psychon. Bull. Rev.* **2021**, *28*, 1281–1288. [CrossRef]
24. Mannan, S.K.; Ruddock, K.H.; Wooding, D.S. Fixation sequences made during visual examination of briefly presented 2D images. *Spat. Vis.* **1997**, *11*, 157–178. [CrossRef] [PubMed]
25. Tal, A.; Bloch, A.; Cohen-Dallal, H.; Aviv, O.; Schwizer Ashkenazi, S.; Bar, M.; Vakil, E. Oculomotor anticipation reveals a multitude of learning processes underlying the serial reaction time task. *Sci. Rep.* **2021**, *11*, 6190. [CrossRef] [PubMed]
26. Wynn, J.S.; Ryan, J.D.; Buchsbaum, B.R. Eye movements support behavioral pattern completion. *Proc. Natl. Acad. Sci. USA* **2020**, *117*, 6246–6254. [CrossRef] [PubMed] [PubMed Central]
27. Wilming, N.; Onat, S.; Ossandón, J.P.; Açık, A.; Kietzmann, T.C.; Kaspar, K.; Gameiro, R.R.; Vormberg, A.; König, P. An extensive dataset of eye movements during viewing of complex images. *Sci. Data* **2017**, *4*, 160126. [CrossRef]
28. Besl, P.J.; McKay, N.D. A Method for Registration of 3-D Shapes. *IEEE Trans. Pattern Anal. Mach. Intell.* **1992**, *14*, 239–256. [CrossRef]
29. Ester, M.; Kriegel, H.P.; Sander, J.; Xu, X. A density-based algorithm for discovering clusters in large spatial databases with noise. In Proceedings of the Second International Conference on Knowledge Discovery and Data Mining, Portland, OR, USA, 2–4 August 1996; pp. 226–231.
30. RStudio Team. *RStudio: Integrated Development for R*; RStudio, PBC: Boston, MA, USA, 2020; Available online: http://www.rstudio.com (accessed on 19 February 2024).
31. Bürkner, P.C. Brms: An R Package for Bayesian Multilevel Models using Stan. *J. Stat. Softw.* **2017**, *80*, 1–28. [CrossRef]
32. Gelman, A.; Hill, J. *Data Analysis Using Regression and Multilevel/Hierarchical Models*, 1st ed.; Cambridge University Press: New York, NY, USA, 2007.
33. Gelman, A.; Rubin, D.B. Inference from iterative simulation using multiple sequences. *Stat. Sci.* **1992**, *7*, 457–472. [CrossRef]
34. Rayner, K. Eye movements in reading and information processing: 20 years of research. *Psychol. Bull.* **1998**, *124*, 372–422. [CrossRef]
35. Lenth, R. Emmeans: Estimated Marginal Means, Aka Least-Squares Means. R Package Version 1.10.4.900001. 2024. Available online: https://rvlenth.github.io/emmeans/ (accessed on 17 May 2024).
36. Torralba, A.; Oliva, A.; Castelhano, M.S.; Henderson, J.M. Contextual guidance of eye movements and attention in real-world scenes: The role of global features in object search. *Psychol. Rev.* **2006**, *113*, 766–786. [CrossRef] [PubMed]
37. Wang, Z.; Bovik, A.C.; Sheikh, H.R.; Simoncelli, E.P. Image quality assessment: From error visibility to structural similarity. *IEEE Trans. Image Process.* **2004**, *13*, 600–612. [CrossRef] [PubMed]
38. Theeuwes, J.; Bogaerts, L.; van Moorselaar, D. What to expect where and when: How statistical learning drives visual selection. *Trends Cogn. Sci.* **2020**, *26*, 860–872. [CrossRef] [PubMed]
39. Follet, B.; Le Meur, O.; Baccino, T. New insights into ambient and focal visual fixations using an automatic classification algorithm. *i-Perception* **2011**, *2*, 592–610. [CrossRef] [PubMed] [PubMed Central]
40. Smith, C.N.; Hopkins, R.O.; Squire, L.R. Experience-dependent eye movements, awareness, and hippocampus-dependent memory. *J. Neurosci.* **2006**, *26*, 11304–11312. [CrossRef]
41. Hannula, D.E.; Ranganath, C.; Ramsay, I.S.; Solomon, M.; Yoon, J.; Niendam, T.A.; Carter, C.S.; Ragland, J.D. Use of eye movement monitoring to examine item and relational memory in schizophrenia. *Biol. Psychiatry* **2010**, *68*, 610–616. [CrossRef]

applied sciences

MDPI

Article

Workload Assessment of Operators: Correlation Between NASA-TLX and Pupillary Responses

Yun Wu, Yao Zhang and Bin Zheng *

Department of Surgery, Faculty of Medicine and Dentistry, University of Alberta, Edmonton, AB T6G 2S2, Canada; yun.wu@ualberta.ca (Y.W.); yao.zhang@ualberta.ca (Y.Z.)
* Correspondence: bin.zheng@ualberta.ca

Abstract: Operators in high-stress environments often face significant cognitive demands that can impair their performance, underscoring the need for comprehensive workload assessment. This study aims to study the relationship between subjective self-reported measures, the NASA task load index (NASA-TLX), objective bio-signal measures, and pupillary responses. The participants engaged in either a visual tracking task or a laparoscopic visuomotor task while their eye movements were recorded using a Tobii Pro Nano eye tracker (Tobii Technology Inc., Stockholm, Sweden). Immediately after completing the tasks, participants provided NASA-TLX scores to assess their perceived workload. The study tested three hypotheses: first, whether increased pupil dilation correlates with higher NASA-TLX scores; second, whether task type affects workload; and third, whether task repetition influences workload. The results showed a moderate positive correlation between pupil size and NASA-TLX scores ($r = 0.513$, $p < 0.001$). The laparoscopic surgery task, which requires visuomotor coordination, resulted in significantly higher NASA-TLX scores ($t = -6.23$, $p < 0.001$), larger original pupil sizes ($t = -22.57$, $p < 0.001$), and more adjusted pupil sizes ($t = -22.57$, $p < 0.001$) than the purely visual task. Additionally, task repetition led to a significant reduction in the NASA-TLX scores ($t = 2.86$, $p = 0.005$), the original mean pupil size ($t = 5.50$, $p < 0.001$), and the adjusted pupil size ($t = 6.34$, $p < 0.001$). In conclusion, the study confirms a positive correlation between NASA-TLX scores and pupillary responses. Task type and repetition were found to influence workload and pupillary responses. The findings demonstrate the value of using both subjective and objective measures for workload assessments.

Keywords: workload assessment; NASA-TLX; pupillary responses; visual-tracking task; laparoscopic task

1. Introduction

Operators in demanding environments such as surgery and driving often face challenging moments that place significant loads on their cognitive and physical resources [1,2]. Workload, also known as cognitive load or mental workload, describes the mental and physical demands placed on an individual during task performance [3]. Excessive workload can impair performance, leading to negative outcomes like errors and accidents [4,5]. This issue is particularly evident in imaging-guided tasks—such as visual tracking and endoscopic surgery—where individuals must process complex visual information in real time while making decisions or performing precise manual actions [6,7].

Given the risk of cognitive overload in these demanding tasks, accurately measuring operator workload is essential for enhancing safety and efficiency. Traditional assessment tools like the NASA task load index (NASA-TLX) have been widely used and validated for measuring subjective perceptions of workload [8]. However, these tools have limitations: they are retrospective, rely on self-reporting, and can be influenced by personal biases. These shortcomings highlight the need for objective, real-time measures.

To address this need, physiological signals offer a promising avenue for objective workload assessment. Among these, pupillary responses based on eye-tracking technologies have gained attention due to their sensitivity to cognitive load, feasibility for real-time monitoring, and cost-effectiveness [9]. Studies have linked pupillary responses with mental effort [10], and latterly with task complexity performed in the hand [11,12]. However, how pupillary responses align with traditional self-report measures in imaging-guided tasks is not well documented. As a multimodal approach that combines subjective assessments with objective data gains popularity, understanding the consistency and correlation between these methods becomes increasingly important [13,14].

Moreover, the task requirements behind different types of tasks and their repetition may affect workload levels differently. For instance, compared to a visual tracking task, performing a surgical procedure in the imaging-guided environment requires processing extensive visual input and delivering motor output with precision, potentially increasing the cognitive load on surgeon [15]. However, with repeated practice of a task, certain movements can become automated, potentially reducing workload [16]. Therefore, if our methods are sensitive in measuring workload changes, this study should yield differences in the workload metrics between different task types and repeated performances.

This study aims to advance the understanding of workload assessment and make several key contributions:

- Reveal a direct correlation between subjective workload measures (NASA-TLX) and objective indicators (pupillary responses) across two different imaging-guided tasks. This will validate pupil responses as a reliable measure of workload applicable to different task types;
- Reveal how tasks requiring both visual and motor coordination differ from those involving only visual tracking in terms of workload, as evidenced by both subjective and objective measures;
- Reveal the influence of repeated task exposure on workload, as evidenced by both subjective and objective measures.

With objectives in mind, this study consisted of three main parts. First, this study examined the relationship between subjective workload scores, as measured by the NASA-TLX, and objective workload indicators derived from pupil responses during two different imaging-guided tasks. This study first hypothesizes that when individuals report higher subjective workload, they will also exhibit increased pupil dilation. Second, this study examines whether the workload differs between a routine visual tracking task and a more demanding visuomotor laparoscopic task that requires precise eye–hand coordination. If both the NASA-TLX and pupil measurements are valid indicators of workload, this study expects to find significant differences in the values between these two tasks. The third goal of this study is to examine how repeated practice affects workload. Previous findings suggest that as individuals become more proficient, their workload decreases because movement control becomes more automatic. If this is the case, both NASA-TLX scores and pupil size should show significant reductions after the repetition of the tasks.

2. Related Work

Existing research has explored the relationship between NASA-TLX scores and surgeons' eye blink rates. It was found that surgeons who blinked less frequently reported higher levels of frustration and overall workload compared to those who blinked more often [17]. Other studies have also investigated how subjective workload measures relate to physiological indicators. For example, heart rate and heart rate variability were linked to self-reported workload on the NASA-TLX during clinical reasoning tasks [18]. Higher workload has been found to potentially cause over-breathing [19]. Additionally, changes in skin conductance, which reflect sweat gland activity driven by the sympathetic nervous system, have been associated with varying levels of workload [20]. These findings indicate that physiological measures can enhance subjective assessments like the NASA-TLX, providing a more comprehensive evaluation of workload.

Previous research has also made some progress in examining the relationship between pupil responses. Pupil size can serve as a bio-signal for occupational risk factors, including stress, fatigue and workload [21,22]. The existing literature on pupil size and subjective workload measures has primarily focused on purely visual (cognitive) tasks [23–25] or visuomotor tasks performed under direct vision, such as manufacturing [25]. Many existing studies did not involve an imaging-guided environment [22,26]. Although a few investigations did consider imaging-guided scenarios, some examined how pupil size varied according to operator expertise [27,28]. Many researchers set predetermined difficulty levels and assessed subjective and objective (pupil size) measures separately, without establishing a direct correlation between the two [29,30]. Moreover, focusing on a single task type does not accurately represent the complexity of real-world contexts, where individuals often switch between purely visual tasks and those requiring both visual perception and motor skills. This study addresses these gaps by integrating both purely visual and visuomotor (laparoscopic) imaging-guided tasks. This work would improve the generalizability of the findings and fill critical knowledge gaps in understanding workload assessments. A comparative table highlighting the main innovations and differences in the current study in relation to the literature can be found in Table 1.

Table 1. Comparative table highlighting the main innovations and differences in the current study in relation to literature.

Study	Objective Measures	Subjective Measures	Task Types Involved	Imaging-Guided Environment	Characterizing Correlation	Compare Visual and Visuomotor Task	Compare Repetition Effect	Reference
Past Literature	Pupil sizes	NASA-TLX SURG-TLX [9]	Pure visual (cognitive) task	No	Yes [23,24] No [19,25,31]	No	No	[23–25,29,31]
			Piloting, manufacturing, open surgery	No	Yes [32] No [33,34]	No	No	[32–34]
			Microscopic and robotic procedures	No	No [15,35] Yes [36]	No	No	[15,35,36]
			Computer games	Yes	Yes [37,38] No [18]	No	No	[37–39]
			Laparoscopic or endoscopic	Yes	No	No	No	[30,40,41]
This study	Pupil size	NASA-TLX	Pure visual and Visuomotor (laparoscopic) hybrid	Yes	Yes	Yes	Yes	

3. Materials and Methods

This study was conducted at the Surgical Simulation Research Laboratory of the University of Alberta between 2022 and 2023. Ethical approval was obtained from the University of Alberta Health Research Ethics Board prior to the research data collection.

3.1. Participants

The participants were recruited from the University of Alberta community. To qualify, the participants needed to be healthy adults with normal or corrected vision and mobility, and they must have had no prior experience in performing visual tracking or laparoscopic tasks. Those who had previously participated in similar studies or simulations were excluded from the study. Informed written consent was obtained from each participant before the experiment. The participants participated in only one of the two tasks to ensure no overlap between the groups.

3.2. Apparatus

To eliminate confounding variables, both of the tasks were conducted using the same computer and eye-tracking setup within a controlled laboratory environment, as demonstrated in Figure 1. The experiment employed a commercially available PC (Intel, Santa Clara, CA, USA) connected to a 476 mm × 370 mm monitor (Asus Computer Inc., Taipei, Taiwan) with a resolution of 1920 × 1080 pixels and a refresh rate of 60 Hz. Monitor parameters such as brightness and distance from the participants' eyes were kept consistent across both tasks.

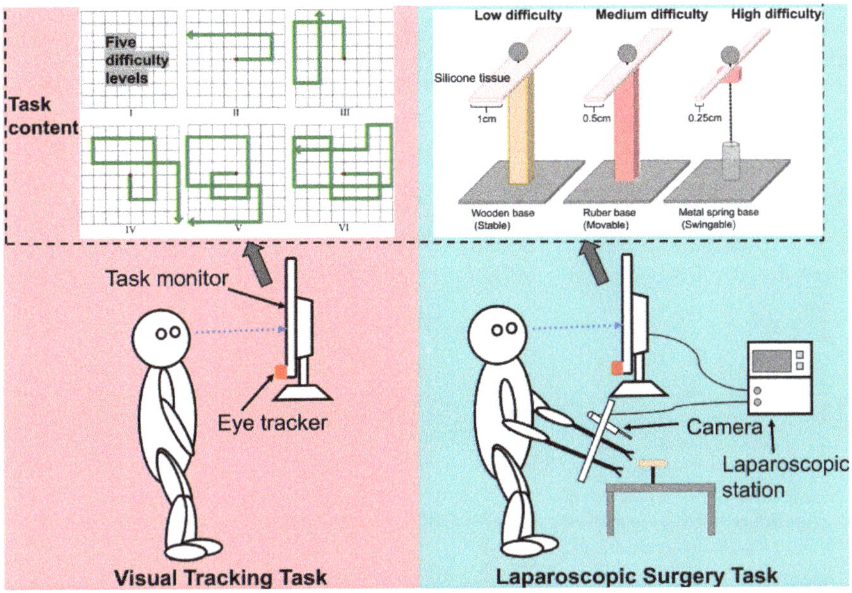

Figure 1. Schematic demonstration of experimental setup and task content.

The participants' eye movements were recorded using a Tobii Pro Nano eye tracker (Tobii Technology Inc., Stockholm, Sweden), a remote, screen-based device with a sampling rate of 60 Hz and a reported accuracy of 0.5 degrees. Before each session, a calibration was performed. This was performed using the Tobii Pro Lab's pre-designed calibration protocol. Data analysis was performed using the Tobii Pro Lab 2021 software, which extracts and reports pupil data for any given task period.

For the second task involving eye–hand coordination, the participants further used a laparoscopic simulation station (Stryker, Kalamazoo, MI, USA) connected to the same computer and monitor. They were required to repeatedly manipulate a size 3–0 needle (Ethicon, Somerville, NJ, USA) using laparoscopic graspers to penetrate target tissue.

3.3. Eye Tracking Calibration

Eye tracking was conducted throughout each formal session. Prior to every experimental session, a calibration process ensured the accuracy of the eye tracking data. The eye tracker calibration began by displaying an animated object to attract the user's eye gaze. The object moved to each calibration point, shrinking and pausing for 0.5 s to focus the user's gaze on the center. Upon focus, the eye tracker collected data for that point and then restored the object's size. This process was animated across all the calibration points, repeating as needed for accuracy. The calibration results were reviewed using plots, and any points with missing or low-accuracy data were recalibrated. Once satisfactory, the calibration was finalized.

3.4. Visual Tracking Task

3.4.1. Task Procedures

Upon entering the lab, the participants received an orientation on how to perform the visual tracking task, and an introduction to the NASA-TLX. After the orientation, they proceeded to watch videos depicting five different movement patterns in the visual tracking task.

These videos all featured an 8 × 8 grid, with each cell measuring 2 cm by 2 cm. An arrow moved within this grid at a constant speed of one cell length per second. The complexity of the movement patterns was manipulated by changing the number of directional changes and the length of the arrow's path, which affected the duration of each video. The task difficulty was structured across five levels, ranging from simple to complex. As the difficulty increased, the arrow made more directional changes and followed longer paths, resulting in longer videos and a greater information load. Specifically, from the simplest videos to the most complex video, the arrow movement direction changes occurred 3, 5, 7, 9, 11 times, respectively, as shown in Figure 1.

To investigate the effect of task repetition on performance measures, each difficulty level of the visual tracking task was presented twice. This allowed us to assess how task repetition influenced participants' workloads. The procedural sequence is demonstrated in Figure 2.

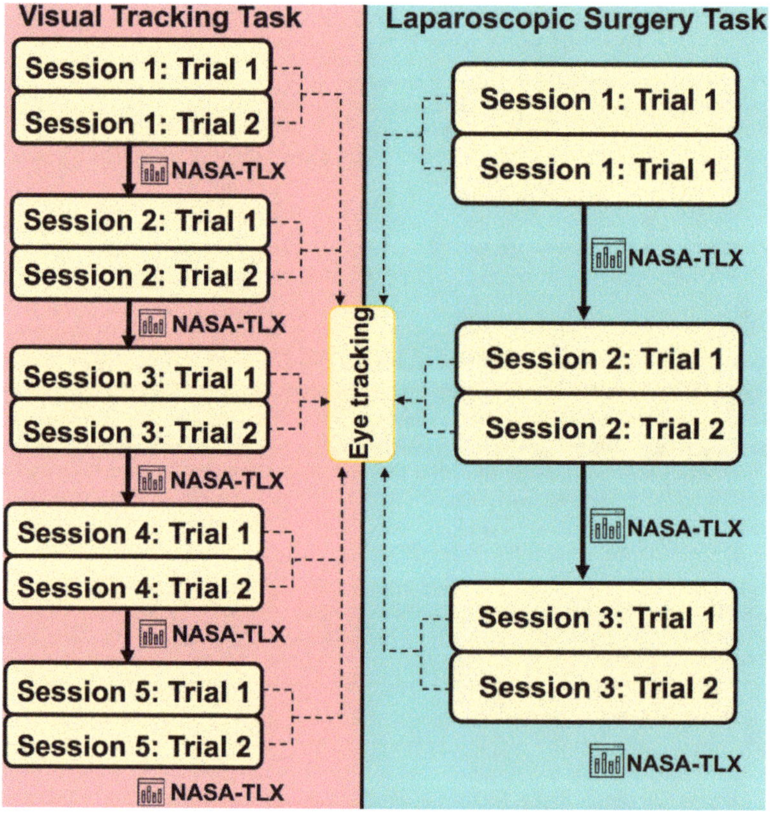

Figure 2. Flowchart on the sequence of experimental procedures.

By systematically varying the number of directional changes and path lengths, we aimed to create measurable task difficulty levels. Following each level, the participants completed a NASA-TLX assessment to evaluate their perceived workload. In the meantime, pupil data taken during their performance was used to report their workloads. They were then provided with ample rest, typically five minutes between blocks to prevent fatigue and maintain alertness. The participants were encouraged to relax and could request additional rest if needed. The entire experimental session lasted approximately 30 min per participant. For detailed experimental procedures and task specifics, please refer to our previous study [42,43].

3.4.2. Tracking Accuracy

The tracking accuracy was assessed by asking the participants to verbally report the movement trajectory immediately after each visual tracking task. The experimenter assessed the tracking accuracy based on the participants' responses during the recall phase. We assessed whether each participant's response matched the correct movement at each turning point, described in terms of "distance-direction" (e.g., "three units up"). The participants received one point for each correct match of both distance and direction; no points were awarded if either component was incorrect. Because the number of turnpoints varied across blocks, we normalized the scores by dividing the total number of points earned by the number of turnpoints in each block. The tracking accuracy was then calculated using the following formula: tracking accuracy = points awarded ÷ total number of turning points. This score ranged from 0 to 1, with 1 indicating perfect tracking accuracy.

3.5. Laparoscopic Surgery Task

3.5.1. Task Procedures

The second task was a laparoscopic surgery task. The participants watched the same display screen while using their hands to manipulate a pair of laparoscopic graspers to perform a needle insertion task. The task was to insert a surgical needle through a silicone pad without tying a knot. They had to complete as many insertions as possible within the allotted time.

This task was structured with three levels of difficulty—easy, medium, and hard—to create varying workloads. The difficulty was adjusted by altering the size of the suturing target and the stability of the platform. In the easy level, the participants sutured a 1 cm-wide silicone strip fixed to a stable wooden platform. The medium level involved a 0.5 cm-wide silicone strip attached to a movable rubber platform. The hard level required suturing a 0.25 cm-wide silicone strip secured to a highly elastic, shaking metal wire platform. The platform height was 6 cm for all three difficulty levels. By modifying the target size and platform stability, we introduced different levels of challenge to assess the participants' operational skills. Knot-tying was not required for this task, and the participants were instructed to perform as many insertions as possible.

Each session was limited to 15 min, a duration determined by a pilot study to accommodate the extended time needed for practice tasks. The three difficulty levels were conducted over three consecutive days, with each day dedicated to one specific level. Additionally, each difficulty level was performed in two separate sessions (first and second trials) to examine the effects of task repetition. Before each session, the eye-tracking devices were calibrated to ensure accurate data collection during their surgical performance. After completing each session, the participants completed a NASA-TLX assessment to evaluate their perceived workload. Adequate rest periods, typically 10–20 min, were provided between sessions to maintain participant well-being and ensure consistent performance throughout the study. The procedural sequence is demonstrated in Figure 2.

3.5.2. Performance Assessment

The number of insertions is used to assess their task performance. Video recordings were made during the task sessions for later analysis. The experimenters conducted a retrospective analysis of these recordings using Behavioral Observation Research Interactive Software (BORIS) version 8.21.10 [4]. BORIS is a software tool that facilitates the conversion of observed behaviors in videos into quantitative data. Observers use it to annotate video footage by tagging predefined behaviors with precise timestamps. The annotation data provides quantitative metrics, such as the frequency of specific events and the duration of tasks captured in the videos, which can be exported for statistical analysis.

3.6. NASA-TLX Score

The workload of the participants from both the visual tracking and surgical tasks was evaluated using the NASA-TLX. This tool measures workload across six dimensions:

mental demands, physical demands, temporal demands, effort, frustration, and performance. The participants used a customized software interface to rate each of these six areas. Subsequently, they responded to a series of paired comparison questions, where each question displayed two of the domains and required participants to select one they deemed more significant. This process determined the relative weight of each domain. The overall NASA-TLX score was then calculated by multiplying each domain's score by its weight, yielding a comprehensive measure of the workload [8].

3.7. Pupil Data Collection and Process

The participants' pupil data were continuously monitored during both tasks using the Tobii Pro Nano eye tracker. The pupil measurements were processed and exported on a frame-by-frame basis using Tobii Pro Lab software (version 1.171.34906). To account for individual variability, the adjusted pupil values were calculated for each participant across all the task conditions and used in subsequent statistical analyses. The adjusted pupil size was determined by subtracting the baseline pupil size from the original mean pupil size, providing a standardized measure of pupil size changes relative to the baseline.

For the arrow tracking task, the baseline pupil size for each condition was calculated by averaging the pupil diameters from the first 10 frames before the onset of the arrow video. The original mean pupil size was calculated by averaging the measurements over all subsequent frames.

Similarly, for the laparoscopic surgery task, the baseline pupil size was determined by averaging the first 100 frames of each session due to the longer session duration. The mean pupil size for all the subsequent frames recorded during each session was calculated as the original mean pupil size. The adjusted pupil size was calculated by subtracting the mean pupil size from baseline in both tasks.

3.8. Statistical Analysis

Statistical analyses were performed using R (version 4.3.2) and RStudio. We first ensured that the data were complete and addressed any missing values or outliers to maintain data integrity. Prior to correlation analysis, we tested the normality of the data using the Shapiro–Wilk test and inspected the scatter plots to confirm linearity. To address the first hypothesis, the trainees' pupil size metrics were correlated with their NASA TLX scores using Pearson's correlation coefficient and simple linear regression. The relationships were analyzed and visualized using scatter plots generated with the GGally package (Barret Schloerke et al., Ann Arbor, MI, USA). For the second and third hypotheses, we assessed the equality of variances using Levene's test. Descriptive analyses were first performed on the original mean pupil size, adjusted pupil size, and NASA TLX scores across the task types and repetitions. Box plots and histograms were used to visualize the distributions. Subsequently, t-tests were performed to compare the outcome variables between the two tasks. Paired t-tests were used to compare outcome variables between two repeated trials. Data were reported as the means and standard deviations for each measure. All the statistical tests were two-tailed with a statistical significance level of $p < 0.05$.

4. Results

4.1. Participants Information

A total of 15 participants voluntarily participated in the visual tracking task, with a mean age of 28 $\pm$ 7 years. The sample included 5 males and 10 females. Each participant completed the five sessions of visual tracking tasks in sequential order from easy to difficult, with each task difficulty repeated twice. A total of 22 subjects participated in and completed the laparoscopic surgery task (10 males and 12 females) with a mean age of 22.8 $\pm$ 4.6 years. The participants completed the task from low to high difficulty, with each task difficulty repeated twice, for a total of six sessions.

4.2. Correlation Between NASA-TLX Scores and Pupil Size

Using the whole dataset, we conducted a correlation analysis to explore the relationships among original mean pupils, adjusted pupil size, and NASA-TLX scores. The results, summarized in a correlation matrix (Figure 3), included Pearson correlation coefficients and scatter plots for visual interpretation. Notably, a moderate positive correlation was found between the original mean pupil and adjusted pupil size ($r = 0.513$, $p < 0.001$), indicating that the adjusted pupil sizes were highly associated with the original mean pupil. A weaker but statistically significant positive correlation was observed between the original mean pupil and the NASA-TLX scores ($r = 0.312$, $p < 0.001$), suggesting that the increased workload was moderately associated with the larger pupil sizes. In contrast, the correlation between the adjusted pupil size and the NASA-TLX scores was weak and not statistically significant ($r = 0.103$, $p = 0.11$). Although not statistically significant, the scatterplots showed positive linear trends between the adjusted pupil size and the NASA-TLX scores.

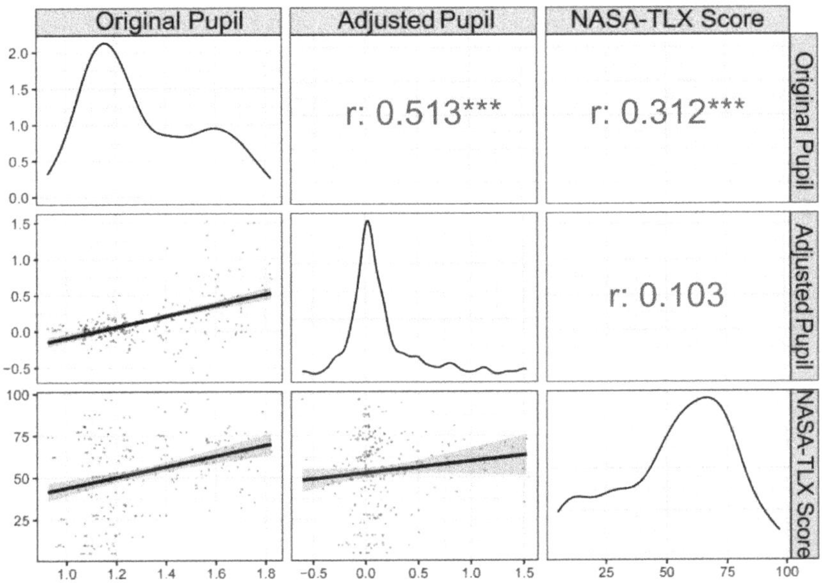

Figure 3. Correlation plots between NASA-TLX scores and pupil size measurements. The diagonal panels display density plots illustrating the distributions of each variable. The lower-left panels present scatter plots with corresponding regression lines, showing the relationships between variables. "r" represents the Pearson correlation coefficient (r value), with "***" indicating a significance level of $p < 0.001$.

4.3. Effect of Task Type on NASA-TLX Scores and Pupil Size Measurements

For the second hypothesis on the effect of task type, the visual representation is shown in Figure 4. In the visual tracking task, the participants reported a NASA-TLX score of 47.5 ± 25.6; their original mean pupil size was 3.20 ± 0.34 mm, and the adjusted pupil size was 0.04 ± 0.12 mm. In contrast, when performing the laparoscopic task, the participants reported a NASA-TLX score of 63.4 ± 14.3; their original mean pupil size was 4.78 ± 0.70 mm, and the adjusted pupil size was 0.31 ± 0.48 mm. We conducted t-tests to examine the difference in the NASA-TLX scores and the pupil size between the visual tracking and the laparoscopic task. Significant group differences were observed on the NASA-TLX scores ($t = -6.23$, $p < 0.001$), original mean pupil ($t = -22.57$, $p < 0.001$), and adjusted pupil size ($t = -22.57$, $p < 0.001$). Specifically, the laparoscopic surgery task resulted in higher NASA-TLX scores and larger pupil sizes—both original and adjusted—compared to the visual tracking task.

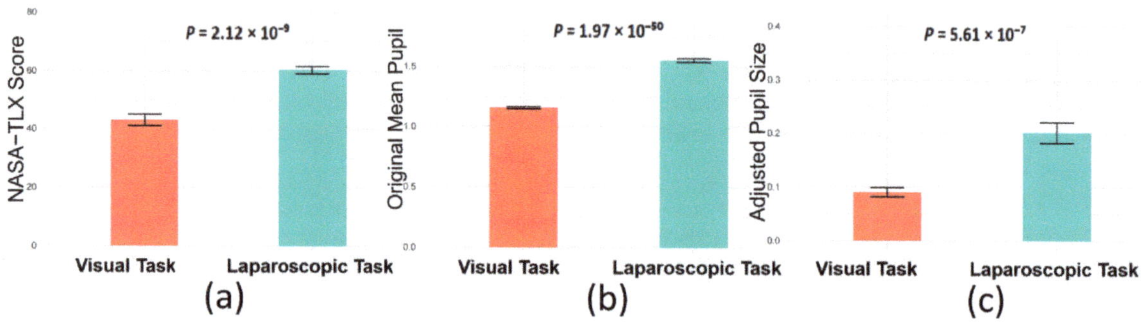

Figure 4. Bar plots across two task types. Bar plots with error bars comparing different task types for: (**a**) NASA-TLX scores, (**b**) original mean pupil size, and (**c**) adjusted pupil size.

4.4. Effect of Task Repetition on NASA-TLX Scores and Pupil Size Measurements

To assess the third hypothesis on the effect of task repetition, we first aggregated data from both the experiments. The visual representation is shown in Figure 5. In the first trial, the participants reported a NASA-TLX score of 54.4 $\pm$ 23.6; their original mean pupil size was 3.84 $\pm$ 0.91 mm, and the adjusted pupil size was 0.19 $\pm$ 0.36 mm. In contrast, during the second trial, the participants reported a NASA-TLX score of 53.0 $\pm$ 22.8; their original mean pupil size was 3.79 $\pm$ 0.94 mm, and the adjusted pupil size was 0.09 $\pm$ 0.31 mm. Paired t-tests were conducted to examine the differences in the NASA-TLX scores and the pupil sizes between the first and second trials. Task repetition was found to have a significant impact on the NASA-TLX (t = 2.86, p = 0.005), on the adjusted pupil size (t = 6.34, p < 0.001), and on the original mean pupil (t = 5.50, p < 0.001). Notably, the workload, and mean and adjusted pupil size were significantly lower during the second trial when the task was repeated.

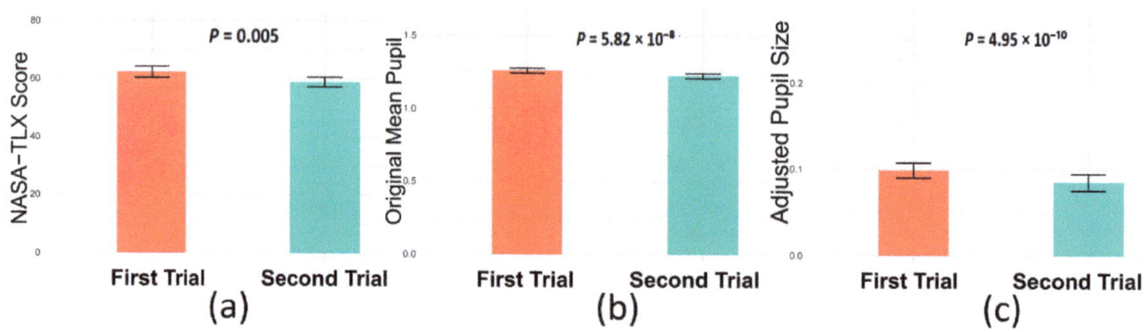

Figure 5. Bar plots across task trials. Bar plots with error bars depicting the effect of task repetition on: (**a**) NASA-TLX scores, (**b**) original mean pupil size, and (**c**) adjusted pupil size.

Analyzing each experiment separately, we found that in the visual tracking task, repetition led to a significant reduction in the adjusted pupil size (t = 11.10, p < 0.001) and original mean pupil (t = 11.57, p < 0.001). Due to the brief nature of the task one sessions and the retrospective design of the NASA-TLX assessment, we did not collect NASA-TLX data between the first and second trials in the visual tracking task. The NASA-TLX was conducted only once for each difficulty level session. Therefore, data on the effect of repetition on the NASA-TLX scores was only available from the laparoscopic surgery task. The laparoscopic task also showed a significant difference in the adjusted pupil size (t = 2.33, p = 0.02), but no significant difference in the original mean pupil (t = 1.01, p = 0.31) after repetition.

4.5. Task Performance

The participants demonstrated significant improvements in both their visual tracking and laparoscopic surgery tasks between the first and second trials. In the visual tracking task, the tracking accuracy increased from 0.71 ± 0.30 in the first trial to 0.90 ± 0.18 in the second trial, indicating a significant improvement (t = 11.57, $p < 0.001$). For the laparoscopic surgery task, the mean number of sutures completed rose from 4.19 ± 3.44 in the first trial to 5.60 ± 3.91 in the second trial, which was a statistically significant difference (t = −4.80, $p < 0.001$). Additionally, the mean suture duration decreased from 36.31 ± 19.20 s in the first trial to 29.51 ± 16.84 s in the second trial, reflecting a significant reduction in performance time (t = 2.83, $p = 0.01$).

5. Discussion

Our study supports our initial three hypotheses. We observed a significantly positive correlation between the NASA-TLX scores and the original mean pupil size from the overall dataset. The tasks involving complex eye–hand coordination in the laparoscopic setting exhibited significantly higher workloads reflected by both the NASA-TLX scores and the pupil size measurements compared to the visual tracking tasks. In addition, task repetition was found to decrease pupil size and the NASA-TLX scores, but not the original mean pupil size in the laparoscopic task.

The synchronized changes between the NASA-TLX scores and pupillary responses provided strong evidence that well-designed paper-based tools, along with pupil size measurements, can reliably assess operator workload when used appropriately. Pupillary response is part of the sympathetic nervous system activation; as the workload increases, the sympathetic activity rises, leading to physiological responses such as increased heart rate, respiration, and pupil dilation [44,45]. Although measuring eye-tracking bio-signals is a logical approach to assessing workload, it requires specialized equipment that must be attached to operators. The interference caused by these devices can limit their use in certain situations, such as during surgery [46]. Therefore, paper-based tools alone still offer a reliable workload assessment in that situation.

We are glad to see that both NASA-TLX and pupil dilation successfully reported significant differences in two tasks with different difficulty. To make the data comparable, we maintained consistent environmental conditions for both the tasks to eliminate confounding variables such as room illumination and environmental interference. This control ensured that the changes in pupil size were due to the tasks rather than external factors. The significantly higher NASA-TLX scores and pupil sizes associated with visuomotor tasks were consistent with cognitive load theory. Motor control requires additional cognitive resources beyond visual processing alone, resulting in higher workload and consequently higher NASA-TLX scores and pupil sizes [47,48].

Task repetition had significant effects on most of the workload measures, except for the original mean pupil size during the laparoscopic task. Repeating the laparoscopic task did not result in noticeable changes in the original mean pupil size. This lack of change may be attributed to the high difficulty inherent in the laparoscopic procedure and suggests that the original pupil size is less sensitive than the baseline-adjusted pupil size. In contrast, repetition of the visual tracking task led to significant changes in both the original and adjusted pupil sizes. This can be explained by the fact that tracking a moving arrow is a common activity, allowing participants to achieve proficiency and automaticity after only one repeated trial. Moreover, performance showed significant improvement in both the tasks, indirectly confirming that repetition reduces trainee workload. These findings are consistent with prior studies demonstrating the effect of practice in reducing workload [49].

A detailed inspection on pupillary response, this study found that the NASA-TLX scores correlated only significantly with the original mean pupil size, showing a trend but not a significant correlation with the adjusted pupil size. However, thee adjusted pupil size was more sensitive in characterizing the effect of task repetition, suggesting that precaution is needed when applying pupillary response to assessing workload, as it can be influenced by many other variations.

One limitation of this study is that NASA-TLX assessments were not conducted between the trials during the visual tracking task. As a result, this study was unable to evaluate how repetition affects NASA-TLX scores in this context. However, previous studies have shown significant improvements in participants' performance between trials, suggesting that trainees did experience reduced workload in the visual tracking task. This supports the reliability of using adjusted pupil size to reflect changes in workload. In addition, this study focused only on the average pupil size; many other eye-related metrics in relation to gaze and pupil oscillation over time are areas for future investigation. The NASA-TLX, while widely used, is a retrospective tool and may not capture moment-to-moment workload variations. Furthermore, although factors like lighting, emotional state, and fatigue were controlled, their influence on pupil size cannot be completely eliminated.

Assessing workload is essential in high-stakes environments. By understanding how objective and subjective measures correlate, we can use eye-tracking technology to more accurately evaluate workload and strengthen its alignment with subjective assessments. Instructors and educators can proactively intervene, pausing tasks or reducing their complexity to prevent cognitive overload during training. Such approaches are especially valuable in the scenarios demanding rapid, imaging-guided decision-making, such as surgery and aviation [50,51]. Additionally, recognizing how task type and repetition influence workload measures can help shape training programs, guiding adjustments when integrating motor components or design training task repetitions.

6. Conclusions

This study contributed to a better understanding of how subjective and objective measures of workload relate to each other in complex imaging-guided task environments and provides a practical basis for future workload assessment using eye tracking.

The main research finding revealed a significant, moderate correlation between NASA-TLX scores and pupil size measurements, supporting the use of eye tracking as an objective tool for assessing workload. Furthermore, this study clarifies how tasks involving eye–hand coordination differ from purely visual tasks in their impact on both subjective and objective workload indicators, and how repetition practice influences these measures.

Future research includes investigating additional eye-tracking metrics such as saccades and pupil size fluctuations, including indices like the Index of Cognitive Activity and the Index of Pupillary Activity [52]. It is also necessary to examine how these biological signals correspond to subjective assessments within more complex human–computer interfaces, such as virtual reality. Finally, exploring the generalizability of the relationship across different tasks, environments, and populations is another valuable future direction.

Author Contributions: Conceptualization, Y.W. and B.Z.; data curation, Y.W.; formal analysis, Y.W.; funding acquisition, Y.W. and B.Z.; investigation, Y.W. and Y.Z.; methodology, Y.W., Y.Z. and B.Z.; project administration, Y.W. and B.Z.; resources, Y.W. and B.Z.; software, Y.W. and Y.Z.; supervision, B.Z.; validation, Y.W. and B.Z.; visualization, Y.W.; writing—original draft, Y.W.; writing—review and editing, Y.W., Y.Z. and B.Z. All authors have read and agreed to the published version of the manuscript.

Funding: Yun Wu acknowledges the financial support from the China Scholarship Council (Grant No. 202106370009).

Institutional Review Board Statement: This study was approved by the University of Alberta Institutional Review Board (Pro00120984 and Pro00128050).

Informed Consent Statement: Written informed consent has been obtained from the patient(s) to publish this paper.

Data Availability Statement: The datasets analyzed in this current study are available from the corresponding author upon reasonable request.

Acknowledgments: We thank our participants for participating in this experiment.

Conflicts of Interest: The authors declare no conflicts of interest.

References

1. Zhang, Z.S.; Wu, Y.; Zheng, B. A review of cognitive support systems in the operating room. *Surg. Innov.* **2024**, *31*, 111–122. [CrossRef] [PubMed]
2. Engström, J.; Johansson, E.; Östlund, J.; Hofmann, U.; Bärgman, J. Effects of cognitive load on driving performance: The cognitive control hypothesis. *Hum. Factors* **2017**, *59*, 734–764. [CrossRef]
3. Sweller, J. Cognitive load theory, learning difficulty, and instructional design. *Learn. Instr.* **1994**, *4*, 295–312. [CrossRef]
4. Howie, E.E.; Coulter, C.; Nguyen, H.; Yan, Y.; Zheng, B. Cognitive load management: An invaluable tool for safe and effective surgical training. *J. Surg. Educ.* **2023**, *80*, 311–322. [CrossRef] [PubMed]
5. Ahmad, R.; Anuar, N.; Md Som, A.P.; Sharif, Z.; Taat, M.S. The impact of workload on job performance among doctors in Malaysian public hospitals: A case study. *Int. J. Bus. Soc.* **2019**, *20*, 1276–1293.
6. Harewood, G.C.; Sharma, V.K.; De Garmo, P. Impact of operator fatigue on endoscopy performance: Implications for procedure scheduling. *Dig. Dis. Sci.* **2009**, *54*, 1656–1661. [CrossRef]
7. Luz, M.; Strauss, G.; Manzey, D. Impact of image-guided surgery on surgeons' performance: A literature review. *Int. J. Hum. Factors Ergon.* **2016**, *4*, 229–263. [CrossRef]
8. Hart, S.G. NASA-Task Load Index (NASA-TLX); 20 Years Later. In Proceedings of the Human Factors and Ergonomics Society Annual Meeting, San Francisco, CA, USA, 16–20 October 2006; Sage Publications: Los Angeles, CA, USA, 2006.
9. Kramer, A.F. Physiological Metrics of Mental Workload: A Review of Recent Progress. In *Multiple Task Performance*; CRC Press: Boca Raton, FL, USA, 2020; pp. 279–328.
10. Van der Wel, P.; Van Steenbergen, H. Pupil dilation as an index of effort in cognitive control tasks: A review. *Psychon. Bull. Rev.* **2018**, *25*, 2005–2015. [CrossRef]
11. Zheng, B.; Jiang, X.; Atkins, M.S. Detection of changes in surgical difficulty: Evidence from pupil responses. *Surg. Innov.* **2015**, *22*, 629–635. [CrossRef]
12. Jiang, X.; Atkins, M.S.; Tien, G.; Bednarik, R.; Zheng, B. Pupil Responses During Discrete Goal-Directed Movements. In Proceedings of the SIGCHI Conference on Human Factors Computing Systems, Toronto, ON, Canada, 26 April–1 May 2014.
13. Herbig, N.; Pal, S.; Vela, M.; Krüger, A.; van Genabith, J. Multi-modal indicators for estimating perceived cognitive load in post-editing of machine translation. *Mach. Transl.* **2019**, *33*, 91–115. [CrossRef]
14. Abbad-Andaloussi, A.; Burattin, A.; Slaats, T.; Kindler, E.; Weber, B. Complexity in declarative process models: Metrics and multi-modal assessment of cognitive load. *Expert Syst. Appl.* **2023**, *233*, 120924. [CrossRef]
15. Bednarik, R.; Bartczak, P.; Vrzakova, H.; Koskinen, J.; Elomaa, A.-P.; Huotarinen, A.; de Gómez Pérez, D.G.; von und zu Fraunberg, M. Pupil Size as an Indicator of Visual-Motor Workload and Expertise in Microsurgical Training Tasks. In Proceedings of the Symposium on Eye Tracking Research & Applications, Warsaw, Poland, 14–17 June 2018.
16. Szulewski, A.; Howes, D.; van Merriënboer, J.J.; Sweller, J. From theory to practice: The application of cognitive load theory to the practice of medicine. *Acad. Med.* **2021**, *96*, 24–30. [CrossRef] [PubMed]
17. Zheng, B.; Jiang, X.; Tien, G.; Meneghetti, A.; Panton, O.N.M.; Atkins, M.S. Workload assessment of surgeons: Correlation between NASA TLX and blinks. *Surg. Endosc.* **2012**, *26*, 2746–2750. [CrossRef] [PubMed]
18. Solhjoo, S.; Haigney, M.C.; McBee, E.; van Merrienboer, J.J.G.; Schuwirth, L.; Artino, A.R.; Battista, A.; Ratcliffe, T.A.; Lee, H.D.; Durning, S.J. Heart rate and heart rate variability correlate with clinical reasoning performance and self-reported measures of cognitive load. *Sci. Rep.* **2019**, *9*, 14668. [CrossRef]
19. Grassmann, M.; Vlemincx, E.; von Leupoldt, A.; Mittelstädt, J.M.; Bergh, O.V.D. Respiratory changes in response to cognitive load: A systematic review. *Neural Plast.* **2016**, *2016*, 8146809. [CrossRef]
20. Buchwald, M.; Kupinski, S.; Bykowski, A.; Marcinkowska, J.; Ratajczyk, D.; Jukiewicz, M. Electrodermal Activity as a Measure of Cognitive Load: A Methodological Approach. In Proceedings of the 2019 Signal Processing: Algorithms, Architectures, Arrangements, and Applications, SPA, Poznan, Poland, 18–20 September 2019.
21. Ferreira, D.F.; Ferreira, S.; Mateus, C.; Barbosa-Rocha, N.; Coelho, L.; Rodrigues, M.A. Advancing the understanding of pupil size variation in occupational safety and health: A systematic review and evaluation of open-source methodologies. *Saf. Sci.* **2024**, *175*, 106490. [CrossRef]
22. Naik, R.; Kogkas, A.; Ashrafian, H.; Mylonas, G.; Darzi, A. The measurement of cognitive workload in surgery using pupil metrics: A systematic review and narrative analysis. *J. Surg. Res.* **2022**, *280*, 258–272. [CrossRef]
23. Jonathan, G.; Rainer, R.; Bernhard, B. Investigation of eye movement measures of mental workload in healthcare: Can pupil dilations reflect fatigue or overload when it comes to health information system use? *Appl. Ergon.* **2024**, *114*, 104150.
24. Mohanty, S.; Kim, J.H.; Pulipati, V.; Wang, F.; Mostowfi, S.; Oprean, D.; Wang, Y.; Seo, K. Measuring Cognitive Workload in Augmented Reality Learning Environments Through Pupil Area Analysis. In Proceedings of the International Conference on Human-Computer Interaction, Washington, DC, USA, 29 June–4 July 2024; Springer: Berlin/Heidelberg, Germany, 2024.
25. Mihelčič, M.; Podlesek, A. Cognitive workload affects ocular accommodation and pupillary response. *J. Optom.* **2023**, *16*, 107–115. [CrossRef]

26. Gorin, H.; Patel, J.; Qiu, Q.; Merians, A.; Adamovich, S.; Fluet, G. A Review of the Use of Gaze and Pupil Metrics to Assess Mental Workload in Gamified and Simulated Sensorimotor Tasks. *Sensors* **2024**, *24*, 1759. [CrossRef]
27. Aljehane, S.D.; Sharif, B.; Maletic, J.I. Studying Developer Eye Movements to Measure Cognitive Workload and Visual Effort for Expertise Assessment. *Proc. ACM Human-Computer Interact.* **2023**, *7*, 1–18. [CrossRef]
28. Li, S.; Duffy, M.C.; Lajoie, S.P.; Zheng, J.; Lachapelle, K. Using eye tracking to examine expert-novice differences during simulated surgical training: A case study. *Comput. Hum. Behav.* **2023**, *144*, 107720. [CrossRef]
29. Schirm, J.; Gómez-Vargas, A.R.; Perusquía-Hernández, M.; Skarbez, R.T.; Isoyama, N.; Uchiyama, H.; Kiyokawa, K. Identification of Language-Induced Mental Load from Eye Behaviors in Virtual Reality. *Sensors* **2023**, *23*, 6667. [CrossRef] [PubMed]
30. Kapp, C.M.; Akulian, J.A.; Yu, D.H.; Chen, A.; Cárdenas-García, J.; Molena, D.; Vachani, A.; Wahidi, M.M.; Maldonado, F.; Fielding, D.; et al. Cognitive load in electromagnetic navigational and robotic bronchoscopy for pulmonary nodules. *ATS Scholar* **2021**, *2*, 97–107. [CrossRef]
31. Mitre-Hernandez, H.; Carrillo, R.C.; Lara-Alvarez, C. Pupillary responses for cognitive load measurement to classify difficulty levels in an educational video game: Empirical study. *JMIR Serious Games* **2021**, *9*, e21620. [CrossRef]
32. Liu, F.; Wang, Y.; Burkhart, T.A.; González Penedo, M.F.; Ma, S.D.; Wanyan, X.; Zhuang, D.; Zhang, H. Improving pilot mental workload evaluation with combined measures. *Bio-Med. Mater. Eng.* **2014**, *24*, 2283–2290.
33. Biondi, F.N.; Saberi, B.; Graf, F.; Cort, J.; Pillai, P.; Balasingam, B. Distracted worker: Using pupil size and blink rate to detect cognitive load during manufacturing tasks. *Appl. Ergon.* **2023**, *106*, 103867. [CrossRef]
34. Tien, T.; Pucher, P.H.; Sodergren, M.H.; Sriskandarajah, K.; Yang, G.-Z.; Darzi, A. Differences in gaze behaviour of expert and junior surgeons performing open inguinal hernia repair. *Surg. Endosc.* **2015**, *29*, 405–413. [CrossRef]
35. Koskinen, J.; Bednarik, R.; Vrzakova, H.; Elomaa, A.-P. Combined gaze metrics as stress-sensitive indicators of microsurgical proficiency. *Surg. Innov.* **2020**, *27*, 614–622. [CrossRef]
36. Wu, C.; Cha, J.; Sulek, J.; Zhou, T.; Sundaram, C.P.; Wachs, J.; Yu, D. Eye-tracking metrics predict perceived workload in robotic surgical skills training. *Hum. Factors* **2020**, *62*, 1365–1386. [CrossRef]
37. Patel, A.N.; Chau, G.; Chang, C.; Sun, A.; Huang, J.; Jung, T.P.; Gilja, V. Affective Response to Volitional Input Perturbations in Obstacle Avoidance and Target Tracking Games. In Proceedings of the 2021 43rd Annual International Conference of the IEEE Engineering in Medicine & Biology Society (EMBC), Virtual Conference, 1–5 November 2021.
38. Marinescu, A.C.; Sharples, S.; Ritchie, A.C.; López, T.S.; McDowell, M.; Morvan, H.P. Physiological parameter response to variation of mental workload. *Hum. Factors* **2018**, *60*, 31–56. [CrossRef] [PubMed]
39. Köles, M.; Szegletes, L.; Forstner, B. Towards a Physiology-Based Difficulty Control System for Serious Games. In Proceedings of the 2015 6th IEEE International Conference on Cognitive Infocommunications (CogInfoCom), Győr, Hungary, 19–21 October 2015.
40. Gao, J.; Liu, S.; Feng, Q.; Zhang, X.; Zhang, J.; Jiang, M.; Wang, L.; Zhang, Q. Quantitative evaluations of the effects of noise on mental workloads based on pupil dilation during laparoscopic surgery. *Am. Surg.* **2018**, *84*, 1951–1956. [CrossRef] [PubMed]
41. Gao, J.; Liu, S.; Feng, Q.; Zhang, X.; Jiang, M.; Wang, L.; Zhang, J.; Zhang, Q. Subjective and objective quantification of the effect of distraction on physician's workload and performance during simulated laparoscopic surgery. *Med. Sci. Monit.* **2019**, *25*, 3127. [CrossRef] [PubMed]
42. Wu, Y.; Zhang, Z.; Aghazadeh, F.; Zheng, B. Early Eye Disengagement Is Regulated by Task Complexity and Task Repetition in Visual Tracking Task. *Sensors* **2024**, *24*, 2984. [CrossRef]
43. Friard, O.; Gamba, M. BORIS: A free, versatile open-source event-logging software for video/audio coding and live observations. *Methods Ecol. Evol.* **2016**, *7*, 1325–1330. [CrossRef]
44. Bradley, M.M.; Miccoli, L.; Escrig, M.A.; Lang, P.J. The pupil as a measure of emotional arousal and autonomic activation. *Psychophysiology* **2008**, *45*, 602–607. [CrossRef]
45. Grassi, G.; Esler, M. How to assess sympathetic activity in humans. *J. Hypertens.* **1999**, *17*, 719–734. [CrossRef]
46. Hermens, F.; Flin, R.; Ahmed, I. Eye movements in surgery: A literature review. *J. Eye Mov. Res.* **2013**, *6*, 1016910. [CrossRef]
47. Schmidt, R.; Lee, T. *Motor Learning and Performance, 6th Edition with Web Study Guide–Loose-Leaf Edition: From Principles to Application*; Human Kinetics Publishers: Champaign, IL, USA, 2019.
48. Godde, B.; Voelcker-Rehage, C. Cognitive resources necessary for motor control in older adults are reduced by walking and coordination training. *Front. Hum. Neurosci.* **2017**, *11*, 156. [CrossRef]
49. Haith, A.M.; Krakauer, J.W. The multiple effects of practice: Skill, habit and reduced cognitive load. *Curr. Opin. Behav. Sci.* **2018**, *20*, 196–201. [CrossRef]
50. Zheng, B.; Rieder, E.; Cassera, M.A.; Martinec, D.V.; Lee, G.; Panton, O.N.M.; Park, A.; Swanström, L.L. Quantifying mental workloads of surgeons performing natural orifice transluminal endoscopic surgery (NOTES) procedures. *Surg. Endosc.* **2012**, *26*, 1352–1358. [CrossRef]

51. Zheng, B.; Cassera, M.A.; Martinec, D.V.; Spaun, G.O.; Swanström, L.L. Measuring mental workload during the performance of advanced laparoscopic tasks. *Surg. Endosc.* **2010**, *24*, 45–50. [CrossRef]
52. Duchowski, A.T.; Krejtz, K.; Gehrer, N.A.; Bafna, T.; Bækgaard, P. The Low/High Index of Pupillary Activity. In Proceedings of the 2020 CHI Conference on Human Factors in Computing Systems, Honolulu, HI, USA, 25–30 April 2020.

applied sciences

MDPI

Article

Studying Pupil-Size Changes as a Function of Task Demands and Emotional Content in a Clinical Interview Situation

Daniel Gugerell [1,*], Benedikt Gollan [2], Moritz Stolte [1] and Ulrich Ansorge [1,3,4]

[1] Faculty of Psychology, University of Vienna, 1010 Vienna, Austria; moritz.stolte@univie.ac.at (M.S.); ulrich.ansorge@univie.ac.at (U.A.)
[2] Research Studios Austria, 1090 Vienna, Austria; benedikt.gollan@researchstudio.at
[3] Vienna Cognitive Science Hub, University of Vienna, 1010 Vienna, Austria
[4] Research Platform Mediatised Lifeworlds, University of Vienna, 1010 Vienna, Austria
* Correspondence: daniel.gugerell@univie.ac.at

Abstract: The human pupil changes size in response to processing demands or cognitive (work)load and emotional processing. Therefore, it is important to test if automatic tracking of cognitive load by pupil-size measurement is possible under conditions of varying levels of emotion-related processing. Here, we investigated this question in an experiment simulating a highly relevant applied context in which cognitive load and emotional processing can vary independently: a clinical interview. Our participants conducted a live clinical interview via computer monitor with a confederate as an interviewee. We used eye-tracking and automatic extraction of participants' pupil size to monitor cognitive load (single vs. dual tasks, between participants), while orthogonally varying the emotional content of the interviewee's answers (neutral vs. negative, between participants). We ensured participants' processing of the verbal content of the interview by asking all participants to report on the content of the interview in a subsequent memory test and by asking them to discriminate if the answers of the interviewee referred to only herself or to somebody else (too). In the dual-task condition, participants had to monitor additionally if the facial emotional expressions of the interviewee matched the content of her verbal responses. Results showed that pupil-size extraction reliably discriminated between high and low cognitive load, albeit to a lower degree under negative emotional content conditions. This was possible with an algorithmic online measure of cognitive load as well as with a conventional pupil-size measure, providing proof of the external validity of the algorithm/online measure.

Keywords: pupil dilation; eye-tracking; interview; task demands; emotional processing

1. Introduction

In research on ergonomics in general and human–machine interactions in particular, but also in many clinical or educational settings, technical applications aim to track humans' current processing demands under varying, real-world conditions. Examples are the presentation of information "just in time" (e.g., when users have a higher likelihood of processing the information), warning the user via visual signals in cases where cognitive load reaches levels that significantly increase the chances of missing crucial information, or monitoring an audience in an educational context to gauge the speed or rate of information presented and processed (cf. [1–5]). Cognitive demands or load and its tracking are therefore a major topic in medical education and applied clinical settings [6]. Clinical interviews, for example, are often conducted under highly distracting conditions (e.g., in an emergency room) and, yet, the correct processing of the information presented is of the highest relevance in this situation (e.g., to decide if a patient needs to be admitted to the hospital, yes or no) [7]. The problem is aggravated by the fact that self-report measures of cognitive load or demands do not necessarily converge. For instance, Naismith et al. [8] used different self-report measures of cognitive load in simulation-based medical training and found little convergence between these measures. For these reasons, researchers seek to identify unobtrusive, pervasive, and affordable technology to track processing demands or cognitive load in areas such as the workspace or educational institutions [9]. One such technology is eye-tracking (e.g., [10–12]). Here, one dependent variable suited to track processing demands or cognitive load is pupil diameter (cf. [13–15]). Higher cognitive load elicits a pupil dilation and, accordingly, research has indicated that measured pupil diameter can be used to successfully discriminate between more or less demanding conditions or conditions requiring more or less mental effort (e.g., [16–19]).

However, a general problem that complicates the read-out of processing demands or cognitive load from measured pupil sizes is that pupillary dilation is a relatively unspecific response elicited by a variety of different conditions, not all of which are diagnostic of cognitive load (e.g., [20]). Among the critical triggers of pupillary responses are changes in the ensuing luminance level (e.g., [21,22]), saccades ([23,24]), felt emotions (e.g., [25–27]), value (e.g., [28,29]), arousal (e.g., [30,31]), psychoactive substances (e.g., [32]), and pathological conditions on the side of the observer (e.g., [33]). Thus, many factors that independently modulate the pupillary response pose a threat to the successful measurement of cognitive load from pupil dilations. For example, if an emotion such as joy or anger, elicits a pupil dilation, there is potentially less room for a load-specific response on top of the one triggered by the emotion [34,35]. This can be a problem, as the valence of the processed content can vary independently of cognitive load. To understand this, think of a clinical interview (e.g., [36,37]). In such a situation, the demands imposed on the interviewer can vary depending on whether the interviewer has to encode only verbal responses of the patient or whether she also pursues additional aims, such as evaluating emotional content by classifying nonverbal expressions, such as mimic and gestures (cf. [38–40]). At the same time, in this situation, the emotional content of the answers may vary (cf. [41]). Sometimes the interviewee might relate emotionally neutral content such as her age or occupation, but, at other times, the interviewee might also talk about personal traumatic or stressful past experiences. The emotional valence of the responses can be entirely independent of the cognitive demands imposed by the task—for example, whether the interviewer processes only verbal or verbal and nonverbal communication. Thus, in such a situation, monitoring the cognitive load imposed on the interviewer by measuring pupillary responses may be reduced. For example, if an emotional response already triggers pupil dilation, there would maybe be less room for a pupil response to the higher demands. Thus, it might be difficult to identify questions or phases of the interview where difficulty increases and where questions possibly have to be rephrased.

Can load be assessed through pupillary responses in all such cases? It should be noted that emotional effects can impact pupil sizes by effects that do not correspond to a cognitive-load effect itself. It is true that, for example, some emotional processing such as an observer's anxiety may draw on her or his limited capacity and, thus, impose a cognitive load itself [42]. In such cases, emotional and load effects converge to increase load, and there is no need to tell the two apart and to derive a pure measure of load. However, other emotional content besides one's own felt stress or anxiety can also elicit pupillary changes, but these are load-independent effects. In these cases, to make use of a pupil-based load measure, it is important to tell apart which pupil-size changes are true reflections of cognitive load and which are due to emotions. For example, it would be important to know if the emotion-specific but load-independent pupillary effect limits the sensitivity of the pupil's response to an additional load effect. It is exactly such a situation, where emotion and load lead to independent effects on pupil size, that we studied in the present research. We did this with a clinical interview in which patients—in fact, a confederate of the experimenter—voiced more neutral or more negative content. Here, the more negative content is independent of the cognitive load per se, and, thus, has the chance to compromise our cognitive-load measure by load-independent means.

As explained, clinical interviews are common practice and an important tool in diagnosis, methods to understand and improve their application are needed [43,44]. Therefore, the present study investigated exactly the question of whether varying processing demands can be successfully and unobtrusively tracked by measurements of pupil dilation under conditions of varying emotional content of the answers of the interviewee. To this end, the current study used four different groups varying with respect to the processing demands or cognitive loads on the one hand, and independently of the load with respect to the emotional content of the answers, on the other hand. All participants acted as interviewers conducting an interview with a confederate of the researchers that gave prepared verbal answers and displayed generic mimics and gestures. Participants in two of the groups had to monitor the verbal content of the answers of the interviewee only, both for current classification of the responses in the course of the interview (by button presses) and for later recall (in a memory test). These were the low-load or low-demand (L) conditions. Participants in the other two groups had to monitor both the verbal and nonverbal content of the interviewee, again for current classification of the responses during the interview and for later recall. These were the high-load or high-demand (H) conditions. In each of the two load conditions, one group of participants heard more neutral, less emotional answers (NEU), while the other group heard less neutral, more negative (NEG), and, thus, more emotional answers. Thus, there were four groups (NEUL, NEUH, NEGL, and NEGH) altogether.

To track the processing demands, we measured the pupillary responses of the interviewers with an eye-tracker and analyzed the data conventionally, and also used an algorithm that automatically subtracts light-elicited pupillary responses [19]. This algorithmic measure of cognitive load could be used online, with little temporal delay between recording and resulting identification of cognitive load experienced. Yet, it stands as external validation by comparison with a more conventional pupil-size measure. This was what we did in the current study: We compared how the algorithm faired in comparison with a standard measure of pupil dilation under the relatively uncontrolled conditions of an interview. In addition, we took various measures of objective task performance, such as the interviewer's correct classifications of both the verbal (in L and H groups) as well as nonverbal responses (H groups only) during the interview, and of the interviewer's performance in a subsequent memory test about the content of the interview. We also collected several measures of the interviewers' self-assessment regarding both the emotions experienced during the interview as well as the felt subjective demands imposed by the interview and tasks.

The following were our hypotheses:

Hypothesis 1. *If the algorithm for automatic extraction of the cognitive load through pupillary responses serves its purpose, we expected to see the above-chance classification of our participants into those that experienced higher processing demand versus those that experienced lower processing demand. If external validity is granted, we also expected a good correspondence between the algorithmic measure and the more conventional measurements of cognitive load based on subtracting pupillary light responses as a baseline from conditions in which the same pupillary light response is likely elicited but on top of this cognitive load triggers a pupil dilation.*

Hypothesis 2. *However, because emotions can elicit pupil dilation, just as the task demands, we also expected this classification to work better under conditions with neutral rather than emotional (here, negative) response content. As a ground truth, we also used raw pupil size measured by the eye-tracker itself, without the preprocessing steps taken by the algorithm such as its automatic modeling of an impulse corresponding to the pupillary light response. To this end, raw pupil sizes during a baseline phase of the interview were subtracted from raw pupil sizes measured during the manipulated parts of the interviews ("during manipulation" in short) that varied according to response content (neutral vs. negative) and task or processing demands (low vs. high). The resulting differences were analyzed with a 2 × 2 analysis of variance (ANOVA), with between-participant variables processing demand (low vs. high) and verbal content (neutral vs. negative).*

Hypothesis 3. *Here, we expected a main effect of task load/processing demand, with larger pupil sizes under high- than low-load conditions (cf. [34]) and possibly a main effect of emotion, with larger pupil sizes under negative than under neutral-content conditions (e.g., [25]; but see [35] for an example of failure to observe emotion-elicited pupil dilations with verbal material of varying task difficulty).*

Hypothesis 4. *In addition, we expected to see an interaction, for example, a compromised cognitive-load effect under emotional conditions compared to neutral conditions if the emotion-elicited pupil dilation can mask or compromise the processing demands.*

Hypothesis 5. *Finally, we looked into the correlation between the algorithmic cognitive-load measure and the ground truth and expected a significant and substantial correlation between the measures, corresponding to external validation of the algorithmic cognitive-load measure.*

Besides these most important analyses, we also analyzed objective task performance and subjective assessments, where we would have expected better objective performance under low than high task-load conditions, lower interviewers' self-assessed effort and demands under low- than under high-load conditions, or more positive ratings under neutral- than under negative-content conditions. In addition, to study the proper functioning of the automatic extraction of the light-elicited pupil response from the algorithmically extracted cognitive-load measure, we correlated these two values and expected no correlation between the two measures. In contrast, if the light-elicited pupillary response can limit the available room for a cognitive-load response, we might see a negative correlation between the two measures, meaning on average fewer pupillary responses to higher demands under darker than under lighter conditions.

2. Methods

2.1. Participants

Forty psychology students from the University of Vienna participated in the experiment (27 female, 13 male, M_{age} = 22.2, range 19–29 years). The sample size was based on an a priori power calculation assuming a small effect size (η^2 = 0.20), striving for a statistical power of 0.8, and allowing an alpha error of 0.05. Participants were randomly assigned to one of four groups: neutral content/low workload (NEUL; nine participants), neutral content/high workload (NEUH; nine participants), negative content/low workload (NEGL; nine participants), and negative content/high workload (NEGH; nine participants). Four participants had to be excluded from the study: Two due to a faulty dataset, where data from the eye-tracker were missing and could not be evaluated correctly. One participant's dataset in the NEUH group had to be excluded since he/she did not perform the Verbal–Visual Task. One participant in the NEGH group had to be excluded due to a very small pupil size at the start. Each participant received partial course credit and had normal or corrected-to-normal visual acuity. Prior to the experiment, informed consent was obtained from all participants. Ethical approval (No. EK_00617) was obtained from the University of Vienna's Ethical Review Board.

2.2. Procedure and Task

Participants wore a head-mounted eye-tracker and were asked to conduct an adapted clinical interview with another person and to additionally ask about the interviewee's experiences with health issues in her family. The interview was conducted via an online video-call using BigBlueButton (2021). Participants were told to read and ask 67 questions (see Supplementary Tables S1–S5) in sequence and wait until the interviewee had responded to each question before asking the next question.

Unbeknownst to the participants, the person to whom they were interviewing (i.e., the interviewee) was a trained confederate of the experimenters, who had been informed about the study's aim beforehand. The confederate answered the questions of the participant in a fixed, prescribed manner, either in an on average more negative (e.g., "My mother died after she was diagnosed with corona") or a more neutral (e.g., "My mother was quarantined after she was diagnosed with corona") manner. In addition, the confederate also used pre-specified and practiced emotional facial expressions and gestures (e.g., frown and/or lower her gaze) during her responses.

Out of the 67 questions, 7 were pre-experimental information questions and another 11 were used as baseline questions. The interviewee's answers to these pre-experimental and baseline questions were the same, regardless of which emotional content group participants were assigned to (i.e., regardless of the manipulation). The participants' pupil responses to the baseline questions served as a reference for comparison with the pupil responses in the later, manipulated part of the experiment (see below), as both the content of the questions and of the answers during baseline and manipulation touched upon less socially desirable personal information about the behavior and or the experiences of the interviewee. After the baseline questions, participants were instructed to ask another 49 questions, the answers to which were manipulated and depended on the group the participants were assigned to. This part of the interview was labeled the "manipulation". In the neutral emotional content condition (low and high cognitive load; NEUL and NEUH), the interviewee answered the participant's questions in a neutral manner (see Supplementary Table S4). In contrast, in the negative emotional content condition (low and high cognitive load; NEGL and NEGH), the interviewee answered participants' questions in a more negative manner (see Supplementary Table S5). The difference between pupil sizes during baseline and manipulation was the conventional measure of demand-elicited pupillary changes during the manipulation, as the lighting conditions during baseline and manipulation were the same and the two phases of the experiment differed only in terms of the workload and the emotional content of the answers. The interview itself lasted between 25 and 30 min.

Regarding the tasks, all participants had to remember as much of the interviewee's responses as they could, since they were tested for their memory of the interview afterward. Participants' mental or cognitive (work)load was manipulated by giving participants one or two tasks. One of these tasks was the same for all participants: Participants had to discriminate if the interviewee only referred to herself while giving an answer, or if she mentioned anybody else as well. If she only talked about herself, participants were to press the "Y" key on the keyboard at the end of an answer, while they were to press the "X" key if the interviewee mentioned someone else ("Verbal Task"). In addition, participants in the high-load conditions (NEUH and NEGH) were asked to discriminate the match between the interviewee's facial expressions and the content of her verbal responses. If they were congruent (e.g., frowning while telling them that her mother had died), participants had to press the "P" key, but if they were incongruent (e.g., smiling while telling them that their mother died), they had to press the "O" key at the end of an answer ("Verbal–Visual Task").

2.3. *Apparatus and Material*

The interview was conducted in a dimly lit laboratory, where the only light source was the monitor. The interview was conducted via a 31 cm × 28.5 cm monitor (resolution, 1920 pixels × 1080 pixels; 60 Hz screen refresh rate). Participants were seated in front of the monitor, with their gaze straight ahead, centered on the screen. Viewing direction and distance (60 cm) were supported by a chinrest. Participants wore a mobile, video-based eye tracker (Pupil Labs, Berlin, Germany; [45]), sampling at 60 Hz, with an estimated gaze accuracy of 0.6° (according to the manufacturer). A PC running Windows 10 (Microsoft, Redmond, WA, USA) and Pupil Labs software (pupil-labs.com/pupil/; version 3.0.7, accessed on the 8 October 2021) was connected to the eye-tracker for recording of pupil size, eye movements, and the external visual surroundings. The Pupil-Player software ran with default settings, and the eye-tracker was calibrated before the start of the interview as instructed by the pupil-labs documentation (https://docs.pupil-labs.com/core/software/pupil-capture/, accessed on the 8 October 2021). Participants also wore headphones (Gindoly PC Headset with microphone). BigBlue-

Button was used for the video-call, in which participants were able to see the interviewee via webcam. PyCharm Community Edition 2020.2.3 was employed for the interview questions and response collection. Participants responded on a standard "QWERTZ"-keyboard. The interview consisted of seven pre-experimental questions, 11 baseline questions, 16 questions regarding the COVID-19 pandemic, and 33 reworded questions from the Diagnostic Interview for Mental Disorders (DIPS; [46]). Following the interview, a questionnaire consisting of 10 items tested whether participants paid attention to and, hence, recollected the content of the interviewee's responses. Additionally, participants were asked to fill out the NASA-Task-Load-Index (NASA-TLX; [47]) and the Self-Assessment Manikin (SAM; [48]) at the end of the interview to measure their self-assessed felt emotional state and mental workload during the interview. Lastly, participants were asked whether they felt that the interview was faked (that they knew that the interviewee was actually a confederate of the researchers) to see if the credibility of the confederate's answers influenced participants' self-assessments and/or their performance.

2.4. Eye-Tracking Data Processing

The pupil data were exported using the Pupil Player v3.0.7 with minimum data confidence of 0.6. Confidence is an assessment by the pupil detector on how sure it is about the measurement. This measurement is taken for each frame and each eye. Pupil labs suggests that any confidence value greater than 0.6 corresponds to useful data. Further data processing was conducted using PyCharm Community Edition 2020.2.3. The whole dataset was reduced to a single eye, which was chosen because of its higher overall average confidence. The extracted pupil data of the Pupil Player provide two types of pupil diameter values: one 3D-corrected diameter of the pupil scaled to millimeters, based on an anthropomorphic average eyeball diameter and corrected for perspective and one raw diameter in image pixels as observed in the eye image frame and not corrected for perspective. We used the latter for our analysis. The data were further restricted to those sections in which the interviewee gave her answers, removing all sections in which the participant read the questions to the interviewee. Artifacts (such as unexplainable spikes and dips in pupil-size measurements).

2.5. Extraction of Cognitive Load

The analysis of cognitive load is obtained via an algorithm for online analysis of cognitive load from pupil data (see [49]) that have been validated for their internal validity [19]. With the conditions of the study not being fully stable in terms of the changing video luminance and with the majority of pupillary activity being associated with reflexes to environmental luminance and not mental activity, we added a component to model and compensate these environmental effects to obtain a reliable estimation of cognitive load.

For this purpose, we implemented a dynamic baseline to compensate for environmental changes by exploiting the first-person camera as a luminance sensor. These luminance data were being fed into empiric models of pupillary reflex behavior [50] to obtain a corresponding score of luminance-related pupil dilation, and temporal adaption of pupillary behavior [51] and to compute realistic temporal adaptions of the pupil (for a detailed description of the modeling of the dynamic baseline, please refer to Gollan [49]).

As a next step, the dynamic baseline is subtracted from the measured pupil dilation score to obtain the effects that are associated with arousal and accordingly cognitive load. These data are processed by the cognitive-load analysis (see [49]), which uses the mathematical modeling of the task-evoked pupillary response (TEPR) [52] to measure cognitive load. This empirical model of pupillary response to cognitive activities has been transferred into an online analysis algorithm [49]. The online deconvolution algorithm performs a curve-matching approach in a frame-wise feedback loop. To be precise, the algorithm measures the difference between the current modeled pupil value and the current actual pupil dilation measure. If this difference exceeds a threshold of 0.25% change in pupil dilation, a new attentional pulse, wi(si, ti), with scale si and temporal onset ti, is dynamically added to the list of attentional impulses.

3. Results

3.1. Conventional Pupil-Size Measures

Looking at the average pupil-size responses, two trends could be observed. First, a very small pupil size at the start of each answer relative to the participant's average pupil size. This was likely caused by the pupillary light reflex in response to the difference in screen luminance from a screen on which participants had read a question to an image of the interviewee while she answered the question. Second, we can see a rise in average pupil size between answers to baseline questions (Questions 8 to 18) and answers to questions differing in emotional content (between groups; Figure 1).

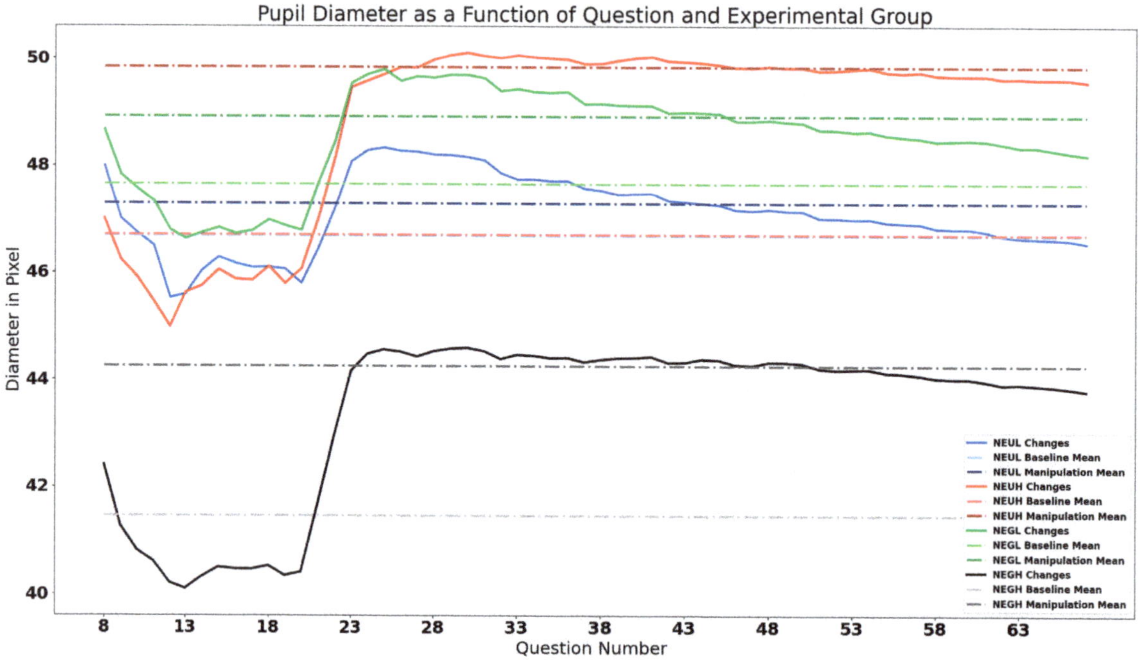

Figure 1. Average pupil diameter in pixels as a function of the answer, separately for different groups (NEUL: neutral content/low load; NEUH: neutral content/high load; NEGL: negative content/low load; NEGH: negative content/high load). The two dotted lines in the colors of each group show the corresponding condition's difference in average pupil size between baseline questions (bottom line) and manipulation questions (top line).

This increase in pupil diameter from the baseline to the manipulation—that is, the interviewee's emotionally more or less negative responses, can be found in every single participant. When comparing the pupillary changes between groups, the largest average change between the baseline and the manipulations occurred in the high-mental-workload conditions (NEUH: with a delta of 3.134 pixels; NEGH: with a delta of 2.786 pixels), followed by participants in the low-workload/negative-content group (NEGL: delta of 1.260 pixels), and lastly the participants in the NEUL group (delta of 0.607 pixels). Although Figure 1 shows that participants in the NEGH Group have had the smallest pupil diameter of all groups, this only means that the average pupil diameter in this group was smaller in the baseline condition to begin with. Figure 2 underlines the difference in pupil diameter change between the baseline and the manipulation in all groups. While participants in the

NEUL group showed the smallest increase in pupil diameter, when the baseline average is subtracted from the average during manipulation, participants in the NEUH group show the biggest increase. Participants in the NEGL group as well as those in the NEGH group showed a similar difference, suggesting that the increase in pupil size due to the emotional content quantitatively limited the influence of mental workload on pupil size.

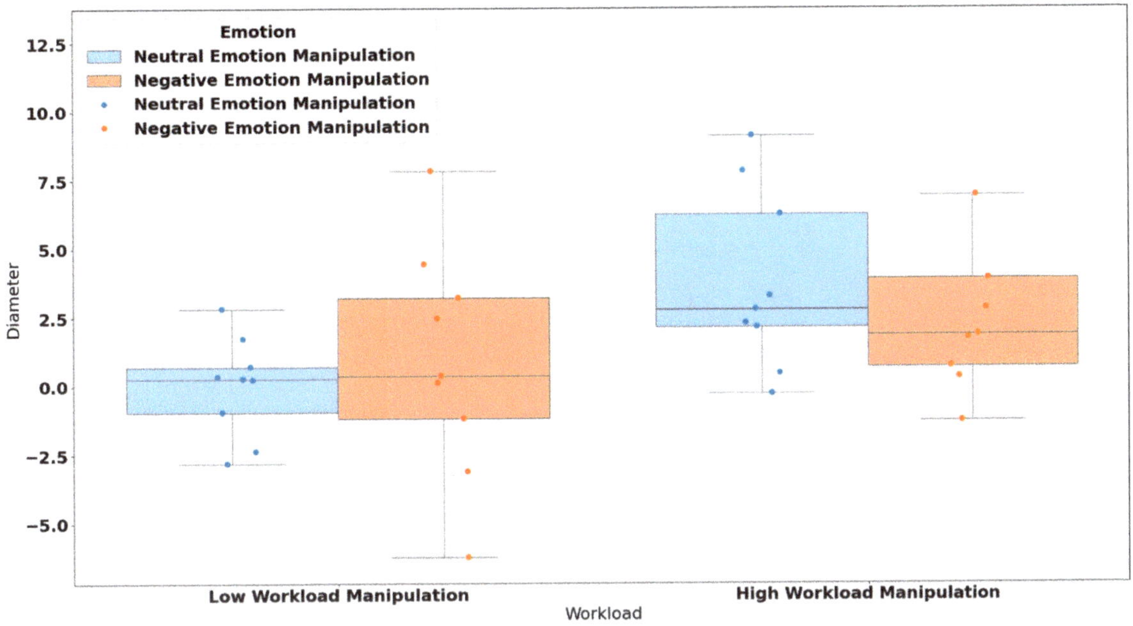

Figure 2. Median diameter changes (black horizontal lines), corresponding quartiles (colored areas above and below the horizontal lines), minima and maxima (whiskers), and individual values (colored dots) of pupil diameter differences between baseline and manipulation on the y axis as a function of condition/Group (low workloads on the left, high workloads on the right; neutral-content conditions in blue, negative-content conditions in red) on the x axis.

An ANOVA yielded a significant main effect of workload, $F(1, 35) = 6.95$, $p = 0.013$, $\eta^2 = 0.17$, but the influence of Emotion, $F(1, 35) = 0.03$, $p = 0.858$, $\eta^2 = 0.002$, and the interaction, $F(1, 35) = 0.32$, $p = 0.576$, $\eta^2 = 0.01$, were not significant. A post hoc t-test between low- ($M = 0.49$, $SD = 3.14$) and high-workload conditions ($M = 3.53$, $SD = 3.68$) showed significantly smaller pupillary changes in low-workload conditions compared to high-workload conditions, $t(35) = -2.667$, $p = 0.012$, $\eta^2 = 0.89$.

Further calculations showed that the main effect of workload was due to a faster average shrinking of pupil sizes of participants in the low-workload conditions over the course of the interview compared to participants in the high-workload group. A point of divergence, from which the pupil size of the low-workload groups significantly differed from those of the high-workload groups, is the 27th question in the interview: An ANOVA using only questions after the 27th question yielded a very similar pattern of results as the initial ANOVA, Workload: $F(1, 35) = 7.73$, $p = 0.009$, Emotion: $F(1, 35) = 0.05$, $p = 0.818$, and Workload × Emotion: $F(1, 35) = 0.39$, $p = 0.537$. The results were similar when only including trials in which participants correctly responded to both the Verbal Task as well as the Verbal–Visual Task (the latter only for participants in the high-workload conditions), Workload: $F(1, 35) = 6.90$, $p = 0.013$, Emotion: $F(1, 35) = 0.15$, $p = 0.699$, and Workload × Emotion: $F(1, 35) = 0.13$, $p = 0.725$.

For the following analyses, participants were divided into two groups based on their self-assessment scores in the NASA-TLX's subscale "Mental-Demand" and the SAM's subscale "Happiness" (see Figures 3 and 4, respectively). Participants who scored equal or below the median score in the respective subscales were assigned to the "low" group, while those who scored above were assigned to the "high" group, and the resulting classifications were tested for their correspondences (with correspondence indicated by the symbol "~" in Figure 3) to the manipulations (i.e., NASA-L/SAM-HAP ~ NEUL, NASA-H/SAM-HAP ~ NEUH, NASA-L/SAM-SAD ~ NEGL, NASA-H/SAM-SAD ~ NEGH).

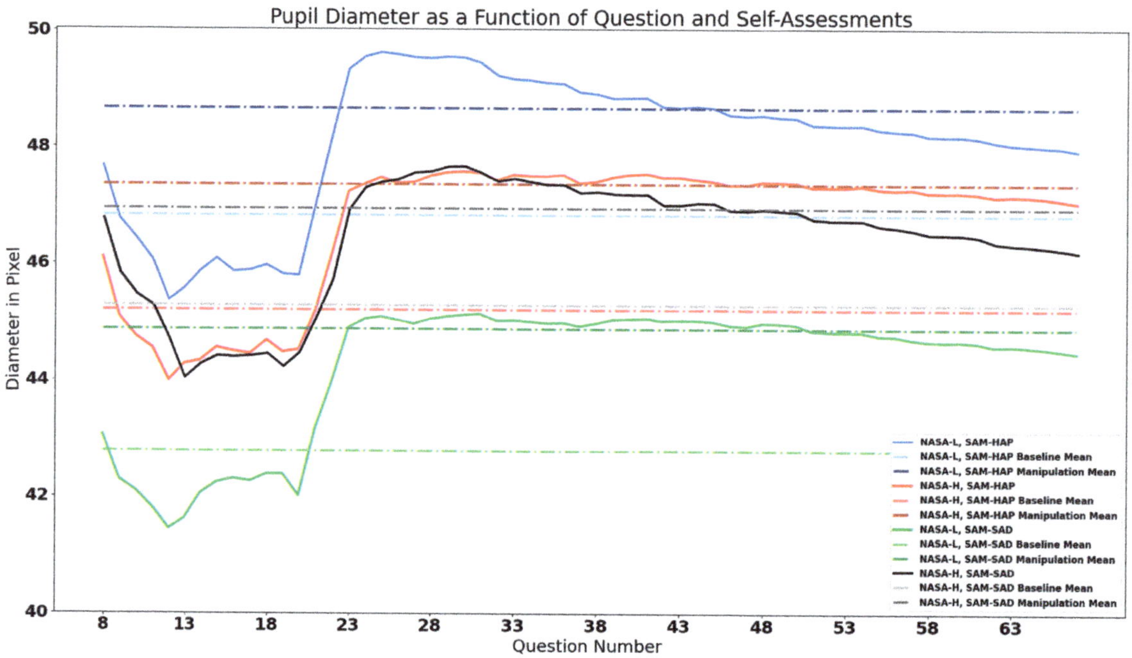

Figure 3. Average pupil diameter in pixels on the y axis as a function of the answer on the x axis, separately for different groups or conditions based on self-assessments (NASA-L, SAM-HAP ~ NEUL: low demand/happy mood; NASA-H, SAM-HAP ~ NEUH: high demand/happy mood; NASA-L, SAM-SAD ~ NEGL: low demand/sad mood; NASA-H, SAM-SAD ~ NEGH: high demand/sad mood). The two dotted lines in the colors of each graph showing the corresponding condition's difference in average pupil size between baseline and manipulation (questions). NASA: NASA Task-Load-Index; SAM: Self-Assessment Manikin; NEU: neutral-content condition; NEG: negative-content condition; H: High; L: Low.

When comparing the pupillary changes between groups based on their self-assessments, the results of the ANOVA were not significant, showing that the self-assessments should be interpreted with caution, all $Fs < 1.00$ (NASA-TLX-Values, F [1, 35] = 0.34, $p = 0.565$, $\eta^2 = 0.01$; SAM-Values, $F[1, 35] = 0.02$, $p = 0.902$, $\eta^2 = 0.001$; (NASA-TLX-Values × SAM-Values, $F[1, 35] = 0.419$, $p = 0.522$, $\eta^2 = 0.01$).

We also collapsed data across self-assessed emotions to check if this improves the picture. However, a two-sample t-test of pupil diameter changes between participants who scored high (which means that participants felt a high mental demand during the task) in the NASA-TLX subtask "Mental-Demand" ($M = 2.37$, $SD = 4.02$) and those who scored low ($M = 1.63$, $SD = 3.51$) showed no significant difference either, $t(35) = -0.589$, $p = 0.559$, $\eta^2 = 0.2$ (see Figure 5).

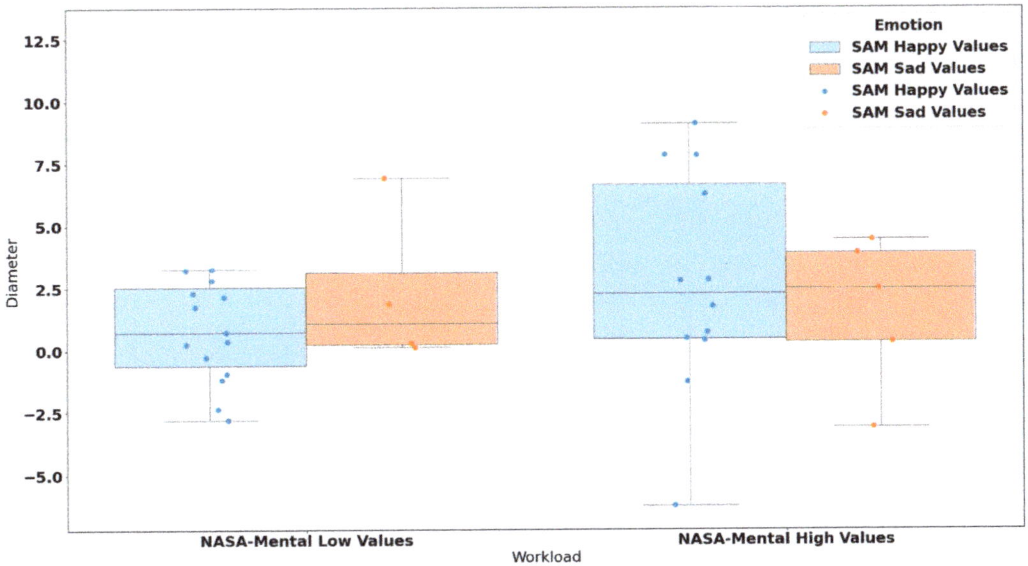

Figure 4. Median diameter changes (black horizontal lines), corresponding quartiles (colored areas above and below the horizontal lines), minima and maxima (whiskers), and individual values (colored dots) of pupil diameter differences between baseline and manipulation on the y axis as a function of condition/Group (low self-assessed mental-demand scores on the left, high self-assessed mental-demand scores on the right; low self-assessed happiness scores in orange, high self-assessed happiness scores in blue) on the x axis. SAM: Self-Assessment Manikin.

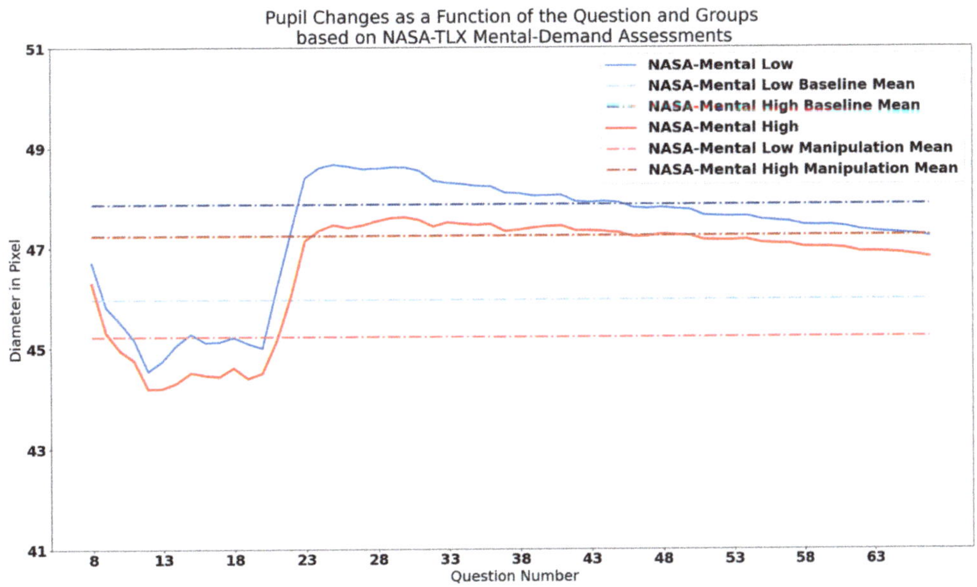

Figure 5. Pupil diameter on the y axis as a function of the question answered on the x axis and whether participants scored high (red graph) on the NASA Task-Load-Index (NASA-TLX "Mental-Demand" subscale) or if they scored low (blue graph). The colored horizontal lines depict the averages of the baseline (lower respective lines) and the manipulation (upper respective lines).

The comparison of pupil diameter changes between participants who believed that the interview was fake ($M = 1.70$, $SD = 4.58$) and those who believed that it was real ($M = 2.17$, $SD = 3.09$) showed no significant difference either, $t(35) = 0.369$, $p = 0.715$, $\eta^2 = 0.12$ (see Figure 6). This is not very surprising given that the workload manipulations were real, such that the beliefs regarding the authenticity of the interviewee should not corrupt the demands imposed by the task(s) at hand.

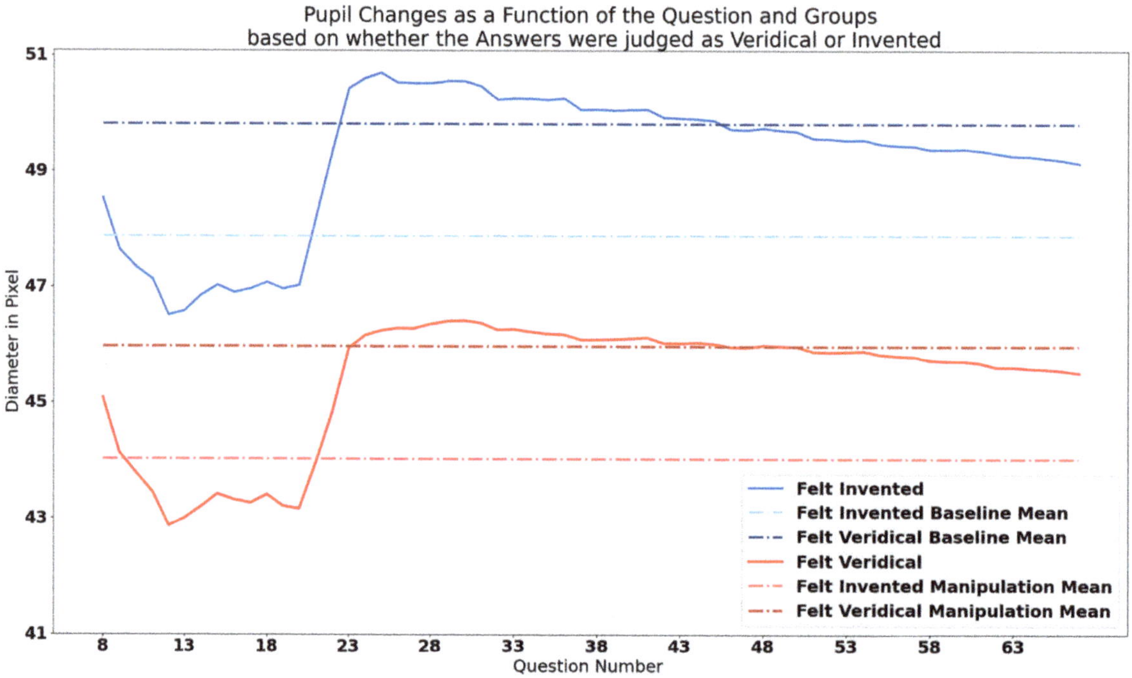

Figure 6. Pupil diameter on the y axis as a function of the question answered on the x axis and whether participants believed the answers were veridical (red graph) or fake (blue graph). The colored horizontal lines depict the averages of the baseline (lower respective lines) and the manipulation (upper respective lines).

3.2. Algorithmically Extracted Workload Measure

The same analysis was conducted using data derived from a "cognitive workload" algorithm, which subtracts the light-elicited pupillary changes ([19]). Quantitatively, the pattern of results is overall similar to that using the raw pupillary changes, with the exception of that of the NEGH Group (see Figure 7).

The increase in cognitive load between the baseline and the manipulation was, again, found in every single participant. When comparing the cognitive-load changes between groups, the largest average change between the baseline and the manipulations occurred in the high NEGL condition (NEGL: with a delta of 6.27), followed by participants in the high-workload/neutral-content group (NEUH: delta of 5.45), and lastly the participants in the NEGH group (delta of 4.47) and the NEUL group (delta of 3.45). Figure 8 illustrates the difference in cognitive-load change between the baseline and the manipulation in all groups. While participants in the NEUL group showed the smallest increase in cognitive load when the baseline average was subtracted from the average during manipulation, participants in the NEUH group as well as those in the NEGL group showed similar increases. Participants in the NEGH group showed a rather small increase in cognitive load as well.

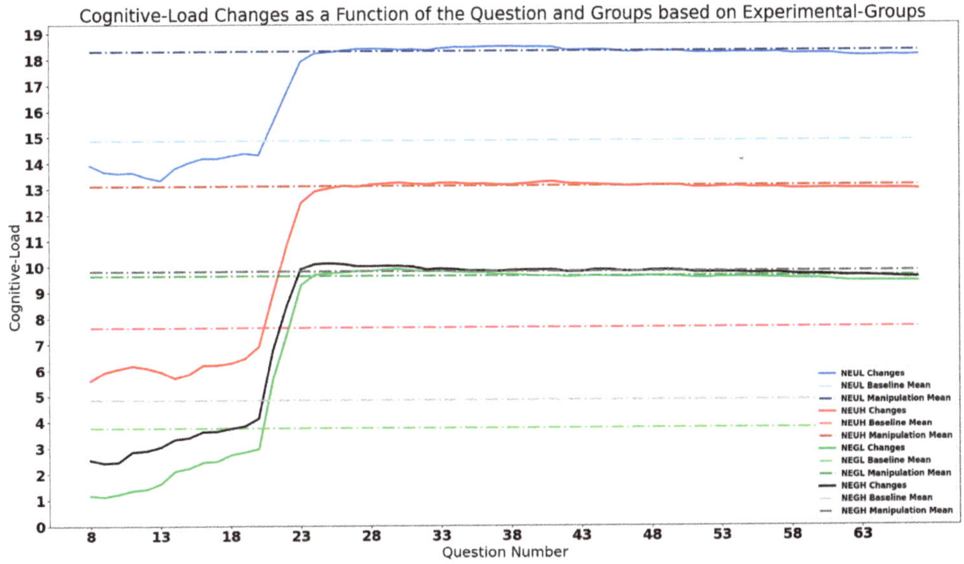

Figure 7. Average cognitive workload (according to the algorithm) on the y axis as a function of the answer on the x axis, separately for different groups or conditions (NEUL: neutral content/low load; NEUH: neutral content/high load; NEGL: negative content/low load; NEGH: negative content/high load). The two dotted lines in the colors of each graph show the corresponding condition's difference in average calculated cognitive workload between baseline and manipulation (questions).

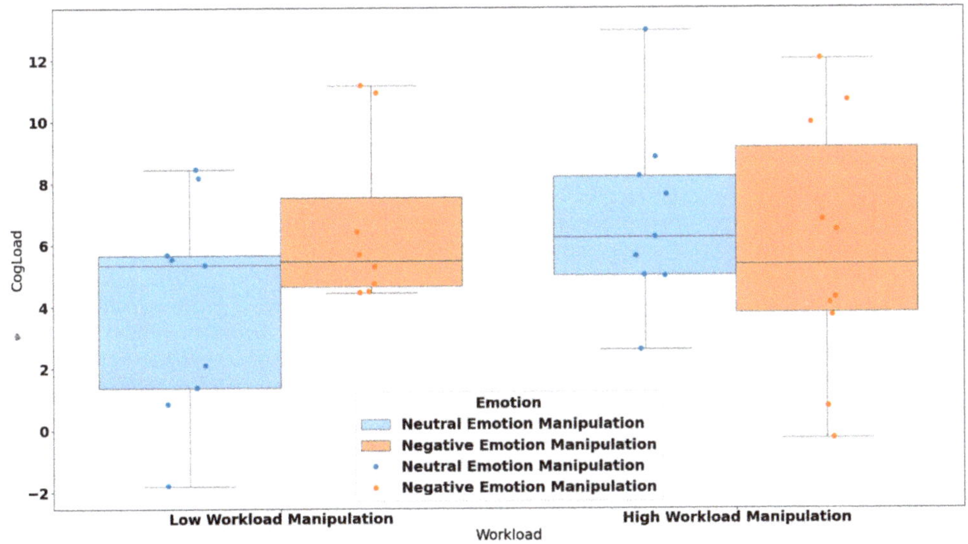

Figure 8. Median cognitive-load changes (black horizontal lines), corresponding quartiles (colored areas above and below the horizontal lines), minima and maxima (whiskers), and individual values (colored dots) of cognitive-load (CogLoad) differences between baseline and manipulation on the y axis as a function of condition/Group (low-workload manipulation on the left, high-workload manipulation on the right; neutral-content conditions in blue, negative-content conditions in red) on the x axis.

An ANOVA yielded a significant interaction between the workload and emotional content manipulation, $F(1, 35) = 4.77$, $p = 0.036$, $\eta^2 = 0.13$, but neither the workload manipulation, $F(1, 35) = 1.38$, $p = 0.248$, $\eta^2 = 0.04$, nor the main effect of emotion, $F(1, 35) = 0.44$, $p = 0.509$, $\eta^2 = 0.01$, were significant. In neutral emotion conditions, a pairwise post hoc t-test between low- ($M = 4.64$, $SD = 3.43$) and high-workload conditions ($M = 6.96$, $SD = 3.21$) showed significantly smaller pupillary changes in low-workload conditions compared to high-workload conditions, $t(35) = -2.108$, $p = 0.047$, $\eta^2 = 0.7$. In contrast, in negative emotion conditions, the t-test between low- ($M = 5.42$, $SD = 3.49$) and high-workload conditions ($M = 6.18$, $SD = 3.52$) did not show a significant change in pupil size, $t(35) = -0.651$, $p = 0.047$, $\eta^2 = 0.7$. This result was due to a load effect in the neutral conditions that numerically even slightly reversed under negative conditions.

3.3. External Validation of Algorithmically Extracted Workload Measure

As explained, the algorithmic cognitive-load measure provides a continuous simulation of the pupil response to the current task that is free of a luminance-created contribution to the same pupillary response. To externally validate the cognitive-load algorithm, we looked at the observed correlations between ground truth and algorithmically derived cognitive-load measures. Ground truth was calculated in the conventional way, by subtracting average baseline pupillary size from the pupillary size at a precisely defined point in time during the manipulation phase. Note: This ground-truth method aims to eliminate the influence of luminance on the pupillary response in the manipulation phases and, thus, corresponds in conventional offline calculations of pupillary responses to what the cognitive-load algorithm does online. To calculate these observed correlations, we first selected the strongest maxima (or highest local peaks) in the modeled cognitive-load responses. In this way, we targeted points in time of assumedly highest signal-to-noise ratio and strongest change in cognitive load across time. In addition, to generalize the insights gained by correlations between the algorithmic model and the ground truth measure, we repeated this method for 20 time points randomly selected from a period of 250 ms before and after a local maximum of the algorithmic impulses.

For the pupil response to each answer of each participant, we correlated the algorithmic cognitive-load measure and baseline corrected pupil-size changes (for example, see Figures 9 and 10).

After calculating a correlation per individual question and for each participant, mean correlations for each participant (see Supplementary Table S6) and for each group were calculated: NEUL: $R^2 = 0.634$; NEUH: $R^2 = 0.652$; NEGL: $R^2 = 0.565$, NEGH: $R^2 = 0.648$. To statistically assess the presence or absence of significant correlations in our observed correlations, we applied a nonparametric resampling procedure. To this end, we randomly reshuffled individual samples of ground truth and algorithmic measures of all used pupillary responses per each participant and, thus, separately per group. We then calculated correlations on the reshuffled data described above for the observed data and repeated this procedure 10,000 times. This created a distribution of 10,000 correlations per participant, from which we determined the statistical thresholds ($p = 0.0125$; i.e., Bonferroni corrected for tests in four separate groups). In other words, only observed correlations exceeding 98.75% of the surrogate correlations were deemed significant. The assessment showed a significant difference between observed measurements and randomly shuffled measurements in each participant and for each answer.

The same nonparametric resampling procedure was conducted to assess a difference in correlation between all four groups. Here, we observed a significantly reduced correlation between ground truth and cognitive load in the NEGL group: (NEUL-NEGL: $p = 0.004$; NEUH-NEGL: $p < 0.001$; NEGH-NEGL: $p = 0.001$). This loss in correlation might also be related to the missing evidence for a clear cognitive-load-dependent pupil-size effect in the NEGL group compared to the NEGH group.

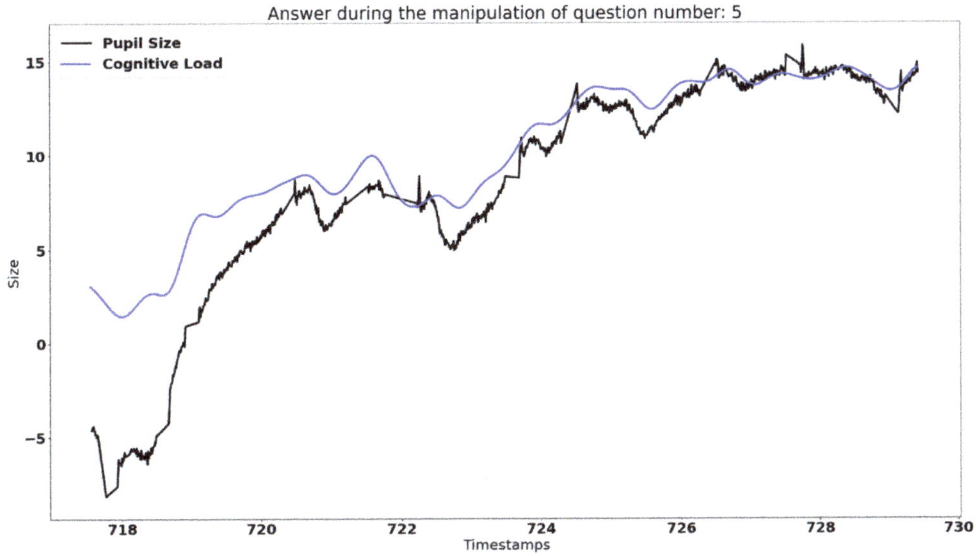

Figure 9. Example data and modeled cognitive load from a single participant's pupil response (size change in pixels) during a single answer of the interviewee (here, to Question No. 5). The black line reflects measured pupil size in pixels, and the blue line reflects modeled cognitive load.

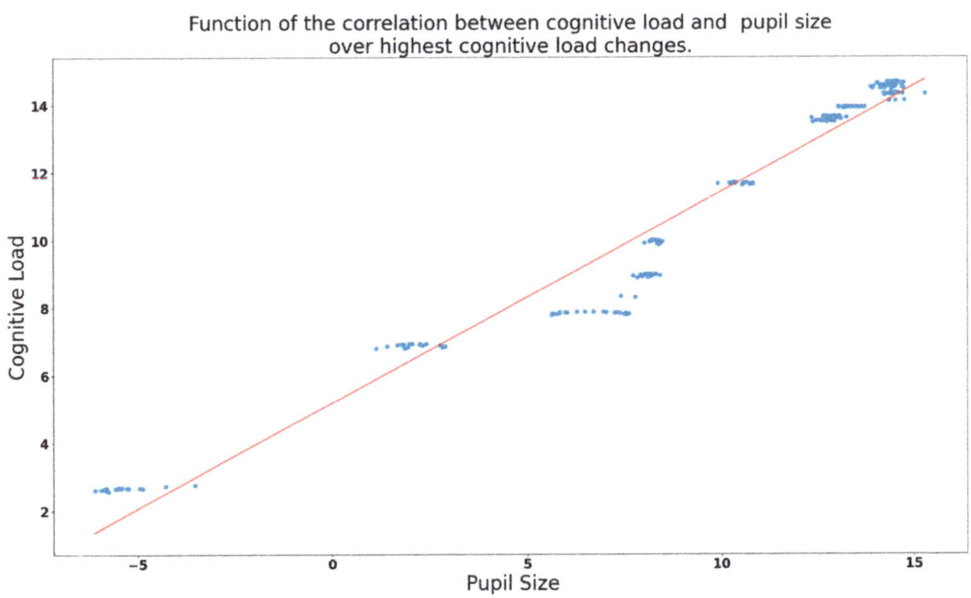

Figure 10. Correlation between modeled cognitive load according to the algorithm (y axis) and ground truth (i.e., pupil size during manipulation minus pupil size during baseline; x axis) of 20 points in a window from 250 ms before to 250 ms after peaks (or local maxima) in the modeled measure of cognitive load during a single answer of one participant. The red line reflects assumption of perfect correlation, the blue dots reflect the actual correlation between measured pupil sizes and cognitive load.

A boxplot diagram with each participant of each group, showing the ranges of all correlation scores per answer provides additional descriptive data on the found correlations (see Figure 11).

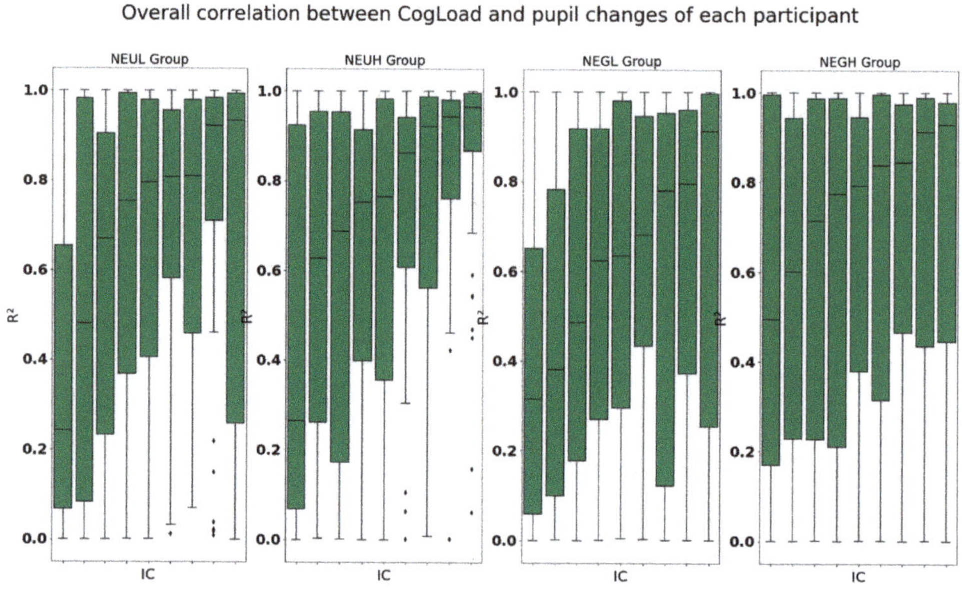

Figure 11. Boxplots showing all correlations (on the y-axis) for each of the nine participants in each group. Each boxplot shows the distribution of R^2 scores of a specific participant, with one score per answer during the manipulation phase. Black horizontal bars indicate mean correlations, green areas depict quartiles of the distributions, whiskers depict maximal and minimal correlations, and the gray diamonds depict outliers that have been removed for the calculations. Boxplots are sorted in ascending order of the mean R^2 scores from left to right. CogLoad: modeled cognitive load according to the algorithm.

In addition to correlating the changes in cognitive load to the changes in pupil size, we conducted a Pearson correlation showing the correlation between the algorithm's extracted cognitive load and the light-elicited pupillary changes. In the first step, we observed that changes in measured luminance, mostly due to the changes between question and answer displays, were negatively correlated, $r(1{,}262{,}808) = -0.55$, $p < 0.001$. This was as expected since higher luminance should lead to pupillary constriction. In contrast, in the second step, though significant due to a huge number of trials, we observed almost no meaningful correlation between changes in luminance and cognitive load, $r(1{,}262{,}808) = 0.02$, $p < 0.001$. Together, results suggested that (1) changes in pupil size due to changes in luminance were successfully captured and that (2) luminance-elicited pupillary responses did not compromise the sensitivity of the algorithm for detection of cognitive load.

3.4. Self-Report Data

The participants' subjective assessment of task difficulty was measured using the NASA-TLX), as can be seen in Figure 12. Consistently lower felt mental demand was observed in neutral relative to negative conditions rather than in low-load relative to high-load conditions (Emotion: $F[1, 35] = 5.69$, $p = 0.023$; Workload: $F[1, 35] = 0.88$, $p = 0.355$; Interaction: $F[1, 35] = 1.89$, $p = 0.179$), where the post hoc t-test revealed significantly higher levels of mental demand in the negative conditions ($M = 3.33$, $SD = 4.23$) than in the neutral conditions ($M = -0.33$, $SD = 5.06$), regardless of workload manipulation, $t(35) = -2.35$, $p = 0.024$, $\eta^2 = 0.78$. A relatively similar difference concerned participants' felt physical demand (Emotion: $F[1, 35] = 6.28$, $p = 0.017$; Workload: $F[1, 35] = 0.25$, $p = 0.619$; Interaction: $F[1, 35] = 1.572$, $p = 0.219$) where the post hoc t-test revealed significantly higher levels of physical demand in the negative conditions ($M = -3.39$, $SD = 4.46$) than in the neutral conditions ($M = -6.72$, $SD = 3.43$), regardless of workload manipulation, $t(35) = -2.51$, $p = 0.017$, $\eta^2 = 0.84$, and felt frustration (Emotion: $F[1, 35] = 7.99$, $p = 0.008$; Workload: $F[1, 35] = 0.20$, $p = 0.658$; Interaction: $F[1, 35] = 0.01$, $p = 0.929$) during the interview, where the post hoc t-test revealed significantly higher levels of frustration in the negative conditions ($M = -2.22$, $SD = 5.81$) than in the neutral conditions ($M = 3.06$, $SD = 5.07$), regardless of the workload manipulation, $t(35) = -2.91$, $p = 0.006$, $\eta^2 = 0.97$. A Tukey-HSD post hoc evaluation revealed a significant difference between groups of different emotional manipulation in the "Mental-Demand" scale ($p = 0.023$, 95%, CI = [0.534, 6.799]), in the "Physical-Demand" scale ($p = 0.017$, 95%, CI = [0.626, 6.041]), and in the "Frustration" scale ($p = 0.008$, 95%, CI = [1.475, 9.080]). All of these were perceived as stronger in the negative emotion condition compared to the neutral condition. All other differences in all other scales, as well as in all pairwise comparisons, were not significant.

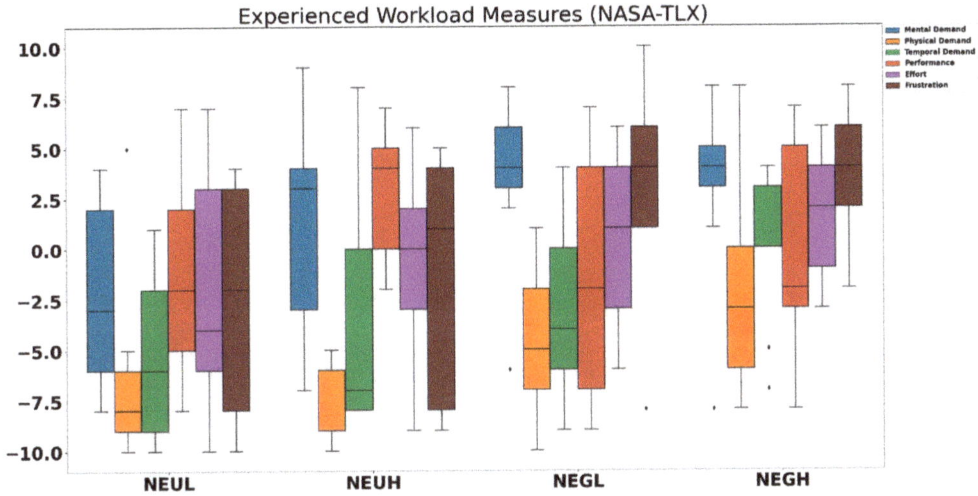

Figure 12. Medians (horizontal lines), corresponding upper and lower quartiles (colored areas above and below the horizontal lines), minima and maxima (lower and upper whiskers), and excluded outliers (gray diamonds) of self-assessments on the NASA-TLX for different groups (NEUL: neutral content/low load; NEUH: neutral content/high load; NEGL: negative content/low load; NEGH: negative content/high load).

The participants' self-assessed performance regarding the verbal and verbal–visual task in the high- and low-workload conditions is backed up by their actual accuracy rates in the verbal task (see Supplementary Figure S2). However, somewhat at odds with an experienced higher demand during the negative than during the neutral conditions, there were no significant performance (accuracy) differences in the verbal–visual task between the neutral and the negative high-workload (see Supplementary Figure S3) groups. There were also no other significant differences between groups. In the memory test at the end of the interview (see Supplementary Figure S1), groups also showed very similar results.

Subjectively felt emotions during the interview were assessed using the Self-Assessment Manikin (SAM), measuring the participants' happiness, arousal, and control (Figure 13). As expected, higher levels of happiness can be observed in the neutral conditions (Emotion: $F[1, 35] = 17.24$, $p < 0.001$; Workload: $F[1, 35] = 1.72$, $p = 0.199$; Interaction: $F[1, 35] = 0.43$, $p = 0.517$), where the post hoc t-test revealed significantly higher levels of happiness in the neutral conditions ($M = 6.17$, $SD = 1.54$) than in the negative conditions ($M = 4.06$, $SD = 1.51$), regardless of workload manipulation, $t(35) = 4.15$, $p < 0.001$, $\eta^2 = 1.38$, while higher levels of arousal can be observed in the negative conditions (Emotion: $F[1, 35] = 5.91$, $p = 0.021$; Workload: $F[1, 35] = 3.01$, $p = 0.092$; Interaction: $F[1, 35] = 1.08$, $p = 0.305$). There was no difference between groups with regard to control (Emotion: $F[1, 35] = 2.29$, $p = 0.140$; Workload: $F[1, 35] < 0.01$, $p = 1.000$; Interaction: $F[1, 35] = 0.89$, $p = 0.351$).

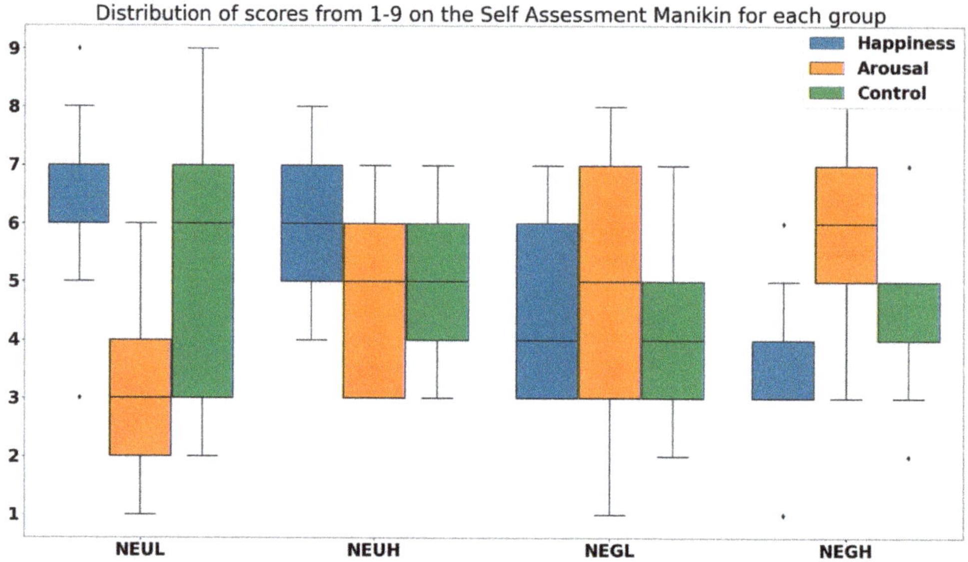

Figure 13. Medians (horizontal lines), upper and lower quartiles (colored areas above and below the horizontal lines),maxima and minima (upper and lower whiskers), and outliers (gray diamonds) of self-assessed experiences (of happiness, arousal, and control) on the y axis as a function of condition/group (NEUL: neutral content/low load; NEUH: neutral content/high load; NEGL: negative content/low load; NEGH: negative content/high load) on the x axis.

4. Discussion

In the current study, we investigated the important question of whether cognitive (work)load or mental demands can be measured by pupil size, regardless of the emotional content of the stimuli used in a task. To this end, we selected a controlled but nonetheless more realistic scenario in which the measurement of cognitive load is of interest: clinical interviews. Under an applied perspective, the measurement of cognitive load in this situation is important for a variety of reasons, not least because the correct usage of data obtained in clinical interviews often also depends on a correct comprehension of the content of the messages conveyed in such an interview, be this through verbal or non-verbal communication [7]. It should be noted that the currently available self-report measures are not necessarily consistent [8], so an unobtrusive and objective measure would be desirable.

With respect to our hypotheses, we found the following results. Regarding Hypothesis 1, we found evidence that the algorithm was partly able to differentiate between low- and high-cognitive workload manipulations.

With respect to Hypothesis 2, however, numerically, although not statistically significant, we found that the effect was larger in neutral- rather than in negative-content conditions. In the negative-content conditions, pupil-size differences between workload conditions were absent.

Concerning Hypothesis 3, nevertheless, we did find evidence that cognitive load can be successfully derived from measured pupillary responses. This was reflected in the main effect of cognitive load on an increased pupillary response. This pattern of results was found using a conventional pupil-size measure of cognitive load—the difference in pupil size between the manipulation (here, the clinical interview proper) and a baseline phase during which the same light-elicited pupillary responses were observed as during the manipulation (but without the critical manipulation). In the current study, the baseline measure of pupillary response was recorded early during the interview, when questions concerned basic information about the interviewee.

Regarding Hypothesis 4, importantly, the observed numerical pattern of an interaction between cognitive workload and emotional content of the interview became significant in an analysis based on arithmetically calculated cognitive-load measures. This was found when we applied an algorithm that uses the currently measured luminance to calculate a pupillary light response, which in turn is subtracted online from the measured pupil size to extract cognitive load. The current study, thus, shows that the algorithmic cognitive-load measure might be more sensitive to the true pattern within the numerical data than the conventional way of calculating a pupillary cognitive-load response (i.e., based on an offline subtraction of pupil sizes during baseline from pupil sizes during manipulation phases).

Finally, with respect to Hypothesis 5, the algorithmic online measure was externally validated by reasonably high and significant correlations with the standard offline measure of cognitive workload in all of the groups and for each of our participants.

In summary, pupil sizes are a promising tool for the relatively unobtrusive measurement of cognitive load in applied scenarios, such as clinical interviews, and it might even be possible to derive estimates of the loads online, during data recording. This is particularly interesting for applied settings, such as clinical interviews, where it is essential that a continuous measure of load is available without delay, so as to allow the interviewer to pay special attention to what was said or to ask the patient to repeat the information. The problem is, however, that the sensitivity of this measure can be decreased by concomitant emotional content as demonstrated in the negative conditions.

Of further interest, the very low correlation between luminance-elicited and load-related pupillary changes indicated that at least the range of luminance values used in the current study does not impose a limit on the sensitivity of the load-dependent pupillary change measured by the algorithm. This was the case despite a significant negative correlation between measured luminance and pupillary responses that indicated that the luminance differences were strong enough to elicit a pupillary response.

From an applied perspective, in the future, the increasingly pervasive presence of cameras in laptops and smartphones may be used to measure pupil sizes without the need for additional equipment and, thus, to measure cognitive load without any additional technical requirements. Technologies such as smart glasses may further boost the widespread availability of the necessary camera or eye-tracking equipment. However, we have to admit that, currently, the resolution of cameras built into laptops and smartphones is insufficient, especially in combination with the frequently suboptimal recording conditions with low luminance, shadows, faces turned away from the screens and cameras, or glasses blocking the detection of the comparatively small pupil.

5. Limitations

Our cognitive-load manipulation was relatively weak. This was not only reflected in similar performance (accuracy) levels in both cognitive-load conditions—high- and low-load conditions but also in participants' subjective self-assessments of cognitive load, which were the same in high- and low-load conditions. In contrast, to our surprise, participants subjectively felt and reported more cognitive and physical demands in negative than neutral conditions; a subjective rating that was not reflected in our pupil-size measures.

Another important limitation concerns doubts about the success of our manipulation of emotional content. Although participants subjectively felt more frustrated and generally reported more negative feelings under negative than under neutral conditions, some participants did not believe in the veridicality of the answers of the confederate/interviewee. This implies that the manipulation of the participants' emotions was probably relatively weak. This problem was potentially exacerbated by the semi-controlled nature of our task as we will discuss next.

The current study used semi-controlled conditions: On the one hand, control itself was, thus, partly limited. For example, emotional content manipulations were confounded with the different exact words or phrases used by the interviewee. Another example is that the exact behavior of the confederate/interviewee varied slightly from participant to participant (e.g., in the form of the exact length of the verbal answer or the exact strength of a facial expression), even within groups, thereby creating additional and unwanted noise in the data. On the other hand, conducting interviews in front of a computer screen allowed some degree of control (e.g., easier calibration of the eye tracker) but might have compromised the perception of nonverbal facial expressions and could have mitigated emotional responses to the stimulus. In this way, control means that our conditions were not entirely the same as those encountered in the field, and our study is, thus, only a step in the direction of closing the gap between laboratory and field conditions.

Another potential limitation concerns insufficient power. It is well known that insufficient power can lead to oversights of existing effects that simply do not turn out to be significant. However, it is probably still less well known that a lack of sufficient power can also compromise the replicability of significant findings [53]. In this respect, our study could also stand improvement in terms of a larger sample besides the aforementioned improvements in terms of the manipulations.

We used a measure of cognitive load that disregards the impact of saccades on pupil dilations. Our research is, thus, silent on how to incorporate saccade-elicited influences on pupil size into pupillary load measures. We think rightly so, as long as it is unclear if and how saccades are related to cognitive load. On the one hand, temporal attention and, thus, a particular cognitive-load factor might temporarily decrease saccade frequencies and related pupillary changes (cf. [54]). On the other hand, spatial attention and, thus, a different factor affecting cognitive load might be accompanied by (micro-)saccades and their associated pupillary changes (e.g., [55,56]), However, we agree that future research would benefit from discriminating between saccade-related pupillary changes and other load-elicited pupil responses in different tasks to understand when saccade-related pupil changes are true reflections of cognitive load and when they are better filtered out as not related to load or other difficulty related measures (cf. [57]).

Since our interview was conducted in front of a computer screen, generalizability to a face-to-face scenario is not granted. Not only might more eye movements under 3D conditions lead to less accurate pupil-size data. Head and body movements of the interviewer reacting to the interviewee would also imply, for example, that light-elicited pupil changes can be more difficult to capture in a less-controlled 3D environment, where the direction of the scene camera and the looking direction can diverge more strongly, leading to less-accurate cognitive-load measurements.

Finally, we conducted our manipulations of emotional content and cognitive load independently of one another, assuming that these two manipulations affected pupil sizes in largely independent ways. However, subjectively, participants assessed the task demands (or load) and their frustration as significantly higher under negative than under neutral emotional conditions. Interestingly, this was not reflected in their performance accuracy. Participants in neutral and negative emotional conditions performed equally well. Thus, the subjective feelings could not be backed up by objective performance indicators. Nonetheless, the assumed independence of the two manipulations—load on the one hand, emotions on the other hand—needs further scrutiny in future studies, too.

6. Conclusions

We showed that pupil-size measures can be used for the online calculation of cognitive load in a clinical interview. The present data, thus, extend prior findings on offline measures and on laboratory measures to show that cognitive load can be estimated in an unobtrusive way and without much temporal delay. Such a measure can now be used in the adaptation of tasks in applied settings, such as training where the speed of instructions could be altered to match the current level of cognitive load, in order not to impose excessive demands on trainees. However, the application still requires the usage of dedicated eye-tracking equipment, and conventional cameras built into smartphones and computers do not provide sufficient resolution. In addition, more research is needed in light of the relatively low power of the current study.

Supplementary Materials: The following supporting information can be downloaded at: https://www.mdpi.com/article/10.3390/app142411714/s1, Table S1: List of questions the confederate has been asked by the participant during the baseline; Table S2: List of questions the confederate has been asked by the participant during the manipulation; Table S3: List of answers the confederate has given to the participant during the baseline; Table S4: List of answers the confederate has given during the manipulation to participants of the "neutral"-groups (Neutral-Low, and Neutral-High); Table S5: List of answers the confederate has given during the manipulation to participants of the "negative"-groups (Negative-Low, and Negative-High); Table S6: Correlations between pupil size changes and cognitive load algorithm for each participant. Figure S1: Performance in the memory test: Number of correct answers given on a questionnaire regarding the answers that have been given by the confederate during the interview (0–10). The black horizontals depict average performance, the colored boxes the upper and lower quartiles, the whiskers the maxima and minima, and the gray diamonds the outliers; Figure S2: Number of correct responses to the verbal task during the manipulation part of the interview (0–49). The black horizontals depict average performance, the colored boxes the upper and lower quartiles, the whiskers the maxima and minima, and the gray diamonds the outliers; Figure S3: Number of correct responses to the visual task during the manipulation part of the interview (0–49). The black horizontals depict average performance, the colored boxes the upper and lower quartiles, the whiskers the maxima and minima, and the gray diamonds the outliers.

Author Contributions: Conceptualization, U.A. and B.G.; Methodology, D.G. and M.S.; Software, B.G.; Validation, U.A.; Formal analysis, D.G.; Resources, B.G.; Data curation, D.G. and M.S.; Writing—original draft, D.G. and U.A.; Writing—review & editing, B.G. and M.S.; Visualization, D.G.; Supervision, U.A.; Project administration, U.A.; Funding acquisition, U.A. All authors have read and agreed to the published version of the manuscript.

Funding: This research was funded by Austrian Research Promotion Agency grant number [880102]. Open Access Funding by the University of Vienna.

Institutional Review Board Statement: The study was conducted in accordance with the Declaration of Helsinki, and approved by the Institutional Review Board (or Ethics Committee) of University of Vienna (protocol code EK_00617 at the 13 April 2021) for studies involving humans.

Informed Consent Statement: Written informed consent was obtained from all subjects involved in the study.

Data Availability Statement: The data presented in this study are openly available in FigShare at https://doi.org/10.6084/m9.figshare.27183681.v3.

Conflicts of Interest: The authors declare no conflict of interest.

References

1. Appel, T.; Scharinger, C.; Gerjets, P.; Kasneci, E. Cross-subject workload classification using pupil-related measures. In Proceedings of the 2018 ACM Symposium on Eye Tracking Research & Applications, Warsaw, Poland, 14–17 June 2018; Bulling, A., Huckauf, A., Eds.; Association for Computing Machinery: New York, NY, USA, 2018; pp. 1–8. [CrossRef]
2. Di Stasi, L.L.; Antolí, A.; Gea, M.; Cañas, J.J. A neuroergonomic approach to evaluating mental workload in hypermedia interactions. *Int. J. Ind. Ergon.* **2011**, *41*, 298–304. [CrossRef]
3. Iqbal, S.T.; Zheng, X.S.; Bailey, B.P. Task-evoked pupillary response to mental workload in human-computer interaction. In Proceedings of the CHI '04 Extended Abstracts on Human Factors in Computing Systems, Vienna, Austria, 24–29 April 2014; Dykstra-Erickson, E., Tscheligi, M., Eds.; pp. 1477–1480. [CrossRef]
4. Ramakrishnan, P.; Balasingam, B.; Biondi, F. Chapter 2—Cognitive load estimation for adaptive human–machine system automation. In *Learning Control*; Zhang, D., Wei, B., Eds.; Elsevier: Amsterdam, The Netherlands, 2021; pp. 35–58; ISBN 978-0-12-822314-7.

5. Scharinger, C.; Kammerer, Y.; Gerjets, P. Pupil Dilation and EEG Alpha Frequency Band Power Reveal Load on Executive Functions for Link-Selection Processes during Text Reading. *PLoS ONE* **2015**, *10*, e0130608. [CrossRef]
6. Young, J.Q.; Van Merrienboer, J.; Durning, S.; Ten Cate, O. Cognitive Load Theory: Implications for medical education: AMEE Guide No. 86. *Med. Teach.* **2014**, *36*, 371–384. [CrossRef]
7. Fava, G.A.; Sonino, N.; Aron, D.C.; Balon, R.; Berrocal Montiel, C.; Cao, J.; Concato, J.; Eory, A.; Horwitz, R.I.; Rafanelli, C.; et al. Clinical Interviewing: An Essential but Neglected Method of Medicine. *Psychother. Psychosom.* **2024**, *93*, 94–99. [CrossRef] [PubMed]
8. Naismith, L.M.; Cheung, J.J.H.; Ringsted, C.; Cavalcanti, R.B. Limitations of subjective cognitive load measures in simulation-based procedural training. *Med. Educ.* **2015**, *49*, 805–814. [CrossRef]
9. Eggemeier, F.T.; Wilson, G.F. Performance-based and subjective assessment of workload in multi-task environments. In *Multiple Task Performance*; CRC Press: Boca Raton, FL, USA, 1991; ISBN 978-1-00-306944-7.
10. Ahlstrom, U.; Friedman-Berg, F.J. Using eye movement activity as a correlate of cognitive workload. *Int. J. Ind. Ergon.* **2006**, *36*, 623–636. [CrossRef]
11. Steinman, R. Gaze control under natural conditions. In *The Visual Neurosciences*; Chalupa, L.M., Werner, J.S., Eds.; MIT Press: Cambridge, MA, USA, 2004; pp. 1339–1356. [CrossRef]
12. Wickens, C.D.; Helton, W.S.; Hollands, J.G.; Banbury, S. *Engineering Psychology and Human Performance*, 5th ed.; Routledge: New York, NY, USA, 2021; ISBN 978-1-00-317761-6.
13. Beatty, J. Task-evoked pupillary responses, processing load, and the structure of processing resources. *Psychol. Bull.* **1982**, *91*, 276–292. [CrossRef] [PubMed]
14. Kahneman, D.; Beatty, J. Pupil diameter and load on memory. *Science* **1966**, *154*, 1583–1585. [CrossRef] [PubMed]
15. van der Wel, P.; van Steenbergen, H. Pupil dilation as an index of effort in cognitive control tasks: A review. *Psychon. Bull. Rev.* **2018**, *25*, 2005–2015. [CrossRef] [PubMed]
16. Chen, S.; Epps, J. Using Task-Induced Pupil Diameter and Blink Rate to Infer Cognitive Load. *Hum.–Comput. Interact.* **2014**, *29*, 390–413. [CrossRef]
17. Krejtz, K.; Duchowski, A.T.; Niedzielska, A.; Biele, C.; Krejtz, I. Eye tracking cognitive load using pupil diameter and microsaccades with fixed gaze. *PLoS ONE* **2018**, *13*, e0203629. [CrossRef] [PubMed]
18. Laeng, B.; Ørbo, M.; Holmlund, T.; Miozzo, M. Pupillary Stroop effects. *Cogn. Process.* **2011**, *12*, 13–21. [CrossRef] [PubMed]
19. Stolte, M.; Gollan, B.; Ansorge, U. Tracking visual search demands and memory load through pupil dilation. *J. Vis.* **2020**, *20*, 21. [CrossRef]
20. Strauch, C.; Wang, C.-A.; Einhäuser, W.; Van der Stigchel, S.; Naber, M. Pupillometry as an integrated readout of distinct attentional networks. *Trends Neurosci.* **2022**, *45*, 635–647. [CrossRef] [PubMed]
21. Heller, P.H.; Perry, F.; Jewett, D.L.; Levine, J.D. Autonomic components of the human pupillary light reflex. *Investig. Ophthalmol. Vis. Sci.* **1990**, *31*, 156–162.
22. Loewenfeld, I.E. *The Pupil: Anatomy, Physiology, and Clinical Applications*; Butterworth-Heinemann: Boston, MA, USA, 1999; Volume 1, ISBN 978-0-7506-7216-0.
23. Benedetto, A.; Binda, P. Dissociable saccadic suppression of pupillary and perceptual responses to light. *J. Neurophysiol.* **2016**, *115*, 1243–1251. [CrossRef] [PubMed]
24. Zuber, B.L.; Stark, L.; Lorber, M. Saccadic suppression of the pupillary light reflex. *Exp. Neurol.* **1966**, *14*, 351–370. [CrossRef] [PubMed]
25. Bradley, M.M.; Miccoli, L.; Escrig, M.A.; Lang, P.J. The pupil as a measure of emotional arousal and autonomic activation. *Psychophysiology* **2008**, *45*, 602–607. [CrossRef] [PubMed]
26. Henderson, R.R.; Bradley, M.M.; Lang, P.J. Emotional imagery and pupil diameter. *Psychophysiology* **2018**, *55*, e13050. [CrossRef] [PubMed]
27. Hess, E.H. Some New Developments in Pupillometrics. In *Die Normale und die Gestörte Pupillenbewegung/Normal and Disturbed Pupillary Movements*; Dodt, E., Schrader, K.E., Eds.; J.F. Bergmann-Verlag: Munich, Germany, 1973; pp. 246–262. ISBN 978-3-642-80488-5.
28. Pauli, W.M.; Larsen, T.; Collette, S.; Tyszka, J.M.; Seymour, B.; O'Doherty, J.P. Distinct Contributions of Ventromedial and Dorsolateral Subregions of the Human Substantia Nigra to Appetitive and Aversive Learning. *J. Neurosci.* **2015**, *35*, 14220–14233. [CrossRef] [PubMed]
29. Seymour, B.; Daw, N.; Dayan, P.; Singer, T.; Dolan, R. Differential Encoding of Losses and Gains in the Human Striatum. *J. Neurosci.* **2007**, *27*, 4826–4831. [CrossRef]
30. Alamia, A.; VanRullen, R.; Pasqualotto, E.; Mouraux, A.; Zenon, A. Pupil-Linked Arousal Responds to Unconscious Surprisal. *J. Neurosci.* **2019**, *39*, 5369–5376. [CrossRef]
31. Joshi, S.; Li, Y.; Kalwani, R.M.; Gold, J.I. Relationships between Pupil Diameter and Neuronal Activity in the Locus Coeruleus, Colliculi, and Cingulate Cortex. *Neuron* **2016**, *89*, 221–234. [CrossRef] [PubMed]
32. Bramness, J.G.; Khiabani, H.Z.; Mørland, J. Impairment due to cannabis and ethanol: Clinical signs and additive effects. *Addiction* **2010**, *105*, 1080–1087. [CrossRef] [PubMed]
33. Peinkhofer, C.; Martens, P.; Grand, J.; Truelsen, T.; Knudsen, G.M.; Kjaergaard, J.; Kondziella, D. Influence of Strategic Cortical Infarctions on Pupillary Function. *Front. Neurol.* **2018**, *9*, 916. [CrossRef] [PubMed]
34. Chen, S.; Epps, J. Automatic classification of eye activity for cognitive load measurement with emotion interference. *Comput. Methods Programs Biomed.* **2013**, *110*, 111–124. [CrossRef]
35. Kuchinke, L.; Võ, M.L.-H.; Hofmann, M.; Jacobs, A.M. Pupillary responses during lexical decisions vary with word frequency but not emotional valence. *Int. J. Psychophysiol.* **2007**, *65*, 132–140. [CrossRef] [PubMed]

36. Doody, O.; Noonan, M. Preparing and conducting interviews to collect data. *Nurse Res.* **2013**, *20*, 28–32. [CrossRef] [PubMed]
37. McGrath, C.; Palmgren, P.J.; Liljedahl, M. Twelve tips for conducting qualitative research interviews. *Med. Teach.* **2019**, *41*, 1002–1006. [CrossRef] [PubMed]
38. Annen, S.; Roser, P.; Brüne, M. Nonverbal behavior during clinical interviews: Similarities and dissimilarities among schizophrenia, mania, and depression. *J. Nerv. Ment. Dis.* **2012**, *200*, 26–32. [CrossRef]
39. Freedman, N.; Hoffman, S.P. Kinetic behavior in altered clinical states: Approach to objective analysis of motor behavior during clinical interviews. *Percept. Mot. Ski.* **1967**, *24*, 527–539. [CrossRef] [PubMed]
40. Robinson, J.D. Nonverbal Communication and Physician-Patient Interaction: Review and New Directions. In *The Sage Handbook of Nonverbal Communication*; Sage Publications, Inc.: Thousand Oaks, CA, USA, 2006; pp. 437–459. ISBN 978-1-4129-0404-9.
41. Donnelly, D.A.; Murray, E.J. Cognitive and emotional changes in written essays and therapy interviews. *J. Soc. Clin. Psychol.* **1991**, *10*, 334–350. [CrossRef]
42. Derakshan, N.; Eysenck, M.W. Anxiety, Processing Efficiency, and Cognitive Performance. *Eur. Psychol.* **2009**, *14*, 168–176. [CrossRef]
43. Fernández-Olano, C.; Montoya-Fernández, J.; Salinas-Sánchez, A.S. Impact of clinical interview training on the empathy level of medical students and medical residents. *Med. Teach.* **2008**, *30*, 322–324. [CrossRef] [PubMed]
44. Ginsburg, H.P. *Entering the Child's Mind: The Clinical Interview in Psychological Research and Practice*; Cambridge University Press: Cambridge, UK, 1997; ISBN 978-0-521-49685-8.
45. Kassner, M.; Patera, W.; Bulling, A. Pupil: An open source platform for pervasive eye tracking and mobile gaze-based interaction. In Proceedings of the 2014 ACM International Joint Conference on Pervasive and Ubiquitous Computing: Adjunct Publication, Washington, DC, USA, 13–17 September 2014; Brush, A.J., Friday, A., Eds.; pp. 1151–1160.
46. Margraf, J.; Cwik, J.C. *Mini-DIPS Open Access: Diagnostisches Kurzinterview bei Psychischen Störungen*; Forschungs- und Behandlungszentrum für Psychische Gesundheit, Ruhr-Universität Bochum: Bochum, Germany, 2020. [CrossRef]
47. Hart, S.G.; Staveland, L.E. Development of NASA-TLX (Task Load Index): Results of empirical and theoretical research. In *Human Mental Workload*; Advances in Psychology, 52; North-Holland: Oxford, UK, 1988; pp. 139–183; ISBN 978-0-444-70388-0.
48. Bradley, M.M.; Lang, P.J. Measuring emotion: The self-assessment manikin and the semantic differential. *J. Behav. Ther. Exp. Psychiatry* **1994**, *25*, 49–59. [CrossRef]
49. Gollan, B.; Ferscha, A. Modeling pupil dilation as online input for estimation of cognitive load in non-laboratory attention-aware systems. In Proceedings of the COGNITIVE 2016, The Eighth International Conference on Advanced Cognitive Technologies and Applications, Rome, Italy, 20–24 March 2016.
50. Watson, A.B.; Yellott, J.I. A unified formula for light-adapted pupil size. *J. Vis.* **2012**, *12*, 12. [CrossRef]
51. Pamplona, V.F.; Oliveira, M.M.; Baranoski, G.V.G. Photorealistic models for pupil light reflex and iridal pattern deformation. *ACM Trans. Graph.* **2009**, *28*, 106. [CrossRef]
52. Hoeks, B.; Levelt, W.J. Pupillary dilation as a measure of attention: A quantitative system analysis. *Behav. Res. Methods Instrum. Comput.* **1993**, *25*, 16–26. [CrossRef]
53. Button, K.S.; Ioannidis, J.P.A.; Mokrysz, C.; Nosek, B.A.; Flint, J.; Robinson, E.S.J.; Munafò, M.R. Power failure: Why small sample size undermines the reliability of neuroscience. *Nat. Rev. Neurosci.* **2013**, *14*, 365–376. [CrossRef]
54. Abeles, D.; Amit, R.; Tal-Perry, N.; Carrasco, M.; Yuval-Greenberg, S. Oculomotor inhibition precedes temporally expected auditory targets. *Nat. Commun.* **2020**, *11*, 3524. [CrossRef] [PubMed]
55. Engbert, R.; Kliegl, R. Microsaccades uncover the orientation of covert attention. *Vis. Res.* **2003**, *43*, 1035–1045. [CrossRef] [PubMed]
56. Poletti, M.; Listorti, C.; Rucci, M. Microscopic eye movements compensate for nonhomogeneous vision within the fovea. *Curr. Biol.* **2013**, *23*, 1691–1695. [CrossRef] [PubMed]
57. Burlingham, C.S.; Mirbagheri, S.; Heeger, D.J. A unified model of the task-evoked pupil response. *Sci. Adv.* **2022**, *8*, eabi9979. [CrossRef] [PubMed]

applied sciences

MDPI

Article

Analysing Virtual Labs Through Integrated Multi-Channel Eye-Tracking Technology: A Proposal for an Explanatory Fit Model

María Consuelo Sáiz-Manzanares [1,*], Raúl Marticorena-Sánchez [2], Javier Sáez-García [3] and Irene González-Díez [4]

1 Department of Health Sciences, GIR DATAHES, UIC JCYL N° 348, Universidad de Burgos, 09001 Burgos, Spain
2 ADMIRABLE Research Group, Department of Computer Engineering, Higher Polytechnic School, Campus Vena, Universidad de Burgos, 09006 Burgos, Spain; rmartico@ubu.es
3 Higher Polytechnic School, Campus Milanera, Universidad de Burgos, 09001 Burgos, Spain; jsg1013@alu.ubu.es
4 Department of Health Sciences, GIR DATAHES, Universidad de Burgos, 09001 Burgos, Spain; igdiez@ubu.es
* Correspondence: mcsmanzanares@ubu.es

Abstract: This study deals with an analysis of the cognitive load indicators produced in virtual simulation tasks through supervised and unsupervised machine learning techniques. The objectives were (1) to identify the most important cognitive load indicators through the use of supervised and unsupervised machine learning techniques; (2) to study which type of task presentation was most effective at reducing the task's intrinsic load and increasing its germane load; and (3) to propose an explanatory model and find its fit indicators. We worked with a sample of 48 health sciences and biomedical engineering students from the University of Burgos (Spain). The results indicate that being able to see the task before performing it increases the germane load and decreases the intrinsic load. Similarly, allowing students a choice of presentation channel for the task respects how they process information. In addition, indicators of cognitive load were found to be grouped into components of position, speed, psychogalvanic response, and skin conductance. An explanatory model was proposed and obtained acceptable fit indicators.

Keywords: eye tracking; galvanic skin response; cognitive load; simulation tasks; machine learning techniques

1. Introduction

This paper will first address the state of the art of the concept of cognitive load during task or problem solving. Secondly, it will address the challenge of interpreting cognitive load indicators through statistical and machine learning algorithms (supervised and unsupervised). Finally, a suggested model of analysis will be analysed as an advance in the knowledge of cognitive load and its application to fieldwork.

1.1. Cognitive Load Definition and Measurement

In recent years, the measurement of cognitive load with respect to task performance or problem solving has become a necessary process, as it helps us understand performance and facilitates an analysis and adjustment of how tasks or problems are presented. Interest arises from a variety of fields and contexts (the educational and instructional sector, marketing, driving, aviation, business, etc.). However, there are many challenges involved in what seems at first to be manageable.

The concept of cognitive load is based on the idea that cognitive resources are limited. Cognitive load develops as a complex interaction between the different demands of tasks and the cognitive processes required to process and execute them, and a subject's performance often decreases when cognitive demands are either very low or very high.

The techniques for measuring cognitive load can be classified into four main categories: (1) performance-based, (2) subjective, (3) behavioural and (4) physiological [1–3]. Performance can be measured through learning outcomes, subjectivity through learning history and the use of scales, and behaviours through behavioural analyses. When it comes to physiological components, various indicators can be applied to measure physiological responses during task performance or problem solving. For example, research using eye tracking technology shows that the frequency of fixations, saccades and pupil diameter appear to be suitable indicators for predicting task difficulty at an early stage [4].

The theory of cognitive load is related to studies in the field of developmental educational psychology and instructional cognitive psychology, specifically the study of the relationship between working memory and long-term memory (LTM) [5]. The theory first appeared in 1976 with Sweller's studies [6]. Subsequently, in the late 1980s, the concept of cognitive load was used to explain performance on different tasks with respect to the use of various forms of information processing. A distinction was made between the intrinsic cognitive load (which refers to the complexity of the knowledge to be processed by the learner and is independent of how the task is presented) [7] and extrinsic cognitive load (referring to how the task is presented) [8,9]. Cognitive load seems to be related to divided attention with respect to information received through different channels (visual, auditory, or both). This information must be integrated in order for the learning agent to gain an understanding of a task [9,10]. The comprehension process also seems to be influenced by the way the task is presented [11]. These results are based on Baddeley's theory [12], according to which memory is divided into different types of storage [short-term memory (STM), working memory, and LTM]. For Baddeley [12], working memory consists of an auditory subsystem and a visual subsystem.

Until the second half of the 1990s, research on cognitive load theory focused on instructional designs seeking to reduce the extrinsic cognitive load, since at that time the intrinsic load was difficult to measure or manipulate. Later, the concept of germane cognitive load was used to refer to the mental resources that learners deploy to deal with the intrinsic cognitive load of learning tasks [8]. It was introduced in reference to the development of cognitive schemas in LTM on knowledge integration. Subsequently, studies by Sweller [13] showed that integrating information through different channels (visual or auditory) does not always lead to better learning. The explanation was that eliminating redundant information can be beneficial in the learning process.

Sweller tried to integrate cognitive load theory into the broader theoretical framework of evolutionary theory [13]. This new view was explained by a representation of the cognitive architecture in which there are different memory stores. More specifically, the working memory store has a limited capacity when dealing with new information and cannot combine or contrast more than four pieces of information at a time [13,14]. In addition, information in working memory is lost in approximately 20 s [15]. In LTM, information is organised into higher-order units called cognitive schemas, according to Baddeley [12]. When the number of information units that are processed simultaneously exceeds the capacity of working memory, these units must be combined into cognitive schemas in order to understand the information [16]. In short, each cognitive process that requires conscious control places a cognitive load on working memory. In this framework, germane load helps in the construction and automation of schemas. Therefore, according to the theory of cognitive load, the germane load should be increased as much as possible.

In addition, explanations of cognitive load should consider Vygotsky's theory of the Zone of Proximal Development (ZDP) [17], where the success of learning new information is related to the cognitive links that the learner makes with respect to similar information. The construction of new information will be based on this interaction [13,17]. Likewise, the theory of cognitive load is currently explained by a theory of instruction [18] that differentiates between primary biological information. This refers to information that we as members of the species have about different abilities (speaking, walking, etc.) that we execute unconsciously. However, when the learner presents certain delays in development or is subjected to socio-environmental deprivation factors, there may be problems in the acquisition of these skills. In addition, this theory of cognitive load includes secondary biological information, which requires conscious effort to successfully learn various skills such as reading, writing, arithmetic, etc. Acquiring this knowledge needs the application of various specific metacognitive strategies (orientation, planning, evaluation, and elaboration of information).

Putting all that together, the starting point is therefore an explanatory scheme that begins with an input of information where the learner first of all perceives information (this can be visual, auditory, tactile, or any combination thereof). The learner then selects some of the information and starts processing it. Some of this information passes to STM and working memory (this store cannot remember more than 7 items of novel information and cannot process more than 3–4 items, and these actions cannot last more than about 20 s). Subsequently, the retained information will pass to LTM (this store has more storage capacity than STM and the information remains more stable). LTM would therefore be the key to the individual difference in human processing, as each learner will relate new information to existing information in a different way. The load here is both intrinsic and extrinsic. In order to achieve good learning, learners' motivation has to be enhanced in the way information is presented by seeking novelty and avoiding redundancy. On the other hand, improving the retention of novel information requires facilitating practice and automation of task resolution procedures. Each of these effects will facilitate the acquisition of information stored in LTM by reducing the load on working memory [18], although all of these processes are subject to individual differences in learners [18].

This means that understanding the human cognitive architecture is important for developmental educational psychology and instructional cognitive psychology, as knowledge of the types of processing will provide the teacher with information for tailoring the design of learning tasks [18]. Therefore, knowledge of the cognitive load in the performance of a given task is important for adjusting the task to each learner's performance possibilities. However, this remains a challenge, as the same task may need to be presented in different ways for each type of learner, and learners may experience different cognitive loads for the same task [19].

Research is currently ongoing about the cognitive load experienced in virtual environments that apply simulation tasks [19]. These studies show that cognitive load is a useful indicator for analysing learning processes in simulated environments. They also differentiate between intrinsic cognitive load, task-specific germane cognitive load, and extraneous cognitive load. If the total cognitive load (the sum of the three types) exceeds the capacity of working memory, performance may suffer [20]. Another classification of cognitive load differentiates between primary cognitive load (referring to domain-specific task performance) and secondary cognitive load (referring to metacognitive performance that supports primary performance). If the secondary cognitive load is activated, performance may be better but the total cognitive load is increased, and so the relationship between cognitive load and performance may be positive or negative depending on the type of task, the type of learner, the task design, and the context of task performance [21].

Research about performance analysis in simulated learning environments has shown psychophysiological indicators of cognitive load to be the most reliable. These include pupil dilation (an involuntary response reflecting noradrenergic activity of the locus coeruleus that regulates arousal, cognitive activity, and emotion), which can indicate the cognitive load related to the emotion that the task produces in the learner [22,23]. However, this measure may be affected by factors other than an individual's processing, such as the type of lighting. Other indicators of cognitive load include the duration and total time of fixation [22], the number of saccades, and the amplitude of the saccades [24,25]. These are all static indicators, but dynamic indicators should also be considered. Dynamic indicators refer to the route that each learner takes in the task resolution process. The route refers to the steps followed in order of priority and the duration of each step in the spatial coordinate space during execution. An analysis of each learner's tracking pattern on the multimedia interface used in the simulation environments can give information about how this information has been retained in working memory and how it has been processed in LTM. The type of interface can increase or decrease a task's cognitive load, depending on the individual learner [26].

In summary, physiological measurements during task performance can provide information on different indicators such as pupil diameter [27] or the galvanic skin response (GSR), which measures skin conductance by recording the sweat secreted from the sweat glands [28]. This can be divided into tonic components, giving information on the level of skin conductance, and phasic components, providing the skin conductance response (SCR), a rapidly changing signal that takes one to five seconds to reflect the response to a stimulus [29]. These two measures, GSR and SCR, are considered reliable indicators for measuring cognitive load levels [27,29], although they can be contaminated by other external factors. As noted above, pupillary activity can be subject to illumination-related factors [30]. Similarly, skin conductance indicators may be affected by the user's physical activity (e.g., GSR-equipped hand movements) [29]. Therefore, to prevent external or internal factors from affecting the physiological measurement of cognitive load, these external variables should be controlled as much as possible [31]. Moreover, current lines of research are aimed at using several indicators to measure cognitive load, understanding this as a multimodal approach [31,32]. Such an approach has many advantages, as using various indicators of cognitive load can mitigate the potential limitations of individual measures and thus provide a more comprehensive analysis of the cognitive load experienced by different users. In addition, recording physiological measures allows real-time data collection, facilitating a more objective analysis of cognitive load [31]. Another factor that is relevant to cognitive load studies is the design of the materials (content, interface, etc.) used in the task to be solved [33]. Table 1 presents a list of possible variables that can influence learners' performance of a task or problem solving. An analysis of these variables may provide researchers with information about each learner's characteristics with respect to the intrinsic and germane cognitive loads experienced in each task.

Table 1. List of elements to be considered in the task performance space and measurement indicators.

Elements to Consider	Measurement
Task characteristics, indicators of the extrinsic cognitive load (form of task presentation (visual—image, text, etc.)) and extrinsic cognitive load written, auditory, all), multimedia learning environment simulation, task execution schedule, etc.	Description of the type of task Analysis of the presentation of the task
Physical space [characteristics of the space in which the task takes place (lighting, space, etc.)].	Indicators for brightness, temperature, noise, time, and time of day
Learner characteristics (age, gender, prior knowledge, motivation towards the task, way of responding, way of solving, type of reasoning applied, etc.).	Variables assigned (age, gender, etc.) Analysis of prior knowledge Cognitive style and learning style
Physiological indicators of the intrinsic cognitive load	Pupil diameter Fixations Saccades GSR amplitude SCR count
Perception of the germane load	Performance indicators in the execution of the task

1.2. Machine Learning Algorithms Used in the Interpretation of Cognitive Load

As discussed above, many different factors can influence the performance of a task or the resolution of a problem. Measuring them provides a huge amount of data to be processed. Various statistical and machine learning techniques can be used to carry out this processing. These techniques are divided into supervised techniques using labelled data to train models (within which we can differentiate between classification and regression techniques); unsupervised techniques that do not require labels and are used to discover hidden patterns in the data; and semi-supervised learning (combines the most appropriate supervised and unsupervised approach in each case of study).

The techniques include algorithms such as logistic regression [34]. These can help in the classification of intrinsic cognitive load into levels (low, medium, and high) [19]. A probabilistic study from a percentile analysis can also be used for this purpose [35]. Similarly, supervised machine learning algorithms can be used—where it is advisable to select features beforehand in order to detect the most influential ones [19]. Next, regression and/or classification techniques can be applied. Regression techniques may include linear regression, Support Vector Machine (SVM), Decision tree, and Neural Network algorithms. Classification techniques that can be used include discriminant analysis and *k*-Nearest neighbour (KNN) algorithms, among others.

Unsupervised learning techniques (clustering) include applying algorithms such as *k*-Means, *k*-Medoids, and s-Means [36]. Therefore, before data processing, it is important for noise to be removed and classifiers to be found. The aim is to determine the best indicators of cognitive load in each task [19]. Principal component analysis can also be applied to find a feature matrix [37]. For example, many workload studies have used the SVM algorithm in multiclass classification [36], whereas other studies have used KNN, Naive Bayes, Random Forest, SVM [38–40], and neural network-based models to infer cognitive load levels [41,42]. Classifier effectiveness can be determined using accuracy, recall, and precision indicators in the different algorithms used [41–43]. Fit can be examined using goodness-of-fit index models, which include various fit tests (NFI = normed-fit-index; RMSEA = root-mean-square error of approximation; SRMR = standardised root-mean-square residual; TLI = Tucker–Lewis index; CFI = comparative fit index; AIC = Akaike information criterion; and ECVI = parsimony index) [44].

In contrast, as already indicated, unsupervised learning clustering techniques can be used to learn the clustering profile with respect to different indicators of cognitive load without applying a prior classifier [45]. Algorithms can also be used to perform pattern analysis. These include distance analysis through string edit methods [46], which are applied in the study of patterns detected with a scan path or gaze point. Similarly, more traditional statistical algorithms such as Analysis of Variance (ANOVA), Analysis of Covariance (ANCOVA), and Multivariate Analysis of Variance (MANOVA) can be used to determine the significant differences in indicators of cognitive load between different types of users [47]. Table 2 presents a summary of the algorithms and their usefulness in the study of cognitive load.

Table 2. Different algorithms and their usefulness for measuring cognitive load.

Contrast Tests	Usefulness in Measuring Cognitive Load
Feature selection [19,37]	The use of feature selection algorithms provides information in a hierarchical order on which features are most relevant.
Logistic regression [34] Probabilistic study (percentile analysis) [36].	Helps in the detection of intrinsic cognitive load levels (low, medium, and high) [19]. Allows a determination of the level of execution (low, medium, and high).
Predictions (linear regression, SVM, Decision tree, Neuronal Networks)	Provide information on cognitive load indicators that may be more relevant in explaining, e.g., learning outcomes.
Classification (analysis, discriminant analysis, KNN, SVM, Naive Bayes, Random Forest [38–40], and Neural networks [41,42])	The classifiers provide information on the most important cognitive load indicators with respect to ranking groups of variables in relation to a reference variable, such as academic performance.

Table 2. *Cont.*

Contrast Tests	Usefulness in Measuring Cognitive Load
Clustering (*k*-Means, *k*-Medoids, s-Means) [45], string-edit Methods [46]	Determines interaction profiles of task performance without prior grouping variables. Identifies route maps in different types of learners.
Adjustment indicators (accuracy, recall, and precision) [41–43], goodness-of-fit indices (NFI, RMSEA, SRMR, TLI, CFI, AIC, ECVI) [44]	Determine the effectiveness, fit and precision when comparing the results obtained with different algorithms.
Parametric statistics (ANOVA, ANCOVA, MANOVA and effect sizes) [47] If the sample does not have a normal distribution, similar tests would be applied in a non-parametric context (Mann–Whitney U test, Kruskal–Wallis, Wilcoxon signed-rank test, etc.)	Determine whether there are significant differences in terms of variables considered as independent variables on variables considered as dependent variables. Also, determine the size of the significant effect.

1.3. Proposed Models for Measuring Cognitive Load

Taking the above into account, the current challenge is to create models to explain cognitive processing in different types of learners during the same task [48]. As noted above, various algorithms have been advanced for measuring the cognitive load of different users during the execution of different tasks [49]. Proposed explanatory and analytical models are based on the construction of multimodal machine learning models that help to estimate people's cognitive loads. However, one potential limitation of these models is that they usually use data extracted from the same scenario. Therefore, the models would need to be tested in learning contexts other than the source application and include information on the execution of different learning tasks. Furthermore, predictive results should be accompanied by estimates of uncertainty (percentage error) in order to reduce misclassification [50].

In summary, the current research focuses on finding robust models that can predict certain situations and facilitate early intervention [51]. These models are based on using machine learning techniques because, as noted above, the assessment of cognitive load is complex. Current research [52] indicates that machine learning techniques such as Deep Learning (DL) are a good resource for finding the most relevant information, allowing a more reliable classification. In particular, hybrid DL architectures that integrate convolution operators and Long Short-Term Memory (LSTM) have been found to produce significantly better results than other models. Similarly, the proposed models should be tested in domains other than the source context, in addition to taking into account individualised differentiation with respect to users [53]. Models that apply inverse reinforcement learning are also proving particularly effective at explaining complex cognitive tasks [54]. Likewise, hyper long-short-term-memory-based modules from time-series multimodal information models have been shown to be effective at explaining the cognitive load in driving tasks [55]. The use of data fusion techniques [56] and Transformer models are likely to be particularly useful resources in explaining the cognitive load experienced by users in the execution of different tasks [57–59]. In recent years, various studies have applied a variety of techniques, such as the Gaussian Mixture Variational-Based Transformer Domain Adaptation Fault Diagnosis method (applying it to diagnosing faults in bearings) [60], and the Prototypical Contrastive-Based Domain Adaptation method to achieve learning-based fault diagnosis under variable working conditions [61].

In summary, modelling is important in different research fields (neurodegenerative diseases, cancer processes, autism spectrum disorders, learning disabilities, etc.) where an early differential diagnosis is essential, as early therapeutic or instructional intervention is critical for a good prognosis. Therefore, cognitive load indicators (biomarkers) need to be identified, and predictive and classification models need to be proposed [62]. Furthermore, the results of current research advocate the use of an approach that includes several indicators of cognitive load collected through different devices. This is because this working methodology increases the reliability of cognitive load detection and the identification of individual differences. Similarly, attention needs to be paid to the design of the task and the development of the interface where it is applied. This area of research faces challenges such as noise reduction with respect to the signal collection devices and the extrapolation of the results produced in laboratories [19].

Based on the state of the art above, the objectives of this study were as follows:

Objective 1. To select characteristics with respect to different indicators of cognitive load related to the performance of simulation tasks in virtual environments.

Objective 2. To determine participants' levels of cognitive load (low, medium, and high).

Objective 3. To apply supervised machine learning prediction techniques to determine how the variables of presentation type 1 (not having previously watched the lab vs. having done so) and presentation type 2 (visual–auditory vs. visual) from the virtual laboratories predict the indicators of cognitive load and learning results.

Objective 4. To apply supervised machine learning techniques for classification in order to determine the most representative classifiers of cognitive load.

Objective 5. To apply unsupervised machine learning techniques to identify possible groupings without a prior assignment variable with respect to cognitive load indicators and learning outcomes.

Objective 6. To determine whether there are significant differences in the indicators of cognitive load and learning results depending on the type of task and the way it is presented (visual–auditory vs. visual) in the participating groups.

Objective 7. To propose an analysis model for the markers of cognitive load.

2. Materials and Methods

2.1. Participants

We worked with a convenience sample of 48 undergraduate students in their final years at the University of Burgos: 21 were studying for a degree in Occupational Therapy, of whom 20 were women and 1 was a man (Group 1); 10 were studying for a degree in Biomedical Engineering (1), of whom 8 were women and 2 were men (Group 2); and 17 were studying for a degree in Biomedical Engineering (2), of whom 9 were women and 8 were men (Group 3). Groups 2 and 3 were different groups, although they shared the same degree and academic level. The mean age of the participants was between 21–23 years old. Participation in the study was voluntary, required written informed consent, and was without financial compensation.

2.2. Instruments

(a) Tobii pro lab (Stockholm, Sweden) version 1.194 with 64 Hz.
It is eye-tracking software that provides a visual user interface and dedicated software functions that support the researcher in all phases of an eye-tracking experiment, from test design and recording to analysis. An image of the device used is shown in Figure 1 below.

(b) Laptop computer. Falkon i7. Screen 15.6"; processor Intel core I7-9750H; hard disk 512 Gb SSD; memory 16 Gb of RAM; Geforce RTX 2070 MAX-Q GDDR6 8 GB; Windows 10 Professional; connections: HDMI/USB-C/USB 3.1/RJ45 × 1/HDMI × 1/3.5 mm Jack/USB-A 3.1 × 4/Thunderbolt × 1—Bluetooth/WiFi/Gigabit Ethernet. Display resolution of 1080 × 1920.

(c) Shimmer3 GSR+ (galvanic skin response) (Europe Research Ltd., Dublin, Ireland) single-channel galvanic skin response data acquisition unit (measures electrodermal resistance (EDR)/electrodermal activity (EDA)). The GSR+ unit is suitable for measuring the electrical characteristics or conductance of skin. Traditional galvanic skin response theory is based on the idea that the electrical resistance of the skin varies with the state of the sweat glands. Sweating in the human body is regulated by the autonomous nervous system (ANS). More specifically, if the sympathetic branch (SNS) of the ANS is excited, the activity of sweat glands increases, which in turn increases skin conductance. This device is compatible with Tobii pro lab version 1.194 and allows for the integration of records. Figure 2 shows the device and how it is placed to record EDA.

(d) Virtual lab for the resolution of clinical cases. This lab was developed within the European project "Specialized and updated training on supporting advance technologies for early childhood education and care professionals and graduates", eEarlyCare-T-No. 2021-1-ES01-KA220-SCH-000032661. In this study, the virtual lab is a representation of a dialogue between two avatars: a student and a teacher.
Two labs were used in this study. In Group 1 and Group 2, a general simulation virtual lab was applied (this referred to content that could be used in both degrees, as they were general and about patient intervention processes). This lab is extensively described in the study by Sáiz-Manzanares et al. [32] (see Figure 1 in that publication). In addition, a specific lab with biomedical engineering content was applied to Group 3. This lab can be found in Supplementary Materials (Figure S1). Both virtual labs can be accessed and used free of charge by logging into the Virtual Classroom in the eEarlyCare-T project https://www2.ubu.es/eearlycare_t/en/project (accessed on 16 August 2024).

(e) Learning Strategies Scales (ACRA) by Román and Gallego [63]. This is an instrument that has been widely tested in research. It identifies 32 strategies at different points in the information processing cycle and has reliability indicators between $\alpha = 0.75$ and $\alpha = 0.90$, and content validity indicators between $r = 0.85$ and $r = 0.88$. In the present study, we applied the metacognitive skills subscale, which includes 17 strategies on the use of metacognitive skills in problem-solving tasks that are distributed over three subscales: Self-knowledge, Planning, and Self-Evaluation.

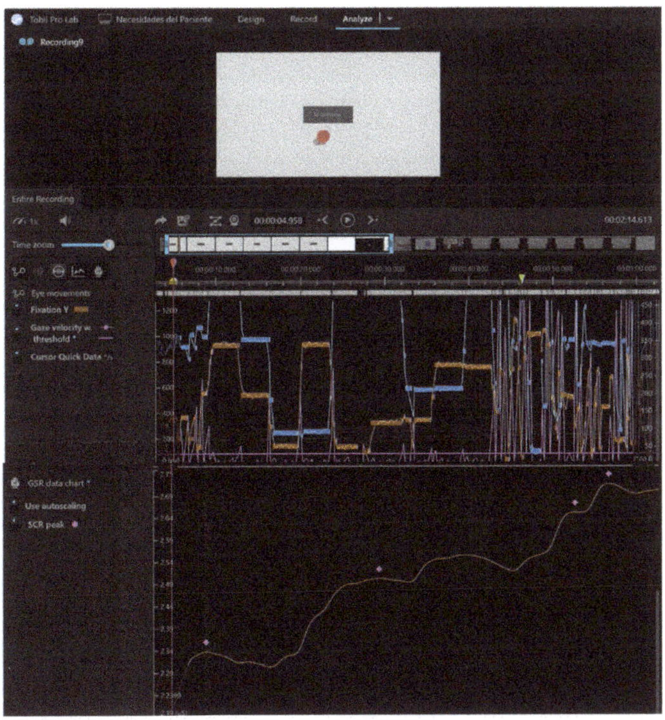

Figure 1. Example of a test performed with Tobii pro lab version 1.194 with 64 Hz, recording eye movements and GSR.

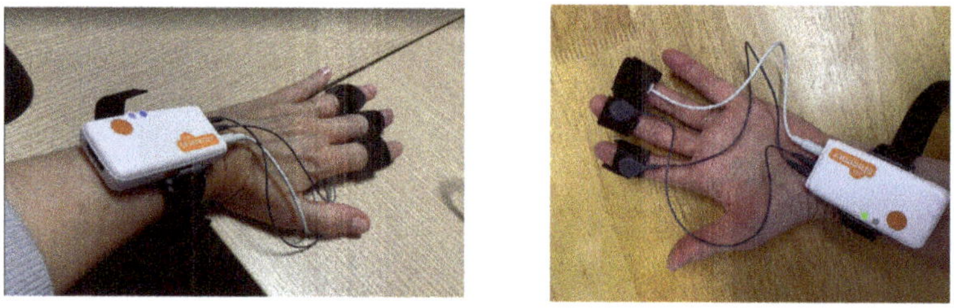

Figure 2. Example of the Shimmer3 GSR+ device.

2.3. Procedure

Prior approval was obtained for the study from the Bioethics Commission of the University of Burgos: No. IO 03/2022 (for carrying out the study) and No. IO 04/2022 (for application of the labs). Written informed consent was then obtained from all participants, their participation was voluntary and without financial compensation. The virtual labs were then implemented for all three groups throughout the second semester. In all cases, lighting and temperature conditions were equivalent (the light and temperature conditions were stable, as the studies were always carried out at the same time with the same light conditions and the temperature was always regulated by a thermostat at 21°). The time each participant spent viewing each lab was discretionary, as they were able to spend as long as they felt necessary to watch the scenes making up each lab. The labs could be watched with audio (an avatar narrating the written text presented in each lab scene) or with only the image of the scene showing the avatars and the text of their dialogue. At the beginning of the lab, each participant was informed that they could voluntarily choose how they wanted to watch (with audio vs. without audio). Group 1 and Group 2 saw the same lab, which included general health content (which can be found in Figure 1 in the paper by Sáiz-Manzanares et al. [32]). The participants in these groups had not previously seen the lab. Group 3 watched a lab that included specific biomedical engineering content (specifically referring to the functionality of using eye-tracking technology and its measurement indicators) (see Figure S1 in Supplementary Materials document). Participants in this group were able to see the lab before watching it again in the in the experimental room. After all three groups had watched the lab in the experimental room, a test was administered to test the knowledge acquired. The test was multiple choice with a single correct answer for each question. The test given to Group 1 and 2 is provided in the Supplementary Materials, Figure S2, and the test given to group 3 is in Figure S3. As the number of questions in the check tests were not equal in order to compare the answers, the answers were categorised into correct versus not correct answers.

2.4. Data Analysis

Table 3 presents the tests and instruments used to test the research objectives.

Table 3. Research objectives, counterfactuals, and analytical tools.

Research Objectives	Test	Test Instrument
Objective 1. To select characteristics with respect to different indicators of cognitive load related to the performance of simulation tasks in virtual environments.	Feature selection	WEKA v.3.8.5 [64]
Objective 2. To determine participants' levels of cognitive load (low, medium, and high).	Percentiles (25, 50, and 75)	SPSS v. 28 [65]
Objective 3. To apply supervised machine learning prediction techniques to determine how the variables of presentation type 1 (not having previously watched the lab vs. having done so) and presentation type 2 (visual–auditory vs. visual) from the virtual laboratories predict the indicators of cognitive load and learning results.	Multiple linear regression	SPSS v. 28 [65]

Table 3. *Cont.*

Research Objectives	Test	Test Instrument
Objective 3 (continued) To apply principal component analysis to determine the groupings of the cognitive load indicators.	Principal component analysis	SPSS v. 28 [65]
Objective 4. To apply supervised machine learning techniques for classification in order to determine the most representative classifiers of cognitive load.	Discriminant analysis KNN	SPSS v. 28 [65]
Objective 5. To apply unsupervised machine learning techniques to find possible groupings without a prior assignment variable with respect to cognitive load indicators and learning outcomes.	*k*-means	SPSS v. 28 [65]
Objective 6. To determine whether there are significant differences in the indicators of cognitive load and learning outcomes depending on the type of task and the way it is presented (visual–auditory vs. visual) in the participating groups.	Two-factor fixed-effect MANOVA	SPSS v. 28 [65]
Objective 7. To propose an analysis model for the indicators of cognitive load.	Structural equation modelling	AMOS v. 29 [66]

3. Results

3.1. Objective 1: To Select Characteristics with Respect to Different Indicators of Cognitive Load Related to the Performance of Simulation Tasks in Virtual Environments

WEKA algorithm [58] Evaluator, weka.attributeSelection.CfsSubsetEval, evaluates the value of a subset of attributes taking into account the individual predictive ability of each feature together with the degree of redundancy between features. More information is available at https://weka.sourceforge.io/doc.dev/weka/attributeSelection/CfsSubsetEval.html (accessed on 20 August 2024). The search is weka.attributeSelection.BestFirst; this searches the attribute subset space using the greedy hill climbing method augmented with a backtracking function. More information is available at https://weka.sourceforge.io/doc.dev/weka/attributeSelection/BestFirst.html (accessed on 20 August 2024). Forty-eight instances and 10 attributes were analysed (mean duration, fixation point X, fixation point Y, average pupil diameter, saccade direction, average velocity, peak velocity, saccade amplitude, average GSR, and SCR count). The results were the merit of best subset found, 0.75, and the features selected were the fixation point X and average GSR.

3.2. Objective 2: To Determine Cognitive Load Levels (Low, Medium, and High) in the Participating Groups

Descriptive statistics, maximum and minimum values, and percentiles were calculated for each of the characteristics. The 25th percentile indicates a low value, the 50th percentile an intermediate value, and the 75th percentile a high value, see Table 4.

Table 4. Descriptive statistics and percentiles for cognitive load indicators.

Indicators of Cognitive Load	Mean	SD	Minimum	Maximum	Percentiles		
					25	50	75
Group 1							
Mean duration	140.72	34.77	105.33	224.75	118.63	140.72	174.36
Fixation point X	994.84	288.42	750.66	1924.86	798.22	994.84	1227.70
Fixation point Y	866.17	195.53	635.68	1247.70	693.81	866.17	1048.09
Average pupil diameter	5311.03	1286.28	3715.67	7991.97	4338.36	5311.03	6558.44
Saccade direction	278,133.97	70,405.30	130,426.22	431,082.91	238,321.58	278,133.97	329,400.50
Average velocity	233,191.56	250,444.66	152,868.39	1,260,057.37	187,812.06	233,191.56	339,436.29
Peak velocity	336,697.42	260,723.55	230,621.84	1,398,529.00	261,396.24	336,697.42	493,273.13
Saccade amplitude	6173.42	2359.82	2020.85	10,407.48	5115.04	6173.42	8437.59
Average GSR	1.47	1.38	0.01	7.44	1.47	1.47	1.47
SCR Count	10.18	5.76	0.00	34.00	10.13	10.18	10.24

Table 4. *Cont.*

Indicators of Cognitive Load	Mean	SD	Minimum	Maximum	Percentiles		
					25	50	75
Group 2							
Mean duration	153.58	35.61	93.25	237.41	136.26	153.58	173.51
Fixation point X	1119.89	527.97	158.17	2424.80	988.74	1119.89	1276.17
Fixation point Y	951.40	438.04	162.18	2023.44	849.83	951.40	1132.69
Average pupil diameter	6371.50	4451.51	1132.41	19,140.53	4432.29	6371.50	7298.63
Saccade direction	319,974.78	115,718.13	41,797.90	397,630.64	267,580.33	319,974.78	375,132.14
Average velocity	229,531.47	94,148.06	49,784.31	382,706.45	199,910.41	229,531.47	266,353.23
Peak velocity	331,430.91	132,544.18	64,173.60	537,397.30	272,005.44	331,430.91	398,652.17
Saccade amplitude	7372.37	2653.55	989.92	9646.79	5714.81	7372.37	8285.77
Average GSR	1.10	1.12	0.01	3.17	0.01	1.10	1.99
SCR count	8.00	11.33	0.00	38.00	0.00	8.00	16.00
Group 3							
Mean duration	250.43	54.94	134.61	399.59	239.68	250.43	286.75
Fixation point X	2919.06	1260.15	1162.95	6185.12	2068.23	2919.06	4091.90
Fixation point Y	2643.52	1307.08	959.91	6157.44	2225.84	2643.52	3557.76
Average pupil diameter	25,055.68	7498.00	11,358.53	40,122.79	19,702.44	25,055.68	29,343.05
Saccade direction	105,956.02	24,936.85	51,991.70	149,285.16	89,490.71	105,956.02	120,311.50
Average velocity	84,465.25	18,890.28	51,017.96	121,059.84	71,890.70	84,465.25	97,068.89
Peak velocity	105,569.23	24,746.30	64,997.00	152,861.24	95,035.23	105,569.23	127,999.83
Saccade amplitude	2006.29	521.67	1300.68	2952.87	1785.50	2006.29	2645.64
Average GSR	0.86	4.11	0.01	14.11	0.18	0.86	4.75
SCR count	11.00	12.14	0.00	39.00	6.00	11.00	24.50

Note: SD = standard deviation; GSR = galvanic skin response, SCR = skin conductance responses.

3.3. Objective 3: To Apply Supervised Machine Learning Prediction Techniques to Determine How the Variables of Presentation Type 1 (Not Having Previously Watched the Lab vs. Having Done So) and Presentation Type 2 (Visual–Auditory vs. Visual) from the Virtual Laboratories Predict the Indicators of Cognitive Load and Learning Results

The presentation type 1 variable (not having previously watched the lab vs. having done so) predicted 76.3% (R^2 = 0.763) of the variance with respect to the cognitive load indicators (see Supplementary Materials, Table S1 for the results of the linear regression analysis). The values of tolerance—the proportion of variance not explained by the rest of the variables (1-R^2), where the higher the Tolerance, the more independent the variable in question—were all less than 1. The variance inflation (VIF) values—which quantifies the intensity of multicollinearity in a normal least squares regression analysis, and the higher the variance of the estimated regression coefficient, the higher the collinearity—were high for the cognitive load indicators average pupil diameter, saccade direction and average velocity.

The presentation type 2 variable (visual–auditory vs. visual) predicted 25.5% (R^2 = 0.255) of the variance with respect to the cognitive load indicators (see Supplementary Materials, Table S2 for the results of the linear regression analysis). In this case, very low values of tolerance were found for the cognitive load indicators fixation point Y, average pupil diameter, saccade direction and average velocity, while there were very high VIF values for the indicators average pupil diameter, saccade direction and peak velocity.

Based on these results, which point to collinearity between the variables, a principal component analysis was carried out to group the cognitive load levels into the most relevant load components (see Supplementary Materials, Figure S4). Three principal components were found (see Supplementary Materials, Table S3 for the matrix of principal components). Component 1 comprised the cognitive load indicators mean duration, fixation point X, fixation point Y, average pupil diameter and saccade direction; Component 2 comprised the indicators average velocity and peak velocity; while Component 3 comprised the indicators average GSR and SCR count.

3.4. Objective 4: To Apply Supervised Machine Learning Classification Techniques to Identify the Most Representative Classifiers of Cognitive Load

A discriminant analysis was carried out to determine the behaviours of the groups with respect to the cognitive load indicators. As Figure 3 shows, the behaviour in cognitive load for Group 1 and Group 2, where the students had not previously watched the lab (which contained general health content), was quite similar to each other, but differed from the behaviour of Group 3, where the students had previously viewed the lab (which was about content specific to the degree). Similarly, all the values of Wilks' lambda (a method for selecting variables by discriminant analysis steps that selects the variables to be added into the equation based on how much they contribute to decreasing Wilks' lambda—at each step, the variable that minimises the overall Wilks' Lambda is added) for the cognitive load indicators was significant except for the average GSR and SCR count indicators (see Table 5).

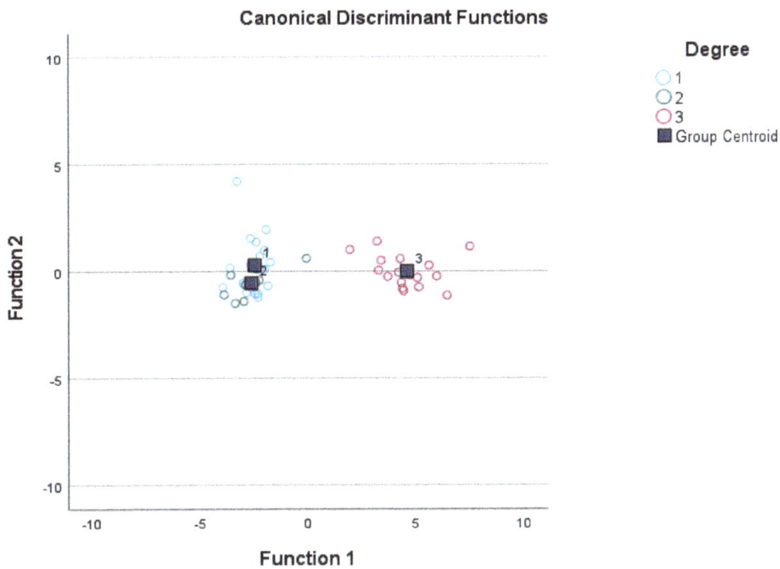

Figure 3. Canonical discriminant functions with respect to the group type variable.

Table 5. Test of equality of group means.

Cognitive Load Indicators	Lambda Wilks	F	df1	df2	pLambda
Mean duration	0.36	40.10	2	45	<0.001 *
Fixation point X	0.38	36.12	2	45	<0.001 *
Fixation point Y	0.39	35.28	2	45	<0.001 *
Average pupil diameter	0.20	92.56	2	45	<0.001 *
Saccade direction	0.35	41.44	2	45	<0.001 *
Average velocity	0.71	9.04	2	45	<0.001 *
Peak velocity	0.62	13.91	2	45	<0.001 *
Saccade amplitude	0.42	31.26	2	45	<0.001 *
Average GSR	0.93	1.79	2	45	0.18
SCR count	0.94	1.43	2	45	0.25

* $p < 0.05$. Note: df = degree of freedom; GSR = galvanic skin response, SCR = skin conductance response.

The KNN algorithm was also applied, giving the most representative cognitive load indicators as fixation point X, fixation point Y, and mean duration.

Figure 4 shows the representation of the focal points with respect to the fixation point X classifier and the distribution of subjects with respect to the other indicators of cognitive load. Figure 5 shows the representation with respect to the fixation point Y classifier and Figure 6 with respect to the mean duration classifier. The numbers in the focal graphs indicate the participant. Thus, it is possible to know the position of each participant with respect to the classifiers and the focal points. This fact is important for being able to design specific tasks for each of them.

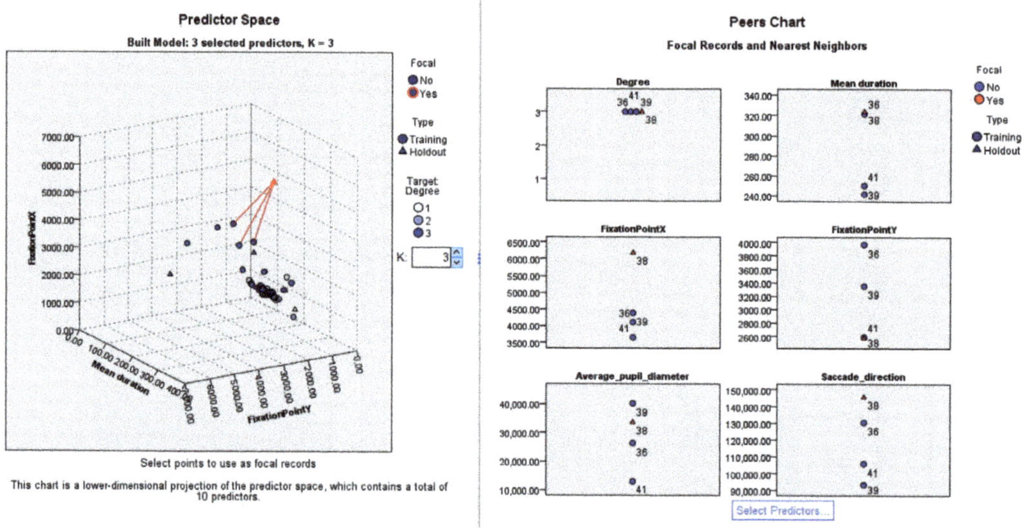

Figure 4. Analysis of the focal points with respect to the fixation point X classifier.

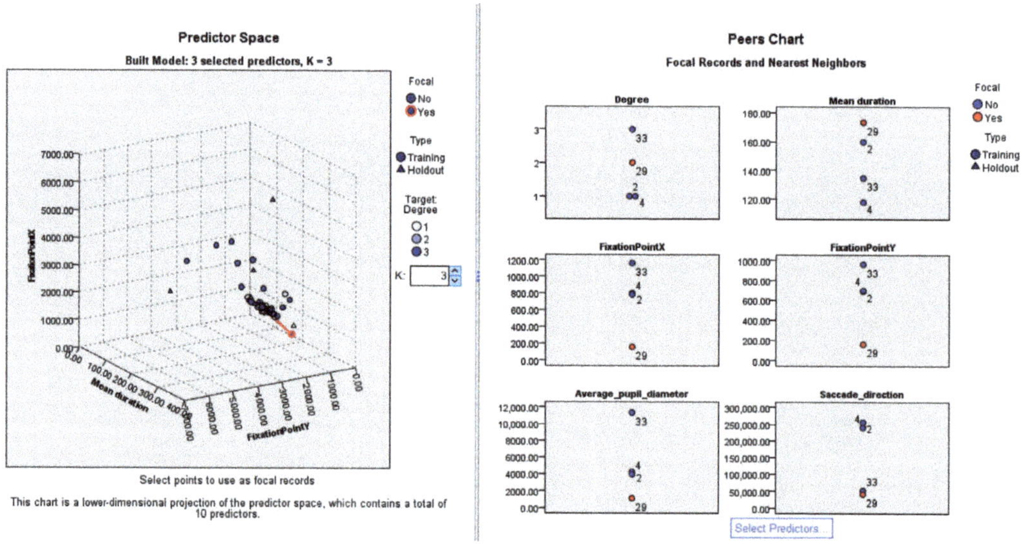

Figure 5. Analysis of the focal points with respect to the fixation point Y classifier.

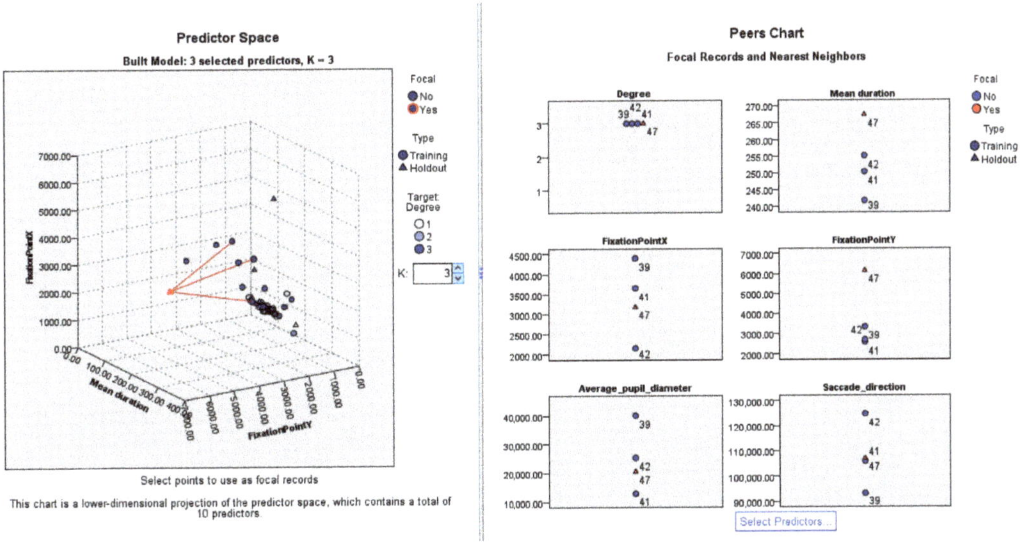

Figure 6. Analysis of the focal points with respect to the mean duration classifier.

3.5. Objective 5: Apply Unsupervised Machine Learning Techniques to Determine Possible Groupings Without a Prior Assignment Variable with Respect to Cognitive Load Indicators and Learning Outcomes

Three clusters were found for the indicators of cognitive load and for the results in the learning check test (see Table 6). ANOVA was also performed to test whether there were significant differences between the clusters in the cognitive load indicators and in the test results (see Table 7). Significant differences were found in all cognitive load indicators except average GSR and SCR count. No significant differences were found in the outcome test.

Table 6. Final cluster centres of *k*-means when $k = 3$ is used.

Indicators of Cognitive Load	Cluster 1 (n = 22)	Cluster 2 (n = 25)	Cluster 3 (n = 1)
Mean duration	150.58	225.34	172.44
Fixation point X	1165.44	2315.30	994.84
Fixation point Y	997.23	2190.93	921.50
Average pupil diameter	6178.29	17,931.22	6156.46
Saccade direction	322,067.82	140,859.37	130,426.22
Average velocity	298,529.77	109,142.61	1,260,057.37
Peak velocity	406,223.39	147,998.56	1,398,529.00
Saccade amplitude	7562.57	3058.55	2861.72
Average GSR	1.64	2.31	1.47
SCR count	11.54	12.86	10.15
Results test	1	1	1

Note: GSR = Galvanic Skin Response, SCR = Skin Conductance Responses.

Table 7. Cluster ANOVA on indicators (cognitive load and test results).

	Cluster		Error		F	*p*
	Mean Square	df	Mean Square	df		
Mean duration	32,856.22	2	3367.96	45	9.756	<0.001 *
Fixation point X	8,035,738.38	2	1,267,468.79	45	6.340	0.004 *
Fixation point Y	8,584,531.38	2	1,266,594.11	45	6.778	0.003 *
Average pupil diameter	827,482,721.38	2	72,932,182.36	45	11.346	<0.001 *

Table 7. *Cont.*

	Cluster		Error		F	p
	Mean Square	df	Mean Square	df		
Saccade direction	196,570,633,958.60	2	4,242,830,417.24	45	46.330	<0.001 *
Average velocity	762,312,710,583.10	2	7,488,341,771.28	45	101.800	<0.001 *
Peak velocity	1,014,921,268,496.76	2	8,945,463,462.81	45	113.457	<0.001 *
Saccade amplitude	121,297,203.11	2	3,563,210.06	45	34.042	<0.001 *
Average GSR	2.76	2	7.55	45	0.365	0.696
SCR count	12.45	2	100.40	45	0.124	0.884
Results test	0.44	2	1.16	45	0.378	0.687

* Note: df = degree of freedom; GSR = galvanic skin response, SCR = skin conductance responses.

Cluster 2 exhibits a longer mean duration compared to the virtual lab display. The mean pupil diameter is also larger, the mean velocity is lower, and the saccade amplitude and maximum velocity have intermediate values.

Cross-tabulation was then performed to test the relationship between cluster membership and group of origin (see Table 8). A contingency coefficient of $C = 0.60$ $p < 0.001$ was found. All members of Group 3 are in Cluster 2. Members of Group 1 and 2 are distributed between Clusters 1 and 2.

Table 8. Cluster number of cases and degree type cross-tabulation.

Cluster Number of Cases		Degrees			Total
		1	2	3	
	1	13	9	0	22
	2	7	1	17	25
	3	1	0	0	1
Total		21	10	17	48

3.6. Objective 6: To Determine Whether There Are Significant Differences in the Indicators of Cognitive Load and Test Results Depending on the Type of Task and the Way the Task Is Presented (Visual–Auditory vs. Visual) in the Participation Groups

First, the normality of the sample was examined with respect to the scores on the ACRA metacognitive strategies scale, a test that was applied before the intervention phase in the three groups. Because the sample size was less than 50 (n = 48), the Shapiro–Wilk test was applied, which, as Table 9 shows, indicated a normal distribution. Therefore, parametric statistics were applied to test Objective 6, which in this case was a two-factor multivariate analysis (MANOVA) with fixed effects and eta-squared effect size (η^2) (presentation type 1 (not having previously watched the lab vs. having done so) and presentation type 2 (visual–auditory vs. visual) (see Table 10). Significant differences from for presentation type 1 (prior watching vs. not) were found for all cognitive load indicators except for the average GSR and SCR count. The effect value was small for the cognitive load indicators average velocity, fixation point X, fixation point Y, and peak velocity, intermediate for the mean duration, saccade direction and saccade amplitude, and high for the average pupil diameter. No differences were found in the outcome test. No significant differences were found for presentation type 2 (visual–auditory vs. visual) or for the interaction of the two factors (presentation type 1 and presentation type 2). Descriptive statistics can be found in the Supplementary Materials in Table S4.

Table 9. Shapiro–Wilk normality test.

Metacognitive Strategies	Groups	Shapiro–Wilk test		
		Statistic	df	p
Self-knowledge	1	0.96	21	0.53
	2	0.90	10	0.23
	3	0.91	17	0.10
Planning	1	0.92	21	0.08
	2	0.86	10	0.08
	3	0.89	17	0.04
Self-Evaluation	1	0.91	21	0.05
	2	0.97	10	0.86
	3	0.791	17	0.09

$p < 0.05$. Note: df = degree of freedom.

Table 10. Multivariate two-factor fixed effects analysis.

Source	Indicators	F	df	p	η^2
Display 1 (no preview vs. display)	Mean duration	25.17	2	0.00 *	0.55
	Fixation point X	17.25	2	0.00 *	0.45
	Fixation point Y	16.90	2	0.00 *	0.45
	Average pupil diameter	54.18	2	0.00 *	0.72
	Saccade direction	30.31	2	0.00 *	0.59
	Average velocity	6.79	2	0.00 *	0.24
	Peak velocity	10.48	2	0.00 *	0.33
	Saccade amplitude	21.68	2	0.00 *	0.51
	Average GSR	0.31	2	0.74	0.01
	SCR count	0.32	2	0.73	0.02
	Results test	0.21	2	0.81	0.01
Visualisation 2 (visual–auditory vs. visual)	Mean duration	0.37	1	0.55	0.01
	Fixation point X	2.21	1	0.15	0.05
	Fixation point Y	2.26	1	0.14	0.05
	Average pupil diameter	0.40	1	0.53	0.01
	Saccade direction	2.27	1	0.14	0.05
	Average velocity	1.35	1	0.25	0.03
	Peak velocity	1.07	1	0.31	0.02
	Saccade amplitude	1.30	1	0.26	0.03
	Average GSR	1.62	1	0.21	0.04
	SCR count	1.54	1	0.22	0.04
	Results test	0.08	1	0.77	0.00
Interaction Visualisation 1 * Visualisation 2	Mean duration	0.90	2	0.41	0.04
	Fixation point X	2.97	2	0.06	0.12
	Fixation point Y	2.21	2	0.12	0.10
	Average pupil diameter	0.52	2	0.60	0.02
	Saccade direction	0.70	2	0.50	0.03
	Average velocity	1.79	2	0.18	0.08
	Peak velocity	2.01	2	0.15	0.09
	Saccade amplitude	0.58	2	0.56	0.03
	Average GSR	0.57	2	0.57	0.03
	SCR count	0.25	2	0.78	0.01
	Results test	0.10	2	0.90	0.00

* $p < 0.05$. Note. GSR = galvanic skin response, SCR = skin conductance responses; df = degree of freedom; η^2 = eta-squared effect size, where values were interpreted according to Cohen [67]: below 0.20 is no effect, between 0.21 and 0.49 is a small effect, between 0.50 and 0.69 is a moderate effect, and $\geq$0.7 is a large effect.

3.7. Objective 7: To Propose a Model of Analysis of the Markers of Cognitive Load

The results found in testing Objectives 1–6 are summarised in Table 11.

We propose a model for analysing cognitive load indicators that considers the following components. Component 1 includes the cognitive load indicators of position, durations and amplitudes of fixations and saccades, and pupil diameter; Component 2 includes cognitive load indicators of velocity; Component 3 includes indicators of psychogalvanic response and skin conductance; and Component 4 includes the pre-task presentation characteristics and single or multi-channel presentation. A representation of the model is shown in Figure S5. A goodness-of-fit index analysis was then performed to check the fit of the model. The proposed model obtained a better fit in the different fit indices than the stand-alone model (see Table 12). However, the fit was not optimal in some indices, such as RMSEA. Figure 7 shows a representation of the model.

Table 11. Results with respect to the research objectives.

Research Objectives	Results
Objective 1. To select characteristics with respect to different indicators of cognitive load related to performance of simulation tasks in virtual environments.	Feature selection Fixation point X, average GSR
Objective 2. To determine participants' levels of cognitive load (low, medium, and high).	Low $P_{25–49}$ Medium $P_{50–74}$ High $P_{75–99}$
Objective 3. To apply supervised machine learning prediction techniques to determine how the variables of presentation type 1 (not having previously watched the lab vs. having done so) and presentation type 2 (visual–auditory vs. visual) from the virtual laboratories predict the indicators of cognitive load and learning results.	Presentation type 1 [form of task presentation (generalist vs. specific content) and preview vs. no preview] explains 76.3% of the results for the cognitive load indicators. Presentation type 2 [form of task display (visual–audio vs. audio)] explains 25.5% of the results for the cognitive load indicators. However, some of the cognitive load indicators may accumulate redundant information. Three main components were found. Component 1 was related to indicators of duration, position, pupil diameter, saccade direction, and saccade amplitude. Component 2 is related to indicators of velocity, and Component 3 is related to indicators of conductance.
Objective 4. To apply supervised machine learning techniques for classification in order to determine the most representative classifiers of cognitive load.	Differences were found for all indicators of cognitive load depending on the way the task was presented, except for those related to skin conductance. Likewise, the classifiers found were related to the gaze position at point X and Y, and the average duration of cognitive load.
Objective 5. To apply unsupervised machine learning techniques to find possible groupings without a prior assignment variable with respect to cognitive load indicators and learning outcomes.	The clusters found showed greater homogeneity in the group that was able to watch the information before the test. Differences were found in all indicators of cognitive load, except those related to skin conductance and in the results of the knowledge test.
Objective 6. To determine whether there are significant differences in the indicators of cognitive load and learning outcomes depending on the type of task and the way it is presented (visual–auditory vs. visual) in the participating groups.	Significant differences were found with respect to the way the task was presented in all cognitive load indicators, except those related to skin conductance; no differences were found in the results of the knowledge test. No significant differences were found for the display type variable.

Table 12. Standardised estimates of the default model.

Goodness-of-Fit Index	Default Model 1	Independence Model 1	Accepted Value
χ^2	85.98, df = 38 p = 0.00 *.	-	$p > 0.05$ $\alpha = 0.95$
RAMSEA	0.16	0.45	>0.05–0.08
RAMSEA intervals (LO 90)	0.12	0.42	>0.05–0.08
SRMR	0.06	-	>0.05–0.08
NFI	0.85	0.00	0.85–0.90<
TLI	0.90	0.00	0.85–0.90<
IFC	0.90	0.00	0.95–0.97<
AIC	163.98	623.68	The lowest value
ECVI	3.49	13.27	The lowest value
ECVI interval (90%)	2.99	11.70	The lowest value

* $p < 0.05$. Note. df = degree of freedom; χ^2 = Chi-squared; RMSEA = root-mean-square error of approximation; SRMR = standardised root-mean-square Residual; NFI = normed fit index; TLI = Tucker–Lewis index; CFI = comparative fit index; AIC = Akaike information criterion; ECVI = parsimony index.

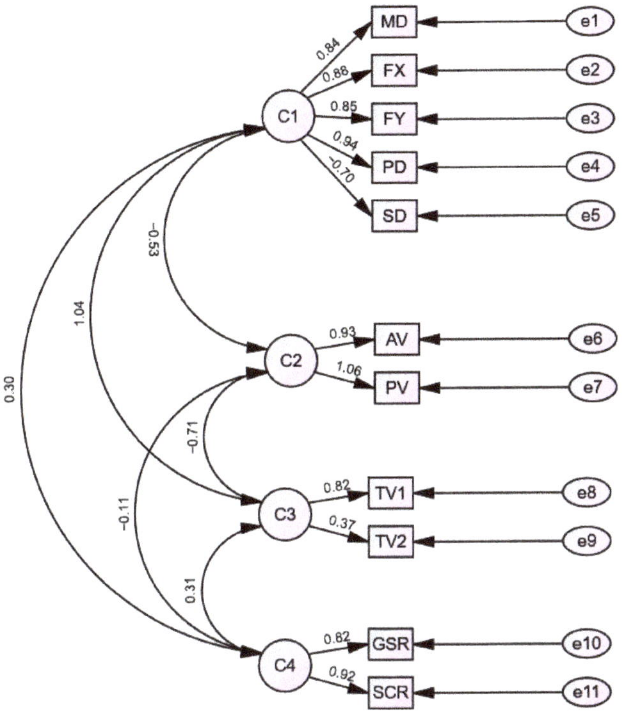

Figure 7. Standardised estimates. Note: Component 1 (C1) includes the following indicators of cognitive load: MD = Mean duration, FX = Fixation point X, FY = Fixation point Y, PD = Average pupil diameter, and SD = Saccade direction. Component 2 (C2) includes the indicators AV = Average velocity and PV = Peak velocity; Component 3 (C3) includes AV = preview, and PV = type of presentation (audio and text vs. audio); and Component 4 (C4) includes the Average GSR and SCR count.

4. Discussion

This study found that the intrinsic cognitive load experienced by students seems to be related to whether they are able to see the task beforehand and to the use of specific learning tasks. However, these conditions did not translate into differences in learning outcomes. Furthermore, these conditions also produced differences with respect to the germane cognitive load, which refers to the construction and automatisation of cognitive schemas in LTM [8,13,17]. Seeing the task beforehand and task-specific content may be related to how the information is processed and the learner's engagement with the task [18,19,21]. One possible explanation is that facilitating knowledge of the information enhances placement of the information in the learner's Proximate Development Zone (PDZ), which helps to improve information processing [17]. At this point, it is assumed that each learner has their own way of processing information [18]; however, by allowing them to see the task before attempting it in the in the simulation lab, it seems that the way information was processed was somehow equalised, with this fact having more explanatory weight than the intrinsic cognitive load that the student might have experienced [20]. The explanation may be related to the fact that the learners' being able to see the information beforehand may help to eliminate redundant information [20] which aids in processing information in working memory and long-term memory [8,9,13,17]. However, this hypothesis will need to be tested in future studies with different tasks and in different groups.

We also found that whether information was presented through one or two channels had an effect on the indicators of cognitive load. This result may be related to the possibility given to students in all groups to voluntarily choose how the task was presented, which may boost the personalisation of the learning process and increase student motivation [18,19,21] as it respects individuality in the way information is processed.

On the other hand, it seems that some of the indicators of cognitive load considered in this study were measuring the same thing with respect to information processing. We found three main components in this study: one related to the cognitive load indicators of duration, position, pupil diameter, saccade direction and saccade amplitude; another related to velocity; and a third related to the psychogalvanic response and skin conductance, i.e., a more emotional component. This result opens the way for future research aimed at studying indicators of cognitive load in terms of explanation from the perceptual and neurological functioning of information processing in different tasks.

Finally, we proposed an explanatory model of cognitive functioning in the execution of simulation tasks. This is intended to highlight the need for applied research to test the generalisability of the research results. More specifically, in this study, we used a data fusion model [56] to analyse the different indicators (cognitive load and type of presentation) and the fit of the results was checked using a goodness-of-fit index. However, future studies will be able to expand on the number of cognitive load indicators, including electroencephalographic recordings; the characteristics of the tasks; and the application of other more powerful models for data processing that include generative artificial intelligence, such as the Transformer model [57–59].

5. Conclusions

Firstly, the results from this study should be considered with caution, as we worked with samples from a single university selected through convenience sampling. In addition, the students voluntarily agreed to participate without receiving any financial compensation, which is a bias in itself. However, the analysis of human processing in learning environments is complex and does not allow for brief tests with very large samples. Therefore, the way to increase generalisability is to replicate research in other contexts with other participants, which will undoubtedly improve the generalisability of the results.

Secondly, bearing in mind that caution is needed when interpreting the results, the study does raise a number of unknowns for future research in this field. It seems that being able to preview tasks in virtual simulation contexts is an important element for homogenising the intrinsic load and increasing the germane load. One possible explanation is that this may help the learner to eliminate the redundant load and improve the processing of information in working memory and long-term memory, which may help increase learning engagement and thus motivation. Another key aspect is to facilitate the learner's choice of information display channel(s). This respects each learner's type of processing (auditory, visual, or both). Therefore, the design of the tasks and the application of the tasks seem to be important elements in improving information processing.

Finally, proposing explanatory models is important. These models must be analysed in terms of their fit and reliability indicators. This is undoubtedly a challenge for developmental educational psychology and cognitive instructional psychology. At this point, the use of generative artificial intelligence similar to what is used in the Transformer models will undoubtedly be of great help in this field of research. Therefore, we stress the importance of working with interdisciplinary teams that include professionals from educational and cognitive psychology, computer science, and information technology in order to effectively address these challenges.

In summary, using characteristic selection techniques can offer researchers important information that helps them understand which of the variables (features in data mining terminology) may be more significant in the study they are conducting. Moreover, an analysis of percentiles with regard to the metrics recorded by multi-channel eye-tracking devices can help researchers to categorise the different types of users (or instances in data mining parlance). Applying supervised machine learning regression techniques gives information about percentage predictions of numerical values, in this case values related to the metrics from multi-channel eye-tracking and GSR devices. Along similar lines, supervised machine learning classification techniques give researchers information about the prediction of previously defined categories. In contrast, non-supervised machine learning clustering techniques help identify hidden patterns, as they do not apply prior labels—unlike supervised learning. Using statistical techniques such as MANOVA allows researchers to determine whether there are significant differences in defined independent variables between the groups being tested. All of these results allow researchers to interpret and develop explanatory models, in this case about information processing during a learning task in a self-regulated virtual lab. Finally, using goodness-of-fit indices gives researchers an idea of how well the proposed model fits.

Supplementary Materials: The following supporting information can be downloaded at https://www.mdpi.com/article/10.3390/app14219831/s1.

Author Contributions: Conceptualisation, all authors; methodology, M.C.S.-M., R.M.-S. and J.S.-G.; software, all authors; validation, all authors; formal analysis, M.C.S.-M., R.M.-S. and J.S.-G.; investigation, M.C.S.-M., J.S.-G. and I.G.-D.; resources, M.C.S.-M.; data curation, M.C.S.-M.; writing—original draft preparation, M.C.S.-M.; writing—review and editing, all authors; visualisation, M.C.S.-M.; supervision, all authors; project administration, M.C.S.-M. and R.M.-S.; funding acquisition, M.C.S.-M. and R.M.-S. All authors have read and agreed to the published version of the manuscript.

Funding: The project "Voice assistants and artificial intelligence in Moodle: a path towards a smart university", SmartLearnUni, Call 2020 R&D&I Projects-RTI Type B, MINISTRY OF SCIENCE AND INNOVATION AND UNIVERSITIES, STATE RESEARCH AGENCY, Government of Spain, grant number PID2020-117111RB-I00". Specifically, in the part concerning the application of multichannel eye-tracking technology with university students and the project "Specialized and updated training on supporting advance technologies for early childhood education and care professionals and graduates" (eEarlyCare-T), grant number 2021-1-ES01-KA220-SCH-000032661 funded by the EUROPEAN COMMISSION. In particular, the funding has enabled the development of the e-learning classroom and educational materials.

Institutional Review Board Statement: This study was conducted in accordance with the Declaration of Helsinki and approved by the Institutional Ethics Committee of University of Burgos (protocol code IO 03/2022 and date of approval 7 April 2022 for the SmartLearnUni project and protocol code IO 04/2022 and date of approval 7 April 2022 for the eEarlyCare-T project).

Informed Consent Statement: Written informed consent was obtained from all subjects involved in this study.

Data Availability Statement: The data presented in this study are available on request from the corresponding author. The data are not publicly available due to the confidentiality of the data and for ethical reasons and to check the use of the data.

Conflicts of Interest: The authors declare no conflicts of interest.

References

1. Brünken, R.; Seufert, T.; Paas, F. Measuring Cognitive Load. In *Cognitive Load Theory*; Plass, J.L., Moreno, R., Brünken, R., Eds.; Cambridge University Press: Cambridge, UK, 2010; pp. 181–202. [CrossRef]
2. Eggemeier, F.T.; Wilson, G.F.; Kramer, A.F.; Damos, D.L. Workload assessment in multi-task environments. In *Multiple-Task Performance*; Damos, D.L., Ed.; Taylor & Francis: London, UK; Washington, DC, USA, 1991; pp. 207–216. [CrossRef]
3. Johannsen, G. Workload and Workload Measurement. In *Mental Workload*; NATO Conference Series; Moray, N., Ed.; Springer: Boston, MA, USA, 1979; Volume 8, pp. 3–11. [CrossRef]
4. Sevcenko, N.; Appel, T.; Ninaus, M.; Moeller, K.; Gerjets, P. Theory-based approach for assessing cognitive load during time-critical resource-managing human-computer interactions: An eye-tracking study. *J. Multimodal User Interfaces* **2023**, *17*, 1–19. [CrossRef]
5. Kirschner, P.A.; Sweller, J.; Kirschner, F.; Zambrano, R.J. From Cognitive Load Theory to Collaborative Cognitive Load Theory. *Intern. J. Comput.-Support. Collab. Learn.* **2018**, *13*, 213–233. [CrossRef]
6. Sweller, J. The effect of task complexity and sequence on rule learning and problem solving. *Br. J. Psychol.* **1976**, *67*, 553–558. [CrossRef]
7. Sáiz-Manzanares, M.C.; Marticorena-Sánchez, R.; Martín-Antón, L.J.; González-Diez, I.; Carbonero-Martín, I. Using eye tracking technology to analyse cognitive load in multichannel activities in university students. *Int. J. Hum.-Comput. Interact.* **2023**, *40*, 3263–3328. [CrossRef]
8. Sweller, J. Cognitive load during problem solving: Effects on learning. *Cogn. Sci.* **1998**, *12*, 257–285. [CrossRef]

9. Sweller, J.; Chandler, P.; Tierney, P.; Cooper, M. Cognitive load as a factor in the structuring of technical material. *J. Exp. Psychol. Gen.* **1990**, *119*, 176–192. [CrossRef]
10. Mayer, R.E. *Multimedia Learning*; Cambridge University Press: New York, NY, USA, 2001. [CrossRef]
11. Tindall-Ford, S.; Chandler, P.; Sweller, J. When two sensory modes are better than one. *J. Exp. Psychol. Appl.* **1997**, *3*, 257–287. [CrossRef]
12. Baddeley, A. Working memory. *Science* **1992**, *255*, 556–559. [CrossRef]
13. Schnotz, W.; Kürschner, C.A. Reconsideration of Cognitive Load Theory. *Educ. Psychol. Rev.* **2007**, *19*, 469–508. [CrossRef]
14. Miller, G.A. The magical number seven, plus or minus two: Some limits on our capacity for processing information. *Psychol. Rev.* **1956**, *63*, 81–97. [CrossRef]
15. Peterson, L.; Peterson, M. Short-term retention of individual verbal items. *J. Exp. Psychol.* **1959**, *58*, 193–198. [CrossRef]
16. Marcus, N.; Cooper, M.; Sweller, J. Understanding instructions. *J. Educ. Psychol.* **1996**, *88*, 49–63. [CrossRef]
17. Vygotski, L.S. Learning and mental development at school age. In *Educational Psychology in the U.S.S.R.*; Simon, B.J., Simon, J., Eds.; Routledge & Kegan Paul: London, UK, 1963; pp. 21–34.
18. Sweller, J. Cognitive load theory and individual differences. *Learn. Individ. Differ.* **2024**, *110*, 102423. [CrossRef]
19. Ahmad, M.I.; Keller, I.; Robb, D.A.; Lohan, K.S. A framework to estimate cognitive load using physiological data. *Pers. Ubiquitous Comput.* **2023**, *27*, 2027–2041. [CrossRef]
20. Lee, Y.; Szulewski, A.; Young, J.Q.; Donkers, J.; Jarodzka, H.; Van Merriënboer, J.J.G. The medical pause: Importance, processes, and training. *Med. Educ.* **2021**, *55*, 1152–1160. [CrossRef]
21. Seufert, T. The interplay between self-regulation in learning and cognitive load. *Educ. Res. Rev.* **2018**, *24*, 116–129. [CrossRef]
22. Holmqvist, K.; Nyström, M.; Andersson, R.; Dewhurst, R.; Jarodzka, H.; Weijer, J.V.D. *Eye Tracking: A Comprehensive Guide to Methods and Measures*; Oxford University Press: Oxford, UK, 2011.
23. Scheutz, M.; Aeron, S.; Aygun, A.; de Ruiter, J.P.; Fantini, S.; Fernandez, C.; Haga, Z.; Nguyen, T.; Lyu, B. Estimating Systemic Cognitive States from a Mixture of Physiological and Brain Signals. *Top. Cogn. Sci.* **2024**, *16*, 485–526. [CrossRef]
24. Mayer, R.E. Unique contributions of eye-tracking research to the study of learning with graphics. *Learn. Instr.* **2010**, *20*, 167–171. [CrossRef]
25. Mutlu-Bayraktar, D.; Ozel, P.; Altindis, F.; Yilmaz, B. Relationship between objective and subjective cognitive load measurements in multimedia learning. *Interact. Learn. Environ.* **2020**, *31*, 1322–1334. [CrossRef]
26. Koć-Januchta, M.; Höffler, T.; Thoma, G.-B.; Prechtl, H.; Leutner, D. Visualizers versus verbalizers: Effects of cognitive style on learning with texts and pictures—An eye-tracking study. *Comput. Hum. Behav.* **2017**, *68*, 170–179. [CrossRef]
27. Duchowski, A.T.; Krejtz, K.; Gehrer, N.A.; Bafna, T.; Bækgaard, P. The low/high index of pupillary activity. In Proceedings of the CHI Conference on Human Factors in Computing Systems, Honolulu, HI, USA, 25–30 April 2020; pp. 1–12.
28. Critchley, H.D. Electrodermal responses: What happens in the brain. *Neuroscientist* **2002**, *8*, 132–142. [CrossRef] [PubMed]
29. Winter, M.; Pryss, R.; Probst, T.; Reichert, M. Towards the applicability of measuring the electrodermal activity in the context of process model comprehension: Feasibility study. *Sensors* **2020**, *20*, 4561. [CrossRef] [PubMed]
30. Kramer, A.F. Physiological metrics of mental workload. In *Multiple Task Performance*; CRC Press: Boca Raton, FL, USA, 1991; pp. 279–328.
31. Abbad-Andaloussi, A.; Burattin, A.; Slaats, T.; Kindler, E.; Weber, B. Complexity in declarative process models: Metrics and multi-modal assessment of cognitive load. *Expert Syst. Appl.* **2023**, *233*, 120924. [CrossRef]
32. Sáiz-Manzanares, M.C.; Marticorena-Sánchez, R.; Escolar-Llamazares, M.C.; González-Díez, I.; Martín Antón, L.J. Using integrated multimodal technology: A way to personalised learning in Health Sciences and Biomedical engineering Students. *Appl. Sci.* **2024**, *14*, 7017. [CrossRef]
33. Liberman, L.; Dubovi, I. The effect of the modality principle to support learning with virtual reality: An eye-tracking and electrodermal activity study. *J. Comput. Assist. Learn.* **2023**, *39*, 547–557. [CrossRef]
34. Li, Q.; Luximon, Y.; Zhang, J.; Song, Y. Measuring and classifying students' cognitive load in pen-based mobile learning using handwriting, touch gestural and eye-tracking data. *Br. J. Educ. Technol.* **2024**, *55*, 625–653. [CrossRef]
35. Raftery, A.E. Use and communication of probabilistic forecasts. Statistical Analysis and Data Mining: The ASA. *Data Sci. J.* **2016**, *9*, 397–410. [CrossRef]
36. Sáiz-Manzanares, M.C. Metacognition and Artificial Intelligence: Beyond the Parallelism of Functioning [Metacognición e Inteligencia Artificial: Más allá del Paralelismo de Funcionamiento]. Ph.D. Thesis, University of Burgos, Burgos, Spain, 2019. Available online: http://hdl.handle.net/10259/5357 (accessed on 12 August 2024).
37. Gogna, Y.; Tiwari, S.; Singla, R. Evaluating the performance of the cognitive workload model with subjective endorsement in addition to EEG. *Med. Biol. Eng. Comput.* **2024**, *62*, 2019–2036. [CrossRef]
38. Gogna, J.; Tiwari, S.; Singla, R. Mental workload assessment of gamers' eeg with multi-domain feature-based cognitive model and its validation. *Biomed. Eng. Appl. Basis Commun.* **2024**, *36*, 2450022. [CrossRef]
39. Sazuka, N.; Katsumata, K.; Komoriya, Y.; Oba, T.; Ohira, H. Association of brain-autonomic activities and task accuracy under cognitive load: A pilot study using electroencephalogram, autonomic activity measurements, and arousal level estimated by machine learning. *Front. Hum. Neurosci.* **2024**, *18*, 1272121. [CrossRef]
40. Minissi, M.E.; Altozano, A.; Marín-Morales, J.; Giglioli, I.A.C.; Mantovani, F.; Alcañiz, M. Biosignal comparison for autism assessment using machine learning models and virtual reality. *Comput. Biol. Med.* **2024**, *171*, 108194. [CrossRef] [PubMed]

41. Karmakar, S.; Kamilya, S.; Dey, P.; Guhathakurta, P.K.; Dalui, T.M.; Bera, T.K.; Halder, S.; Koley, C.; Pal, T.; Basu, A. Real time detection of cognitive load using fNIRS: A deep learning approach. *Biomed. Signal Process. Control* **2023**, *80*, 104227. [CrossRef]
42. Shaposhnyk, O.; Yanushkevich, S.; Babenko, V.; Chernykh, M.; Nastenko, I. Inferring Cognitive Load Level from Physiological and Personality Traits. In Proceedings of the 2023 International Conference on Information and Digital Technologies (IDT), Zilina, Slovakia, 20–22 June 2023; pp. 233–242. [CrossRef]
43. Murdaca, G.; Banchero, S.; Casciaro, M.; Tonacci, A.; Billeci, L.; Nencioni, A.; Pioggia, G.; Genovese, S.; Monacelli, F.; Gangemi, S. Potential Predictors for Cognitive Decline in Vascular Dementia: A Machine Learning Analysis. *Processes* **2022**, *10*, 2088. [CrossRef]
44. Sáiz-Manzanares, M.C.; Escolar-Llamazares, M.-C.; Arnaiz González, Á. Effectiveness of Blended Learning in Nursing Education. *Int. J. Environ. Res. Public Health* **2020**, *17*, 1589. [CrossRef]
45. Sáiz-Manzanares, M.C.; Ramos Pérez, I.; Arnaiz-Rodríguez, Á.; Rodríguez-Arribas, S.; Almeida, L.; Martin, C.F. Analysis of the learning process through eye tracking technology and feature selection techniques. *Appl. Sci.* **2021**, *11*, 6157. [CrossRef]
46. Sáiz-Manzanares, M.C.; Rodríguez-Díez, J.J.; Marticorena, R.; Zaparaín, M.J.; Cerezo, R. Lifelong Learning from Sustainable Education: An Analysis with Eye Tracking and Data Mining Techniques. *Sustainability* **2020**, *12*, 1970. [CrossRef]
47. Sáiz-Manzanares, M.C.; Marticorena-Sánchez, R.; Martín-Antón, L.J.; Almeida, L.; Carbonero-Martín, I. Application and challenges of eye tracking technology in Higher Education. *Comunicar* **2023**, *76*, 35–46. [CrossRef]
48. Langner, M.; Toreini, P.; Maedche, A. Cognitive state detection with eye tracking in the field: An experience sampling study and its lessons learned. *i-com* **2024**, *23*, 109–129. [CrossRef]
49. Wang, A.; Huang, C.; Wang, J.; He, D. The association between physiological and eye-tracking metrics and cognitive load in drivers: A meta-analysis. *Transp. Res. Part F Psychol. Behav.* **2024**, *104*, 474–487. [CrossRef]
50. Foltyn, A.; Deuschel, J.; Lang-Richter, N.R.; Holzer, N.; Oppelt, M.P. Evaluating the robustness of multimodal task load estimation models. *Front. Comput. Sci.* **2024**, *6*, 1371181. [CrossRef]
51. Havugimana, F.; Moinudin, K.A.; Yeasin, M. Deep Learning Framework for Modeling Cognitive Load From Small and Noisy EEG Data. *IEEE Trans. Cogn. Dev. Syst.* **2024**, *16*, 1006–1015. [CrossRef]
52. Khan, M.K.; Asadi, H.; Zhang, L.; Qazani, M.R.C.; Oladazimi, S.; Loo, C.K.; Lim, C.P.; Nahavandi, S. Application of artificial intelligence in cognitive load analysis using functional near-infrared spectroscopy: A systematic review. *Expert Syst. Appl.* **2024**, *249*, 123717. [CrossRef]
53. Hijazi, H.; Gomes, M.; Castelhano, J.; Castelo-Branco, M.; Praça, I.; de Carvalho, P.; Madeira, H. Dynamically predicting comprehension difficulties through physiological data and intelligent wearables. *Sci. Rep.* **2024**, *14*, 13678. [CrossRef] [PubMed]
54. Gong, J.; Cao, S.; Korivand, S.; Jalili, N. Reconstructing human gaze behavior from EEG using inverse reinforcement learning. *Smart Health* **2024**, *32*, 100480. [CrossRef]
55. Yang, H.; Wu, J.; Hu, A.; Lv, C. Real-Time Driver Cognitive Workload Recognition: Attention-Enabled Learning With Multimodal Information Fusion. *IEEE Trans. Ind. Electron.* **2024**, *71*, 4999–5009. [CrossRef]
56. Chango, W.; Lara, J.A.; Cerezo, R.; Romero, C. A review on data fusion in multimodal learning analytics and educational data mining. Wiley Interdiscip. *Rev. Data Min. Knowl. Discov.* **2022**, *12*, e1458. [CrossRef]
57. Vaswani, A.; Shazeer, N.; Parmar, N.; Uszkoreit, J.; Jones, L.; Gomez, A.N.; Kaiser, L.; Polosukhin, I. Attention Is All You Need. *arXiv* **2017**, arXiv:1706.03762. [CrossRef]
58. Ozdel, S.; Rong, Y.; Albaba, B.M.; Kuo, Y.-L.; Wang, X.; Kasneci, E. A Transformer-Based Model for the Prediction of Human Gaze Behavior on Videos. *arXiv* **2024**, arXiv:2404.07351v1. [CrossRef]
59. Toy, J.; MacAdam, J.; Tabor, P. Metacognition is all you need? Using Introspection in Generative Agents to Improve Goal-directed Behavior. *arXiv* **2024**, arXiv:2401.10910. [CrossRef]
60. An, Y.; Zhang, K.; Chai, Y.; Zhu, Z.; Liu, Q. Gaussian Mixture Variational-Based Transformer Domain Adaptation Fault Diagnosis Method and Its Application in Bearing Fault Diagnosis. *IEEE Trans. Ind. Inform.* **2024**, *20*, 615–625. [CrossRef]
61. An, Y.; Li, Z.; Li, Y.; Zhang, K.; Zhu, Z.; Chai, Y. Few-Shot Learning-Based Fault Diagnosis Using Prototypical Contrastive-Based Domain Adaptation Under Variable Working Conditions. *IEEE Sens. J.* **2024**, *24*, 25019–25029. [CrossRef]
62. Suzuki, Y.; Wild, F.; Scanlon, E. Measuring cognitive load in augmented reality with physiological methods: A systematic review. *J. Comput. Assist. Learn.* **2024**, *40*, 375–393. [CrossRef]
63. Román Sánchez, J.M.; Gallego Rico, S. *ACRA Escalas de Estrategias de Aprendizaje [Learning Strategy Scales]*; TEA: Madrid, Spain, 2008; Available online: http://www.web.teaediciones.com/Ejemplos/ACRA_extracto_web.pdf (accessed on 21 October 2024).
64. Eibe, F.; Hall, M.A.; Witten, I.H. The WEKA Workbench. In *Online Appendix for "Data Mining: Practical Machine Learning Tools and Techniques"*, 4th ed.; Morgan Kaufmann: Burlington, MA, USA; San Francisco, CA, USA, 2016.
65. IBM Corp. *SPSS Statistical Package for the Social Sciences (SPSS)*, 28th ed.; IBM Corp.: Armonk, NY, USA, 2024.
66. IBM Corp. *Structural Equation Modeling (AMOS SPSS)*, 29th ed.; IBM Corp.: Armonk, NY, USA, 2024.
67. Cohen, J.; Cohen, P.; West, G.; Aiken, L.S. *Applied Multiple Regression/Correlation Analysis for the Behavioral Sciences*, 3rd ed.; Routledge: New York, NY, USA, 2002. [CrossRef]

applied sciences

MDPI

Article

Evaluating Pupillometry as a Tool for Assessing Facial and Emotional Processing in Nonhuman Primates

Xinhe Liu [1,2,3], Zhiting Zhang [1,2,3] and Ji Dai [1,2,3,4,*]

1 Shenzhen Technological Research Center for Primate Translational Medicine, Shenzhen-Hong Kong Institute of Brain Science, Shenzhen Institutes of Advanced Technology, Chinese Academy of Sciences, Shenzhen 518055, China; xh.liu@siat.ac.cn (X.L.); zt.zhang@siat.ac.cn (Z.Z.)

2 CAS Key Laboratory of Brain Connectome and Manipulation, The Brain Cognition and Brain Disease Institute, Shenzhen Institutes of Advanced Technology, Chinese Academy of Sciences, Shenzhen 518055, China

3 Guangdong Provincial Key Laboratory of Brain Connectome and Behavior, Shenzhen Institutes of Advanced Technology, Chinese Academy of Sciences, Shenzhen 518055, China

4 University of Chinese Academy of Sciences, Beijing 100049, China

* Correspondence: ji.dai@siat.ac.cn

Abstract: Non-human primates (NHPs) are extensively utilized to investigate the neural mechanisms underlying face processing; however, measuring their brain activity necessitates a diverse array of technologies. Pupillometry emerges as a convenient, cost-effective, and non-invasive alternative for indirectly assessing brain activity. To evaluate the efficacy of pupillometry in assessing facial and emotional processing in NHPs, this study designed a face fixation task for experimental monkeys (Rhesus macaque) and recorded variations in their pupil size in response to face images with differing characteristics, such as species, emotional expression, viewing angles, and orientation (upright vs. inverted). All face images were balanced with luminance and spatial frequency. A sophisticated eye-tracking system (Eye-link 1000 plus) was employed to observe the pupils and track the viewing trajectories of monkeys as they examined images of faces. Our findings reveal that monkeys exhibited larger pupil sizes in response to carnivore faces (versus human faces, $p = 0.035$), negative conspecific faces (versus human faces, $p = 0.018$), and profile viewing angles (versus frontal view angles, $p = 0.010$). Notably, pupil size recorded during the 500–1000 ms post-stimulus interval was negatively correlated with their gaze durations directed at those images ($r = -0.357$, $p = 0.016$). Overall, this study demonstrates that pupillometry effectively captures subtle differences in facial and emotional processing, underscoring its potential as a valuable tool in future cognitive research and the diagnosis of disorders.

Keywords: pupillometry; nonhuman primate; pupil; face processing; eye-tracking

1. Introduction

Facial expressions are pivotal in social interactions, serving as essential indicators of identity, species, gender, emotion, and intention [1]. Accurately interpreting facial cues is vital for effective social engagement, positioning face processing as a significant area of interest within psychology and neuroscience. Both humans and nonhuman primates (NHPs) exhibit comparable behaviors in face observation [2,3], making NHPs an excellent model for exploring the neural mechanisms involved in facial recognition. Research utilizing functional magnetic resonance imaging (fMRI) and multi-channel electrophysiological recording has shown that NHPs' brain activity is modulated by facial cues. For example, neurons in the prefrontal cortex encode species-specific information [4,5], while the temporal cortex contains neurons responsive to both face orientation [6] and viewing angles [7–10].

Moreover, pupil size in primates is influenced by neural activity, offering a unique "window" into brain function [11–13]. Key neural substrates that regulate pupil size, primarily located in the brain stem, include the locus coeruleus (LC) and pretectal olivary nucleus (PON), which receive input from retinal ganglion cells [11,14]. In NHPs, the LC receives projections from both the prefrontal cortex (PFC) and anterior cingulate cortex (ACC) [15,16], influencing PFC activity through noradrenergic pathways [17]. The PON connects to the inferior temporal and visual cortices [18,19]. Electrophysiological studies have revealed that variations in pupil size correspond to neural activity in the PFC and ACC during passive fixation and unexpected events in NHPs [20]. Additionally, microstimulation of the PFC has been shown to affect the pupil light reflex in these animals [21,22]. In humans, similar MRI studies have identified correlations between blood oxygenation level-dependent (BOLD) activity in the ACC and orbitofrontal cortex and changes in pupil size [23]. Collectively, these findings suggest that changes in pupil size can reflect underlying brain activity in primates.

Given this context, pupillometry has emerged as a valuable tool for assessing emotional and cognitive processing in NHPs [11,24]. This study aims to evaluate whether pupillometry can be applied to investigate facial processing in NHPs. Compared to traditional methods for studying primate brain activity—such as multi-channel electrophysiological recording and fMRI—pupillometry offers a cost-effective, convenient, and non-invasive alternative [20]. As NHPs lack the ability to communicate verbally with humans, pupillometry presents an ideal approach for uncovering their internal states [24]. The study developed a face fixation task (FFT) featuring various face images, differing in species, expressions, orientation, and viewpoints—all known to affect neural activity in NHPs [4–6,25–30]. By employing an eye-tracking system, it will monitor the pupil area and gaze location of NHPs while they observe these facial stimuli. Through this research, it aims to determine whether pupillometry effectively reveals psychological differences in facial processing among NHPs.

The key contributions of this research can be outlined as follows:

(1) This study investigates the potential of pupil size variations as indicators of emotional responses in NHPs. Furthermore, it aims to evaluate the effectiveness of changes in pupil size as a method for interpreting the emotional states of animals.
(2) By analyzing the pupillary reactions of NHPs to facial imagery, it can differentiate their emotional responses to faces from those of humans.
(3) Ultimately, this research seeks to explore the viability of pupil responses as a biomarker for deficits in face processing, as such impairments are a significant hallmark of several neurological disorders.

The structure of this paper is organized as follows: Section 1 (Introduction) offers a review of recent related literature in this field. Section 2 (Related Works) reviews recent related works. Section 3 (Materials and Methods) outlines the animal subjects, experimental procedures, and materials utilized throughout the experiments, while also covering data and statistical analysis methods. Section 4 (Results) presents the analysis of the study's findings. Finally, Section 5 (Discussion) discusses the key conclusions drawn from the research and suggests directions for future investigations.

2. Related Works

Pupillometry has garnered increasing interest among psychologists and neuroscientists due to its non-invasive nature and cost-effectiveness (Table 1). For example, research led by Chang YH et al. has revealed that pupillometry correlates with various physiological measures, including heart rate, skin conductance, pulse wave amplitude, and respiratory signals. This suggests its potential as a reliable real-time indicator of autonomic arousal [31]. Additionally, studies have demonstrated that emotional stimuli—both visual and auditory—such as those prompting feelings of sadness, happiness, or fear, result in significant pupil dilation in humans, while neutral stimuli do not elicit such a response [32–34]. Further investigations indicate that the pupil dilation associated with arousal is more pronounced within social contexts [35,36]. This effect has also been documented in rodents, although it may be suppressed under anesthesia (specifically with dexmedetomidine-isoflurane) [37]. Moreover, findings from human studies have shown that increased cognitive workload and demands for attentional control can lead to greater pupil size [38,39], suggesting that cognitive effort is a contributing factor to pupil dilation. Therefore, it is hypothesized that facial characteristics may influence variations in pupil size. However, there is a lack of research exploring the responses of non-human primates (NHPs) to facial images through pupillometry.

Table 1. Main differences between the current study and related works in the past three years.

Study	Objective Measures	Subjects	Experimental Materials	Explore Cross-Species Effects?	Control Low-Level Properties	Compare Effects of View Point	Compare Correlation with Gaze Duration
Chang YH et al., 2024 [31]	Pupil size	Human	Images and words	No	Not mention	No	No
Lee CL et al., 2023 [32]	Pupil size	Human	Videos	No	Not mention	No	No
Pan J et al., 2024 [33]	Pupil size	Human	Auditory stimuli	No	Yes	No	No
Yuan T. et al., 2024 [34]	Pupil size	Human	Motion visual stimuli	No	Yes	No	No
Yu P et al., 2024 [35]	Pupil size	Human	Images	Yes	Yes	No	No
Bonino G et al., 2024 [36]	Pupil size	Human	Touch	No	No mention	No	No
Current study	Pupil size	NHP	Images	Yes	Yes	Yes	Yes

In contrast to previous studies conducted over the past three years, this research is innovative in several ways: (1) it uses NHPs as subjects; (2) it rigorously controls the lower-level properties of images and minimizes the influence of extraneous factors on pupil size; and (3) it investigates the relationship between gaze duration and pupil variation (Table 1). While prior reviews have indicated that pupillometry can effectively measure arousal levels [11,24], we contend that it also serves as an indicator of cognitive effort and attention during facial processing.

3. Materials and Methods

The entire experimental procedure unfolded as follows: it commenced by assembling the necessary materials, specifically images of faces, along with the required equipment. Subsequently, all macaques participated in the behavioral task to facilitate data collection. Finally, a thorough analysis was conducted of the experimental data and a discussion regarding the findings ensued.

3.1. Animals

Seven male Rhesus macaques (age: 8–12 years; weight: 7.2–11.2 kg) were utilized in this study. All monkeys were housed individually in an AAALAC-accredited NHP facility with controlled temperature (20–26 °C) and humidity (40–70%). They lived in the same colony room, which maintained a 12:12 h light–dark cycle (light onset at 7 a.m.) [40,41]. To ensure stable eye tracking, each monkey was surgically implanted with a titanium headpost [42]. All experimental procedures adhered strictly to the Guide for the Care and Use of Laboratory Animals (Eighth Edition) and were approved by the Institutional Animal Care and Use Committee (IACUC) at the Shenzhen Institute of Advanced Technology, Chinese Academy of Sciences.

3.2. Experimental Procedures

The monkeys sat comfortably in a custom-designed chair in a dark, quiet recording room with their heads fixed via the titanium headpost. An LED monitor (ASUS, 24 inches, 1920 × 1080 resolution, 144 Hz refresh rate) was positioned 57 cm in front of the subjects. Eye positions and pupil sizes were monitored using an eye-tracking system (Eyelink 1000 Plus; sampling rate: 1000 Hz, Figure 1). Prior to each recording, subjects underwent a nine-point calibration.

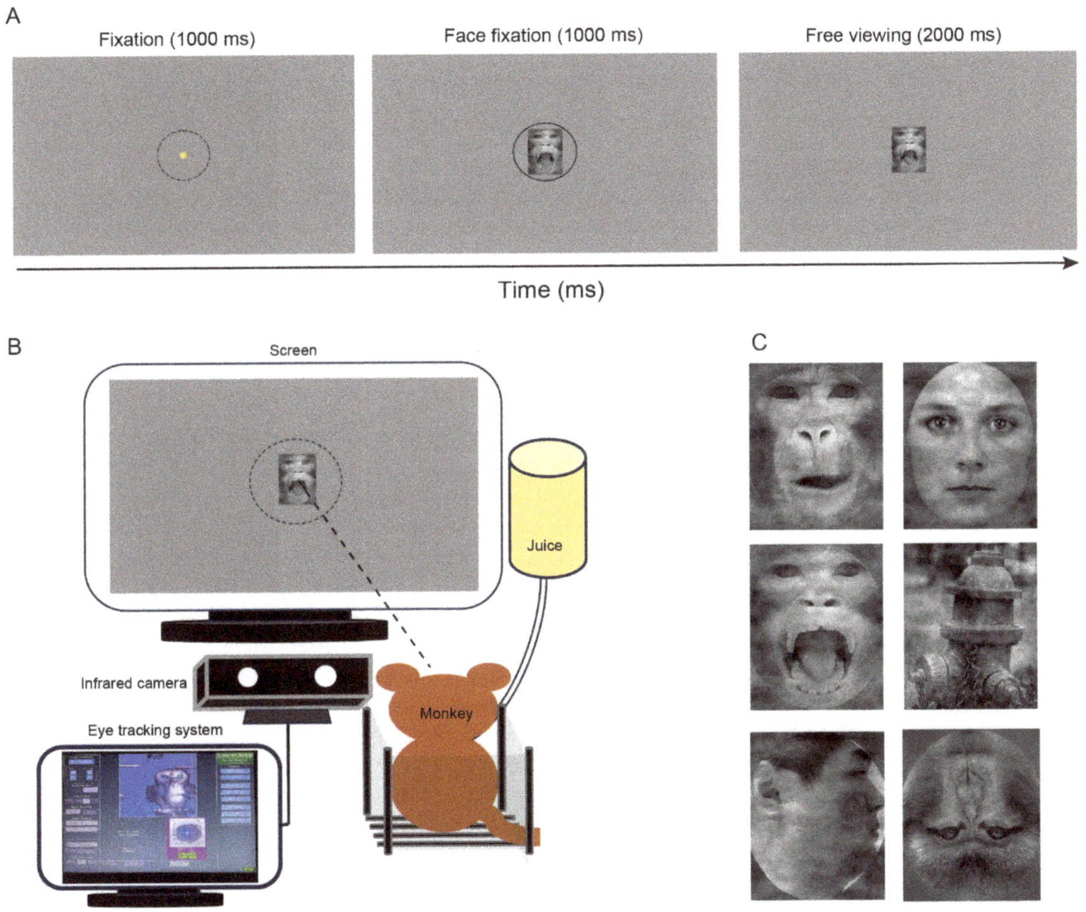

Figure 1. The schematic of the behavioral task and examples of stimulus. (**A**) The flow of the task. (**B**) Illustration of the experimental setup. (**C**) The stimuli set includes faces of monkeys, humans, animals, and objects, with different emotions, viewing angles, and upright or inverted.

All monkeys were trained to perform a face fixation task (FFT; see Figure 1), adapted from previous studies involving humans and NHPs [43–45]. Each trial commenced with a fixation point at the center of the screen, which subjects had to focus on for 1 s to trigger the presentation of face stimuli, which lasted for 3 s. Subjects needed to maintain fixation within four visual degrees during the initial second; breaking this fixation would terminate the trial. Successful completion of a trial resulted in a juice reward. The background of the screen was set to the average RGB of all visual stimuli used (gray) to minimize contrast. The inter-trial interval (ITI) was set to 5–7 s, and all stimuli were presented in a random order. Overall, four distinct experiments utilizing the FFT paradigm were conducted.

3.3. Experiment 1: Responses to Different Animals and Objects

In this experiment, 12 images of faces and three images of objects were used as visual stimuli. The face stimuli included four categories: human, monkey, carnivore, and snake, with three images per category. Object images included a fire hydrant, lamp, and clock. Human, snake, and object images were sourced from the International Affective Picture System (IAPS) [46], while monkey and predator images (tiger, lion, cheetah) were obtained from publicly available sources. All subjects in the images were looking sideways. Monkeys completed two FFT sessions, each consisting of 360 trials.

3.4. Experiment 2: Responses to Monkey and Human Facial Expressions

For this experiment, six face images featuring different facial expressions were employed as visual stimuli, encompassing two species (human and monkey) and three expression types (negative, positive, and neutral). Human faces were obtained from the NimStim database [47], while monkey face images were sourced from one lab monkey displaying open-mouth, lip-smacking, and relaxed expressions, classified as negative, positive, and neutral, respectively [25,48–51]. All faces were oriented straight ahead. This experiment involved one FFT session comprising 120 trials.

3.5. Experiment 3: Responses to Faces of Different Viewing Angles

This experiment employed 12 face images captured from two rhesus macaques and two young male humans at three viewing angles (0°, 45°, and 90°), termed frontal view, middle profile, and profile view, respectively. All faces displayed neutral expressions. Monkeys performed two FFT sessions, each containing 360 trials.

3.6. Experiment 4: Responses to Faces of Different Orientations

In the final experiment, 12 face images featuring upright and inverted orientations were used as visual stimuli. This included three upright and three inverted human faces (female), as well as three upright and three inverted monkey faces. Human face images were acquired from IAPS [46], while monkey images were sourced from publicly available platforms. All faces displayed neutral expressions, oriented straight ahead. Monkeys completed two FFT sessions, each comprising 360 trials.

All images were processed using Photoshop (2017 version). The images were trimmed for consistent sizing, and face-irrelevant information was removed or obscured. Images within each experiment were balanced for luminance and spatial frequency utilizing the ShineToolbox in MATLAB (2016a version) [52]. Briefly, the luminance distributions of each image were normalized and adjusted to align with the corresponding mean and standard deviations. Additionally, each image's spatial frequency was subjected to a fast Fourier transform to generate a spectrum, with the average spectrum established as the reference target. Subsequently, the phase information from the original image was merged with the target amplitude spectrum, and an inverse fast Fourier transform was applied to convert the data from the frequency domain back into the spatial domain.

3.7. Data Analysis

Eye tracking data underwent preprocessing through the utilization of custom scripts in Matlab (Version 2016a). For the analysis of pupil size, only those trials were considered where the pupil sizes remained stable and were continuously recorded for 1000 ms prior to and 1000 ms following the visual stimulus. To obtain the mean pupil size change curve, pupil areas were averaged across trials aligned to 200 ms prior to visual presentation. Notably, it observed that monkey pupils began to constrict approximately 200 ms post visual presentation, reaching minimum levels between 400 and 500 ms. The mean pupil sizes were calculated during three phases: early (1–200 ms), middle (201–500 ms), and late (501–1000 ms) after stimulus onset.

In conducting the gaze duration analysis, data from the free viewing interval, specifically between 1000 ms and 3000 ms following visual stimulation, were extracted for evaluation. To assess the relationship between pupil size and gaze duration for each image, we first determined the average pupil size during three distinct phases and recorded the gaze duration for each image presented to the four monkeys. Subsequently, we computed the z-scores for both pupil size and gaze duration within the context of the experiment. Finally, the Pearson product-moment correlation coefficient was calculated to quantify the correlation between the z-scores of gaze duration and pupil size. All figures were produced using Prism (Version 8.0).

3.8. Statistical Analysis

In Experiment 1 and 2, a Friedman test (a nonparametric test) was employed to assess the main effects between groups, followed by Dunn's multiple comparisons. For Experiments 3 and 4, repeated measures two-way ANOVA (RM two-way ANOVA) was utilized to evaluate the effects of facial attributes (viewing angle, orientation) and species on pupil size changes. Data conformed to the sphericity assumption according to the Mauchly sphericity test. A Greenhouse–Geisser correction was applied if the data violated the sphericity assumption. A Shapiro–Wilk test was used to test the normality of data. Bonferroni correction for multiple comparisons was applied when differences reached significance ($\alpha = 0.05$) or marginal significance ($\alpha = 0.06$). All statistical analyses were conducted using SPSS (version 25, IBM Corporation).

4. Results

4.1. Experiment 1: Pupil Change in Response to Face Images of Different Species

It began by examining whether the pupil responses of monkeys varied when viewing images of different animals and objects. The experiment involved six monkeys, and the results concerning pupil size changes are illustrated in Figure 2A–C. Although no significant differences were detected in the early phase across conditions, a marginally significant effect of species on pupil size emerged in the middle phase ($p = 0.059$, Friedman test, Figure 2B), followed by a significant effect in the late phase ($p = 0.048$, Figure 2C). Further analysis through multiple comparisons revealed that pupil size in response to carnivores was significantly larger than to human faces in the late phase ($p = 0.035$, Figure 2C). Given that the physical features of the images were balanced, these variations in pupil responses likely reflect the innate responses of the subjects to different species. Thus, these findings indicate that pupil size can serve as a sensitive indicator of a monkey's perceptual state.

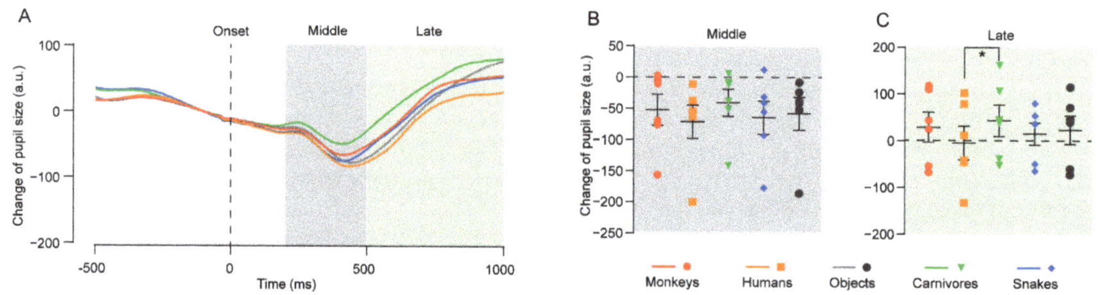

Figure 2. Changes in pupil size in response to images of various animals and objects. (**A**) Pupil alterations observed 500 ms before and 1000 ms post the time of stimulus presentation. Light gray and green zones indicate the middle (201–500 ms) and late phases (501–1000 ms), respectively. (**B**,**C**) Averaged pupil changes during the mid and late phases in reaction to different animals and objects. Color coding: red for monkeys; yellow for humans; gray for objects; green for carnivores; and blue for snakes. Error bar: SEM. * $p < 0.05$.

4.2. Experiment 2: Pupil Change to Facial Expressions of Humans and Monkeys

Recognizing emotional states in facial expressions is crucial for primates. Therefore, it investigated whether eye tracking could yield insights by employing the FFT paradigm. This experiment involved six monkeys and presented human and monkey facial expressions as visual stimuli. The pupil area responses to these expressions are presented in Figure 3A–C, featuring two species (human/monkey) and three expressions (negative/positive/neutral). Consistent with Experiment 1, no significant differences in mean pupil size were observed during the early phase. Notably, facial expressions exhibited significant effects on mean pupil size in the middle ($p = 0.028$, Friedman test) and later phases ($p = 0.040$, Friedman test). Dunn's multiple comparisons indicated that pupil size in response to monkey negative expressions was greater than that for human negative expressions in the middle phase ($p = 0.018$, Figure 3B) and late phase ($p = 0.051$, Figure 3C). These results imply that pupil size is a sensitive measure for detecting nuanced differences in responses to conspecific facial expressions.

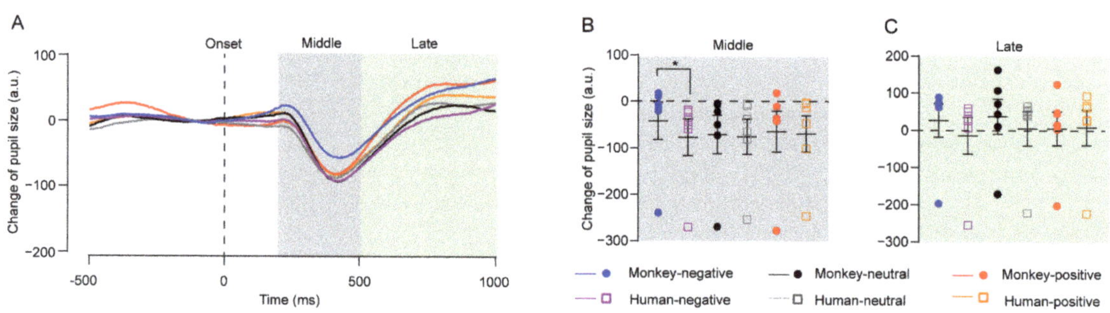

Figure 3. Changes in pupil size in response to facial expressions of humans and monkeys. (**A**) Pupil alterations observed 500 ms before and 1000 ms post the time of stimulus presentation. Light gray and green zones indicate the middle (201–500 ms) and late phases (501–1000 ms), respectively. (**B**,**C**) Averaged pupil changes during the mid and late phases in reaction to different facial expressions of humans and monkeys. Color coding: blue for monkey—negative; dark gray for monkey—neutral; red for monkey—positive; purple for human—negative; light gray for human—neutral; and yellow for human—positive. Error bar: SEM. * $p < 0.05$.

4.3. Experiment 3: Pupil Size Change in Response to Face Images with Different Viewing Angles

Next, it aimed to explore how viewing angles (frontal view, mid-profile view, profile view) affect pupil size. Pupil size changes were examined in response to human and monkey faces presented at various angles (Figure 4A–C), involving five monkeys. While no significant effects of viewing angle were seen in the early phase, significant main effects emerged in the middle phase (repeated-measures two-way ANOVA: interaction $F(2, 8) = 0.883$, $p = 0.450$; viewing angle $F(2, 8) = 5.031$, $p = 0.038$) and late phase (interaction $F(2, 8) = 0.098$, $p = 0.908$; viewing angle $F(2, 8) = 8.744$, $p = 0.010$). Post hoc analysis revealed that pupil size in the late phase was larger for the profile view compared to the frontal view ($p = 0.030$, Figure 4C). These findings suggest that pupil size may reflect the cognitive processing related to viewing angles.

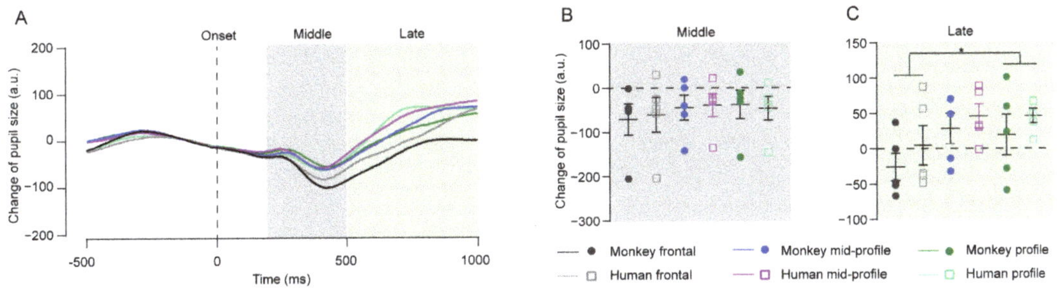

Figure 4. Changes in pupil size in response to face images with different viewing angles. (**A**) Pupil alterations observed 500 ms before and 1000 ms post the time of stimulus presentation. Light gray and green zones indicate the middle (201–500 ms) and late phases (501–1000 ms), respectively. (**B**,**C**) Averaged pupil changes during the mid and late phases in reaction to face images with different viewing angles. Color coding: dark gray for monkey frontal; blue for monkey mid-profile; green for monkey profile; light gray for human frontal; purple for human mid-profile; and light green for human profile. Error bar: SEM. * $p < 0.05$.

4.4. Experiment 4: Pupil Size Change in Response to Upright and Inverted Faces

Investigating face orientation is a common approach in studies of facial processing. It further sought to assess the efficacy of pupil tracking in gauging differences in processing upright versus inverted faces. This experiment involved six monkeys, utilizing both human and monkey faces as stimuli. The results of pupil size changes in response to upright and inverted faces are displayed in Figure 5A–C. No significant differences were found across face orientations at any phase (early, middle, late). These findings suggest that pupil size may not be particularly sensitive to variations in face orientation.

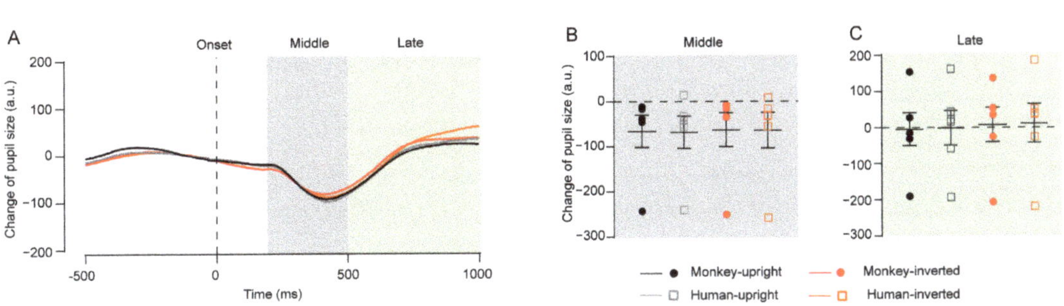

Figure 5. Changes in pupil size in response to upright and inverted faces. (**A**) Pupil alterations observed 500 ms before and 1000 ms post the time of stimulus presentation. Light gray and green zones indicate

the middle (201–500 ms) and late phases (501–1000 ms), respectively. (**B**,**C**) Averaged pupil changes during the mid and late phases in reaction to upright and inverted faces. Color coding: dark gray for monkey upright; red for monkey inverted; light gray for human upright; and yellow for human inverted. Error bar: SEM.

4.5. Correlation Between Gaze Duration and Mean Pupil Size

Lastly, variability in gaze duration was observed among monkeys fixating on different stimuli, ranging from 2500 to 2900 ms, despite a consistent presentation duration of 3000 ms. This study aimed to identify any potential correlation between gaze duration and pupil dynamics. The z-scores were obtained for both gaze duration and pupil size for each visual stimulus across the three phases, followed by Pearson correlation analysis for all phases with data from four monkeys who participated in all experiments. Notably, a significant negative correlation was found between gaze duration and mean pupil size in the late phase ($r = -0.357$, $p = 0.016$, Figure 6A,B, with Figure 6A depicting results for each stimulus). This suggests that pupil size in the late phase may predict a monkey's interest in specific images; for example, smaller pupil sizes correspond to longer gaze durations towards those images in the FFT paradigm. This correlation was absent in the early ($r = -0.063$, $p = 0.679$) and middle phases ($r = -0.255$, $p = 0.091$, Figure 6C).

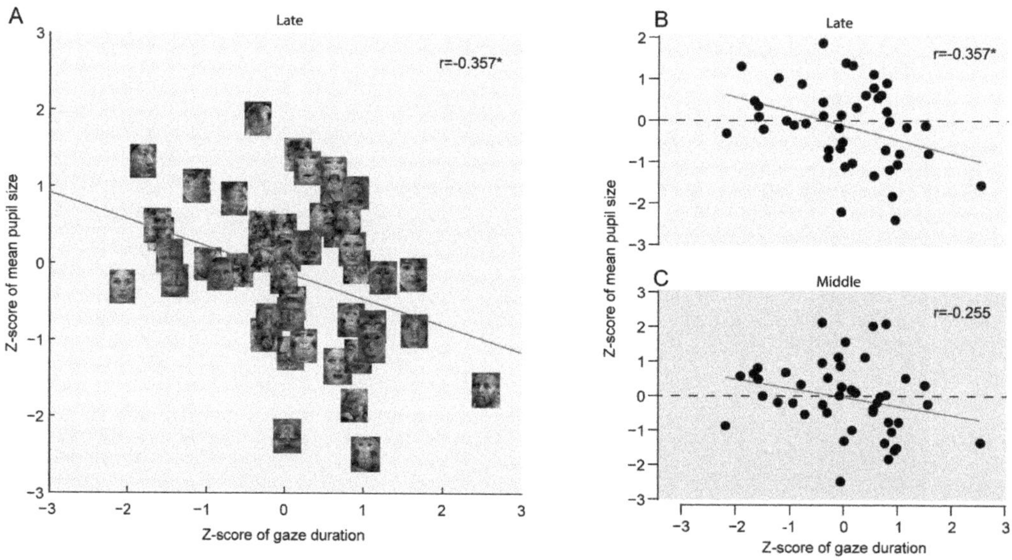

Figure 6. Correlation between gaze duration and mean pupil size. (**A**) The z-scores for gaze duration (x-axis) and pupil size (y-axis) for each visual stimulus. (**B**) The correlation between gaze duration and pupil size during the late (**B**) and middle (**C**) phases. * $p < 0.05$.

5. Discussion

In this study, it was demonstrated that pupillometry could effectively capture subtle differences in facial and emotional processing in NHPs. Our findings indicate that pupil size increases when NHPs view images of carnivores, negative conspecific facial expressions, and profile views. Notably, a significant negative correlation was also observed between mean pupil size in the late phase and gaze duration for each visual stimulus. These results suggest that pupil size may serve as a sensitive, indirect biomarker to gauge face processing in NHPs.

5.1. Pupil Size Changes in Face Processing

Firstly, it was found that NHPs exhibited larger pupil sizes in the late phase when viewing images of carnivores (tigers, leopards, and lions) compared to human faces. The subject monkeys had been trained to interact with humans, receiving food, water, and fruits from them, and had visual exposure to both monkeys and humans within their environments. As a result, these monkeys can be considered familiar with human faces. In contrast, they had no prior exposure to carnivores or snakes, which are recognized as primary predators of primates [53,54]. Din et al. [4] reported that neurons in the prefrontal cortex show stronger responses to images of snakes and carnivores than to humans in NHPs. Significant pupil size changes were not observed in response to snake images; this may be attributed to our control of low-level image properties, such as color, spatial frequency, and luminance. Snakeskin is characterized by a distinctive color and texture (smoothy and hairless) that differs from that of common mammals. By manipulating these low-level attributes of the images, it may have complicated the monkeys' recognition of snakes, ultimately reducing their salience.

Secondly, it identified that NHPs displayed larger pupil sizes in response to negative facial expressions of monkeys, but not humans. The ability to recognize negative emotions from faces is crucial for survival [1], and imaging studies have shown that multiple brain regions in NHPs exhibit increased BOLD signals when viewing conspecific negative expressions. Electrophysiological research suggests that the orbitofrontal cortex and medial prefrontal cortex contain neurons responsive to these expressions [4,5,28]. Our findings align with this biological evidence regarding pupil size modulation by conspecific negative expressions. However, a similar pupil response to negative human expressions was not observed, potentially because NHPs and humans communicate emotional facial expressions differently [41,49,55], making it challenging for NHPs to infer negative emotions from human faces.

Thirdly, viewing angle influenced pupil size, with NHPs responding with larger pupil sizes to profile views compared to frontal views. Research has shown that neurons in the temporal cortex of NHPs are sensitive to viewing angles [8,10]. The recognition of emotions and identities from frontal views occurs more quickly and accurately than from profile views [56–59]. Therefore, we propose that viewing faces from a frontal perspective allows NHPs to more efficiently gather social information, such as identity and emotions, thereby reducing uncertainty in facial recognition [56–59].

These findings indicate that changes in pupil size during the middle and late phases (200–1000 ms post-stimulus presentation) reflect the processing of facial information. Human studies have shown that pupil sizes increase under high working memory load and attentional control [38,39]. It is plausible that NHPs also require increased cognitive effort to recognize threatening or uncertain face information quickly. Moreover, our discovery of a negative correlation between gaze duration and pupil size in the late phase (500–1000 ms post-stimulus presentation) suggests that smaller pupils correspond to longer gaze durations, indicating that NHPs might allocate less attentional control and engage in prolonged viewing to comprehend facial information.

5.2. Limitations of Pupillometry in Exploring Face Processing

While our study demonstrates that pupillometry can reveal differences in face processing among NHPs, certain limitations must be acknowledged. Compared to commonly used brain research techniques, such as MRI, multi-channel electrophysiological recordings, and EEG, pupillometry offers several advantages, including shorter preparation times, no need for consumable materials, absence of required anesthesia, and non-invasive measurements.

Therefore, it posits that pupillometry is a valuable and suitable tool for studying NHPs as experimental models.

However, pupillometry does present some challenges. Firstly, it typically requires a longer latency to reveal differences in face processing. EEG and MEG studies have shown that the N170 and M170 components reflect differences in face processing in humans around 150–200 ms post-stimulus [60–62]. In NHPs, similar components appear approximately 112 ms after face presentation [63], while pupillometry may need 200–1000 ms to reveal processing differences. Secondly, controlling low-level properties of visual input, such as luminance, spatial frequency, and color, is essential in pupillometry studies, as these factors can significantly impact pupil size [64]. However, controlling these properties can affect subjective experiences with the images [65,66], which may also influence pupil size. Thirdly, the neural mechanisms modulating pupil size changes are not fully elucidated; thus, pupillometry serves as an indirect method to assess brain activity [11]. It is essential to consider these limitations prior to implementing pupillometry in a particular study. Integrating artificial intelligence with pupillometry may further enhance its application in future cognitive research and the diagnosis of disorders [67].

5.3. Conclusions and Directions for Future Research

In summary, our findings indicate that pupillometry is a valuable tool for uncovering variations in how non-human primates (NHPs) process facial expressions. The levels of arousal, along with cognitive control, appear to influence this mechanism. While pupillometry does have certain limitations, it presents an economical and practical approach for indirectly assessing brain activity. This research reinforces the notion that changes in pupil size can be an effective measure of emotional face processing in primates, potentially aiding in the diagnosis of neurological conditions such as autism.

Looking ahead, we aim to create a specialized task and formulate a mathematical model that will enable us to interpret the emotional states and cognitive demands of NHPs based on data derived from pupil size fluctuations [68].

Author Contributions: Conceptualization, X.L. and J.D.; methodology, X.L. and Z.Z.; software, X.L.; validation, X.L., Z.Z., and J.D.; formal analysis, X.L.; investigation, X.L.; resources, X.L. and Z.Z.; data curation, X.L.; writing—original draft preparation, X.L.; writing—review and editing, J.D.; visualization, X.L.; supervision, J.D.; project administration, J.D.; funding acquisition, J.D. All authors have read and agreed to the published version of the manuscript.

Funding: This research was funded by the National Natural Science Foundation of China (32371066), the Guangdong Basic and Applied Basic Research Foundation (2022A1515010134, 2022A1515110598), the Shenzhen-Hong Kong Institute of Brain Science–Shenzhen Fundamental Research Institutions (NYKFKT2019009), and Shenzhen Technological Research Center for Primate Translational Medicine (XMHT20220104005). The APC was funded by the National Natural Science Foundation of China (32371066).

Institutional Review Board Statement: The animal study protocol was approved by the Institutional Animal Care and Use Committee (IACUC) at the Shenzhen Institute of Advanced Technology, Chinese Academy of Sciences. (SIAT-IACUC-201123-NS-DJ-A1483, approved on 12-07-2020).

Informed Consent Statement: Not applicable.

Data Availability Statement: The datasets presented in this article are not readily available because the data are part of an ongoing study. Requests to access the datasets should be directed to the corresponding author.

Acknowledgments: We appreciate the assistance of the SIAT MRI facility and NHP veterinaries.

Conflicts of Interest: The authors declare no conflicts of interest.

References

1. Hesse, J.K.; Tsao, D.Y. The macaque face patch system: A turtle's underbelly for the brain. *Nat. Rev. Neurosci.* **2020**, *21*, 695–716. [CrossRef] [PubMed]
2. Dahl, C.D.; Wallraven, C.; Bülthoff, H.H.; Logothetis, N.K. Humans and macaques employ similar face-processing strategies. *Curr. Biol.* **2009**, *19*, 509–513. [CrossRef] [PubMed]
3. Shepherd, S.V.; Steckenfinger, S.A.; Hasson, U.; Ghazanfar, A.A. Human-monkey gaze correlations reveal convergent and divergent patterns of movie viewing. *Curr. Biol.* **2010**, *20*, 649–656. [CrossRef]
4. Dinh, H.T.; Nishimaru, H.; Matsumoto, J.; Takamura, Y.; Le, Q.V.; Hori, E.; Maior, R.S.; Tomaz, C.; Tran, A.H.; Ono, T.; et al. Superior Neuronal Detection of Snakes and Conspecific Faces in the Macaque Medial Prefrontal Cortex. *Cereb. Cortex* **2018**, *28*, 2131–2145. [CrossRef]
5. Gothard, K.M.; Battaglia, F.P.; Erickson, C.A.; Spitler, K.M.; Amaral, D.G. Neural responses to facial expression and face identity in the monkey amygdala. *J. Neurophysiol.* **2007**, *97*, 1671–1683. [CrossRef]
6. Sugase-Miyamoto, Y.; Matsumoto, N.; Ohyama, K.; Kawano, K. Face inversion decreased information about facial identity and expression in face-responsive neurons in macaque area TE. *J. Neurosci.* **2014**, *34*, 12457–12469. [CrossRef]
7. Desimone, R.; Albright, T.; Gross, C.; Bruce, C. Stimulus-selective properties of inferior temporal neurons in the macaque. *J. Neurosci.* **1984**, *4*, 2051–2062. Available online: https://www.jneurosci.org/content/jneuro/4/8/2051.full.pdf (accessed on 1 August 2024). [CrossRef]
8. Perrett, D.I.; Oram, M.W.; Harries, M.H.; Bevan, R.; Hietanen, J.K.; Benson, P.J.; Thomas, S. Viewer-centred and object-centred coding of heads in the macaque temporal cortex. *Exp. Brain Res.* **1991**, *86*, 159–173. Available online: https://www.scopus.com/inward/record.uri?eid=2-s2.0-0025734432&doi=10.1007/BF00231050&partnerID=40&md5=f116a3a935f8853eb7f2b018e3bbf82a (accessed on 1 January 2024). [CrossRef]
9. Perrett, D.I.; Mistlin, A.J.; Chitty, A.J. Visual neurones responsive to faces. *Trends Neurosci.* **1987**, *10*, 358–364. Available online: https://www.sciencedirect.com/science/article/pii/0166223687900713 (accessed on 1 September 2024). [CrossRef]
10. Perrett, D.I.; Rolls, E.T.; Caan, W. Visual neurones responsive to faces in the monkey temporal cortex. *Exp. Brain Res.* **1982**, *47*, 329–342. [CrossRef]
11. Joshi, S.; Gold, J.I. Pupil Size as a Window on Neural Substrates of Cognition. *Trends Cogn. Sci.* **2020**, *24*, 466–480. [CrossRef] [PubMed]
12. Ferencová, N.; Višňovcová, Z.; Bona Olexová, L.; Tonhajzerová, I. Eye pupil—A window into central autonomic regulation via emotional/cognitive processing. *Physiol. Res.* **2021**, *70*, S669–S682. [CrossRef] [PubMed]
13. Ebitz, R.B.; Moore, T. Both a Gauge and a Filter: Cognitive Modulations of Pupil Size. *Front. Neurol.* **2019**, *9*, 1190. [CrossRef] [PubMed]
14. Gamlin, P.D.; McDougal, D.H.; Pokorny, J.; Smith, V.C.; Yau, K.W.; Dacey, D.M. Human and macaque pupil responses driven by melanopsin-containing retinal ganglion cells. *Vis. Res.* **2007**, *47*, 946–954. [CrossRef]
15. Arnsten, A.F.; Goldman-Rakic, P.S. Selective prefrontal cortical projections to the region of the locus coeruleus and raphe nuclei in the rhesus monkey. *Brain Res.* **1984**, *306*, 9–18. [CrossRef] [PubMed]
16. Porrino, L.J.; Goldman-Rakic, P.S. Brainstem innervation of prefrontal and anterior cingulate cortex in the rhesus monkey revealed by retrograde transport of HRP. *J. Comp. Neurol.* **1982**, *205*, 63–76. [CrossRef] [PubMed]
17. Morrison, J.H.; Foote, S.L.; O'Connor, D.; Bloom, F.E. Laminar, tangential and regional organization of the noradrenergic innervation of monkey cortex: Dopamine-beta-hydroxylase immunohistochemistry. *Brain Res. Bull.* **1982**, *9*, 309–319. [CrossRef] [PubMed]
18. Steele, G.E.; Weller, R.E. Subcortical connections of subdivisions of inferior temporal cortex in squirrel monkeys. *Vis. Neurosci.* **1993**, *10*, 563–583. Available online: https://www.scopus.com/inward/record.uri?eid=2-s2.0-0027603502&doi=10.1017/S0952523800004776&partnerID=40&md5=11acec2b634f23922c4ae5c7d37ebb22 (accessed on 31 May 2024). [CrossRef]
19. Leichnetz, G.R. Preoccipital cortex receives a differential input from the frontal eye field and projects to the pretectal olivary nucleus and other visuomotor-related structures in the rhesus monkey. *Vis. Neurosci.* **1990**, *5*, 123–133. Available online: https://www.scopus.com/inward/record.uri?eid=2-s2.0-0025477761&doi=10.1017/S095252380000016X&partnerID=40&md5=2faad4f2ef1de2448a4727adec881e17 (accessed on 15 August 2024). [CrossRef]
20. Joshi, S.; Li, Y.; Kalwani, R.M.; Gold, J.I. Relationships between Pupil Diameter and Neuronal Activity in the Locus Coeruleus, Colliculi, and Cingulate Cortex. *Neuron* **2016**, *89*, 221–234. Available online: https://www.sciencedirect.com/science/article/pii/S089662731501034X (accessed on 6 January 2024). [CrossRef]
21. Ebitz, R.B.; Moore, T. Selective Modulation of the Pupil Light Reflex by Microstimulation of Prefrontal Cortex. *J. Neurosci.* **2017**, *37*, 5008–5018. Available online: https://www.jneurosci.org/content/jneuro/37/19/5008.full.pdf (accessed on 10 May 2024). [CrossRef] [PubMed]
22. Lehmann, S.J.; Corneil, B.D. Transient Pupil Dilation after Subsaccadic Microstimulation of Primate Frontal Eye Fields. *J. Neurosci.* **2016**, *36*, 3765–3776. [CrossRef]

23. Murphy, P.R.; O'Connell, R.G.; O'Sullivan, M.; Robertson, I.H.; Balsters, J.H. Pupil diameter covaries with BOLD activity in human locus coeruleus. *Hum. Brain Mapp.* **2014**, *35*, 4140–4154. [CrossRef]
24. Kuraoka, K.; Nakamura, K. Facial temperature and pupil size as indicators of internal state in primates. *Neurosci. Res.* **2022**, *175*, 25–37. [CrossRef] [PubMed]
25. Liu, N.; Hadj-Bouziane, F.; Jones, K.B.; Turchi, J.N.; Averbeck, B.B.; Ungerleider, L.G. Oxytocin modulates fMRI responses to facial expression in macaques. *Proc. Natl. Acad. Sci. USA* **2015**, *112*, E3123–E3130. [CrossRef]
26. Xu, P.; Peng, S.; Luo, Y.J.; Gong, G. Facial expression recognition: A meta-analytic review of theoretical models and neuroimaging evidence. *Neurosci. Biobehav. Rev.* **2021**, *127*, 820–836. [CrossRef]
27. Adolphs, R.; Tranel, D.; Damasio, H.; Damasio, A. Impaired recognition of emotion in facial expressions following bilateral damage to the human amygdala. *Nature* **1994**, *372*, 669–672. [CrossRef]
28. Barat, E.; Wirth, S.; Duhamel, J.R. Face cells in orbitofrontal cortex represent social categories. *Proc. Natl. Acad. Sci. USA* **2018**, *115*, E11158–E11167. [CrossRef]
29. Moeller, S.; Crapse, T.; Chang, L.; Tsao, D.Y. The effect of face patch microstimulation on perception of faces and objects. *Nat. Neurosci.* **2017**, *20*, 743–752. [CrossRef] [PubMed]
30. Tsao, D.Y.; Freiwald, W.A.; Tootell, R.B.; Livingstone, M.S. A cortical region consisting entirely of face-selective cells. *Science* **2006**, *311*, 670–674. [CrossRef] [PubMed]
31. Chang, Y.H.; Yep, R.; Wang, C.A. Pupil size correlates with heart rate, skin conductance, pulse wave amplitude, and respiration responses during emotional conflict and valence processing. *Psychophysiology* **2024**, *62*, e14726. [CrossRef] [PubMed]
32. Lee, C.L.; Pei, W.; Lin, Y.C.; Granmo, A.; Liu, K.H. Emotion Detection Based on Pupil Variation. *Healthcare* **2023**, *11*, 322. [CrossRef] [PubMed]
33. Pan, J.; Sun, X.; Park, E.; Kaufmann, M.; Klimova, M.; McGuire, J.T.; Ling, S. The effects of emotional arousal on pupil size depend on luminance. *Sci. Rep.* **2024**, *14*, 21895. [CrossRef] [PubMed]
34. Yuan, T.; Wang, L.; Jiang, Y. Multi-level processing of emotions in life motion signals revealed through pupil responses. *Neuroscience* **2024**, *12*, RP89873. [CrossRef]
35. Yu, P.; Yu, L.; Li, Y.; Qian, C.; Hu, J.; Zhu, W.; Liu, F.; Wang, Q. Emotional and visual responses to trypophobic images with object, animal, or human body backgrounds: An eye-tracking study. *Front. Psychol.* **2024**, *15*, 1467608. [CrossRef] [PubMed]
36. Bonino, G.; Mazza, A.; Capiotto, F.; Berti, A.; Pia, L.; Dal Monte, O. Pupil dilation responds to the intrinsic social characteristics of affective touch. *Sci. Rep.* **2024**, *14*, 24297. [CrossRef] [PubMed]
37. Sun, J.; Zhu, L.; Fang, X.; Tang, Y.; Xiao, Y.; Jiang, S.; Lin, J.; Li, Y. Pupil dilation and behavior as complementary measures of fear response in Mice. *Cogn. Neurodyn.* **2024**, *18*, 4047–4054. [CrossRef]
38. Keene, P.A.; deBettencourt, M.T.; Awh, E.; Vogel, E.K. Pupillometry signatures of sustained attention and working memory. *Atten. Percept. Psychophys.* **2022**, *84*, 2472–2482. [CrossRef] [PubMed]
39. Zokaei, N.; Board, A.G.; Manohar, S.G.; Nobre, A.C. Modulation of the pupillary response by the content of visual working memory. *Proc. Natl. Acad. Sci. USA* **2019**, *116*, 22802–22810. Available online: https://www.pnas.org/doi/abs/10.1073/pnas.1909959116 (accessed on 5 November 2024). [CrossRef]
40. Zhang, Z.; Shan, L.; Wang, Y.; Li, W.; Jiang, M.; Liang, F.; Feng, S.; Lu, Z.; Wang, H.; Dai, J. Primate preoptic neurons drive hypothermia and cold defense. *Innovation* **2023**, *4*, 100358. Available online: https://www.ncbi.nlm.nih.gov/pubmed/36583100 (accessed on 5 December 2023). [CrossRef]
41. Liu, X.H.; Gan, L.; Zhang, Z.T.; Yu, P.K.; Dai, J. Probing the processing of facial expressions in monkeys via time perception and eye tracking. *Zool. Res.* **2023**, *44*, 882–893. Available online: https://www.ncbi.nlm.nih.gov/pubmed/37545418 (accessed on 18 September 2023). [CrossRef]
42. Shan, L.; Yuan, L.; Zhang, B.; Ma, J.; Xu, X.; Gu, F.; Jiang, Y.; Dai, J. Neural Integration of Audiovisual Sensory Inputs in Macaque Amygdala and Adjacent Regions. *Neurosci. Bull.* **2023**, *39*, 1749–1761. Available online: https://www.ncbi.nlm.nih.gov/pubmed/36920645 (accessed on 15 March 2023). [CrossRef]
43. Wang, C.A.; Baird, T.; Huang, J.; Coutinho, J.D.; Brien, D.C.; Munoz, D.P. Arousal Effects on Pupil Size, Heart Rate, and Skin Conductance in an Emotional Face Task. *Front. Neurol.* **2018**, *9*, 1029. [CrossRef] [PubMed]
44. Finke, J.B.; Deuter, C.E.; Hengesch, X.; Schächinger, H. The time course of pupil dilation evoked by visual sexual stimuli: Exploring the underlying ANS mechanisms. *Psychophysiology* **2017**, *54*, 1444–1458. [CrossRef]
45. Suzuki, T.W.; Kunimatsu, J.; Tanaka, M. Correlation between Pupil Size and Subjective Passage of Time in Non-Human Primates. *J. Neurosci.* **2016**, *36*, 11331–11337. [CrossRef] [PubMed]
46. Lang, P.J.; Bradley, M.M.; Cuthbert, B.N. *International Affective Picture System (IAPS): Affective Ratings of Pictures and Instruction Manual*; University of Florida: Gainesville, FL, USA, 2008.
47. Tottenham, N.; Tanaka, J.W.; Leon, A.C.; McCarry, T.; Nurse, M.; Hare, T.A.; Marcus, D.J.; Westerlund, A.; Casey, B.J.; Nelson, C. The NimStim set of facial expressions: Judgments from untrained research participants. *Psychiatry Res.* **2009**, *168*, 242–249. [CrossRef] [PubMed]

48. Maréchal, L.; Levy, X.; Meints, K.; Majolo, B. Experience-based human perception of facial expressions in Barbary macaques (*Macaca sylvanus*). *PeerJ* **2017**, *5*, e3413. [CrossRef] [PubMed]
49. Micheletta, J.; Whitehouse, J.; Parr, L.A.; Waller, B.M. Facial expression recognition in crested macaques (*Macaca nigra*). *Anim. Cogn.* **2015**, *18*, 985–990. [CrossRef]
50. Morozov, A.; Parr, L.A.; Gothard, K.; Paz, R.; Pryluk, R. Automatic Recognition of Macaque Facial Expressions for Detection of Affective States. *eNeuro* **2021**, *8*, 1–16. [CrossRef]
51. Wang, L.; Zhang, B.; Lu, X.; Wang, R.; Ma, J.; Chen, Y.; Zhou, Y.; Dai, J.; Jiang, Y. Genetic and Neuronal Basis for Facial Emotion Perception in Humans and Macaques. *Natl. Sci. Rev.* **2024**, *11*, nwae381. [CrossRef]
52. Willenbockel, V.; Sadr, J.; Fiset, D.; Horne, G.O.; Gosselin, F.; Tanaka, J.W. Controlling low-level image properties: The SHINE toolbox. *Behav. Res. Methods* **2010**, *42*, 671–684. [CrossRef] [PubMed]
53. Isbell, L.A. Snakes as agents of evolutionary change in primate brains. *J. Hum. Evol.* **2006**, *51*, 1–35. [CrossRef] [PubMed]
54. Isbell, L.A. Predation on primates: Ecological patterns and evolutionary consequences. *Evol. Anthropol. Issues News Rev.* **1994**, *3*, 61–71. [CrossRef]
55. Parr, L.A.; Waller, B.M.; Vick, S.J.; Bard, K.A. Classifying chimpanzee facial expressions using muscle action. *Emotion* **2007**, *7*, 172–181. [CrossRef] [PubMed]
56. Swystun, A.G.; Logan, A.J. Quantifying the effect of viewpoint changes on sensitivity to face identity. *Vis. Res.* **2019**, *165*, 1–12. [CrossRef] [PubMed]
57. Lee, Y.; Matsumiya, K.; Wilson, H.R. Size-invariant but viewpoint-dependent representation of faces. *Vis. Res.* **2006**, *46*, 1901–1910. [CrossRef] [PubMed]
58. Hill, H.; Schyns, P.G.; Akamatsu, S. Information and viewpoint dependence in face recognition. *Cognition* **1997**, *62*, 201–222. [CrossRef] [PubMed]
59. Guo, K.; Shaw, H. Face in profile view reduces perceived facial expression intensity: An eye-tracking study. *Acta Psychol.* **2015**, *155*, 19–28. [CrossRef] [PubMed]
60. Ewbank, M.P.; Smith, W.A.; Hancock, E.R.; Andrews, T.J. The M170 reflects a viewpoint-dependent representation for both familiar and unfamiliar faces. *Cereb. Cortex* **2008**, *18*, 364–370. [CrossRef] [PubMed]
61. Magnuski, M.; Gola, M. It's not only in the eyes: Nonlinear relationship between face orientation and N170 amplitude irrespective of eye presence. *Int. J. Psychophysiol.* **2013**, *89*, 358–365. [CrossRef]
62. Jacques, C.; Rossion, B. Misaligning face halves increases and delays the N170 specifically for upright faces: Implications for the nature of early face representations. *Brain Res.* **2010**, *1318*, 96–109. [CrossRef]
63. Orczyk, J.; Schroeder, C.E.; Abeles, I.Y.; Gomez-Ramirez, M.; Butler, P.D.; Kajikawa, Y. Comparison of Scalp ERP to Faces in Macaques and Humans. *Front. Syst. Neurosci.* **2021**, *15*, 667611. [CrossRef] [PubMed]
64. Kardon, R. Pupillary light reflex. *Curr. Opin. Ophthalmol.* **1995**, *6*, 20–26. [CrossRef]
65. Menzel, C.; Hayn-Leichsenring, G.U.; Langner, O.; Wiese, H.; Redies, C. Fourier power spectrum characteristics of face photographs: Attractiveness perception depends on low-level image properties. *PLoS ONE* **2015**, *10*, e0122801. [CrossRef]
66. Barros, F.; Soares, S.C.; Rocha, M.; Bem-Haja, P.; Silva, S.; Lundqvist, D. The angry versus happy recognition advantage: The role of emotional and physical properties. *Psychol. Res.* **2023**, *87*, 108–123. [CrossRef] [PubMed]
67. Huang, T.; Xu, H.; Wang, H.; Huang, H.; Xu, Y.; Li, B.; Hong, S.; Feng, G.; Kui, S.; Liu, G.; et al. Artificial intelligence for medicine: Progress, challenges, and perspectives. *Innov. Med.* **2023**, *1*, 100030. Available online: https://www.the-innovation.org/medicine//article/doi/10.59717/j.xinn-med.2023.100030 (accessed on 15 September 2023). [CrossRef]
68. Feng, G.; Xu, H.; Wan, S.; Wang, H.; Chen, X.; Magari, R.; Han, Y.; Wei, Y.; Gu, H. Twelve practical recommendations for developing and applying clinical predictive models. *Innov. Med.* **2024**, *2*, 100105. Available online: https://www.the-innovation.org/medicine//article/doi/10.59717/j.xinn-med.2024.100105 (accessed on 6 December 2024). [CrossRef]

applied sciences

MDPI

Article

An Eye-Tracking Study on Text Comprehension While Listening to Music: Preliminary Results

Georgia Andreou and Maria Gkantaki *

Department of Special Education, University of Thessaly, 38221 Volos, Greece; andreou@sed.uth.gr
* Correspondence: mariagkantaki90@gmail.com

Abstract: The aim of the present study was to examine the effect of background music on text comprehension using eye-tracking technology. Ten Greek undergraduate students read four texts under the following four reading conditions: preferred music, non-preferred music, café noise, and in silence. Eye movements were tracked to assess visual patterns, while reading performance and attitudes were also evaluated. The results showed that fixation measures remained stable across conditions, suggesting that early visual processing is not significantly influenced by auditory distractions. However, reading performance significantly declined under non-preferred music, highlighting its disruptive impact on cognitive processing. Participants also reported greater difficulty and fatigue in this condition, consistent with an increased cognitive load. In contrast, preferred music and silence were associated with enhanced understanding, confidence, and immersion, café noise also had a moderate but manageable effect on reading outcomes. These findings underscore the importance of tailoring reading environments to individual preferences in order to optimize reading performance and engagement. Future research studies should focus on the effects of different musical attributes, such as tempo and genre, and use more complex reading tasks, in order to better understand how auditory stimuli interact with cognitive load and visual processing.

Keywords: adult students; background music; eye movements; text comprehension; reading attitudes

1. Introduction

Reading constitutes a cognitively demanding process which is influenced by multiple factors that may impact the assimilation of written information [1]. Students often engage in reading and studying within environments where various distractions—whether voluntary or involuntary—compete for their cognitive resources. One such common scenario is reading while simultaneously listening to music or watching television or in the presence of people talking in a café [2–4]. Research has demonstrated that listening to music can influence performance in cognitively demanding tasks. However, the magnitude of this effect varies depending on multiple factors, including the nature of the task one is engaged in [5,6], the type of music played [1,7,8], as well as the personality of the individual [6,9]. Therefore, it can be inferred that physiological and psychological factors affect the readers' text comprehension and have an impact on their reading performance.

Visual patterns, as captured through eye movement measurements, constitute a well-established physiological indicator for assessing the cognitive and perceptual load during the reading process [10,11]. The eye-tracking technology is considered one of the methods that has started to attract the interest of educational researchers [12,13], as it is a method that offers the distinct advantage of providing real-time measures of the reading process while capturing moment-to-moment cognitive control during reading [14]. Furthermore, numerous studies have demonstrated that eye-tracking is an effective research tool for elucidating the readers' processing strategies during reading [15–22]. One of the most significant general eye movement measures relevant to reading is fixation duration [12]. Fixation duration refers to the period during which the eye remains focused on a specific word. In adult readers, fixations typically range from 60 to 500 milliseconds (ms), with an average duration of approximately 250 ms [12]. A longer fixation duration is typically linked to increased cognitive demands during task performance [23], such as tasks performed while listening to music or under the conditions of café noise.

The distribution of visual attention is susceptible to be influenced by a multitude of cognitive processes, including, but not limited to perception, memory, language, and decision-making. Additionally, the duration of focus can be influenced by factors such as the point at which the eyes focus. Despite the inaccuracy of the eye-brain connection, a substantial body of research has demonstrated that the eyes reflect the cognitive processing of the object being looked at in that particular moment [23–26]. This makes eye-tracking a widely used tool in research investigating cognitive processes, as it provides data on the moment-to-moment evolution of cognition over time, rather than only revealing the final outcome. Although each individual is capable of selecting what to look at and when to look at it, the subtleties of that eye movement are largely involuntary. This is due to the fact that each individual lacks memory regarding the specific point in time at which they looked at a particular object [13]. Consequently, eye-tracking can capitalize on the processing of data that occurs outside conscious awareness.

A second key aspect in the examination of text comprehension is reading performance, which is assessed through comprehension and reading time. Comprehension reflects the reader's ability to process textual information, understand it, and integrate objective content, whereas reading time denotes the duration required for a participant to complete a given text [10,27]. Previous research has investigated the influence of background music on reading comprehension, with findings indicating that certain types of music, such as Mozart's compositions and highly repetitive music, can have a positive effect [28]. In addition, there are studies which suggest that background music may enhance the learners' emotional states by helping to regulate their valence and arousal levels during reading [1,6]. On the other hand, other studies have reported negative effects on reading comprehension when background music includes genres such as hip-hop, slow-tempo music, fast and loud music, or familiar non-lyrical music [29]. The discrepancies in these findings may be attributed to variations in musical characteristics, including loudness, tempo, structure, complexity, genre, familiarity, and whether the music contains lyrics or is purely instrumental [30]. Moreover, individual differences among the participants, such as musical preferences and expertise, may further contribute to these inconsistencies [1,6]. Furthermore, one possible explanation for this discrepancy is that the reading comprehension is influenced by a multitude of factors. As a complex cognitive process, reading is shaped by elements such as word recognition, reading strategies, comprehension skills, and motivation [31].

Finally, understanding the readers' attitudes towards music offers a psychological perspective on reading conditions that may differ from objective reading outcomes, such as text comprehension. From a cognitive science standpoint, an individual's perception or self-assessment (e.g., perceived levels of understanding or difficulty) does not always align with objective behavioral measures, including visual patterns and reading comprehension [10]. Therefore, beyond assessing reading comprehension, evaluating reading attitudes—by exploring perceived aspects such as difficulty, confidence, and preference—

may provide additional insights into text comprehension. Previous research on the subject has highlighted the impact of reading attitudes on comprehension. Specifically, intrinsic motivation, perceived difficulty, and self-assessed confidence in comprehension have been found to influence reading performance and achievement [10].

Given the significance of investigating visual patterns, reading performance, and reading attitudes, this study examines these variables to gain a deeper understanding of how text comprehension is affected by different types of music and a certain type of background noise, using the eye-tracking technology. In this study, in order to investigate whether the claim that certain types of music improve text comprehension is true, we have chosen to focus not on a specific type of music or certain aspects of it, but on personal preferences: reading while listening to preferred music and reading while listening to non-preferred music. In addition, as students often choose to study in environments where there is noise from people talking, text comprehension will be studied in the presence of recorded noise from a café. Finally, text comprehension is studied under the condition of silence. In order to study the reading process as it unfolds, we used eye-tracking technology. In particular, the research hypotheses are proposed in the following sequence, which reflects the process from visual patterns to reading performance and to reading attitudes. This sequence is consistent with the structure and the aims of the study:

Visual Patterns Hypotheses:

Hypothesis 1 (H1): *Time to First Fixation (TTFF) is expected to be longer in the non-preferred music and café noise conditions than in the preferred music and silence conditions.*

Hypothesis 2 (H2): *Participants are expected to exhibit longer fixation duration and more fixations count when reading under the non-preferred music and café noise conditions than under the preferred music and silence conditions.*

Hypothesis 3 (H3): *First Fixation Duration (FFD) is not expected to differ significantly across the four reading conditions.*

Reading Performance Hypotheses:

Hypothesis 4 (H4): *Participants are expected to achieve higher reading comprehension scores in the preferred music and silence conditions than in the non-preferred music and café noise conditions.*

Hypothesis 5 (H5): *Reading time is expected to be longer in the non-preferred music and café noise conditions than in the preferred music and silence conditions.*

Reading Attitudes Hypotheses:

Hypothesis 6 (H6): *Participants are expected to report higher perceived difficulty and fatigue when reading under the non-preferred music and café noise conditions than reading under the preferred music and silence conditions.*

Hypothesis 7 (H7): *Participants are expected to report higher perceived understanding, confidence, and immersion when reading under the preferred music and silence conditions than under the non-preferred music and café noise conditions.*

2. Materials and Methods

The objective of the experiment was to investigate the impact of different auditory environments on visual patterns, reading performance and reading attitudes. Four reading conditions were employed: preferred music, non-preferred music, café noise, and silence. The experimental design was focused on measuring the extent to which readers could comprehend different texts in different reading conditions. Visual patterns were measured by fixation duration, fixations count, TTFF and FFD. Reading performance was measured by reading comprehension and reading time, and reading attitudes were assessed by five measures of reader attitude: perceived difficulty, perceived understanding, perceived confidence, perceived fatigue, and perceived immersion.

2.1. Experimental Set Up

The reading materials were selected from a reading comprehension section of the Greek language texts of the Panhellenic Examinations, which are conducted on an annual basis and serve to assess the comprehension and knowledge of the Greek language of students seeking admission to Greek universities. In order to ensure consistency across the texts, the reading materials were selected in accordance with two criteria. Firstly, it was essential that all the selected texts were of an identical length and readability. The texts were of a similar length, with an average of 480 words (ranging from a minimum of 407 to a maximum of 564) and comprised five to six paragraphs (sentences) of approximately 90 words each (with a minimum of 83 and a maximum of 100). Secondly, it was ensured that the selected texts were different from those used in the national university entrance examinations of the participants. The texts were displayed on a computer screen in a large Arial font (21 points) with a line spacing of 1.5. Each text was followed by five true/false questions. The subject areas of the texts and the corresponding questions were sociology, philosophy and education.

The eye-tracking device employed in the experiment was the Tobii Pro X2-60 Hz system, which records the user's gaze data in real time. The system operates at a sampling rate of 60 Hz and provides a spatial accuracy of <0.5°, making it well-suited for reading research. The computer utilized was a 15.6-inch Dell Vostro 15,500 series monitor, with a resolution of 1024 × 768 pixels and a physical size of 380 × 300 mm. The musical accompaniment was delivered via speakers positioned in front of the participant, adjacent to the monitor. The volume of the background music was standardized to ensure that each participant was able to read the text comfortably. The café noise was a recorded sound from a café, which was then normalized using Adobe Audition 3.0 software. The average volume of the noise was between 60 and 70 dB, with no instances of exceeding 80 dB. The sound was measured while the participants were seated.

2.2. Experimental Protocol

The experiment employed a within-subject experimental design, whereby each participant was exposed to four distinct texts under four different conditions: preferred music, non-preferred music, café noise, and silence. In order to minimize the possibility of carry-over effects, each participant was assigned to receive the reading conditions and texts in a specific order. The particulars of the experimental protocol are described below.

The participants were invited to the Laboratory of Bilingual Education of the Department of Special Education of the University of Thessaly, where the experiment leader provided an explanation of the experiment's purpose and the process of the study (see Figure 1 for an overview of the study design). Ensuring confidentiality and privacy is a standard practice to protect research study participants. Therefore, this process involves concealing the participants' real names and other identifiable characteristics to minimize the risk of exposing their personal information [32]. Participants were de-identified and assigned a unique ID number (for student male and female participants, e.g., SM1, SF2). They were assured that their personal information would be protected and thus each participant signed a consent form, indicating their understanding of the research aims and confidentiality policies, as well as their agreement to participate in the study. After that, the participants were required to complete a reading comprehension ability test and a pre-survey questionnaire concerning their demographics, as well as their music and reading habits. All participants were instructed to bring a 20–30 min long piece of music that they preferred to listen to while reading. The participants selected a diverse range of musical genres, including instrumental classical music, various forms of modern pop and jazz, and vocal music. Additionally, the participants indicated to the experiment leader which genres of music they did not prefer to listen to while reading. The most commonly cited examples of such music were metal and heavy metal. A comprehensive list of all preferred and non-preferred music genres is provided in Table A1. The experiment leader selected 20–30 min long pieces of music from these categories that the participants indicated they did not prefer to listen to while reading. These pieces were then brought to the Laboratory for the experiment.

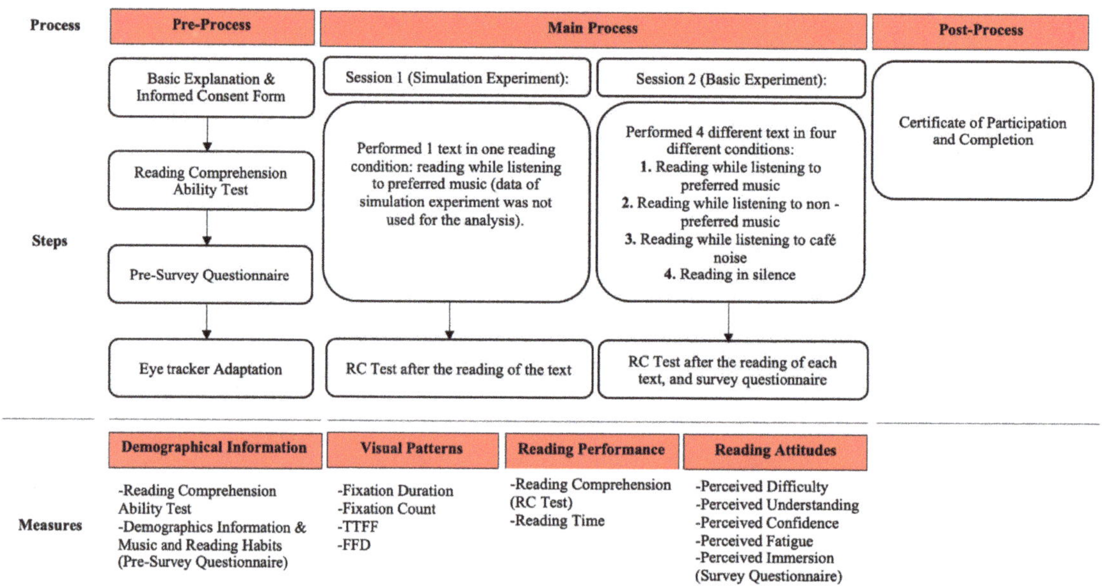

Figure 1. Overview of the study design.

Subsequently, the participants were instructed to assume a seated position in front of the computer and the eye-tracking device. The participants were seated at a distance of 0.40 m from the computer screen, in a manner that was as comfortable as possible. The experiment leader informed the participants that they would be reading four texts while the eye tracker would record their eye movements. Subsequently, they were required to answer to a series of questions of a reading comprehension test (RC Test) designed to evaluate their comprehension of the text they had just read. The experiment leader informed the participants that their eye movements would not be recorded during the responses to the questions. A one-minute intermission was permitted between each reading task. The participants were instructed to read one of the texts while listening to their preferred music, one while listening to non-preferred music, one to recorded noise from a café, and one in complete silence. The participants were not informed of the sequence in which the four reading conditions would be presented. The participants were informed that they could access the texts by using the mouse, and that they were permitted as much time as they required to complete each reading task. To prevent significant discrepancies in overall completion times between the participants, the completion time for each reading task should not exceed 20 min. However, the participants were not informed that their time was being recorded in order to facilitate a sense of comfort and avoid any undue pressure throughout the experiment. Following the completion of the four reading tasks, participants were invited to answer in a survey questionnaire designed to assess their attitudes towards reading.

In regard to the eye-tracking procedure, the participants were informed that a 9-point calibration would be conducted prior to the commencement of each reading task. Calibration accuracy was assessed using deviation errors, with a threshold of <0.5° for acceptance. If the accuracy exceeded this threshold, recalibration was conducted. Additionally, the participants were instructed to maintain a fixed position of the head during the reading tasks. However, they were permitted to move their head freely during the response phase of each text, during which the eye tracker would not record their eye movements. The experiment was conducted in two sessions, with a five-minute rest period between them. In order to facilitate the participants' comfort with the eye-tracking technology, the calibration process, the reading procedure, and the RC Test, a simulation experiment was conducted during the Session 1. The participants were required to read an example text in one reading condition: reading while listening to preferred music and after that, they were asked to complete a RC test. In the second session, the participants completed the basic experiment as described above. After that, the experiment leader presented each participant with a certificate of participation in the research project and provided them with a summary of the project's conclusion.

2.3. Subjects

A total of 10 undergraduate students from the University of Thessaly in Greece ($n = 10$, comprising 9 females and 1 male) participated voluntarily in this pilot study. Table A2 presents the participants' demographic details, as well as their music and reading habits. The mean age of the participants was 22.8 years (SD = 7.2), and they were majoring at the time in the subject of special education. All participants were native speakers of the Greek language and reported having normal or corrected-to-normal vision, with or without the use of contact lenses or glasses.

2.4. Measurements

Prior to the reading tasks, participants completed a survey comprising questions about their demographics, as well as their music and reading habits. The pre-reading survey, comprising 10 questions, was designed to assess eight variables: demographics, reading environment, reading habits, reading performance with background noise and music, self-perception of reading ability with music, reading frequency with music, and music preferences while reading. In the experiment, the visual patterns of the participants were measured in terms of four eye movement measures: fixation duration, fixations count, TTFF and FFD. Moreover, the experiment resulted in the acquisition of two measures of reading performance, which included both reading comprehension and reading time, is evaluated in conjunction with five distinct measures of reading attitudes: perceived difficulty, perceived understanding, perceived confidence, perceived fatigue, and perceived immersion.

2.4.1. Visual Patterns Measurements

The visual patterns were assessed in terms of Fixation Duration (ms), Fixations Count (number of fixations), Time To First Fixation (TTFF) (ms) and First Fixation Duration (FFD) (ms). The measurements were calculated by computing the average value of those performed while the participant read a text on the computer screen. Fixation Duration is the average time spent fixating on a word or area of interest (AOI). Longer fixation duration indicate increased cognitive processing demands. Fixations Count is the total number of fixations recorded per reading task, reflecting the overall level of visual engagement with the text. Time To First Fixation is the time elapsed from text onset until the participant's first fixation on the text, serving as a measure of attentional allocation efficiency, and First Fixation Duration is the duration of the initial fixation on a word, which is associated with early lexical processing. Fixations falling outside the specified cutoff range were excluded from the analysis. The cutoff range was defined based on the standard deviation, specifically two standard deviations above and below the mean. It ranged from below 100 ms to above 500 ms.

2.4.2. Reading Performance Measurements

The reading performance of the participants was evaluated in terms of their reading comprehension and reading time. To assess reading comprehension, participants were required to answer to an RC Test responding to five questions for each text. The questions were in a true/false format and were designed to assess the general understanding of the text by the participants. A score of one was awarded for each correct answer, with a maximum score of five. Given that each reading task was based on a single reading text, the maximum score for each task was five points. The reading time was calculated as the number of seconds (s) required to read each text, beginning with the onset of eye movement and concluding when the reading stopped. The data pertaining to the reading activities were recorded via the eye-tracking device, while the reading time was recorded and analyzed using the eye-tracking software. The data were validated by the research team.

2.4.3. Reading Attitudes Measurements

In the experiment, participants were invited to answer a survey questionnaire designed to assess their attitudes towards reading. The experiment leader employed a fourth-grade Likert scale to calculate the scores for five perceived reading attitudes: difficulty, understanding, confidence, fatigue, and immersion. The following section provides an explanation of the methodology used to calculate each reader's attitude score. Firstly, perceived difficulty refers to the level of difficulty associated with each reading task, as evaluated by the participant. This was assessed through the following question: "How difficult did you perceive this reading task to be?". A score was calculated for the perceived difficulty of reading task. A score was calculated to quantify the perceived difficulty of each reading task, ranging from 1 (very easy) to 4 (very difficult). Secondly, perceived understanding was assessed through the question: "How clearly did you understand the reading text?". Responses were recorded on a scale from 1 (poorly understood) to 4 (completely understood), providing a measure of the participants' self-assessed comprehension of the reading task. Thirdly, the levels of perceived confidence were evaluated through the following question: "How confident are you about your text comprehension?". The scale ranged from 1 (lack of confidence) to 4 (complete confidence). Fourth, perceived fatigue was assessed using two statements: "My eyes became tired as I read the text" and "I felt comfortable reading the text" (reverse-scored). Responses were recorded on a scale from 1 (very unlikely) to 4 (very likely), capturing the extent to which participants experienced fatigue during reading. Similarly, perceived immersion was evaluated through two statements: "I became really involved in reading the text" and "I could not focus on reading the text" (reverse-scored). This scale, ranging from 1 (strongly disagree) to 4 (strongly agree), was employed to measure the degree of engagement participants experienced while reading. The content of the survey questionnaire was based on the relevant literature concerning reading attitudes [10,33].

2.5. Data Preprocessing and Analysis

Raw gaze data were collected and processed using Tobii Pro Eye-Tracker Manager (Tobii Pro X2-60) and iMotions 7.1 software to ensure data integrity and consistency. The preprocessing procedure involved multiple steps, including blink removal, where data points containing blinks or missing values were excluded from fixation calculations, and saccade filtering, where eye movements exceeding a velocity of 30°/s were classified as saccades and removed to isolate fixations. Additionally, fixation aggregation was performed by computing and averaging fixation durations and fixations count for each participant and reading condition. The Statistical Package for Social Sciences (SPSS) was utilized to analyze statistical differences across four visual patterns and two reading performance measures under the four auditory conditions: preferred music, non-preferred music, café noise, and silence. Statistical tests were conducted to examine how these four reading conditions (independent variable) influenced the measured dependent variables, which included fixation duration, fixations count, (TTFF), (FFD), reading time, reading comprehension, perceived difficulty, perceived understanding, perceived confidence, perceived fatigue, and perceived immersion. In order to decide on the suitable test to be performed, the assumptions of Analysis of Variance (ANOVA) were initially examined for all the related dependent variables. For each variable, the assumptions assessed included the independence of observations, normality (observations within each status should be normally distributed) and homogeneity of variance (variance of the observations within each status should be equal). In case all the assumptions were met, the ANOVA test is used, otherwise the non-parametric statistical test of Kruskal–Wallis is used to determine if there were statistical differences for the dependent variable between the independent reading statuses. The results of the ANOVA assumptions assessment are summarized in Table 1. For the variables fixation duration, fixations count and reading time, all the assumptions were met and ANOVA is used. For the rest of the variables, the normality assumption was violated, therefore the Kruskal–Wallis test was used. All analyses were corrected using the post hoc tests—the Bonferroni post hoc test after an ANOVA has been performed and the 'Dwass, Steel, Critchlow-Fligner' test after a Kruskal–Wallis test has been performed.

Table 1. ANOVA assumptions assessment.

	Independence of Observations	**Normality**	**Homogeneity of Variance**	**ANOVA**	**Kruskal–Wallis Test**
Fixation Duration (ms)	Yes	Yes	Yes	Yes	No
Fixation Counts	Yes	Yes	Yes	Yes	No
TTFF (ms)	Yes	No	No	No	Yes
FFD (ms)	Yes	No	Yes	No	Yes
Reading Comprehension (0–10 point scale)	Yes	No	No	No	Yes
Reading Time (s)	Yes	Yes	Yes	Yes	No
Perceived Difficulty	Yes	No	Yes	No	Yes
Perceived Understanding	Yes	No	No	No	Yes

Table 1. *Cont.*

	Independence of Observations	Normality	Homogeneity of Variance	ANOVA	Kruskal–Wallis Test
Perceived Confidence	Yes	No	No	No	Yes
Perceived Fatigue	Yes	No	Yes	No	Yes
Perceived Immersion	Yes	No	Yes	No	Yes

3. Results

This section presents the results from the statistical analysis of the data obtained from the experiment regarding our hypotheses on visual patterns, reading performance, and reading attitudes.

The mean values and the standard deviations of the examined variables, as well as the F-values and *p*-values of the ANOVA analyses, are presented in Table 2 below. For the visual patterns, ANOVA analysis results for the fixation duration (F(3,36) = 0.16, *p*-value = 0.924), fixations count (F(3,36) = 0.66, *p*-value = 0.585) and Kruskal–Wallis test for TTFF (X2(3) = 6.02, *p*-value = 0.110) and FFD (X2(3)= 3.60, *p*-value = 0.309) indicated, as shown in Figure 2, that there is no statistically significant difference in the variables' means across the four different study conditions.

Table 2. Mean values for visual patterns, reading performance, and reading attitudes across preferred music, non-preferred music, café noise, and silence conditions.

Variable	Measure (Unit)	Preferred Music (SD)	Non-Preferred Music (SD)	Café Noise (SD)	Silence (SD)	ANOVA		Kruskal–Wallis	
						F (df)	Sig.	X^2 (df)	*p*-Value
Visual patterns	Fixation Duration (ms)	156.2 (24.5)	160.9 (26.4)	155.6 (24.2)	153.4 (24.9)	0.16 (3.36)	0.924	-	-
	Fixations Count	489.8 (345.1)	392.9 (232.6)	388.4 (291.6)	315.9 (229.6)	0.66 (3.36)	0.585	-	-
	TTFF (ms)	668.5 (1024.1)	1999.3 (2517.3)	4581.1 (10,284.0)	1394.9 (1682.4)	-	-	6.02 (3)	0.110
	FFD (ms)	136.7 (33.1)	164.0 (101.1)	203.6 (120.1)	146.7 (76.2)	-	-	3.60 (3)	0.309
Reading performance	Reading Comprehension (0–10 point scale)	9.2 (1.4)	8.8 (1.7)	9.6 (0.8)	9.2 (1.0)	-	-	1.47 (3)	0.690
	Reading Time (s)	220.9 (79.0)	170.5 (46.8)	189.6 (49.5)	154.4 (60.5)	2.25 (3.36)	0.099	-	-
Reading Attitudes	Perceived Difficulty (0–4 point scale)	2.0 (0.5)	2.9 (0.3)	2.4 (0.5)	1.8 (0.4)	-	-	20.80 (3)	0.000 *
	Perceived Understanding (0–4 point scale)	3.1 (0.2)	1.8 (0.6)	2.8 (0.4)	3.1 (0.3)	-	-	25.67 (3)	0.000 *

Table 2. *Cont.*

Variable	Measure (Unit)	Preferred Music (SD)	Non-Preferred Music (SD)	Café Noise (SD)	Silence (SD)	ANOVA		Kruskal–Wallis	
						F (df)	Sig.	χ^2 (df)	*p*-Value
Reading Attitudes	Perceived Confidence (0–4 point scale)	3.7 (0.5)	2.3 (0.7)	2.8 (0.4)	3.9 (0.3)	-	-	27.62 (3)	0.000 *
	Perceived Fatigue (0–4 point scale)	1.7 (0.7)	2.3 (0.5)	1.6 (0.5)	1.6 (0.5)	-	-	8.80 (3)	0.032 *
	Perceived Immersion (0–4 point scale)	2.2 (0.3)	1.2 (0.2)	1.8 (0.5)	2.2 (0.4)	-	-	21.00 (3)	0.000 *

* $p \leq 0.05$.

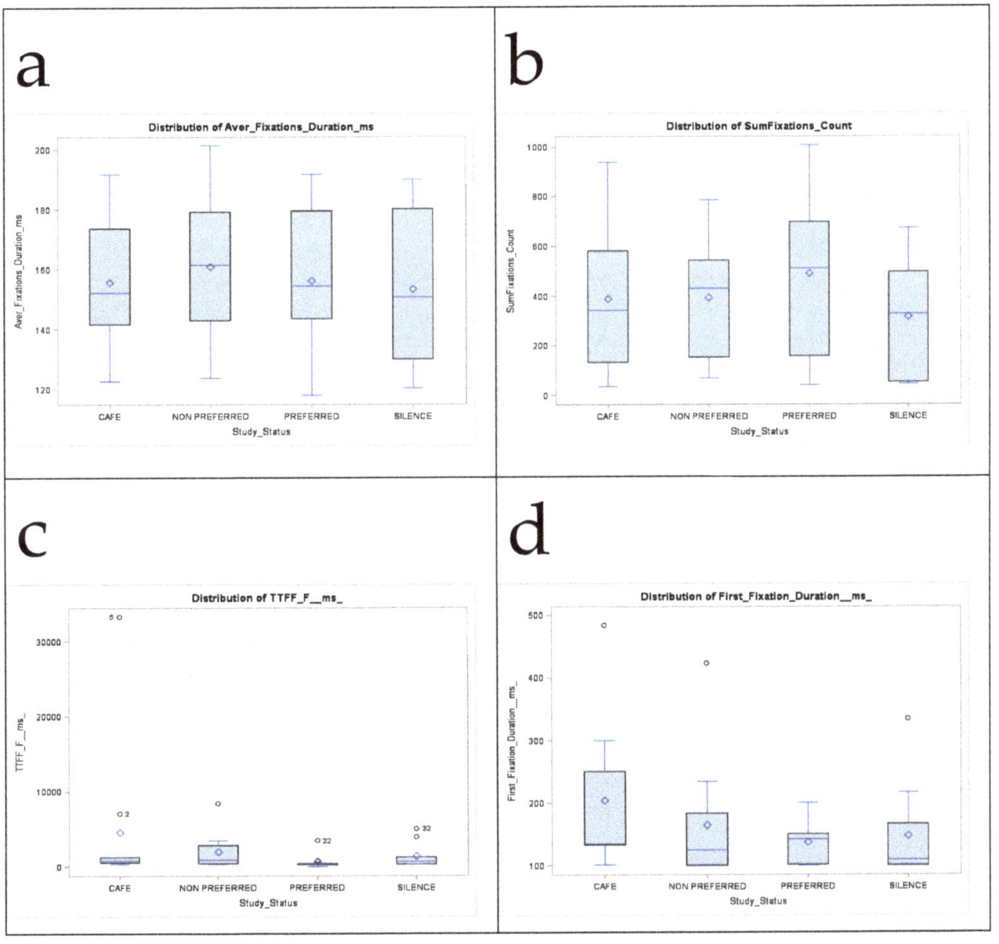

Figure 2. Mean measures for visual patterns during the four reading conditions. Boxplot-distribution of Fixation Duration (**a**), Fixations Count (**b**), TTFF (**c**), and FFD (**d**) for each reading condition.

Regarding reading performance, for both reading comprehension (X2(3) = 1.47, p-value = 0.690) and reading time (F(3,36) = 2.25, p-value = 0.099), the Kruskal–Wallis and ANOVA tests, respectively, revealed, as shown in Figure 3, that there was not a statistically significant difference between at least two groups.

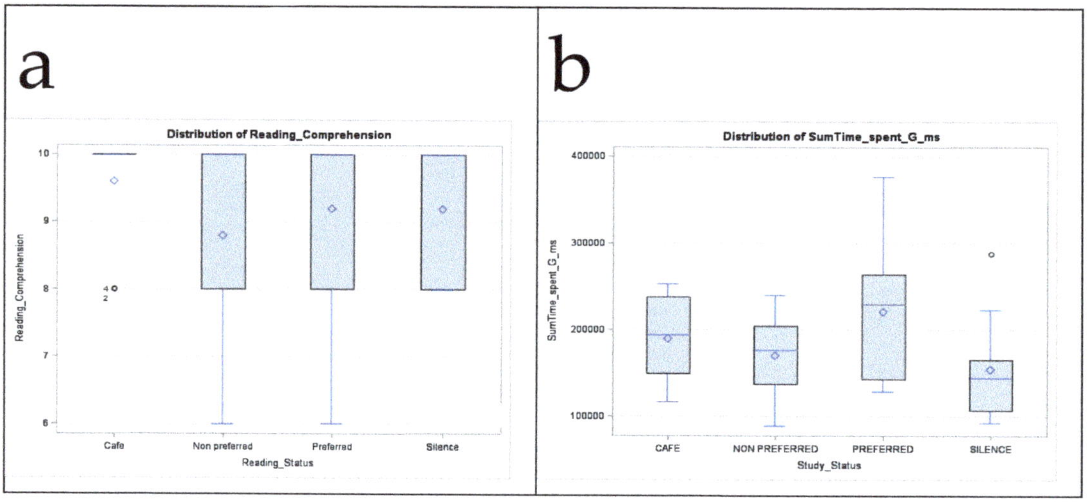

Figure 3. Mean measures for reading performance during the four reading conditions. Boxplot-distribution of Reading Comprehension (**a**), and Reading Time (**b**) for each reading condition.

For the five examined reading attitudes, perceived difficulty (X2(3) = 20.80, p-value = 0.000), perceived understanding (X2(3) = 25.67, p-value = 0.000), perceived confidence (X2(3) = 27.62, p-value = 0.000), perceived fatigue (X2(3) = 8.80, p-value = 0.032) and perceived immersion (X2(3) = 21.00, p-value = 0.000), the Kruskal–Wallis test indicated, as shown in Figure 4, that there is statistically significant difference across preferred music (P), non-preferred music (NP), café noise (C), and silence (S) (Figure 4).

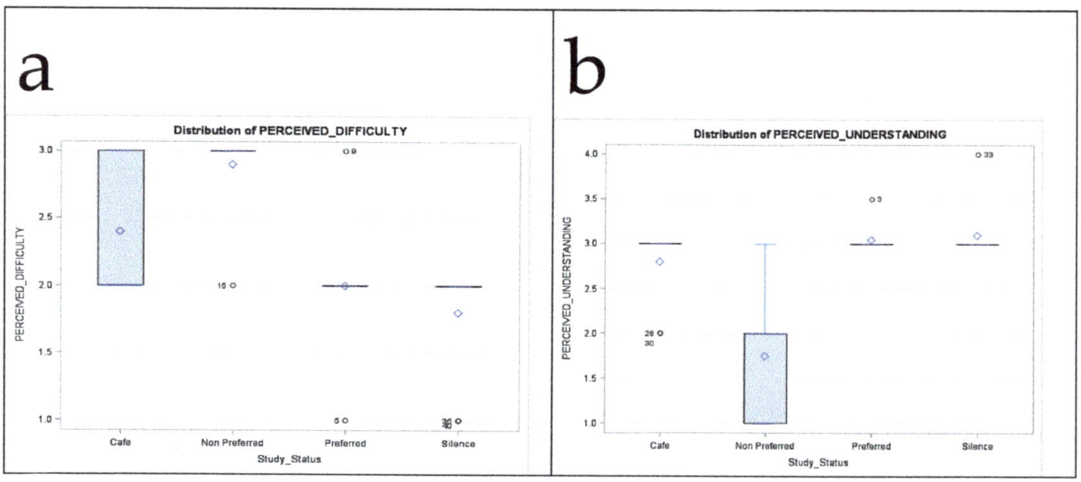

Figure 4. *Cont.*

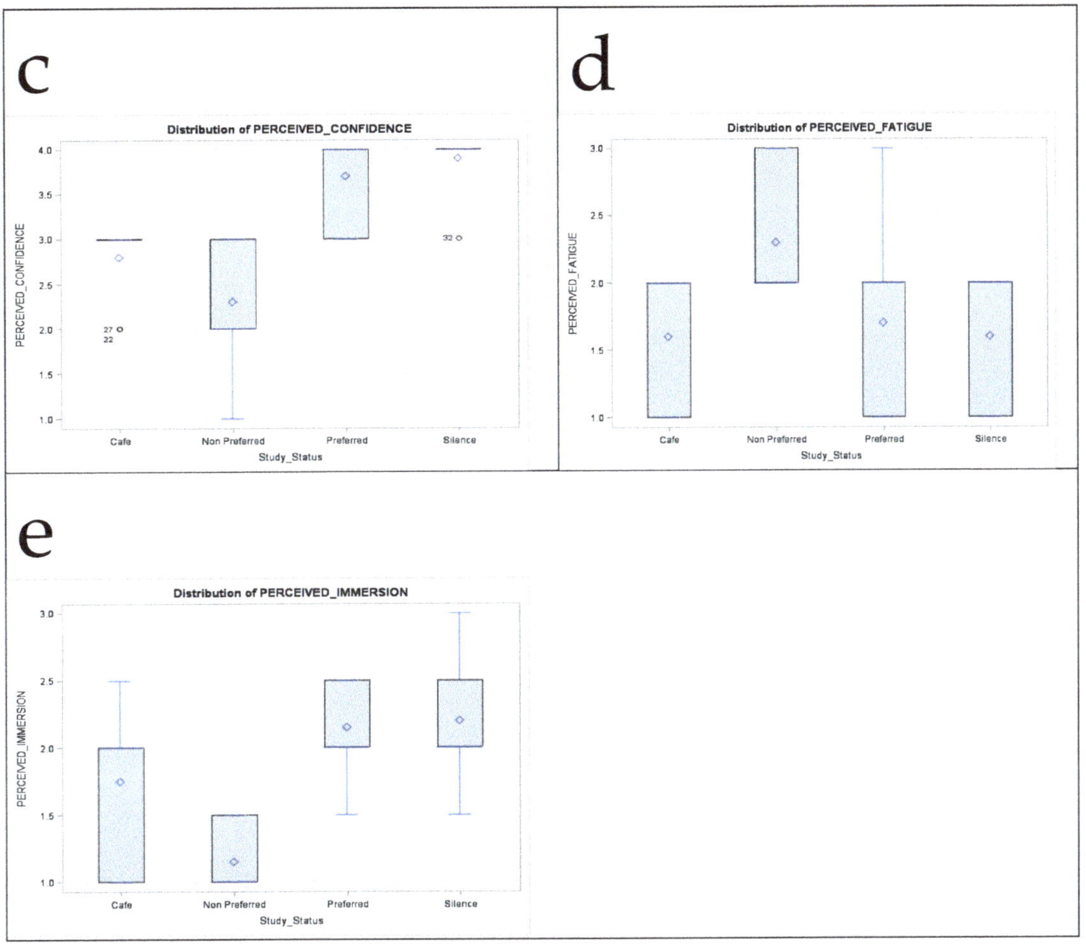

Figure 4. Mean measures for reading attitudes during the four reading conditions. Boxplots-distribution of Perceived Difficulty (**a**), Perceived Understanding (**b**), Perceived Confidence (**c**), Perceived Fatigue (**d**), and Perceived Immersion (**e**) for each reading condition.

The "Dwass, Steel, Critchlow-Fligner" (DSCF) post hoc tests for each of the reading attitudes are presented in Table 3. For perceived difficulty, the non-preferred music had significantly higher mean value compared to silence and preferred music. The larger difference was observed between silence and non-preferred music (DSCF = 5.43, $p = 0.000$), followed by the difference between preferred and non-preferred music (DSCF = 4.89, $p = 0.000$). For perceived understanding, the significant differences were observed between preferred and non-preferred music, silence and non-preferred music and café noise and non-preferred music. The higher level of understanding was observed for preferred music and silence (mean = 3.1), then for café noise (mean = 2.8) and finally for non-preferred (mean = 1.8). The larger difference was observed between silence and non-preferred music and preferred and non-preferred music (DSCF = 5.25, $p = 0.000$). For perceived confidence, the silence and preferred music had the higher levels of confidence, while a lower mean confidence was observed for non-preferred (mean = 2.3). The largest difference was observed between silence and non-preferred music (DSCF = 5.47, $p = 0.000$). For perceived fatigue, non-preferred music (mean = 2.3) has significantly higher mean fatigue values compared to silence (mean = 1.6) and café noise (mean = 1.6). The differences are the same for both comparisons ($p = 0.047$). For perceived immersion, the significant differences were observed between preferred and non-preferred music, silence and non-preferred music and café noise and non-preferred music. The highest level of immersion was observed for preferred music and silence (mean = 2.2), then for café noise (mean = 1.8), and for non-preferred music (mean = 1.2). The largest difference was observed between preferred and non-preferred music (DSCF = 5.38, $p = 0.001$). In conclusion, it is noted that for all reading attitudes, the DSCF post hoc tests indicated no significant difference between preferred music and silence, while the most significant and larger differences were observed between silence and non-preferred music.

Table 3. 'Dwass, Steel, Critchlow-Fligner' post hoc tests for each of the reading attitudes under the preferred music (P), non-preferred music (NP), café noise (C), and silence (S) conditions.

	P-S		P-C		P–NP		S-C		S–NP		C-NP	
Reading Attitude	**DSCF**	***p*-Value**	**DSCF**	***p*-Value**	**DSCF**	***p*-Value**	**DSCF**	***p*-Value**	**DSCF**	***p*-Value**	**DSCF**	***p*-Value**
Perceived Difficulty (0–4 point scale)	0.98	1.381	2.40	0.325	4.89 *	0.003	3.45	0.070	5.43 *	0.001	3.23	0.102
Perceived Understanding (0–4 point scale)	0.10	1.000	2.41	0.322	5.25 *	0.001	2.41	0.322	5.25 *	0.001	4.48 *	0.008
Perceived Confidence (0–4 point scale)	1.54	0.696	4.57 *	0.007	4.96 *	0.003	5.44*	0.001	5.47 *	0.001	2.60	0.254
Perceived Fatigue (0–4 point scale)	0.37	0.994	0.37	0.994	2.92	0.164	0.00	1.000	3.66 *	0.047	3.66 *	0.047
Perceived Immersion (0–4 point scale)	0.23	0.998	2.48	0.294	5.38 *	0.001	2.53	0.278	5.37 *	0.001	3.54	0.059

* $p \leq 0.05$.

4. Discussion

In this study, we used the eye-tracking technology to examine how text comprehension is affected by different reading conditions (preferred music, non-preferred music, café noise and in silence) across the three variables of visual patterns, reading performance and reading attitudes.

4.1. Visual Patterns in Text Comprehension

According to H1, Time To First Fixation (TTFF) is expected to be longer in the non-preferred music and café noise conditions than in the preferred music and silence conditions. This hypothesis was based on the assumption that non-preferred music and café noise would delay visual attention allocation due to increased cognitive load. Contrary to the hypothesis, there was no statistically significant difference in TTFF across the four reading conditions: preferred music, non-preferred music, café noise, and silence. Participants initiated the reading at a consistent pace regardless of the auditory condition, indicating their ability to adapt promptly to the auditory stimuli.

This finding contradicts studies indicating that auditory distractions, especially unpredictable or aversive noise, can delay attention allocation and increase TTFF [22,34,35]. For instance, Tsai et al. [22] found that a noise-induced cognitive load prolongs fixation time as readers filter out irrelevant stimuli. However, the findings align with studies suggesting that visual attention processes, particularly initial fixation, are largely automatic and less affected by auditory distractions [13,23]. Furthermore, the consistent TTFF observed in this study lends support to the notion that visual attention remains stable in the presence of familiar or predictable distractions, as evidenced by prior research in reading in café noise and preferred music conditions [23].

In conclusion, the observed discrepancy between the results and our initial hypothesis may be attributed to the participants' familiarity with background noise, as indicated by the 40% response rate indicating frequent use of music while studying (see Appendix A, Table A2). Furthermore, the predictable nature of café noise is likely to have reduced its cognitive load, resulting in a reduced disruptive impact. Therefore, H1 was not confirmed.

According to H2, the participants were expected to exhibit longer fixation duration and more fixations count when reading under the non-preferred music and café noise conditions than under the preferred music and silence conditions. This was based on the assumption that non-preferred music and café noise would increase cognitive load, leading to more frequent and prolonged fixations as readers make an effort to compensate for auditory distractions. However, the findings did not support this hypothesis as no statistically significant differences were observed in the fixation duration and fixations count across all four conditions (preferred music, non-preferred music, café noise, and silence). Participants demonstrated consistent visual patterns irrespective of the auditory environment, indicating effective management of distractions and sustained visual focus.

The findings contradict the predictions of the cognitive load theory, which states that auditory distractions increase cognitive demands, resulting in longer fixation duration and higher fixations count as readers process text more carefully [13,35,36]. For example, Rayner [13] found that challenging tasks, particularly those involving distractions, often require additional visual resources to overcome external challenges.

However, the results are consistent with studies suggesting that predictable noise, such as café noise, exerts minimal impact on visual patterns compared to unpredictable or fluctuating auditory distractions [34]. The findings further corroborate the notion that individuals with experience in handling background noise or environments devoid of distractions possess the capacity to regulate their attention effectively [13]. The participants' inclination towards instrumental or low-distraction music (55.6%) and their frequent engagement in study activities accompanied by background music (40%), as shown in Table A2, are likely to have contributed to their ability to maintain consistent visual patterns.

The observed discrepancy between the results and our initial hypothesis could be attributed to the relatively uncomplicated reading tasks used in this study, which did not impose a significant cognitive load on the participants. Additionally, the participants' familiarity with the auditory environments may have also reduced the impact of the distraction caused by non-preferred music and café noise. Therefore, H2 was not confirmed.

According to H3, First Fixation Duration (FFD) is not expected to differ significantly across the four reading conditions. This prediction was based on the assumption that FFD, as a measure of initial visual engagement, reflects automatic, low-level processing that is less susceptible to external auditory distractions. FFD did not vary significantly across the four reading conditions, indicating that auditory distractions, such as non-preferred music and café noise, do not affect the participants' initial engagement with the text.

These findings are consistent with those of the existing research. Rayner [13] claimed that FFD primarily represents automatic processes such as word recognition and attention allocation, which are robust against environmental distractions. In addition, Tsai et al. [22] found that FFD remains stable in predictable environments, such as those of café noise, where distractions are less intrusive. Furthermore, Haapakangas et al. [34] observed that low-level auditory distractions minimally influence early visual processing stages, though they may affect later cognitive engagement. Additionally, Reichle et al. [23] observed that FFD remains stable unless readers encounter significantly complex or unfamiliar text. Our results align with the broader consensus that FFD reflects low-level, automatic processing unaffected by auditory distractions in familiar or predictable environments. Therefore, H3 was confirmed and the findings suggest that FFD is a robust measure of early visual processing that remains stable across auditory conditions, particularly when tasks are relatively simple and readers are familiar with the auditory environment.

4.2. Reading Performance in Text Comprehension

According to H4, participants are expected to achieve higher reading comprehension scores in the preferred music and silence conditions than in the non-preferred music and café noise conditions. This prediction was based on the idea that non-preferred music and café noise would increase cognitive load, impairing text comprehension, while preferred music and silence would allow for better cognitive focus and processing. The participants' reading comprehension scores were statistically significantly lower in the non-preferred music condition than the other conditions, supporting the hypothesis that non-preferred music increases cognitive load and has a negative impact on text comprehension. However, no statistically significant difference was observed between comprehension scores obtained for the preferred music, café noise, and silence conditions. This suggests that preferred music and silence allow for comparable reading performance, while also café noise did not impair comprehension at a statistically significant level.

The findings align with those of several studies that highlight the disruptive effects of non-preferred music and the cognitive benefits of silence or preferred music. Many studies reported that non-preferred music impairs cognitive performance, while silence and preferred music enhance comprehension [1,8,28]. Haapakangas et al. [34] and Sweller et al. [35] claimed that silence reduces cognitive load, enabling better focus and comprehension. Johansson et al. [1] observed better comprehension under preferred music or silence compared to non-preferred or loud music. Haapakangas et al. [34] and Que et al. [6] suggested that moderate, predictable noise, such as café noise, has less detrimental effects than irregular, disruptive noise, which aligns with this study's finding of no significant impact from café noise.

While this study found no statistically significant difference in comprehension between café noise, preferred music, and silence conditions, some prior studies indicate that café noise can impair cognitive performance. Perham and Sykora [8] emphasized that background noise, including café noise, often disrupts cognitive tasks by increasing cognitive load. In the same vein, Sweller et al. [35] suggested that auditory distractions, including café noise, could overwhelm cognitive systems. The lack of a statistically significant effect of the café noise in this study may be explained by the fact that a number of participants (20% as shown in Table A2) reported familiarity with studying in environments with background noise. Familiarity is likely to have reduced the disruptive potential of café noise, allowing participants to adapt and maintain comprehension. Therefore, H4 was partially confirmed.

According to H5, reading time is expected to be longer in the non-preferred music and café noise conditions than in the preferred music and silence conditions. Our results showed that although reading time in the non-preferred music condition was slightly longer than in other conditions, the difference was not statistically significant. This means that participants were able to maintain a consistent reading pace despite the presence of non-preferred music.

The findings of this study partially align with the existing literature. A number of studies have shown that background music, particularly music that individuals do not prefer, can impede cognitive performance, leading to longer reading times [3,37]. However, other research studies suggest that the effect of noise, including non-preferred music, may vary depending on the individual's ability to concentrate or the complexity of the reading task [38]. In this study, the non-statistically significant difference in reading time among preferred music, café noise, and silence conditions indicates that the participants may have employed effective strategies to alleviate the potential distractions. Therefore, H5 was not confirmed.

4.3. Reading Attitudes in Text Comprehension

According to H6, participants are expected to report higher perceived difficulty and fatigue when reading under the non-preferred music and café noise conditions than reading under the preferred music and silence conditions. This hypothesis is based on the premise that non-preferred music and café noise impose a higher cognitive load than the other three reading conditions preferred music and silence, making the reading task more mentally demanding. Participants perceived the non-preferred music and café noise conditions as significantly more challenging and fatiguing than the preferred music and silence conditions. This indicates that non-preferred music and café noise are associated with increased cognitive load, causing participants to exert more mental effort during the reading task. Descriptive statistics revealed that the non-preferred music conditions frequently involved genres such as metal and heavy metal, characterized by high tempo, loud volume, and unpredictable rhythm. These attributes are likely to have disrupted focus and heightened cognitive demands, leading to increased fatigue and difficulty. In addition, as shown by the descriptive statistics, the café noise condition included background sounds typically found in public environments, such as background chatter, clinking utensils and intermittent auditory distractions, which had a negative impact on the participants' studying/reading. The unpredictable nature of these sounds may have increased the participants' cognitive load and disrupted their reading focus, contributing to their increased perceptions of perceived difficulty and fatigue. Conversely, preferred music and silence created more conducive environments for maintaining focus and reducing perceived mental effort.

The findings align with those of the existing literature. Cognitive load theory suggests that distractions such as non-preferred music and café noise impose additional cognitive demands, diverting attention from the primary task and causing increased fatigue and difficult [35]. Perham and Sykora [8] reported that non-preferred music disrupts cognitive tasks by requiring individuals to exert more effort to filter out auditory distractions, leading to perceived fatigue and difficulty. The present findings that low levels of fatigue and difficulty did present in some of the participants under the preferred music condition align with those of several other studies that demonstrated that preferred music fosters positive emotional states, reduces stress, and alleviates perceived task difficulty of lower fatigue and difficulty in the preferred music condition [28]. Similarly, café noise, although often perceived as ambient and less intrusive, can still disrupt reading performance, especially when it involves background chatter, clinking sounds and variable auditory inputs [34]. Studies suggest that the inconsistency and unpredictability of café noise forces readers to allocate additional cognitive resources to maintain focus, leading to increased perceptions of difficulty and fatigue [39]. Despite this, café noise may be less detrimental than non-preferred music due to its more consistent acoustic profile and participants' potential familiarity with such environments. Therefore, H6 was confirmed.

According to H7, the participants are expected to report higher perceived understanding, confidence, and immersion when reading under the preferred music and silence conditions than under the non-preferred music and café noise conditions. The results of the study showed that the participants reported higher levels of perceived understanding, confidence, and immersion in the preferred music and silence conditions. These environments supported cognitive engagement and task mastery, as evidenced by higher ratings in these areas. Perceived understanding, confidence and immersion were statistically significantly greater in the preferred music and silence conditions than non-preferred music and café noise conditions, reflecting the participants' ability to focus and engage deeply with the reading task in familiar or quiet auditory settings. Conversely, non-preferred music and café noise were associated with low ratings across all three measures, highlighting its disruptive effects on focus and task engagement.

The findings align with those in the existing literature supporting the positive impact of preferred music and silence on cognitive performance and subjective task engagement. Hallam et al. [28] demonstrated that preferred music facilitates task engagement and immersion by fostering positive emotional states and reducing anxiety. The present findings align with those of several studies which demonstrated that the participants felt more confident and immersed when listening to their preferred music. Sweller et al. [35] proposed that silence reduces extraneous cognitive load, enabling a better allocation of cognitive resources to the task. This enhances understanding, confidence, and immersion, as observed in this study. The disruptive effects of non-preferred music are consistent with the findings of Perham and Sykora [8], who demonstrated that unpredictable rhythms and a high intensity increase cognitive load, hindering focus and task confidence. In addition, Haapakangas et al. [34] support the notion that café noise, while distracting, is less disruptive than non-preferred music but does not foster the same level of task engagement as silence or preferred music. Therefore, H7 was confirmed.

5. Conclusions

The findings of this study showed that visual measures, such as fixation duration, fixations count, TTFF and FFD, remained stable across the experimental conditions. This finding aligns with prior research that has highlighted the stability of early-stage visual processing in predictable environments [13,22]. The familiarity with background noise may have enabled the participants to maintain their focus, even when faced with non-preferred or noisy conditions [40].

Furthermore, text comprehension was significantly impaired in the non-preferred music condition, in accordance with the cognitive load theory proposed by Sweller et al. [35]. This theory posits that stimuli perceived as negative and aversive result in increased cognitive demands. In contrast, comprehension scores remained stable in the preferred music, café noise, and silence conditions, indicating that predictable or familiar auditory environments can reduce the disruptive effects of noise [34]. For example, participant SF1 used to listen to pop music while reading, a genre that was familiar to her and did not hinder the reading process, as reflected in her high performance on the comprehension test. On the contrary, metal music, which she chose to listen to in the non-preferred music condition, was a genre she was not familiar with and did not listen to while reading in the past. As a result, her performance was low under this condition. These findings are in accordance with prior studies which have demonstrated that preferred music and silence foster conditions conducive to learning [1].

Moreover, the participants of this study indicated that the non-preferred music condition was perceived as significantly more challenging and fatiguing, thereby confirming the negative impact of distracting noise on task engagement [8]. In contrast, the participants reported improved understanding, confidence, and immersion when exposed to preferred music and silence, which is in accordance with the positive effects of familiar auditory environments on emotional and cognitive states [28]. Although café noise was found to be distracting, it was observed to have a reduced negative impact, which is likely attributable to its consistent and predictable characteristics [39]. These findings highlight the significance of aligning the study environments with individual preferences. It can be concluded that preferred music and silence represent optimal auditory environments for reading, promoting both cognitive engagement and emotional well-being. In contrast, the presence of non-preferred music was found to significantly increase cognitive load and fatigue, underscoring the necessity to avoid aversive auditory stimuli during study sessions. The ability of participants to adapt to café noise indicates that consistent background sounds may be tolerable, particularly for those who are used to studying in public spaces.

This study contributes to the existing literature on the subject in a number of ways. Firstly, the reading conditions of preferred music, café noise, and silence selected in this study are commonly used by students, thus the findings offer insight into how reading habits affect reading performance. Moreover, while previous studies have focused on a single type of music or noise background, contrasting it only with reading sessions conducted in silence [41,42], this study presents four reading conditions, each of which is compared with the silent reading condition. Additionally, this study specifically examined the role of personal music preferences in reading, distinguishing between preferred and non-preferred music conditions, while previous research has primarily focused on factors such as the nature of the reading tasks, the type of music played, and the personality of the individuals involved [5,7,8]. Furthermore, the eye-tracking methodology employed in this study, in which reading comprehension performance was assessed across four reading conditions, represents a significant contribution to the existing literature on the subject, since there is a lack of research investigating text comprehension while listening to music and tracking readers' eye movements simultaneously.

However, the results of this study are not in accordance with the body of research that claims that text comprehension is enhanced in the presence of music rather than in its absence [1,6]. This discrepancy may be attributed to the several limitations of our study. To date, no studies have been conducted examining text comprehension and eye movements while listening to different kinds of music and noise in the context of the Greek language. Unlike many previous studies that examined reading comprehension in English, this research focused on Greek-language texts, which may involve different syntactic structures, lexical properties, and cognitive processing demands. Furthermore, cultural reading habits and educational background may influence how Greek students engage with texts in the presence of background music.

Cultural background plays a significant role in reading habits, cognitive engagement, and text interpretation. Different cultures emphasize distinct reading strategies, such as rote memorization, inferential reasoning, or critical analysis. In the context of this study, Greek university students may approach reading differently from students in English-speaking countries, where previous research on background music and reading comprehension has been conducted. Additionally, exposure to background noise varies across cultures-some students may be accustomed to studying in noisy environments (e.g., cafés, communal study spaces), while others prefer quiet, isolated settings. These factors could influence how participants respond to auditory distractions like music or café noise during reading tasks.

Linguistic structure plays a crucial role in reading comprehension, influencing how individuals process and extract meaning from text. Greek and English differ significantly in morphology, syntax, and orthographic depth, which may affect reading strategies and cognitive load. Greek is a highly inflected language, where word endings change based on grammatical case, number, and tense, requiring readers to engage in more detailed morphological processing than in English. Additionally, Greek has a shallow orthography, meaning there is a strong correspondence between letters and sounds, whereas English has a deep orthography, with irregular spelling rules requiring greater reliance on whole-word recognition. These differences suggest that Greek readers may depend more on phonological processing, while English readers rely on lexical familiarity and contextual cues. Furthermore, Greek syntax allows for a flexible word order, which may require greater cognitive effort in sentence parsing, potentially influencing reading comprehension under auditory distractions. These linguistic differences highlight the need for cross-linguistic comparisons to determine whether background music and noise affect reading comprehension in a language-specific or universal manner.

Furthermore, the present study was conducted with a limited number of participants, namely university students in Greece and an underrepresentation of males in the sample. As a result, the findings cannot be generalized to readers of other languages or cultural contexts. This study was conducted as a pilot investigation with a small sample size ($n = 10$), which limits the generalizability of the findings. While no formal power analysis was performed prior to data collection, the sample size was determined based on feasibility constraints and the exploratory nature of the research. While we acknowledge its limitations, the primary objective was to test the experimental procedure and to refine the methodology for a subsequent main study. This future study will include a larger sample with a balanced number of males and females and a more comprehensive analysis, which will enhance the generalizability of the findings.

It is therefore recommended that future research in this field should focus on a number of additional details concerning the type of music played while reading, as well as how specific characteristics of music (e.g., tempo, lyrics, genre) and noise variability affect reading performance and cognitive load. For example, future research could examine the effects of progressively increasing noise levels during the reading process, as well as the differential impact of exposure to speech-based versus instrumental music. Furthermore, future studies could investigate the impact of eye movements on the syntactic and semantic aspects of the texts in greater detail and employ larger sample sizes in order to increase statistical power and improve the generalizability of findings. A within-subject experimental design would allow for a more controlled comparison across different auditory conditions, reducing variability due to individual differences in reading ability and music sensitivity.

Additionally, controlled exposure to varying noise levels could help determine whether background noise intensity (e.g., low, moderate, or high decibel levels) influences reading performance differently. Manipulating specific acoustic properties—such as music tempo, complexity, familiarity, or instrumental versus vocal content—could provide a more detailed understanding of how auditory input interacts with reading comprehension. Furthermore, cross-linguistic comparisons across languages with different orthographic and syntactic structures would help clarify whether the effects of background music and noise are language-specific or universal. These methodological refinements would contribute to a more comprehensive model of how auditory environments influence reading processes.

Author Contributions: Conceptualization, G.A. and M.G.; methodology, G.A. and M.G.; software, G.A. and M.G.; validation, G.A. and M.G.; formal analysis, G.A. and M.G.; investigation, G.A. and M.G.; resources, G.A. and M.G.; data curation, G.A. and M.G.; writing—original draft preparation, G.A. and M.G.; writing—review and editing, G.A. and M.G.; visualization, G.A. and M.G.; supervision, G.A. and M.G.; project administration, G.A. and M.G.; funding acquisition, G.A. and M.G. All authors have read and agreed to the published version of the manuscript.

Funding: This research received no external funding.

Institutional Review Board Statement: Not applicable.

Informed Consent Statement: Not applicable.

Data Availability Statement: Data is contained within the article.

Acknowledgments: The authors would like to thank the students from the University of Thessaly in Greece, who participated in the study for their contribution to our research.

Conflicts of Interest: The authors declare no conflicts of interest.

Appendix A

Table A1. Participants' choices of preferred and non-preferred music.

Participant	Preferred Music	Non-Preferred Music
SF1	Adele—Hello (Pop)	Pantera—Walk (Metal)
SF2	Dave Brubeck—Take Five (Jazz)	Snik, Ivan, Greko—Gaza (Trap)
SF3	Miles Davis—Freddie Freeloader (Jazz)	Pantera—Walk (Metal)
SF4	Dave Brubeck—Take Five (Jazz)	Metallica—The four horsemen (Heavy Metal)
SF5	Lana del Ray—Born to die (Pop)	Metallica—The four horsemen (Heavy Metal)
SF6	Chopin—Nocturne Op. 9 No. 2 (Classic)	Metallica—Whiskey in the jar (Heavy Metal)
SF7	Adele—Hello (Pop)	Metallica—Whiskey in the jar (Heavy Metal)
SF8	Chopin—Nocturne Op. 9 No. 2 (Classic)	Pantera—Walk (Metal)
SM9	Miley Cyrus—Flowers (Pop)	Eminem—Mockingbird (Rap)
SF10	Taylor Swift—Wildest Dreams (Pop)	Led Zeppelin—Immigrant song (Rock)

Table A2. Descriptive statistics of participants' demographics information, and music and reading habits.

		Experiment (n = 10)	
Variables		n	%
Gender	Female	9	90
	Male	1	10
Preferred reading/studying enviroment	Home	8	80
	Library	2	20
	Café	0	
	Other	0	
Study in silence	Yes	8	80
	No	1	10
	Does not matter	1	10
Noise affects studying/reading	Yes	10	100
	No	0	
	Do not know	0	
Music helps comprehension during the reading of a text	Yes	4	40
	No	2	20
	Do not know	4	40
Study better with music	Yes	2	20
	No	7	70
	Do not know	1	10
Study while listening to music	Often	2	20
	Sometimes	4	40
	Rarely	3	30
	Never	1	10
Study with music volume	High	0	
	Moderate	1	11, 11
	Low	8	88, 89

Table A2. *Cont.*

		Experiment (n = 10)	
Preferred reading/studying music	Instrumental	5	55, 56
	With Vocals	4	44, 44
Preferred kind of reading/studying music	Pop/Rock	2	22, 22
	Jazz	2	22, 22
	Classical	3	33, 33
	Other	2	22, 22

References

1. Johansson, R.; Holmqvist, K.; Mossberg, F.; Lindgren, M. Eye movements and reading comprehension while listening to preferred and non-preferred study music. *Psychol. Music* **2012**, *40*, 339–356. [CrossRef]
2. Calderwood, C.; Ackerman, P.L.; Conklin, E.M. What else do college students "do" while studying? An investigation of multitasking. *Comput. Educ.* **2014**, *75*, 19–29. [CrossRef]
3. Furnham, A.; Trew, S.; Sneade, I. The distracting effects of vocal and instrumental music on the cognitive test performance of introverts and extraverts. *Personal. Individ. Differ.* **1999**, *27*, 381–392. [CrossRef]
4. Ransdell, S.E.; Gilroy, L. The effects of background music on word-processed writing. *Comput. Hum. Behav.* **2001**, *17*, 141–148. [CrossRef]
5. Banbury, S.P.; Macken, W.J.; Tremblay, S.; Jones, D.M. Auditory distraction and short-term memory: Phenomena and practical implications. *Hum. Factors* **2001**, *43*, 12–29. [CrossRef]
6. Que, Y.; Zheng, Y.; Hsiao, J.H.; Hu, X. Studying the effect of self-selected background music on reading task with eye movements. *Sci. Rep.* **2023**, *13*, 1704. [CrossRef]
7. Cassidy, G.; Macdonald, R. The effects of music choice on task performance: A study of the impact of self-selected and experimenter-selected music on driving game performance and experience. *Music Sci.* **2009**, *13*, 357–386. [CrossRef]
8. Perham, N.; Sykora, M. Disliked music can be better for performance than liked music. *Appl. Cogn. Psychol.* **2012**, *26*, 550–555. [CrossRef]
9. Goltz, F.; Sadakata, M. Do you listen to music while studying? A portrait of how people use music to optimize their cognitive performance. *Acta Psychol.* **2021**, *220*, 103417. [CrossRef]
10. Jeong, Y.J.; Gweon, G. Advantages of print reading over screen reading: A comparison of visual patterns, reading performance, and reading attitudes across paper, computers, and tablets. *Int. J. Hum. Comput. Interact.* **2021**, *37*, 1674–1684. [CrossRef]
11. Rayner, K.; Slattery, T.J.; Bélanger, N.N. Eye movements, the perceptual span, and reading speed. *Psychon. Bull. Rev.* **2010**, *17*, 834–839. [CrossRef]
12. Rayner, K. Eye movements in reading and information processing: 20 years of research. *Psychol. Bull.* **1998**, *124*, 372–422. [CrossRef] [PubMed]
13. Rayner, K. Eye movements and attention in reading, scene perception, and visual search. *Q. J. Exp. Psychol.* **2009**, *62*, 1457–1506. [CrossRef] [PubMed]
14. Just, M.A.; Carpenter, P.A. A theory of reading: From eye fixations to comprehension. *Psychol. Rev.* **1980**, *87*, 329–354. [CrossRef] [PubMed]
15. Andreou, G.; Gkantaki, M. Tracking adults' eye movements to study text comprehension: A review article. *Languages* **2024**, *9*, 360. [CrossRef]
16. Jian, Y.C. Reading instructions influence cognitive processes of illustrated text reading not subject perception: An eye-tracking study. *Front. Psychol.* **2018**, *9*, 2263. [CrossRef]
17. Jian, Y.C. Reading instructions facilitate signaling effect on science text for young readers: An eye-movement study. *Int. J. Sci. Math. Educ.* **2019**, *17*, 503–522. [CrossRef]
18. Jian, Y.C. Influence of science text reading difficulty and hands-on manipulation on science learning: An eye-tracking study. *J. Res. Sci. Teach.* **2021**, *59*, 358–382. [CrossRef]
19. Kim, Y.-S.G.; Vorstius, C.; Radach, R. Does online comprehension monitoring make a unique contribution to reading comprehension in beginning readers? Evidence from eye movements. *Sci. Stud. Read.* **2018**, *22*, 367–383. [CrossRef]
20. Liao, C.-N.; Chang, K.-E.; Huang, Y.-C.; Sung, Y.-T. Electronic storybook design, kindergartners' visual attention, and print awareness: An eye-tracking investigation. *Comput. Educ.* **2020**, *144*, 103703. [CrossRef]
21. Mason, L.; Pluchino, P.; Tornatora, M.C. Effects of picture labeling on science text processing and learning: Evidence from eye movements. *Read. Res. Q.* **2013**, *48*, 199–214. [CrossRef]

22. Tsai, M.J.; Wu, A.H.; Chen, Y. Static and dynamic seductive illustration effects on text-and-graphic learning processes, perceptions, and outcomes: Evidence from eye-tracking. *Appl. Cogn. Psychol.* **2019**, *33*, 109–123. [CrossRef]
23. Reichle, E.D.; Pollatsek, A.; Fisher, D.L.; Rayner, K. Toward a model of eye movement control in reading. *Psychol. Rev.* **1998**, *105*, 125–157. [CrossRef]
24. Anderson, J.R.; Bothell, D.; Byrne, M.D.; Douglass, S.; Lebiere, C.; Qin, Y. An integrated theory of the mind. *Psychol. Rev.* **2004**, *111*, 1036–1060. [CrossRef] [PubMed]
25. Pickering, M.J.; Ferreira, V.S. Structural priming: A critical review. *Psychol. Bull.* **2008**, *134*, 427–459. [CrossRef]
26. Steindorf, L.; Rummel, J. Do your eyes give you away? A validation study of eye movement measures used as indicators for mindless reading. *Behav. Res. Methods* **2020**, *52*, 162–176. [CrossRef]
27. Sage, K.; Augustine, H.; Shand, H.; Bakner, K.; Rayne, S. Reading from print, computer, and tablet: Equivalent learning in the digital age. *Educ. Inf. Technol.* **2019**, *24*, 2477–2502. [CrossRef]
28. Hallam, S.; Price, J.; Katsarou, G. The effects of background music on primary school pupils' task performance. *Educ. Stud.* **2014**, *28*, 111–122. [CrossRef]
29. Zhang, H.; Miller, K.; Cleveland, R.; Cortina, K. How listening to music affects reading: Evidence from eye-tracking. *J. Exp. Psychol. Learn. Mem. Cogn.* **2018**, *44*, 1778–1791. [CrossRef]
30. Perham, N.; Currie, H. Does listening to preferred music improve reading comprehension performance? *Appl. Cogn. Psychol.* **2014**, *28*, 279–284. [CrossRef]
31. McLaughlin, M. Reading comprehension: What every teacher needs to know. *Read. Teach.* **2012**, *65*, 432–440. [CrossRef]
32. Cohen, L.; Manion, L.; Morrison, K. *Research Methods in Education*, 8th ed.; Routledge: London, UK, 2018. [CrossRef]
33. Hou, J.; Rashid, J.; Lee, K.M. Cognitive map or medium materiality? Reading on paper and screen. *Comput. Hum. Behav.* **2017**, *67*, 84–94. [CrossRef]
34. Haapakangas, A.; Hongisto, V.; Hyönä, J.; Koskela, H. Effects of unattended speech on performance and subjective distraction: The role of acoustic design in open-plan offices. *Appl. Acoust.* **2018**, *134*, 70–81. [CrossRef]
35. Sweller, J.; Ayres, P.; Kalyuga, S. *Cognitive Load Theory*; Springer: New York, NY, USA, 2011. [CrossRef]
36. Crawford, M.; Rutkowski, D.; Rutkowski, L. Improving reading abilities, attitudes and practices: A home-based intervention of supplementary texts for young readers in Cambodia. *Int. J. Educ. Dev.* **2023**, *103*, 102906. [CrossRef]
37. Schellenberg, E.G.; Peretz, I. Music, language and cognition: Unresolved issues. *Trends Cogn. Sci.* **2008**, *12*, 45–46. [CrossRef]
38. Ragert, M.; Fairhurst, M.T.; Keller, P.E. Segregation and integration of auditory streams when listening to multi-part music. *PLoS ONE* **2014**, *9*, e84085. [CrossRef]
39. Hongisto, V.; Haapakangas, A.; Hyönä, J. Effect of sound masking on concentration in open-plan offices: Acoustic and perceptual aspects. *J. Environ. Psychol.* **2020**, *72*, 101533. [CrossRef]
40. Mehta, R.; Zhu, R.; Cheema, A. Is noise always bad? Exploring the effects of ambient noise on creative cognition. *J. Consum. Res.* **2012**, *39*, 784–799. [CrossRef]
41. Hygge, S. Classroom experiments on the effects of different noise sources and sound levels on long-term recall and recognition in children. *Appl. Cogn. Psychol.* **2003**, *17*, 895–914. [CrossRef]
42. Iwanaga, M.; Ito, T. Disturbance effect of music on processing of verbal and spatial memories. *Percept. Mot. Ski.* **2002**, *94*, 1251–1258. [CrossRef]

applied sciences

MDPI

Article

Translators' Allocation of Cognitive Resources in Two Translation Directions: A Study Using Eye-Tracking and Keystroke Logging

Yifang Wang [1], Saihong Li [2,*] and Yubo Zhou Rasmussen [3]

[1] China-Africa Institute, Chinese Academy of Social Sciences, Beijing 100101, China; xyfwangyf@cass.org.cn
[2] Division of Literature and Languages, University of Stirling, Stirling FK9 4LA, UK
[3] Department of Actuarial Mathematics and Statistics, Maxwell Institute for Mathematical Sciences, Heriot-Watt University, Edinburgh EH14 4AS, UK; yr2001@hw.ac.uk
* Correspondence: saihong.li@stir.ac.uk

Abstract: This study investigates how novice translators distribute their cognitive resources during translation between English and Chinese in both directions, with particular attention paid to the role of translation direction and the divergence between empirical findings and participants' introspective reports. A combination of eye-tracking and keystroke logging was used to quantify cognitive effort, incorporating participant variation, attention unit type (ST, TT, parallel), gaze event duration, and average pupil dilation. A Generalized Linear Model (GLM) was applied, with average pupil dilation as the response variable and gaze event duration, AU type, and participant as covariates. An interaction term between gaze event duration and AU type was included in the E-C GLM but omitted from the C-E GLM due to non-significance. The results reveal distinct cognitive demands across translation directions. In English–Chinese (E-C) translation, ST processing significantly reduces pupil dilation (by 3.56%, $p < 0.001$), whereas TT processing leads to increased cognitive load, particularly during prolonged fixations, with pupil dilation increasing by 1.4% ($p = 0.033$). In Chinese–English (C-E) translation, ST processing does not significantly differ from parallel processing ($p = 0.285$), and TT processing reduces pupil dilation by 4.75% ($p < 0.001$), suggesting that it involves a lower cognitive effort than E-C translation. Gaze event duration significantly affects pupil dilation in C-E translation ($p < 0.001$); however, its influence in E-C translation varies according to the types of cognitive processing involved. Moreover, a significant gap is observed between the participants' self-reported reflections and the quantitative data, a disparity that is strongly shaped by the direction of translation. These findings contribute to a deeper understanding of cognitive effort in translation and raise implications for translator training, assessment, and cognitive translation studies, particularly in contexts where translation direction and processing mode interact to shape cognitive demand.

Keywords: translators; eye-tracking; keystroke logging; cognitive resources; translation direction; self-reflection

1. Introduction

The concept of translation directionality—whether translation occurs from a second language into the translator's native language (L1) or vice versa—has been a longstanding and central focus in translation studies [1–5]. Researchers have examined this topic through multiple lenses, such as 'inward' versus 'outward' translation practices, differences across regions and nations, and the impact of political, economic, and sociocultural contexts [3] (p. 898). Despite these advancements, a crucial question remains unanswered: does translation directionality influence the allocation of cognitive resources, and if so, how?

While previous studies have utilized eye-tracking to examine cognitive load in written translation [1–4], conference interpretation [2,6], simultaneous interpretation [7–12], translation competence [5,6], and translator training [8] between L1 and L2 translations [10], the findings remain inconsistent and often inconclusive. A key shortcoming in existing research is the limited attention given to the alignment—or lack thereof—between objective, data-driven indicators (such as those captured through eye-tracking and keystroke logging) and translators' introspective accounts of their own mental processes. This disconnect obscures a fuller understanding of how cognitive resources are allocated in real time and how translators perceive their own mental effort during the task.

To address this gap, the present study adopts an integrative methodology combining eye-tracking and keystroke logging techniques to investigate cognitive resource distribution among novice translators working in both English-to-Chinese and Chinese-to-English directions. The research focuses on the role that translation direction plays in shaping mental effort and aims to examine how observed behavior aligns—or contrasts—with participants' self-assessments. Cognitive effort in this study is quantified using four core indicators, developed by Hvelplund [12]: Total Attentional Duration (TA duration), AU count, AU duration, and pupil dilation. These measures offer a multifaceted view of how translators attend to the source text (ST), produce the target text (TT), or engage in both actions simultaneously. We aim to test the following three primary hypotheses:

(1) Translators engage in overlapping processing of the source text (ST) and target texts (TT) in both translation directions.
(2) Translators allocate more cognitive resources to TT processing than to ST processing or simultaneous (parallel) processing in both translation directions.
(3) Translators may underestimate the cognitive resources devoted to first-language production during L2-L1 translation.

To test these hypotheses, data were collected from 24 Chinese translator trainees and analyzed use Generalized Linear Models (GLMs) to assess the relationship between directionally and cognitive effort. By combining behavioral data with retrospective self-reflection, this research seeks to provide a more nuanced perspective on how cognitive processes function during translation. It aims to refine theoretical models and offer practical implications for translator training and assessment, especially in relation to translation directionality and the mental demands it entails.

2. Theoretical and Methodological Considerations

2.1. Processing Types and Allocation of Cognitive Resources in Translators

Translation engages multiple layers of cognitive effort, involving both intentional focus and more automatic mental functions. Scholars have approached the division and description of these resources in various ways. Translation researchers have described these cognitive demands in various ways. Traditionally, the translation workflow has been categorized into two main phases, comprehending the source text (ST) and generating the target text (TT), e.g., [12–14]. Recent theoretical and empirical developments, however, have introduced a third dimension, "parallel processing", highlighting how translators often switch between or simultaneously handle ST understanding and TT formulation [15–18].

Insights from neurolinguistics and cognitive psychology suggest that multitasking is fundamental to how translation unfolds. Rather than progressing linearly from reading to writing, translators frequently adopt overlapping strategies, integrating comprehension and production [19–21]. Evidence from keystroke logging and eye-tracking studies reinforces this view, showing that there is frequent simultaneous attention to both ST and TT, challenging linear models of the translation process [6,22–26].

To better capture this complexity, Hvelplund [12] introduced the concept of attention units (AUs), mapping this activity more precisely. These include (1) ST AUs—focused on interpreting the source material; (2) TT AUs—concerned with constructing the translation; and (3) parallel AUs—capturing moments when both tasks occur concurrently. These distinctions have enabled more precise measurements of attention distribution and cognitive effort during translation.

Building on this framework, the present study employs four empirical indicators—Total Attentional Duration, AU count, AU duration, and pupil dilation—to quantify cognitive effort for different translation directions and processing types. By adopting Hvelplund's AU categorization, we aim to capture the dynamic nature of attention allocation in translation and provide deeper insights into how cognitive resources are managed during ST comprehension, TT production, and their concurrent execution.

2.2. Defining Cognitive Units in the Translation Process

To capture the way translators allocate cognitive resources, the concepts of translation unit (TUs) and attention units (AUs) have been proposed to quantify continuous cognitive processing over time. Originally defined in linguistics as the smallest meaningful segment in a translation [27–36], the translation unit (TU) has evolved within cognitive translation studies to represent cognitive processing effort. Specifically, it denotes the focus of attention during translation [28], encapsulating the segment that is actively processed [12] and serving as a key unit of cognitive activity [29] (p. 953). This expanded definition integrates real-time keystroke and eye-tracking data, enabling a detailed analysis of cognitive resource allocation during translation [28].

Carl and Kay [34] describe AUs as segments of text that attract a translator's attention at any given moment. Their research suggests that AU boundaries shift depending on cognitive demands and translator experience. Hvelplund [12] formalized these observations by identifying three AU types: ST (comprehension), TT (production), and parallel (combined). This study adopts Hvelplund's AU framework to measure real-time cognitive effort across ST, TT, and parallel processing [12] (p. 116). These units provide a precise means of evaluating cognitive allocation patterns, enhancing models of translation as a cognitively regulated process. To achieve this, we categorize our data into four groups—Group A, Group B, Group C and Group D—based on the distinct macro- and micro-AU classifications of Hvelplund [12], as presented in Table 1.

Table 1. Macro- and micro-AU classifications used in this study.

AUs Classification	Macro AUs	Micro AUs	Details
Group A	ST AU (Source Text Attention Unit)	ST Gaze	Corresponds to ST comprehension; includes eye gaze on ST and typing without gaze.
Group B	TT AU (Target Text Attention Unit)	Gaze Off + Typing No Gaze + Typing TT Gaze + Typing TT Gaze	Corresponds to TT production; includes gaze on TT and typing activity.
Group C	Parallel AU	ST Gaze + Typing	Reflects simultaneous ST and TT processing.
Group D	No data	Gaze Off	Inactive moments, pauses, or unclassified activities.

2.3. Empirical Approaches to Modeling the Translation Process

Eye-tracking technologies allow for the precise observation of how translators read and allocate attention, providing valuable temporal and spatial data. While these tools are highly effective, they still face methodological limitations, particularly in distinguishing between early comprehension and deeper interpretive phases. The boundary between understanding and production can also become blurred in authentic translation settings [31] (p. 129). Nonetheless, the combined use of eye-tracking and keystroke logging has emerged as one of the most efficient methods for investigating cognitive processes in translation.

Several empirical models have been developed to account for the complexities of translation cognition. Tirkkonen-Condit [37] proposed that literal translation serves as a default strategy until a problem is detected, which then triggers a shift to conscious decision-making [31] (pp. 407–408). Building on this, Schaeffer et al. [38] suggested that translators initially form provisional representations that mirror the source text's structure and meaning, adjusting these only when contextual or linguistic constraints intervene.

Alves and Vale [33], drawing on Relevance Theory [39–41], introduced a dual-mode model comprising a stimulus mode (s-mode) and an interpretive mode (i-mode), which are thought to unfold along distinct cognitive timelines [33] (p. 256). Carl [31] expanded on this framework through the Monitor Model, which conceptualizes translation as a process driven by both automatized priming routines and higher-order monitoring strategies. Translation, according to this model, progresses in translation units, with oversight mechanisms regulating the degree of interlingual similarity based on task-specific goals, stylistic norms, or quality expectations [19] (p. 257). Moreover, this Monitor Model emphasizes the interplay between horizontal (within-level) and vertical (cross-level) monitoring processes, as outlined by De Groot [27]. This theoretical framework is particularly relevant to our study, as it offers a robust basis for analyzing cognitive effort using eye-tracking and keylogging. It provides a nuanced lens through which to understand the coordination of automatic and controlled processes, especially in relation to variations in cognitive demand across translation directions.

2.4. Cognitive Resource Allocation Across Translation Directions

The role of translation direction—whether from L2 to L1 or vice versa—has been investigated across various dimensions, including metaphor processing [3], translation accuracy [3,39–42] and cognitive load [17,43–45]. For example, Chang [42] utilized eye-tracking and fMRI to evaluate whether cognitive asymmetries found at the word level persist in full-text translation. His findings supported the views of the Revised Hierarchical Model. Expanding on this work, Wang [3] investigated how directionality influences metaphor translation and attention distribution. Later, Wang et al. [17] examined the differential cognitive demands of automatic versus controlled processing and found that translation into L2 required greater attentional resources. Interestingly, translations from L2 into L1 were often more accurate, likely due to greater familiarity with the target language's structure.

While such studies have yielded valuable insights, they tend to underrepresent translators' subjective experiences and self-perceptions of cognitive load. To address this gap, we adopt a multimethod design that integrates eye-tracking, keystroke logging, and retrospective self-reflection. Keystroke logging captures behavioral indicators of cognitive effort, including typing speed, revision patterns, and pauses during target text (TT) production. In parallel, participants' introspective accounts provide qualitative insights into their perceived mental workload. This triangulated methodology enables a more comprehensive analysis of cognitive resource allocation in translation. Specifically, our study aims to examine (1) how attention is distributed between source text (ST) comprehension and target text (TT) production; (2) variations in attentional metrics such as total attentional (TA) duration, AU count, and pupil dilation; and (3) the extent to which introspective reflections align with empirical data. By bridging objective measurement and subjective experience, this study contributes to a more nuanced understanding of how translation directionality shapes cognitive demand.

3. Materials and Research Design

3.1. Participants

The study involved 24 postgraduate students enrolled in translation studies programs at various UK universities. All participants were native speakers of Chinese (L1) and used English as their second language (L2), with IELTS scores of 7.0 or above, ensuring a consistent level of English proficiency. They reported prior experience with computer-assisted translation tools and had previously taken part in experiments utilizing eye-tracking, keystroke logging, and cue-based Retrospective Think-Aloud Protocols (RTA), which contributed to the smooth implementation of the current research.

The decision to recruit only native Chinese speakers was based on our research focus: examining cognitive effort allocation in different translation directions (L1 $\rightarrow$ L2 and L2 $\rightarrow$ L1) among novice translators with a shared linguistic background. This design helps reduce variability stemming from differences in language proficiency, translation training, or linguistic typology. Including a mixed population of native English and native Chinese translators could introduce confounding factors, potentially obscuring the results. While we acknowledge that a cross-linguistic comparison between native English and native Chinese translators would provide additional insights, such an approach falls beyond the scope of this study and is highlighted as a future research direction.

Although the sample size of 24 participants may appear limited, our study adopts a within-subject design, where each participant performed multiple translation tasks. This approach resulted in a substantial dataset of eye-tracking and keystroke-logging records and enhanced statistical power by allowing for direct comparison across conditions within the same individuals, thereby reducing the variability that can arise in a between-subject design.

To ensure the adequacy of our sample size, we conducted a post hoc power analysis using G*Power 3.1.9.7 (statistical power analysis to compute effect sizes), based on effect sizes observed in previous eye-tracking studies on cognitive translation processing (e.g., [1,12]). The analysis focused on detecting differences in Total Attentional Duration (TA duration), AU count, AU duration, and pupil dilation across translation directions. The power analysis yielded the following results:

Effect size (Cohen's f^2) = 0.35 (large effect, based on prior research estimates);

Alpha level = 0.05;

Power $(1-\beta) = 0.85$;

Required minimum sample = 19 participants (for within-subject comparisons).

Since our sample size (N = 24) exceeds the minimum required for detecting statistically significant effects at a power level of 0.85, we are confident that our dataset is sufficient for drawing meaningful conclusions.

Prior to participation, all individuals provided informed consent, in accordance with established ethical research guidelines. Data from twenty-one participants met the inclusion criteria for analysis. These criteria were based on established benchmarks and included (1) the total gaze duration on the screen, (2) the proportion of gaze samples successfully classified as fixations, and (3) the average fixation duration. We followed the standards proposed by Rayner [43] and Hvelplund [12].

3.2. Materials

The source texts used in this study comprised simulated dialogs situated within a business context, depicting interactions between Chinese and British speakers. These dialogs were displayed onscreen and deliberately constructed to align with the study's research parameters. Both the English and Chinese texts were informal, conversational in tone, and free from poetic elements, complex sentence structures, or specialized terminology. The aim was to mirror authentic professional communication, thereby enhancing the relevance of the materials for translation tasks in business environments.

The English text consisted of 125 words, while the Chinese version contained 151 words, with both comprising 9 sentences, labeled S1–S9 for analytical consistency. Each sentence was rigorously evaluated and aligned at multiple linguistic levels—including syllables, words, and sentence structures—to ensure cross-linguistic equivalence. Alignment criteria included word count, stylistic features, genre, sentence type and structure, sentence length, word frequency, lexical difficulty, and character or word length.

This meticulous standardization of linguistic and contextual variables ensured equivalency between the two texts, thereby supporting valid comparisons in cognitive resource allocation during translation. Additionally, the simulated healthcare dialog incorporated into the design enhances the study's applicability, particularly in translation scenarios relevant to medical and professional interpretation contexts.

3.3. Experimental Procedure

The entire experimental session lasted for approximately 35–40 min per participant, depending on task completion speed (see Figure 1—workflow flowchart).

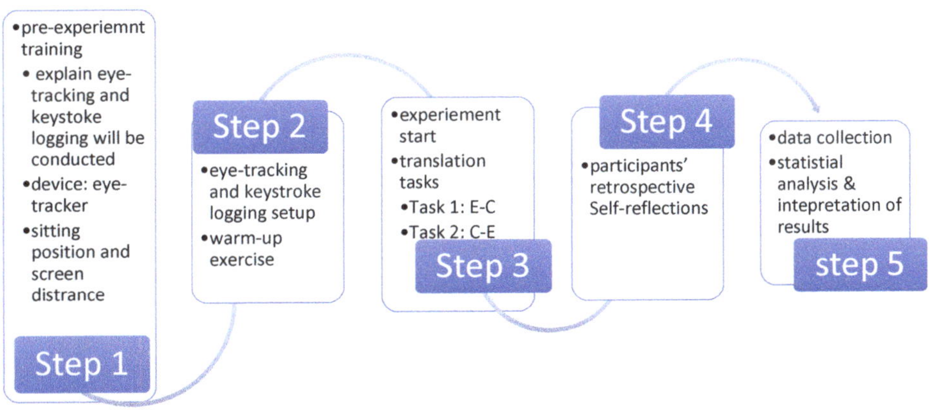

Figure 1. Step-by-step workflow of the experimental procedure.

Step 1: Before commencing with the main experiment, participants completed a preliminary training session designed to acquaint them with the experimental setup and procedures. During this session, they were given instructions on how to use the tools and were provided with examples of the tasks to ensure understanding. Participants were asked to seat at a distance of 60–65 cm from the screen throughout the session and the Tobii TX300 (Tobii Technology AB, Stockholm, Sweden) remote eye tracker was set up, allowing calibration for each participant to ensure the accurate recording of eye movements.

Step 2: Participants completed two sets of 50-word translation tasks, one translating from English to Chinese and the other from Chinese to English, as a warm-up exercise. This was performed to ensure participants' comfort and consistency in their use of the equipment.

Step 3: Participants began their translation tasks, during which their eye movements and keystroke logging data were recorded.

Step 4: Follow-up participants' retrospective self-reflections.

Step 5: Data collection and statistical analysis. To analyze the collected data, a Generalized Linear Model was employed, considering relevant co-variables. For the English-to-Chinese (E-C) dataset, variables included the size of the Area of Interest (AOI) for eye fixations, word frequency, average syllables per word, and the average number of letters per word. In the Chinese-to-English (C-E) models, AOI size and word frequency were used as key predictors. These linguistic variables were selected based on widely recognized readability metrics, such as the Flesch Reading Ease, Gunning Fog Index, Flesch–Kincaid Grade Level, Coleman–Liau Index, SMOG Index, Automated Readability Index, and Linsear Write Formula. Core factors evaluated included sentence count, average sentence length (in characters, aligned with AOI size), syllabic complexity, the proportion of complex words, and character length per word.

This carefully structured analytical approach enhances the reliability and validity of the findings, supporting the study's objective to investigate how cognitive resources are allocated during translation tasks.

4. Results and Discussions

4.1. Eye Tracking and Keystroke Logging Data Analysis

Table 2 provides a detailed description of the 10 categorical and continuous variables collected throughout the experiment in both English-to-Chinese (E-C) and Chinese-to-English (C-E) translation directions. These variables were recorded using eye-tracking and keystroke logging technology, capturing key metrics related to fixations, gaze behavior, and pupil dilation.

Table 2. Description of variables collected during experimental process.

Variable	Description	Comment	Categories
FixationIndex	Unique ID for fixation	[1:936]	-
Participant	Unique ID for participant	Factor with 18 levels	P01, P02, P03, P04, P05, P08, P09, P10, P11, P12, P13, P14, P16, P17, P18, P19, P20, P22
AU_type	The task which is currently being performed	Factor with 3 levels	1: parallel, 2: source text (ST), 3: target text (TT)
FixationPointY (MCSpx)	Point on the screen of gaze along the y-axis	[0:900]	-
GazeEventDuration	The amount of time spent gazing at a single point, recorded in (ms)	[0:9500]	-
KeyDefinition	The label generated by the eye tracking software	This takes the form of a string	-
FixationPointX (MCSpx)	Point on the screen of gaze along the x-axis	[0:1400]	-
PupilLeft	The dilation of the left eye pupil	Continuous	-
PupilRight	The dilation of the right eye pupil	Continuous	-
AVGpupildillation	The average pupil dilation of both pupils	Continuous	-

The dataset includes categorical variables such as participant ID, which identifies individual participants, and AU_type, which specifies the translation task being performed (parallel, source text (ST), or Target Text (TT)). Additionally, the dataset contains continuous variables related to gaze behavior, including FixationPointX and FixationPointY (screen coordinates of gaze), GazeEventDuration (time spent fixating on a point in milliseconds), and pupil dilation measures (PupilLeft, PupilRight, and their average, AVGpupildilation).

Table 3 presents summary statistics for average pupil dilation in both English-to-Chinese (E-C) and Chinese-to-English (C-E) translation directions. Pupil dilation values were obtained by averaging left and right eye pupil measurements, providing a single measure of cognitive effort. The results indicate that the median pupil dilation is slightly lower in C-E translation (2.799) compared to E-C translation (2.853), suggesting potential differences in cognitive load between translation directions.

Table 3. Summary of statistics for average pupil dilation.

English-to-Chinese (E-C) Translation Direction						
Variable	Min	1st Qu.	Median	Mean	3rd Qu.	Max
PupilLeft	1.690	2.658	2.885	2.853	3.012	3.912
PupilRight	1.466	2.645	2.811	2.852	3.043	4.197
AVGpupildillation	1.812	2.655	2.853	2.853	3.029	4.033
Chinese to English (C-E) Translation Direction						
Variable	Min	1st Qu.	Median	Mean	3rd Qu.	Max
PupilLeft	1.800	2.606	2.827	2.817	3.004	4.000
PupilRight	1.899	2.594	2.773	2.838	3.072	4.184
AVGpupildillation	1.927	2.602	2.799	2.828	3.032	4.090

To visualize the distribution of pupil dilation, Figure 2 displays histograms for both translation directions. The distributions exhibit a slight right skew, with a peak around 3.0. This suggests that most pupil dilation values fall within a central range, with higher dilation values appearing less frequently. Instances of elevated pupil dilation may indicate moments of increased cognitive load, potentially due to difficult translation segments or higher processing demands. The presence of extreme dilation values further supports the idea that certain translation tasks induce cognitive overload, requiring increased attentional resources.

The boxplots in Figures 3 and 4 provide a broad overview of participant variation in cognitive effort across both English-to-Chinese (E-C) and Chinese-to-English (C-E) translation directions, without distinguishing between AU types. In each boxplot, the dots represent individual data points that fall outside the interquartile range (IQR), indicated by the box, and are referred to as outliers. These points highlight instances where measurements such as pupil dilation or gaze event duration notably deviate from the median (marked by the horizontal line within each box), reflecting moments of particularly elevated or reduced cognitive load during translation tasks. Figure 3 illustrates overall pupil dilation per participant, while Figure 4 presents gaze event duration across participants, capturing fixation behavior during translation tasks. In Figure 3, some participants, such as P01 and P09, exhibit consistently higher median pupil dilation, whereas others, such as P18, show lower median pupil dilation. These variations may reflect differences in cognitive load, translation expertise, or individual processing strategies. The presence of numerous outliers in both translation directions suggests that certain translation segments required significantly greater cognitive effort, potentially due to challenging lexical choices, ambiguous structures, or increased decision-making demands. Similarly, Figure 4 highlights variation in gaze event duration across participants, providing insights into fixation behavior during translation. Some participants, such as P10 and P20, display longer gaze event durations, indicating that they spend more time fixating on specific areas before making translation decisions. Conversely, others exhibit shorter fixation durations, suggesting a more rapid or fluent translation approach. The large variance in fixation durations suggests that some participants engage in frequent, shorter fixations, while others rely on longer fixations to process information.

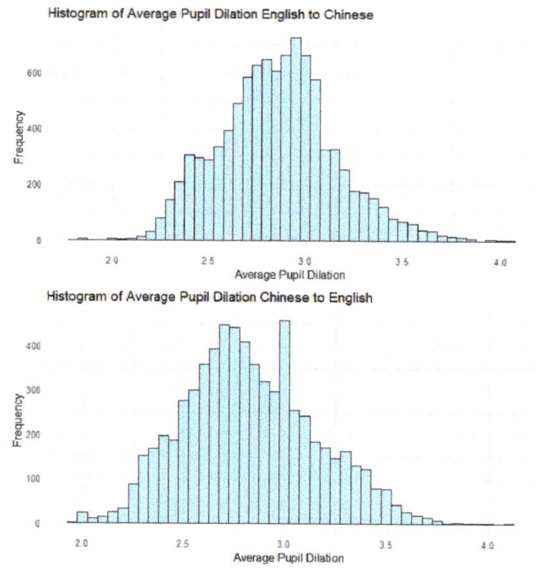

Figure 2. Histogram of average pupil dilation.

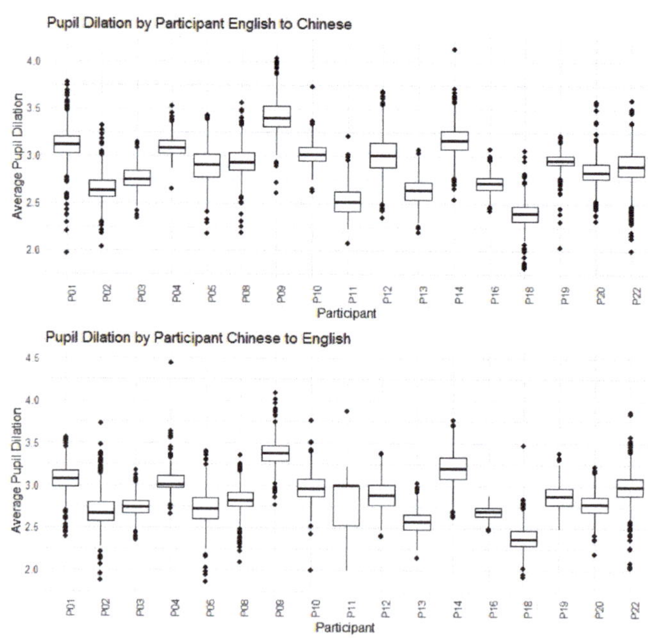

Figure 3. Pupil dilation by participant.

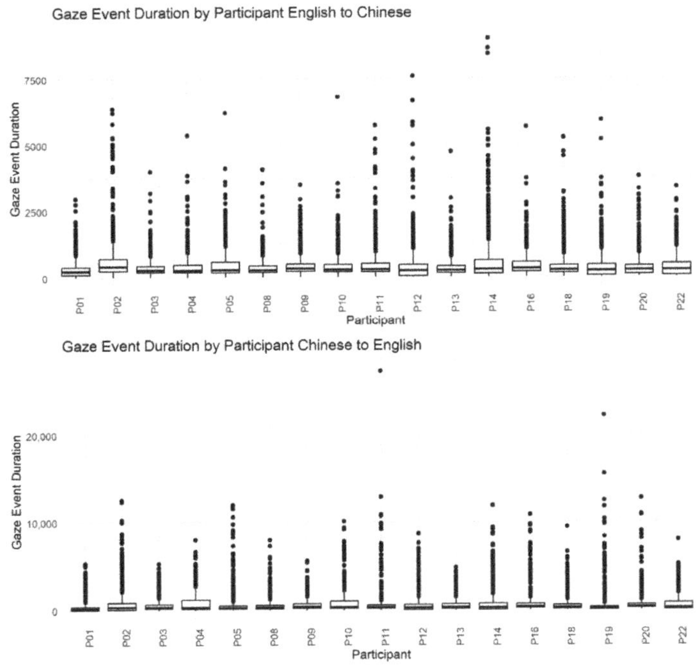

Figure 4. Gaze event duration by participant.

Figure 5 presents boxplots of pupil dilation across AU types (parallel, ST, and TT) for both translation directions (E-C and C-E), highlighting differences in cognitive effort across translation tasks. The results indicate that TT processing exhibits a slightly higher median pupil dilation than ST and parallel processing, suggesting that text production (TT) demands greater cognitive effort than comprehension (ST). In both translation directions, ST processing has a slightly lower median pupil dilation than parallel processing, aligning with the expectation that comprehension (ST) is less cognitively demanding than multitasking (parallel). The spread of dilation values appears consistent across AU types, indicating that cognitive effort varies within each condition but follows a similar distribution. The presence of outliers above 3.5 suggests that some translation segments elicited substantially higher cognitive load, potentially due to increased difficulty, ambiguous sentence structures, or complex lexical choices. Figure 6 displays the distribution of gaze event duration across AU types for both translation directions, offering insights into how visual attention is allocated during different translation tasks. The results show that TT processing has the longest median gaze event duration, reinforcing the idea that text production requires prolonged fixations, likely due to increased cognitive demands. In contrast, ST processing exhibits the shortest gaze durations, suggesting that comprehension is a more fluid process that requires less sustained attention. Parallel processing falls between ST and TT, indicating that multitasking demands more visual attention than ST but less than TT. The presence of substantial outliers in TT processing, with some gaze durations exceeding 6000 ms (E-C) and 20,000 ms (C-E), suggests that certain translation segments require significantly more cognitive effort, possibly due to lexical retrieval challenges, decision-making processes, or syntactic complexity. These findings also support the necessity of including an interaction term between gaze event duration and AU type in the statistical models. Given that gaze event duration appears to influence pupil dilation differently depending on the processing condition, failing to model this interaction could overlook important cognitive differences between ST, TT, and parallel processing across translation directions. By incorporating this interaction into our analysis, we can gain deeper insights into how translation processing types modulate the relationship between fixation duration and cognitive load, further refining our understanding of cognitive effort in bilingual translation tasks.

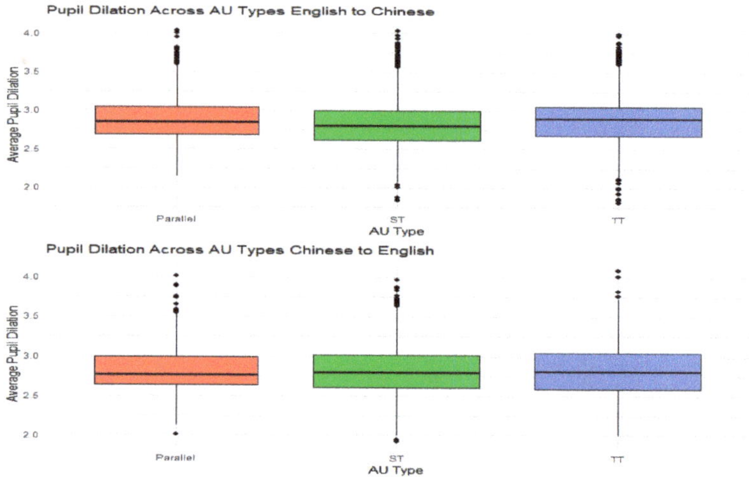

Figure 5. Average pupil dilation across AU_type.

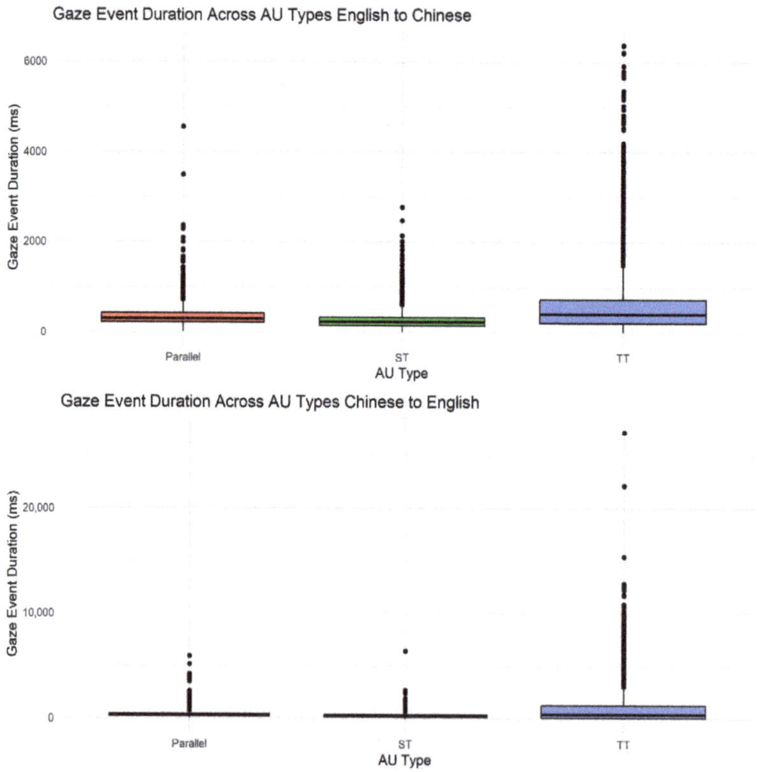

Figure 6. Gaze event duration across AU_type.

Figure 7 illustrates the relationship between pupil dilation and gaze event duration for both English-to-Chinese (E-C) and Chinese-to-English (C-E) translation directions. The blue curve in each scatterplot represents a smoothed trend line, obtained using a locally weighted regression (LOESS),indicates a weak overall relationship, with minimal variation in pupil dilation across gaze durations. In the E-C direction, pupil dilation remains largely stable, suggesting that longer fixations do not significantly impact the cognitive load. In contrast, the C-E direction shows a slight increase in pupil dilation for shorter fixations (~250 ms), followed by a decline for longer durations (~750+ ms). This may indicate higher cognitive effort during initial fixations, followed by adaptation over prolonged gazes. Most fixations are brief (<250 ms), with greater variability at longer durations, particularly in the C-E condition, suggesting increased cognitive demand in specific translation segments. These findings highlight the importance of modeling the interaction between gaze event duration and AU type, as cognitive load appears to vary by translation direction and fixation length.

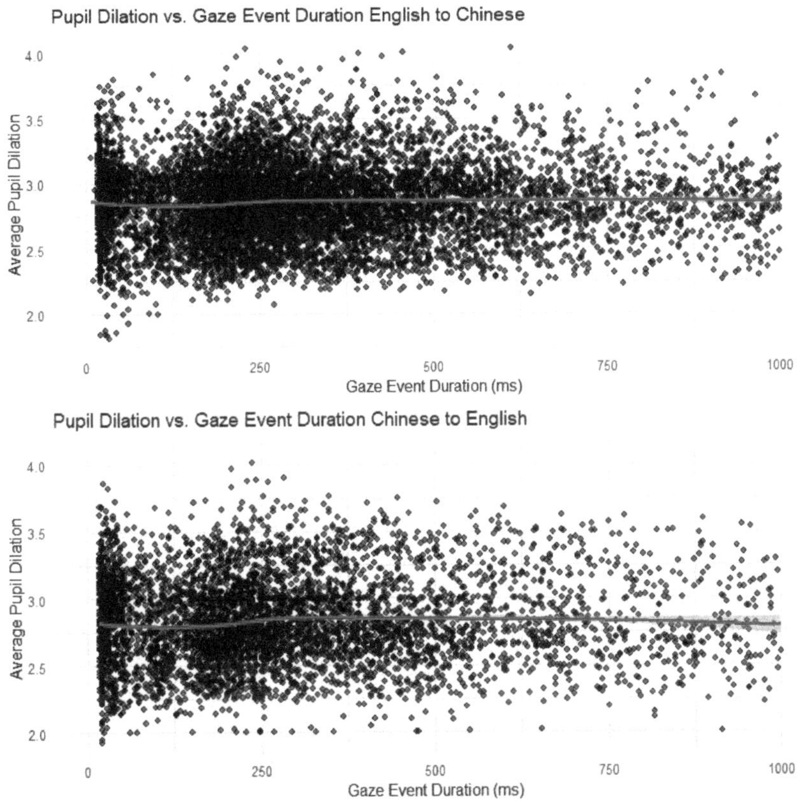

Figure 7. The relationship between pupil dilation and gaze event duration.

We now proceed to visualize gaze event duration. As an example, we illustrate through Figure 8 the cognitive processing throughout Participant 13's complete English–Chinese translation task. The horizontal axis represents a typing event, and the vertical axis represents duration of eye gaze for that event, also known as AU. TT (red) dominates throughout, indicating the translator's sustained cognitive efforts directed predominantly toward text production. Longer-duration fixations (higher bars) in TT segments suggest points of greater cognitive difficulty or decision-making complexity in target text production. Parallel processing (green bars) appears frequently, showing the translator often simultaneously comprehends the source text and produces the target text. Frequent occurrences of parallel processing indicate moments when the translator actively integrates the comprehension of the source text with simultaneous text production. This aligns clearly with prior studies suggesting that translation often involves parallel cognitive effort as in Hvelplund [12]. ST processing (blue bars) occurs more frequently, but with a shorter duration, suggesting that the translator reads whole sentences to completion before starting their translation task. This aligns with Carl and Dragsted's [29] findings on parallel processing, where translators gather information from the ST while producing the TT.

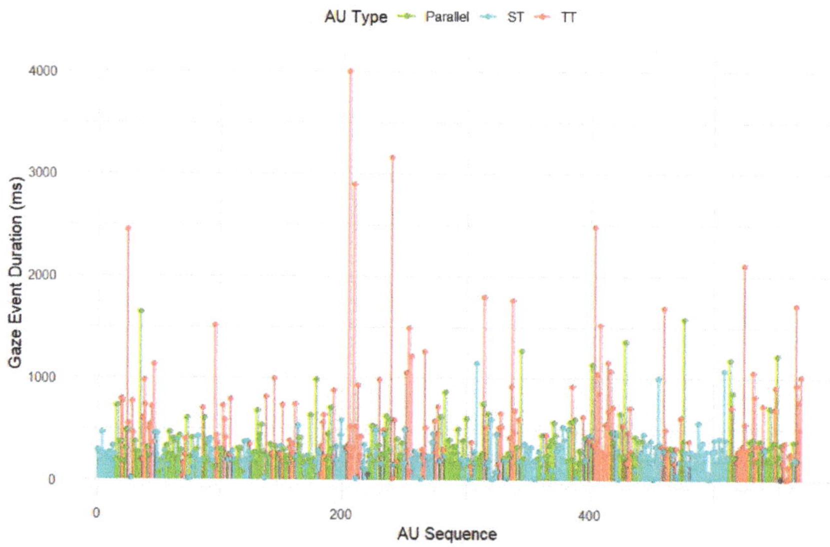

Figure 8. Timeline of attention units for the entire experiment for Participant 13: English–Chinese Translation. Parallel is displayed in green, source text (ST) in light blue, and target text (TT) in red. Each AU sequence has not been labeled for reliability.

To interpret and make sense of what is going on during the translation, we display the first 75 events of for Participant 13 in Figure 9. Early in the translation task (AU 1–20), Participant 13 mainly engaged with ST comprehension (blue bars). This clearly indicates an initial orientation phase, dedicated primarily to un-derstanding source content before typing. After initial comprehension, TT (red) and Parallel (green) processing clearly dominate, demonstrating frequent simultaneous comprehension-production activities. Clear instances of parallel processing highlight ongoing engagement with the source text even during active typing activities, reflecting cognitive multitasking during translation. Pronounced peaks such as "BackYOU" around AU 30 and indicate heightened cognitive effort possibly due to complex linguistic decisions, in particular on self-correction or during translation. The annotations like "Return", "Space", and specific characters provide evidence of active typing and editing processes, clearly linking cognitive activities (gaze duration) to concrete keystroke events.

Parallel coordination of ST comprehension and TT production was frequently observed. Keystroke logging activity often overlapped with eye fixations on the ST, indicating concurrent comprehension and production processes. Keystroke logging activity itself is evidence of cognitive effort in TT production, as participants explicitly reported ST comprehension while typing, corroborated by eye-tracking data.

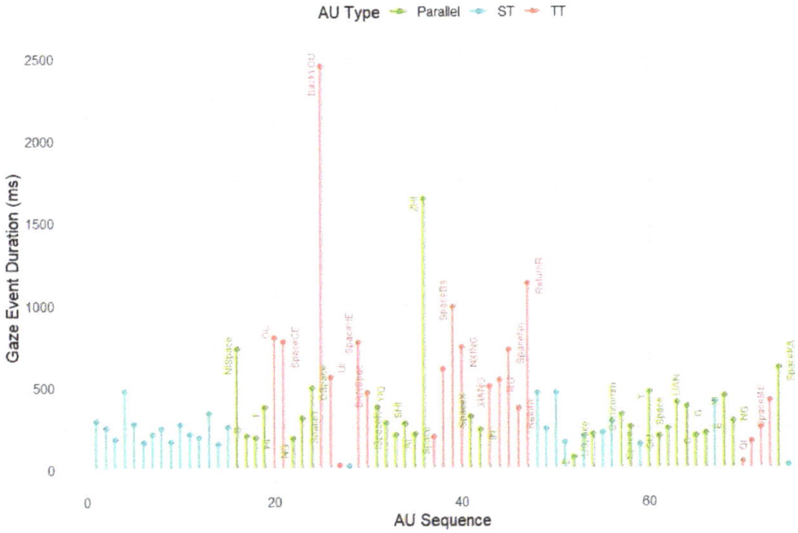

Figure 9. A timeline of attention units for the first 75 events for Participant 13: English–Chinese Translation. Parallel is displayed in Green, Source Text (ST) in Light Blue, and Target Text (TT) in Red. Each AU is labeled.

4.2. *Building a Generalized Linear Model to Assess the Difficulty of E-C and C-E Translation Tasks*

To analyze average pupil dilation (AVGpupildilation) in English-to-Chinese (E-C) translation, we employed Generalized Linear Models (GLMs) to examine the effects of AU_type (attention unit type: ST, TT, parallel [reference]), GazeEventDuration (fixation duration), and participant (individual differences, with P01 as the reference). Given that pupil dilation was a continuous, positive-value response variable, we compared three different GLM families to identify the best-fitting model:

1. Gaussian GLM (normal distribution): assumes that pupil dilation is normally distributed with constant variance.
2. Gamma GLM (Log-link Gamma distribution): accounts for the skewness of physiological measures like pupil dilation, which is strictly positive.
3. Inverse Gaussian GLM: models highly skewed data where variance increases quadratically with the mean.

To assess whether the impact of gaze duration varies across AU types, we included an interaction term (AU_type × GazeEventDuration). Model selection was based on the Akaike Information Criterion (AIC), which balances model fit and complexity. The results indicate that the Gaussian GLM achieved the lowest AIC (-11048), followed by the Gamma GLM (-11033) and Inverse Gaussian GLM (-10876). Given the comparable AIC values, we prioritize the use of the Gaussian model for interpretation due to its intuitive linear relationship between predictors and pupil dilation.

The results, as presented in Table 4, suggest that translation processing type significantly impacts the cognitive load in English-to-Chinese translation. Specifically, ST processing leads to significantly lower pupil dilation than parallel processing ($p < 0.001$), while TT processing also reduces pupil dilation, though to a lesser extent ($p = 0.033$). These findings align with research indicating that text comprehension (ST) requires less cognitive effort than multitasking (parallel), whereas text production (TT) involves similar or slightly lower cognitive demands. Furthermore, gaze event duration influences cognitive load differently across processing conditions. Overall, longer fixations are associated with a slight decrease in pupil dilation ($p = 0.026$), potentially indicating adaptation or cognitive efficiency. However, in TT processing, longer fixations significantly increase pupil dilation ($p = 0.0026$), suggesting that sustained fixations during text production impose higher cognitive demands. This finding is consistent with prior research showing that fixations during writing require prolonged cognitive processing, leading to increased cognitive loads. In contrast, gaze duration does not significantly affect pupil dilation in ST processing ($p = 0.265$), suggesting that cognitive effort remains stable regardless of fixation duration during text comprehension. Additionally, participant differences were highly significant across all models ($p < 0.001$), reflecting individual variability in cognitive processing strategies. Some participants, such as P10, exhibited higher pupil dilation, suggesting increased cognitive effort, whereas others, such as P18 and P12, showed reduced pupil dilation, indicating differences in cognitive load management. However, given the complexity of these variations and their secondary relevance to the study's primary objectives, we omit a detailed discussion of participant differences.

Table 4. Interpretation of Gaussian model for English-to-Chinese translatiom, with a corresponding estimate and *p*-value with Signif. codes: '***' 0.001 '**' 0.01 '*' 0.05.

Predictor	Estimate	*p*-Value	Interpretation
Intercept	3.138	<0.001 ***	Baseline pupil dilation (parallel processing, reference participant).
AU_typeST	−0.0356	<0.001 ***	ST processing significantly reduces pupil dilation compared to parallel processing.
AU_typeTT	−0.0140	0.033 *	TT processing slightly reduces pupil dilation relative to parallel processing.
GazeEventDuration	−0.000028	0.026 *	Longer gaze durations are associated with a small but significant decrease in pupil dilation.
AU_typeST:GazeEventDuration	−0.000018	0.265	Not significant: gaze duration does not significantly alter pupil dilation in ST processing.
AU_typeTT:GazeEventDuration	0.000039	0.0026 **	Significant: longer gaze durations increase pupil dilation in TT processing.

To analyze the average pupil dilation (AVGpupildilation) in Chinese-to-English (C-E) translation, we employed a Generalized Linear Model (GLM) as displayed above, but without an interaction term between AU_type and GazeEventDuration. Importantly, interaction terms between AU_type and GazeEventDuration were tested in previous models but were found to be non-significant. This indicates that the effect of gaze duration on pupil dilation does not vary meaningfully across translation processing types in C-E translation. As a result, we report the final model without interaction terms for a more parsimonious interpretation.

The results, presented in Table 5, indicate that translation processing type impacts cognitive load differently in Chinese-to-English translation. Unlike in E-C translation, where ST significantly reduced pupil dilation, in C-E translation, ST processing does not significantly differ from parallel processing ($p = 0.285$). However, TT processing leads to a significant decrease in pupil dilation ($p < 0.001$), suggesting lower cognitive effort during text production compared to multitasking (parallel processing). A key distinction from the E-C results is that gaze event duration significantly influences pupil dilation in C-E translation ($p < 0.001$). Longer fixation durations are associated with increased pupil dilation, suggesting that sustained fixations demand higher cognitive effort in this translation direction. However, no significant interaction effects between gaze event duration and AU_type were observed, indicating that the impact of fixation duration on pupil dilation remained stable across all translation processing types. Additionally, participant differences were highly significant across all models ($p < 0.001$), demonstrating individual variability in cognitive processing strategies. Some participants, such as P09, exhibited higher pupil dilation, whereas others, such as P18 and P13, showed reduced pupil dilation, indicating variability in cognitive load management. However, due to the complexity and secondary relevance of these variations, we omit a detailed discussion of participant differences.

Table 5. Interpretation of the Gaussian model for Chinese-to-English translation, with a corresponding estimate and *p*-value with Signif. codes: '***' 0.001.

Predictor	Estimate	*p*-Value	Interpretation
Intercept	3.092	<0.001 ***	Baseline pupil dilation (parallel processing, reference participant).
AU_typeST	−0.0077	<0.285 ***	Not significant: ST processing does not significantly differ from parallel processing.
AU_typeTT	−0.0475	<0.001 ***	TT processing significantly reduces pupil dilation compared to parallel processing.
GazeEventDuration	0.0000076	<0.001 ***	Longer gaze durations are significantly associated with increased pupil dilation.

The findings from our Generalized Linear Models (GLMs) highlight key differences in cognitive load between English-to-Chinese (E-C) and Chinese-to-English (C-E) translation. In E-C translation, ST processing significantly reduces pupil dilation, indicating that comprehension requires less cognitive effort than multitasking (parallel processing). Additionally, longer gaze durations generally decrease pupil dilation, except in TT processing, where sustained fixations increase cognitive load. In contrast, C-E translation exhibits a different pattern, where ST does not significantly differ from parallel processing, and TT processing still reduces cognitive effort. Notably, longer gaze event durations significantly increase pupil dilation across all processing types, suggesting that sustained fixations in C-E translation may indicate greater cognitive effort.

4.3. Retrospective Self-Reflections

In this study, participants' introspective accounts were gathered using cue-based Retrospective Think-Aloud (RTA) protocols. They were asked to articulate how they distributed cognitive resources between source text (ST) and target text (TT) processing in both translation directions—that is, to compare the effort invested in ST versus TT during English-to-Chinese (E-C) and Chinese-to-English (C-E) tasks. Comparisons between these subjective reflections and the objective data were conducted at a macro level. Participants' self-reflections are summarized below in Tables 6 and 7:

Table 6. Self-reflection: E-C translation.

	Allocation of Cognitive Resources to ST and TT				Allocation of Cognitive Resources to ST and TT		
	S1–3	S4–6	S7–9		S1–3	S4–6	S7–9
P01	60/40	50/50	50/50	P12	60/40	70/30	80/20
P02	40/60	40/60	40/60	P13	60/40	60/40	65/35
P03	50/50	60/40	60/40	P14	50/50	60/40	60/40
P04	50/50	50/50	30/70	P15	50/50	50/50	50/50
P05	30/70	40/60	50/50	P16	50/50	50/50	50/50

Table 6. *Cont.*

	Allocation of Cognitive Resources to ST and TT				Allocation of Cognitive Resources to ST and TT		
	S1–3	S4–6	S7–9		S1–3	S4–6	S7–9
P06	60/40	60/40	60/40	P17	70/30	70/30	70/30
P07	50/50	60/40	60/40	P18	40/60	40/60	50/50
P08	50/50	60/40	60/40	P19	70/30	70/30	80/20
P09	50/50	60/40	60/40	P20	50/50	40/60	50/50
P10	60/40	70/30	70/30	P21	60/40	70/30	80/20
P11	50/50	50/50	50/50	P22	50/50	50/50	50/50

Table 7. Self-reflection: C-E translation.

	Allocation of Cognitive Resources to ST and TT				Allocation of Cognitive Resources to ST and TT		
	S1–3	S4–6	S7–9		S1–3	S4–6	S7–9
P01	40/60	30/70	30/70	P12	30/70	30/70	30/70
P02	40/60	40/60	40/60	P13	50/50	40/60	40/60
P03	40/60	30/70	30/70	P14	30/70	30/70	30/70
P04	20/80	30/70	50/50	P15	30/70	20/80	20/80
P05	20/80	S	S	P16	40/60	40/60	35/65
P06	40/60	50/50	50/50	P17	50/50	40/60	40/60
P07	40/60	50/50	50/50	P18	30/70	30/70	20/80
P08	20/80	20/80	10/90	P19	50/50	45/55	40/60
P09	40/60	30/70	20/80	P20	50/50	50/50	40/60
P10	40/60	30/70	30/70	P21	50/50	40/60	30/70
P11	20/80	20/80	20/80	P22	50/50	45/55	30/70

In the case of C-E translation, participants' retrospective reflections obtained, through RTA, closely aligned with the eye-tracking and keystroke logging results. Most participants recognized that producing content in their second language required more time overall, a greater number of attention units (AUs), and longer durations per AU.

However, when reflecting on their E-C translation processes, many participants believed that they devoted more cognitive effort to understanding the English source text than to producing the Chinese target text. Of the 22 participants, only 4 (P02, P04, P05, and P18) reported allocating greater cognitive resources to TT production than to ST comprehension at any point. A small number (P11, P15, P16, and P22) perceived their efforts to be evenly divided between the two processing types. The majority, however, appeared unaware of the dominant cognitive load associated with TT production. Even among those who acknowledged investing more effort in TT production, their estimates were significantly lower than what the empirical data suggested—typically ranging between 30% and 40% of total cognitive effort.

For the majority of participants, the main difficulty encountered during L2-to-L1 translation stemmed from comprehending the English source text. Their strong confidence in producing text in their native language—rooted in familiarity and habitual use—appears to be a key factor contributing to the notable gap between their subjective assessments and the objective findings obtained through eye-tracking and keystroke logging. As participant P19 observed, translating from Chinese into English requires careful consideration of linguistic and cultural appropriateness. In contrast, during English-to-Chinese tasks, she experienced no hesitation during production and rarely reviewed her Chinese target text, trusting that it would naturally sound appropriate. Similarly, P22 noted that reviewing her output was unnecessary when translating into Chinese, describing the language as "too familiar" to pose any challenge.

Interestingly, even though most participants were unfamiliar with the formal concept of parallel processing in translation, several reflected on the overlapping nature of source text comprehension and target text production. Some also referred to the automatic nature of certain translation behaviors, particularly when working in their first language. These observations align with the findings of Hvelplund [12], who identified instances of automatic and parallel processing in English–Danish translation tasks, suggesting that either the input is passively retained in sensory memory during typing, or that keystrokes themselves occur without conscious monitoring.

Similar patterns emerged in the present study. Several participants reported that typing in their native language, especially for simple and frequently used expressions such as "所以" (so) or "但是" (but), occurred with minimal cognitive effort. As P12 described, the act of typing felt entirely automatic: "I just feel my fingers moving and typing without thinking". Notably, no such comments were made regarding English target text production during L1-to-L2 translation. Furthermore, in the context of C-E translation, participant P16 acknowledged the difficulty in separating ST comprehension from TT production, noting that the processes often overlapped—particularly when the Chinese source material was clear and easily understood.

To summarize, the eye-tracking and keystroke logging data reveal that participants exhibit diverse patterns of resource allocation across both translation directions, highlighting significant variations in translation strategies and cognitive approaches among individuals. Translation into English generally requires greater focus on the target text, likely due to the linguistic and stylistic challenges of producing fluent English, especially since the interpreters are not native English speakers. In contrast, translating into Chinese demonstrates a more balanced allocation of cognitive resources between the source text and target text, potentially reflecting lower cognitive demands for TT production in this direction. Both directions show an increased focus on the TT as tasks progress, but this trend is more pronounced in C-E translation.

This study identified instances of parallel coordination between source text (ST) comprehension and target text (TT) production among translators. Evidence from both the quantitative data and participants' self-reports supports the existence of a strong association between cognitive effort allocation and the type of processing involved. Eye-tracking and keystroke logging results indicated that TT processing consistently accounted for the largest share of cognitive resources. While many participants demonstrated awareness of the differing cognitive demands across processing types, their estimates regarding the distribution of attention units (AUs) often diverged from the empirical findings. When compared to the study's hypotheses, the results show that two of the three proposed hypotheses were fully supported, while the third received partial confirmation. The third hypothesis proposed that "there are differences between descriptive data and participants' self-reflection. For instance, participants may have a tendency to be unaware of the cognitive resources invested in first-language production during L2-to-L1 translation". This was partially validated by the findings. Additionally, some participants reported experiencing multitasking, parallel processing, and instances of automatic processing during translation tasks.

This study relies on eye-tracking and keylogging data to investigate cognitive resource allocation during translation. While these methods provide valuable insights, they inherently only capture a subset of the complex cognitive and physiological responses involved in the translation process. As a result, certain dimensions of translator cognition, such as emotional regulation and stress, may remain underexplored. Future research could address this limitation by adopting a multimodal methodology, incorporating additional tools such as facial expression analysis, respiratory monitoring, heart rate (HR) tracking, galvanic skin response (GSR), and cue-based Retrospective Think-Aloud (RTA) protocols. Such an integrated approach would allow for a more holistic understanding of the interplay between cognitive and emotional processes during translation tasks. Another limitation lies in the study's exclusive focus on novice translators. While this provides valuable insights into the translation behaviors of this specific cohort, the findings cannot be directly generalized to more experienced or professional translators, whose cognitive strategies and resource allocation may differ significantly. Investigating variations across translator groups with differing levels of expertise and professional backgrounds represents an important avenue for future research. Lastly, this study is constrained by its focus on the English–Chinese language pair. To enhance the broader applicability of the findings, future research could expand to other language pairs or replicate the study with a focus on Chinese-to-English translation. Such comparative studies would provide deeper insights into how language directionality and typological differences influence translation processes.

5. Conclusions

This study examined how novice translators coordinate source text (ST) and target text (TT) processing, revealing strong correlations between cognitive effort allocation and processing types. Our findings indicate that TT processing consumes the largest share of cognitive resources, as discussed regarding the GLMs. While most translators were aware of processing differences, their assumptions about effort distribution often diverged from the observed data. The purpose of this study was not to suggest that translators were 'wrong' in their perceptions but to explore how objective indicators of cognitive effort aligned with subjective experiences. Objective measures, such as eye-tracking, provide quantifiable data, while self-reports capture unique insights into translators' conscious attention and perceived difficulty. By integrating these methods, we aim to present a holistic understanding of cognitive effort during the translation process.

Our findings highlight that translators' subjective perceptions may not fully reflect underlying cognitive processes, emphasizing the importance of using multimodal approaches to study cognitive effort. The combination of objective measures and self-reports bridges the gap between externally observable cognitive effort and translators' subjective experiences, offering a richer perspective on attention allocation and perceived challenges at various stages of translation. These insights have broader implications for translator training and research. Developing metacognitive awareness of cognitive effort allocation could help translators manage their workload better. This study further contributes to debates on the reliability and validity of self-reports when assessing cognitive effort.

Supplementary Materials: The following supporting information can be downloaded at: https://www.mdpi.com/article/10.3390/app15084401/s1, Supplementary File S1: Experiment Source Texts; Supplementary File S1: Experiment data.

Author Contributions: Y.W. and S.L. were responsible for drafting the original manuscript. Y.Z.R. was responsible for data processing, data analysis and writing up data analysis and discussion section of the paper. All authors have read and agreed to the published version of the manuscript.

Funding: This research is co-funded by "A Study on the Influence of Historical and Cultural Empathy between China and Africa on the Cognitive Process of Translators", Project no: 2424018, the Chinese Academy of Social Sciences and the UCCL (Project No: 409164).

Institutional Review Board Statement: The study was conducted in accordance with the code of conduct at the Chinese Academy of Social Science and the University of Stirling, and the Project no: 2424018 was approved by CASS Institutional Review Board in 2022.

Informed Consent Statement: Informed consent was obtained from all subjects involved in the study and the documentation is provided in the Supplementary Materials.

Data Availability Statement: The data presented in this study are available in the Supplementary Material.

Acknowledgments: We would like to thank William Hope for proofreading the final draft of our article. We would like to express our thanks to the anonymous reviewers for their valuable comments and feedback.

Conflicts of Interest: The authors declare no conflicts of interest.

References

1. Zhan, J.; Jiang, Y. Testing the unique item hypothesis with phrasal verbs in Chinese–English translations of Lu Xun's short stories: The perspective of translation directionality. *Humanit. Soc. Sci. Commun.* **2024**, *11*, 344. [CrossRef]
2. Alves, F.; Jakobsen, A.L. *The Routledge Handbook of Translation and Cognition*; Routledge: London, UK, 2021.
3. Wang, Y. The Impact of Directionality on Cognitive Patterns in the Translation of Metaphors. In *Advances in Cognitive Translation Studies*; Muñoz Martín, R., Sun, S., Li, D., Eds.; New Frontiers in Translation Studies; Springer: Singapore, 2021; pp. 201–228.
4. Chang, V.; Chen, I. Translation directionality and the Inhibitory Control Model: A machine learning approach to an eye-tracking study. *Front. Psychol.* **2023**, *14*, 1196910. [CrossRef]
5. Martín, R.M.; Sun, S.; Li, D. *Advances in Cognitive Translation Studies*; Springer: Singapore, 2021.
6. Latif, M.M. *Researching and Modelling the Translation Process*; Cambridge University Press: Cambridge, UK, 2025.
7. Lv, Q.; Liang, J. Predicting Fluency with Language Proficiency, Working Memory, and Directionality in Simultaneous Interpreting. *Front. Psychol.* **2018**, *9*, 1543. [CrossRef]
8. Gile, D. *Basic Concepts and Models for Interpreter and Translator Training*; John Benjamins: Amsterdam, The Netherlands, 1995.
9. Hu, T.; Wang, X.; Xu, H. Eye-Tracking in Interpreting Studies: A Review of Four Decades of Empirical Studies. *Front. Psychol.* **2022**, *13*, 872247. [CrossRef] [PubMed]
10. Jia, J.; Wei, Z.; Cheng, H.; Wang, X. Translation directionality and translator anxiety: Evidence from eye movements in L1-L2 translation. *Front. Psychol.* **2023**, *14*, 1120140. [CrossRef]
11. Pavlović, N. What Were They Thinking?! Students' Decision Making in L1 and L2 Translation Processes. *Hermes J. Lang. Commun. Stud.* **2010**, *44*, 63–87. [CrossRef]
12. Hvelplund, K. Allocation of Cognitive Resources in Translation: An Eye-Tracking and Key-Logging Study. Ph. D Thesis, Copenhagen Business School, Copenhagen, Denmark, 2011.
13. Balling, L.; Hvelplund, K.; Sjørup, A. Evidence of parallel processing during translation. *Meta* **2014**, *2*, 234–259. [CrossRef]
14. Pokorn, N.K.; Blake, J.; Reindl, D.; Pisanski Peterlin, A. The influence of directionality on the quality of translation output in educational settings. *Interpret. Transl. Train.* **2019**, *14*, 58–78. [CrossRef]
15. Godijns, R.; Hinderdael, M. (Eds.) Directionality in Interpreting. In *The 'Retour' or the Native?* Communication and Cognition: Ghent, Belgium, 2005.
16. Bartlomiejczyk, M. Strategies of Simultaneous Interpreting and Directionality. *Interpret. Int. J. Res. Pract. Interpret.* **2006**, *6*, 149–174. [CrossRef]
17. Wang, X.; Wang, L.; Zheng, B. The Influence of Translation Direction on the Process and Quality of Information Processing. *Foreign Lang. Teach. Res.* **2022**, *54*, 128–139.
18. Frenchk-Mestre, C. Eye-movement Recording as a Tool for Studying Syntactic Processing in a Second Language: A Review of Methodologies and Experimental Findings. *Second. Lang. Res.* **2005**, *21*, 175–198. [CrossRef]
19. Zou, D.; Guo, J. Parallel translation process in consecutive interpreting: Differences between beginning and advanced interpreting students. *Acta Psychol.* **2024**, *248*, 104358. [CrossRef] [PubMed]
20. Monti, C.; Bendazzoli, C.; Sandrelli, A.; Russo, M. Studying Directionality in Simultaneous Interpreting through an Electronic Corpus: EPIC (European Parliament Interpreting Corpus). *Meta Transl. J.* **2005**, *50*, 2–16. [CrossRef]
21. Li, S.; Wang, Y.; Rasmussen, Y.Z. Studying Interpreters' Stress in Crisis Communication Evidence from Multimodal Technology of Eye-tracking, Heart Rate and Galvanic Skin Response. *Translator* **2022**, *28*, 468–488. [CrossRef]
22. Carl, M. Empirical Translation Process Research: Past and Possible Future Perspectives. *Transl. Cogn. Behav.* **2023**, *6*, 252–274. [CrossRef]
23. Olive, T. Working Memory in Writing: Empirical Evidence from the Dual-task Technique. *Eur. Psychol.* **2004**, *9*, 32–42. [CrossRef]

24. Jakobsen, A.L. Logging time delay in translation. *LSP Texts Transl. Process* **1998**, *1*, 73–101.
25. Jakobsen, A.; Jensen, K. Eye Movement Behaviour across Four Different Types of Reading Task. In *Looking at Eyes: Eye-Tracking Studies of Reading and Translation Processing, Copenhagen Studies in Language 36*; Göpferich, S., Jakobsen, A.L., Mees, I., Eds.; Samfundslitteratur: Copenhagen, Denmark, 2008; pp. 103–124.
26. De Groot, A. The Cognitive Study of Translation and Interpretation: Three Approaches. In *Cognitive Processing in Translation and Interpreting*; Danks, H., Shreve, G., Fountain, S., McBeath, M., Eds.; Sage: Thousand Oaks, CA, USA, 1997; pp. 25–56.
27. Alves, F.; Gonçalves, J.L. A Relevance Theory approach to the investigation of inferential processes. In *Triangulating Translation: Perspectives in Process-Oriented Research*; Alves, F., Ed.; John Benjamins: Amsterdam, The Netherlands, 2003; Volume 45, pp. 3–24.
28. Dragsted, B. Segmentation in translation: Differences across level of expertise and difficulty. *Target* **2005**, *17*, 49–70. [CrossRef]
29. Carl, M.; Dragsted, B. Inside the Monitor Model: Process of Default and Challenged Translation Production. *Transl. Corpora Comput. Cogn.* **2012**, *2*, 127–145.
30. Jääskeläinen, R. Think-Aloud Protocol Studies into Translation- an Annotated Bibliography. *Target* **2002**, *14*, 107–136. [CrossRef]
31. Carl, M. Towards a classification of translator profiles based on eye-tracking and keylogging data. *J. Writ. Res.* **2013**, *5*, 133–158. [CrossRef]
32. Dragsted, B. Segmentation in Translation and Translation Memory Systems. An Empirical Investigation of Cognitive Segmentation and Effects of Integrating a TM System into the Translation Process. Ph.D Thesis, Copenhagen Business School, Amsterdam, The Netherlands, 2004.
33. Alves, F.; Vale, D. Probing the Unit of Translation in Time: Aspects of the Design and Development of a Web Application for Storing, Annotating, and Querying Translation Process Data. *Across Lang. Cult.* **2009**, *10*, 251–273. [CrossRef]
34. Carl, M.; Kay, M. Gazing and typing activities during translation: A comparative study of translation units of professional and student translators. *Meta Transl. J.* **2011**, *56*, 952–975.26. [CrossRef]
35. Alves, F.; Gonçalves, J. Investigating the Conceptual-procedural Distinction in the Translation Process: A Relevance-theoretic Analysis of Micro and Macro translation units. *Target* **2013**, *25*, 107–124. [CrossRef]
36. Seleskovitch, D. Interpretation: A Psychological Approach to Translating. In *Translation: Applications and Research*; Brislin, R., Ed.; Gardner: New York, NY, USA, 1976; pp. 92–116.
37. Tirkkonen-Condit, S. The monitor model revisited: Evidence from process research. *Meta Transl. J.* **2005**, *50*, 405–414. [CrossRef]
38. Schaeffer, M.; Dragsted, B.; Hvelplund, K.T.; Balling, L.W.; Carl, M. Word Translation Entropy: Evidence of Early Target Language Activation During Reading for Translation. In *New Directions in Empirical Translation Process Research: Exploring the CRITT TPR-DB*; Carl, M., Bangalore, S., Schaeffer, M., Eds.; Springer: Cham, Switzerland, 2016; pp. 183–210.
39. Ernst, A.G. *Translation and Relevance: Cognition and Context*; St. Jerome: Manchester, UK, 2000.
40. Ernst, A.G. On the Significance of the Cognitive Core of Translation. *Translator* **2005**, *11*, 25–49.
41. Carl, M.; Schaeffer, M. Outline for a Relevance Theoretical Model of Machine Translation Post-editing. In *Researching Cognitive Processes of Translation*; Li, D., Lai, V., Lei, C., He, Y., Eds.; Springer: Singapore, 2019; pp. 49–67.
42. Chang, V.C. Testing Applicability of Eye-tracking and fMRI to Translation and Interpreting Studies: An Investigation into Directionality. Ph.D Thesis, Imperial College, London, UK, 2009.
43. Rayner, K. Eye movements in reading, and information processing. *Psychol. Bull.* **1998**, *124*, 372–422. [CrossRef]
44. Komkova, Y.N.; Bezrukikh, M.M.; Babanova, K.U.; Ryabkova, T.S.; Ustsova, A.G. Eye Movements and Autonomic Regulation of Cognitive Activity During Reading in Adolescence. Part I. Functional "Cost" of Cognitive Activity when Reading Text from the Screen in Adolescence. *Hum. Physiol.* **2024**, *50*, 127–136. [CrossRef]
45. Ferreira, A.S.; Gries, T.; Schwieter, J.W. Assessing indicators of cognitive effort in professional translators: A study on language dominance and directionality. In *Translation, Interpreting, Cognition: The Way Out of the Box*; Tra&Co, Group, Ed.; Language; Science Press: Berlin, Germany, 2021; pp. 115–143. [CrossRef]

applied sciences

MDPI

Article

Eye Movement Parameters in Children with Reading Difficulties

Ilze Ceple *, Linda Krauze, Evita Serpa, Aiga Svede, Viktorija Goliskina, Sofija Vasiljeva, Evita Kassaliete, Angelina Ganebnaya, Liva Volberga, Renars Truksa, Tomass Ruza and Gunta Krumina

Department of Optometry and Vision Science, Faculty of Science and Technology, University of Latvia, Jelgavas Street 1, LV-1004 Riga, Latvia; linda.krauze@lu.lv (L.K.); evita.serpa@lu.lv (E.S.); aiga.svede@lu.lv (A.S.); evita.kassaliete@lu.lv (E.K.); renars.truksa@lu.lv (R.T.); gunta.krumina@lu.lv (G.K.)
* Correspondence: ilze.ceple@lu.lv

Abstract: Children with dyslexia have previously been demonstrated to perform more corrective saccades in sequential non-reading saccade tasks, having trouble with maintaining a stable fixation, as well as reduced performance in smooth pursuit tasks. The current study aims to provide a comprehensive analysis of different eye movements (saccades, smooth pursuit eye movements and fixation stability) in children with and without reading difficulties who were tested by a validated reading performance test applied in the school environment. Altogether, 377 children (6–13 years old; 182 boys and 196 girls) were voluntarily involved in the study. Reading performance was determined by the Acadience® Reading literacy assessment tool. The Tobii Pro Fusion Eye-tracker was applied to determine oculomotor performance in children with and without reading difficulties. The results of saccadic eye movement, smooth pursuit eye movement, and fixation stability in a non-reading task did not reveal a statistically significant difference in oculomotor performance in children with and without reading difficulties. The results of the current study indicate that the oculomotor system in children with reading difficulties (not excluding dyslexia) is intact. The results propose that the observed differences with previous studies exploring eye movements in dyslexic children might be attributed to the participant selection.

Keywords: eye movements; eye tracking; reading difficulties; saccadic eye movements; smooth pursuit eye movements; fixation stability

1. Introduction

Reading is a complex visuomotor task involving both a precise performance of the visual and oculomotor system when guiding the reader throughout the text, and a success in the principal components of reading, i.e., phonological awareness, phonics, vocabulary, fluency, and comprehension [1]. During reading, the eyes perform continuous movements from word to word called saccades, as well as regressions that are backward saccades aimed to read a previously read part of the text and refixations that are small-amplitude saccades within the same word. In between saccades, a relatively stable eye position called fixation is acquired during which the visual information, i.e., letters and words in reading, is obtained. Fixations usually last about 200–300 ms, but can last longer depending on the text complexity and other parameters [2].

The relation between saccadic eye movement programming and the higher information processes in reading has been widely discussed in the context of how the perceptual, cognitive, and motor systems interact during the reading process [3–5]. Based on different models of eye movement control in reading, saccadic eye movement programming is considered to be driven either by the completion of the lexical processing of a certain word [6,7] or by a random timer that initiates saccadic programming independently from the lexical processing [8,9]. There are other models of saccadic eye movement programming during reading (e.g., SERIF); however, they are out of the scope of the current research.

Early reading development is associated with the executive cognitive functions, i.e., shifting attention, inhibition of dominant responses, working memory, and cognitive flexibility, all of which begin to develop in infancy [10–12]. It has been demonstrated that executive function performance as early as in infancy can be related to reading performance developing later in life [13,14]. Reading acquisition has also been attributed to a stage-like process during which children learn visual discriminations of letters and words, grapheme-phoneme correspondences and identifying words (orthographic stage) [15]. Different stages of reading development and performance have also been reflected in eye tracking studies, indicating that as reading proficiency increases, shorter fixations, larger-amplitude saccades and less regressions and refixations are observed [16,17]. Reading patterns in children are also subjected to the language and its orthographic complexity and predictability [18].

Children with reading difficulties not only present with lower reading rates, accuracy, fluency, and reading comprehension [19], but also demonstrate different performance in the objective measurement of eye movements during reading tasks, i.e., higher number of fixations, increased average fixation duration, reduced average saccadic duration, as well as increased total number of saccades and average number of fixations [20–25]. Rayner et al. [26] suggested that the reduced eye movement performance in reading tasks might reflect the comprehension processes in reading. But, since reading is a complex cognitive task involving a precise coordination between the processes of visual perception, attention, phonological awareness and other cognitive processes [27], as well as the oculomotor system guiding the fixation from a word to the next word, the analysis of reading eye movements does not reveal whether the differences in the eye movement performance are related to oculomotor dysfunction or if they serve as an indicator of the cognitive load.

Pavlidis [28] proposed that patients with dyslexia present with more corrective saccades in sequential non-reading saccade tasks, and had troubles with maintaining a stable fixation comparing to the control group, therefore indicating that patients with dyslexia might also present with oculomotor dysfunction. Following the results of Pavlidis [28], several other studies have also indicated reduced performance in different eye movement tasks in children with reading difficulties (including dyslexia). Biscaldi et al. [29] demonstrated that children with dyslexia have lower performance in anti-saccade tasks. Fukushima demonstrated that children with learning difficulties (including reading and mathematical disorders) have longer latencies in memory-guided and anti-saccade tasks, as well as lower smooth pursuit gain than the control group. Eden et al. [30] demonstrated that dyslexic children not only have more jerking movements in the smooth pursuit tasks, but also present with reduced fixation stability comparing to the control group. The study by Tiadi et al. [31] also indicated on impaired fixation stability in children with dyslexia, demonstrating that children with dyslexia exhibit more saccadic eye movements during fixations compared to non-dyslexic children, suggesting potential deficits in cortical structures responsible for saccade suppression and fixation control, as well as attentional deficiencies in children with dyslexia. Caldani et al. [32] found that pursuit performance significantly improved with age in non-dyslexic children, while the number of catch-up saccades during pursuit movements was higher in dyslexic children. Finally, a more recent study by Bilbao and Pinero [33] also indicated lower saccadic eye movement performance (more hypometric saccades) in children with neurodevelopmental disorders (including dyslexia).

Significant differences in oculomotor performance in children with and without reading difficulties have also been demonstrated in the DEM test, which is an objective method designed to evaluate eye movements in horizontal and vertical directions, including reading-like conditions [29,34]. Children with reading difficulties demonstrate poorer results in all standard metrics of the horizontal DEM test (both horizontal and vertical), indicating on oculomotor deficiencies [35–37].

The previously mentioned studies clearly demonstrate reduced oculomotor performance in children with dyslexia; however, the general take on whether children with reading difficulties of nonspecific cause present with oculomotor abnormalities is still under debate. A recent study by Vinuela-Navarro [38] did not demonstrate any difference in saccadic main sequence parameters, nor in the number of saccadic eye movements during the fixation task between children with and without delayed reading skills. And as opposed to above, Black et al. [39,40] demonstrated that there are no significant differences in saccadic eye movement latency, accuracy, and velocity parameters, nor a difference in smooth pursuit parameters, in children with and without dyslexia.

It is clear that all of the above-mentioned studies have applied different research designs, including the selection of the eye movement parameters, for their analyses, different methodologies for data analysis, and have applied different criteria for subject selection, which is a crucial part of these studies. Could the observed differences in previous studies be attributed to different methods applied for determining reading dysfunctions and dyslexia? The current study aims to provide a comprehensive eye movement analysis in children with and without reading difficulties determined by the Acadience® Reading literacy assessment tool. Therefore, determining whether reduced oculomotor performance observed in children with dyslexia also applies to children with reading difficulties of nonspecific cause.

2. Method

2.1. Participants

Altogether, 379 children (6–13 years old; 183 boys and 196 girls) from four basic education schools of Latvia were voluntarily involved in the study. Since the number of participants who were 6 years old (n = 6) and 13 years old (n = 1) was low, only the data from 7–12-year-old children were involved in further data analysis, leading to 371 children (7–179 boys and 192 girls) as the final number of participants. The demographic statistics of the participant data are represented in Table 1. All participants underwent general visual function evaluation, including visual acuity at near (65 cm) and eye dominance evaluation (Dolman test). Only the participants with the uncorrected binocular and monocular visual acuity of at least 0.6 decimal units (with or without contact lens correction) were involved in the eye tracking study. None of the children reported any previous eye diseases or other functional disorders that could affect eye movement parameters.

Table 1. Demographic statistics of the two study groups—children with reading difficulties (RD) and the control group (CG).

Task	Demographics	Reading Difficulties (RD)	Control Group (CG)
Saccadic eye movements	Gender *	M n = 49 F n = 41	M n = 119 F n = 137
	Age	9.71 ± 1.65	9.16 ± 1.61
Smooth pursuit eye movements	Gender *	M n = 48 F n = 44	M n = 120 F n = 136
	Age	9.74 ± 1.68	9.17 ± 1.62
Fixation stability	Gender *	M n = 42 F n = 32	M n = 98 F n = 106
	Age	9.39 ± 1.76	8.90 ± 1.62

* M—male, F—female.

2.2. Reading Skills

Reading skills were determined with the Acadience® Reading (formerly DIBELS Next) literacy assessment tool, which had been adapted and validated for Latvian language [41,42]. Reading performance was evaluated by professional speech therapists certified for the Acadience® Reading test performance. Children with reading performance below the 20th percentile were regarded as with reading difficulties (RDs), and children whose performance was above the 20th percentile were regarded as the control group (CG). Of all participants, 97 children were included in the RD group and 275 children were included in the CG.

The research was carried out in accordance with the requirements of the Declaration of Helsinki. The study was approved by the Life and Medical Sciences Research Ethics Committee of the University of Latvia (No. 13.06.2022). Children were involved in the study on a voluntary basis, based on a written consent provided by their parents or their legal guardians. The children and the parents/legal guardians were informed about the aim and the full process of the study, the equipment, and their responsibilities and rights throughout the study.

2.3. Eye Tracking

Eye movements were recorded with Tobii Pro Fusion (120 or 250 Hz) video-oculograph (Tobii, Sweden) in dark pupil illumination mode. Sampling frequency of 120 Hz was applied in recordings of saccadic and smooth pursuit eye movements, while the 250 Hz sampling frequency was applied for fixation stability measurements. Stimuli were demonstrated on DELL P2419H 23.8″ computer screen (52.7 × 29.6 cm; 1920 × 1080 px). Data recording for each participant was performed only once, and the approximate recording time was around 10 min. The participants were seated at a distance of 65 cm from the computer screen, the head position was fixed with chin and forehead rests, the height of the chair was adjusted so that the visual stimuli were at the participant's eye level. The participants were instructed to keep their heads as still as possible and to change their gaze only with their eyes. All data were collected during a single session, and participants were allowed to take a break when necessary and continue the eye movement task afterwards.

2.4. Saccadic Eye Movement Evaluation

Saccadic eye movement evaluation started with the calibration (5 points) and validation (4 points) of the eye tracker. If the calibration accuracy did not reach 0.5 degrees, a repeated calibration and validation process was performed. The calibration and validation stimulus and background color settings were identical to the experimental settings—black stimuli (5 cd/m^2; RGB (0, 0, 0)) demonstrated on a light grey background (90 cd/m^2; RGB (179, 179, 179)) [43].

After the calibration and validation, a fixation cross appeared at the center of the screen (3 s), after which the saccadic eye movement stimulus appeared in one of four positions (up, down, left, or right) relative to the central fixation object. Saccadic eye movement stimuli were either a circle or a star (depending on the task) 0.7 degrees in size. The distance from the center of the fixation cross to the center of the saccadic eye movement target was 11.3 degrees in the vertical positions and 14.0 degrees in the horizontal positions.

Two saccadic eye movement tasks were presented to the participants—(1) a reflexive saccade task and (2) an anti-saccade task (Figure 1). In the reflexive saccade task, participants were instructed to shift their gaze (make a saccade) from the fixation cross to the stimulus (circle) presented in the periphery as soon as soon as the stimulus appeared in either of the positions (left, right, up or down from the fixation cross). After the saccadic stimulus disappeared (5 s), the participants were instructed to return to the fixation stimulus. Note that during the presentation of the saccadic eye movement stimulus, the fixation cross was no longer visible. In the anti-saccade task, the participants were instructed to make a gaze transfer (anti-saccade) in the opposite direction from the saccadic eye movement stimulus (star), but with the same amplitude as the distance of the presented stimulus. When the anti-saccade stimulus disappeared, the participants were instructed to return to the fixation stimulus.

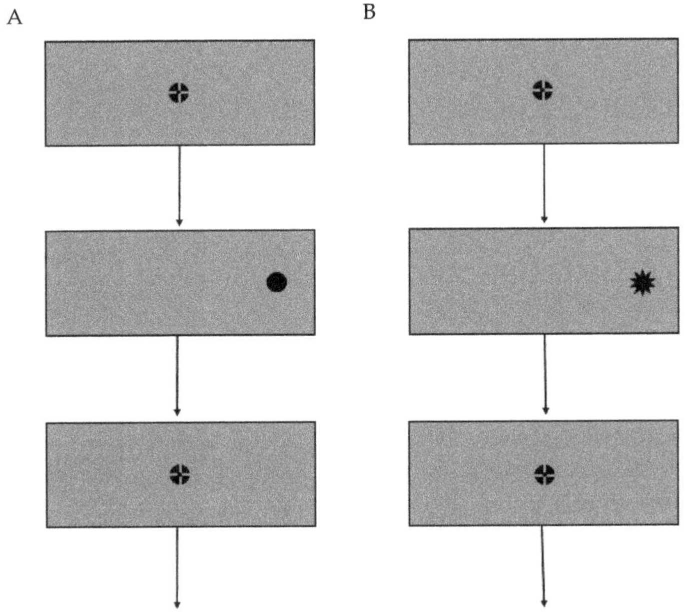

Figure 1. Saccadic eye movement stimulus for analysis of reflexive saccades (**A**) and anti-saccades (**B**).

Saccadic eye movement data were extracted and analyzed using Tobii Pro Lab software, working in accordance with the Tobii Pro Fusion eye tracker. Tobii I-VT and ClearView filters were applied to determine fixations, saccades, and blinks [44]. Only monocular data for the dominant eye were used in further data analysis. The data for saccadic eye movement analysis included 352 (171 boys and 181 girls) participants who managed to record at least one saccade or more with the Tobii Pro Fusion device. Of all participants whose saccadic eye movements were analyzed, 90 had reading difficulties (RD), and

256 children were included in the control group (CD). In order to determine the minimum sample size required for the study, power analysis (G*Power version 3.1.9.7) was conducted (Faul et al., 2007 [45]). The results of the power analysis indicated that the required sample size to achieve 80% power for detecting a medium effect (f = 0.25), at a significance criterion of $\alpha = 0.05$, was N = 60 for ANOVA: repeated-measures within–between interaction. Thus, the obtained sample size of N = 346 is sufficient to test the study hypothesis. Data analysis was performed using the Microsoft Excel, RStudio, and SPSS programs.

To obtain information about the first saccade towards the target object, the Tobii Pro Lab function for AOI (Area of Interest) analysis was applied. The area of interest (an area invisible to the observer) was placed in the center of the screen, where the fixation stimulus was located before the presentation of saccadic eye movement stimulus. In order to obtain the necessary saccadic eye movement parameters, the peak velocity of the exit saccade (saccade from the AOI towards the target) was determined. The peak velocity of saccadic eye movements was determined using a single-parameter I-VT filter with a peak velocity threshold of 30°/s [44].

The saccadic eye movement amplitude was calculated based on raw data analysis. The amplitude of saccadic eye movements was determined in Microsoft Excel using the following equation (Equation (1)):

$$Amplitude = \sqrt{(x_2 - x_1)^2 + (y_2 - y_1)^2}, \quad (1)$$

where x_1 and x_2 are the horizontal gaze coordinates before and after the saccade, respectively (px), and y_1 and y_2 are the vertical gaze coordinates before and after the saccade, respectively (px).

In order to estimate saccadic eye movement accuracy, only the first saccade towards the saccadic target was taken into account. Saccadic eye movement accuracy was calculated in Microsoft Excel using the following equation (Equation (2)):

$$Accuracy = 1 - \frac{S_e}{T_e}, \quad (2)$$

where S_e represents the landing error (cm) and T_e represents the distance between the gaze position after the saccade and the center of the saccadic target (cm).

Landing error S_e represents the vector size between the coordinates of the gaze direction after the saccade ($\vec{b}$—the actual amplitude of saccadic eye movements) and the coordinates of the center of the saccadic target ($\vec{a}$—representing the distance from the fixation point to the center of the target). Therefore, the magnitude of the landing error depends on the location of the saccadic target, including the direction and distance. The following equation based on the Euclidean distance was applied to calculate the landing error (Equation (3)):

$$S_e = \sqrt{(a - b)^2}, \quad (3)$$

where a represents the distance between the center of the fixation point and the center of the target (px) and b represents the amplitude of saccadic eye movements (px).

Since the experiment for each participant was performed only once (without replications), a portion of the data (saccades at specific directions) were not obtained due to blinking, distractions, or other factors. These data were referred as "*Unclassifide*" or "*EyesNotFound*" by Tobii Pro Fusion. In order to ensure a successful data analysis without excluding the already obtained results from children who had some of the data missing, missing data replacement was performed. In order to evaluate the amount and the possible source of missing data, a *missing value analysis* was performed in SPSS. The results of the missing value analysis were not statistically significant ($p = 0.386$), therefore enabling the substitution of missing values [46]. Scheffer [47] concluded that mean value or median substitution can be only applied if less than 10% of the data is missing, and if more than 10% is missing, the regression or ML algorithm should be applied. In order to include all of the children in the further data analysis, the K-Nearest Neighbors Algorithm (KNN) with k = 10 was applied for handling the missing data. KNN is a simple classification method or algorithm that determines the class of a target object based on the class most frequently found in its nearest neighbors [48].

In order to compare the saccadic eye movement performance in children with and without reading difficulties, mixed-model ANOVA was performed. The study employed a between-subjects design with three factors—reading performance (two levels: with and without reading difficulties), age (six levels: 7, 8, 9, 10, 11 and 12 years), and direction—that was analyzed separately for horizontal directions (two levels: right and left) and vertical directions (with two levels: up and down). A Bonferroni post hoc analysis was applied for pairwise comparisons. Horizontal saccades were analyzed separately from the vertical saccades, since the distance between the fixation point and the saccade target was different.

2.5. Smooth Pursuit Evaluation

Prior to the smooth pursuit evaluation, repeated calibration (5 points) and validation (4 points) was performed. If the calibration accuracy did not reach 0.5 degrees, a repeated calibration and validation process was performed. The stimulus color remained black (5 cd/m^2; RGB (0, 0, 0)), and the background color of the calibration and validation stimuli was grey (33 cd/m^2; RGB (114, 114, 114)), adjusted to the experimental settings in smooth pursuit evaluation.

The experiment started with a fixation cross demonstrated at the center of the screen. After 3 s the fixation cross disappeared, and a black stimulus (circle, 0.7 degrees) was presented at the same position and started moving to the right at a speed of 6 degrees per second. Once the stimulus position reached 10.8 degrees from the initial position, the object motion reversed coming to 10.8 degrees to the left from the initial fixation and returned to the center of the screen. After completing three cycles of movement (moving to the right, left and returning to the initial position), the stimulus disappeared and a fixation cross was once again presented at the center of the screen. A similar procedure was performed in the vertical direction, with 3 cycles of movement up (10.8 degrees) and down (10.8 degrees) from the initial position.

Eye movement data were extracted and analyzed in Tobii Pro Lab software, working in accordance with the Tobii Pro Fusion eye tracker. The data for smooth pursuit eye movement analysis included 348 (horizontal movement) and 346 (vertical movement) children who managed to record at least one smooth pursuit movement in one direction with the Tobii Pro Fusion device. Of all participants whose smooth pursuit eye movements were analyzed, 92 (horizontal) and 91 (vertical) had reading difficulties (RD), and 256 (horizontal) and 255 (vertical) children were included in the control group (CD). In order to determine the minimum sample size required for the study, power analysis (G*Power version 3.1.9.7) was conducted (Faul et al., 2007 [45]). The results of the power analysis indicated that the required sample size to achieve 80% power for detecting a medium effect (f = 0.25), at a significance criterion of $\alpha = 0.05$, was N = 60 for ANOVA: repeated measures within–between interaction. Thus, the obtained sample size of N = 346 is sufficient to test the study hypothesis.

Data analysis was performed on Microsoft Excel and SPSS programs. Data analysis was performed based on the raw data file provided by Tobii Pro Lab. The following data were excluded from the data analysis: (1) the latency period and the first 100 ms of the smooth pursuit corresponding to the open-loop phase of smooth pursuit eye movements; (2) blinks detected by the Tobii Pro Fusion and classified as "*EyesNotFound*"; (3) saccades (eye movements reaching velocity > 100 degrees/second); and (4) fixations corresponding to a relative stable gaze position. From the filtered data, eye movement velocity (degrees/second) and velocity gain were calculated, and the number of saccades performed during the smooth pursuit task was recorded.

The smooth pursuit eye movement velocity was calculated based on raw data analysis in Microsoft Excel using the following equation (Equation (4)):

$$v_{eye} = \frac{(x_n - x_{n-1}) * 1000}{(t_n - t_{n-1})} * dpp * 1000, \quad (4)$$

where v_{eye} is eye tracking velocity (deg/s), $(x_n - x_{n-1})$ are the difference of two adjacent x-axis coordinates (px), $(t_n - t_{n-1})$ are the difference of two adjacent time (μs), and dpp represents degrees per pixel (deg/px).

Equation (4) represents the velocity calculations for horizontal smooth pursuit eye movements. Velocity in the vertical direction was calculated using the same equation by applying the vertical coordinates (y instead of x).

Smooth pursuit velocity gain refers to the ratio of eye velocity to the stimulus velocity. The gain of 1 represents an exact match of eye and stimulus velocity, a gain of less than 1 indicates that the eye velocity was slower than the stimulus velocity, and a gain that is larger than 1 indicates that the eye moved with a larger speed than the stimulus. Smooth pursuit velocity gain was calculated in Microsoft Excel using the following equation (Equation (5)):

$$Gain = \frac{v_{eye}}{v_{stimulus}}, \quad (5)$$

where v_{eye} is the eye movement velocity (deg/s) and $v_{stimulus}$ is the stimulus velocity (deg/s) (6 deg/s).

In the study, a mixed-model ANOVA was conducted to compare smooth pursuit eye movement performance between children with and without reading difficulties. The analysis included three between-subject factors: reading performance (two levels: children with reading difficulties and children without reading difficulties) and age (6 levels: 7-, 8-, 9-, 10-, 11- and 12-year-old) and stimulus movement direction (with 4 levels: left, right, up, and down). The within-subject factor was velocity gain, with average values calculated from 3 cycles of stimulus movements back and forth. To control for multiple comparisons and reduce the risk of Type I errors, a Bonferroni post hoc analysis was applied to pairwise comparisons, helping to identify specific differences between groups.

2.6. Fixation Stability Evaluation

The experiment for fixation stability evaluation was generated using the Titta Master toolbox developed by Niehorster et al. [49]. Before eye-movement recording, a 5-point monocular calibration and 4-point validation procedure was performed. If the calibration accuracy did not reach 0.5 degrees, a repeated calibration and validation process was performed.

In order to assess fixation stability, a fixation task was generated in which participants were required to maintain their gaze on a fixation stimulus for 10 s. The demonstrated fixation stimulus consisted of a combination of black circles (5 cd/m^2, RGB (0, 0, 0)) and a white cross (180 cd/m^2; RGB * (255, 255, 255)). The stimulus was displayed on a grey background (90 cd/m^2; RGB (180, 180, 180) and had an overall size of 0.6 degrees (see Figure 2). The fixation task stimulus was chosen based on the study by Thaler et al. [50], where this stimulus was found to provide a relatively stable fixation. After validation, a black dot appeared in the center of the screen for 3 s, after which a fixation task was presented at the same screen position. The fixation stimulus was displayed for 10 s, but the analysis of fixation included only the middle 9 s, excluding the first and last 0.5 s of the stimulus presentation. The fixation task was performed under binocular viewing conditions, meaning both eyes fixated on the stimulus simultaneously.

The data from 279 children (140 boys and 139 girls) were included in the analysis of the fixation stability. Children who did not have a valid eye movement recording due to insufficient participation or technical issues, e.g., one eye not being recorded, were excluded from the further data analysis. Of all the children, 75 had reading difficulties (RDs) and 204 were referred as the control group (CG). In order to determine the minimum sample size required for the study, a power analysis (G*Power version 3.1.9.7) was conducted (Faul et al., 2007 [45]). The results of the power analysis indicated that the required sample size to achieve 80% power for detecting a medium effect (f = 0.25), at a significance criterion of α = 0.05, was N = 60 for ANOVA: repeated measures within–between interaction. Thus, the obtained sample size of N = 346 is sufficient to test the study hypothesis. The results of the power analysis indicated that the required sample size to achieve 80% power for detecting a medium effect (f = 0.25), at a significance criterion of α = 0.05, was N = 279 for [ANOVA: Fixed effects, special, main effects and interactions]. Thus, the obtained sample size of N = 279 is adequate to test the study hypothesis".

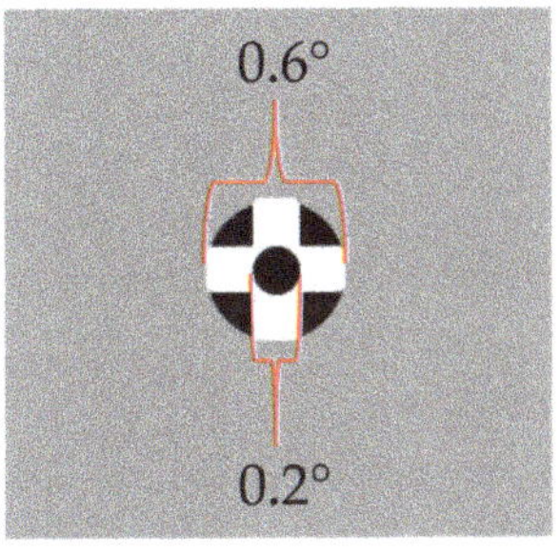

Figure 2. Fixation target.

In order to extract gaze fixations from the eye movement recordings, the I2MC (identification by two-means clustering) algorithm was applied. This algorithm has been designed for analyzing data with high levels of noise and missing data, making it particularly suitable for analyzing children's data [51]. Fixation stability was expressed by the bivariate contour ellipse area (BCEA), which was calculated using the following equation (Equation (6)). Note that a smaller BCEA value indicates a more stable fixation.

$$BCEA = 2k\pi\sigma_H\sigma_V\left(1-\rho^2\right)^{\frac{1}{2}}, \tag{6}$$

where σ_H represents the standard deviation of fixations in the horizontal meridian; σ_V represents the standard deviation of fixations in the vertical meridian; ρ represents the Pearson product–moment correlation coefficient between two directions; and the k value is 1.14, as the chosen probability area is 68% [52].

To compare gaze fixation stability between children with and without reading difficulties, a mixed-model ANOVA was conducted. The analysis included two between-subject factors: reading performance (levels: children with reading difficulties and children without reading difficulties) and age (6 levels: 7-, 8-, 9-, 10-, 11-, and 12-year-old). The within-subject factor was fixation stability.

3. Results

3.1. Saccadic Eye Movements

3.1.1. Reflexive Saccades

The results of mixed repeated measures ANOVA determining the effects of the reading performance, age, and direction on the peak velocity of saccadic eye movements only indicated a significant direction effect when comparing the horizontal (right and left) saccadic eye movements (F(1; 668) = 16.733, $p < 0.001$, $\eta_p^2 = 0.024$) (see Figure 3). Post hoc analysis revealed that the peak velocity of leftwards-oriented saccades was significantly lower than in saccades directed to the right side (Bonferroni post hoc: $p < 0.001$). Mixed ANOVA analysis for horizontally directed saccades did not demonstrate a significant age or reading performance effect on reflexive saccade peak velocity measurements.

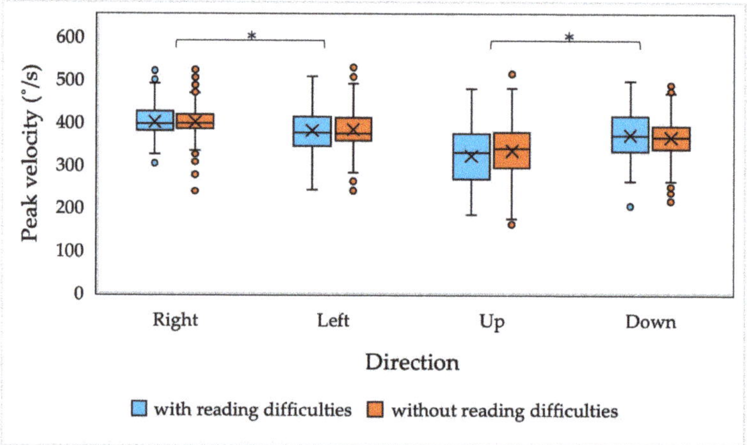

Figure 3. Peak velocity of reflexive saccades in children with and without reading difficulties. Data outliers significantly different from the rest of the data are marked as dots. * Statistically significantly difference (mixed ANOVA, Bonferroni post hoc analysis $p < 0.05$).

Similar results were obtained when assessing the peak velocity in vertically directed saccades—mixed repeated measures ANOVA only indicated a significant direction effect on saccadic peak velocity (F(1; 668) = 62.250, $p < 0.001$, $\eta_p^2 = 0.085$). Post hoc analysis revealed that the peak velocity for upwards-oriented saccades was significantly lower than for downwards-oriented saccades. (Bonferroni post hoc: $p < 0.001$). No significant age or reading performance effect on the peak velocity measurements was observed.

The results of the mixed repeated measures ANOVA, which determined the effects of the reading performance, age, and direction on the accuracy of reflexive saccades, only indicated a significant direction effect when comparing the horizontal (F(1; 668) = 11.182, $p < 0.001$, $\eta_p^2 = 0.016$) and vertical (F(1; 668) = 48.658, $p < 0.001$, $\eta_p^2 = 0.068$) saccadic eye movements (see Figure 4), but there was no significant effect of age or reading performance. Post hoc analysis revealed significantly lower accuracy results for leftward saccades than for rightward ones (Bonferroni post hoc: $p < 0.001$), as well as lower accuracy results for downward saccades than for upward-oriented saccades (Bonferroni post hoc: $p < 0.001$).

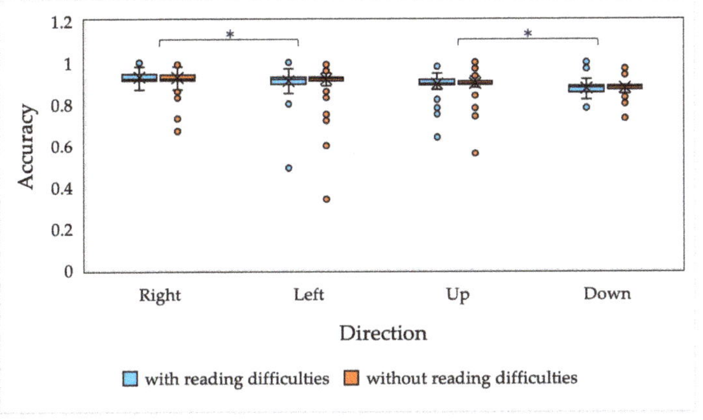

Figure 4. Reflexive saccade accuracy in children with and without reading difficulties. Data outliers significantly different from the rest of the data are marked as dots. * Statistically significantly difference (mixed ANOVA, Bonferroni post hoc analysis $p < 0.05$).

3.1.2. Anti-Saccades

Further data analysis explored the effect of saccadic eye movement direction, participant age, and reading performance on the peak velocity measurements in the anti-saccade tasks (Figure 5). The results of the mixed repeated measures ANOVA did not reveal a significant direction or reading performance effect on the peak velocity of horizontally oriented anti-saccades. However, a significant age effect was observed ($F(5, 668) = 3.085$, $p = 0.009$, $\eta_p^2 = 0.023$). Post hoc analysis revealed a significant difference in the peak velocity measurements between 7- and 11-year-old children (Bonferroni post hoc: $p = 0.047$), with 7 year olds demonstrating higher peak velocities in the anti-saccade tasks compared to 11 year olds. A significant direction effect was observed for vertically oriented anti-saccades ($F(1; 668) = 29.651$, $p < 0.001$, $\eta_p^2 = 0.043$), indicating a lower peak velocity for downwards-oriented saccades compared to saccades performed upwards (Bonferroni post hoc: $p < 0.001$). No significant effect of age or reading performance on the vertically oriented anti-saccades was observed.

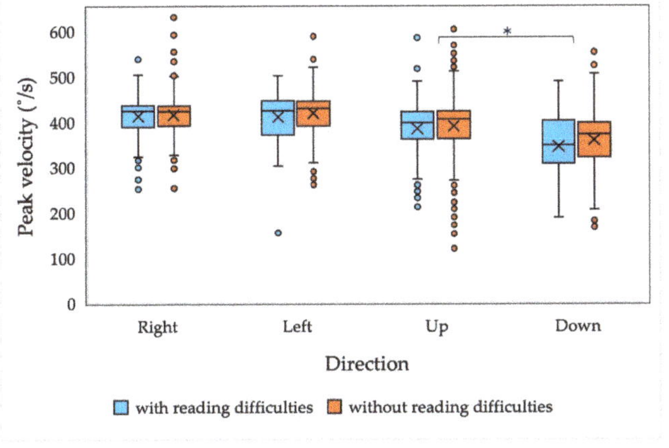

Figure 5. Peak velocity of anti-saccades in children with and without reading difficulties. Data outliers significantly different from the rest of the data are marked as dots. * Statistically significantly difference (mixed ANOVA, Bonferroni post hoc analysis $p < 0.05$).

The results of the mixed repeated measures ANOVA revealed a significant direction effect on anti-saccade accuracy measurements for both horizontally (F(1; 668) = 63.754, $p < 0.001$, $\eta_p^2 = 0.087$) and vertically (F(1; 668) = 462.540, $p < 0.001$, $\eta_p^2 = 0.409$) oriented anti-saccades (Figure 6). Post hoc analysis revealed that saccades oriented to the right were less accurate than saccades oriented to the left (Bonferroni post hoc: $p < 0.001$), and saccades oriented downwards were less accurate than saccades oriented upwards (Bonferroni post hoc: $p < 0.001$). No significant age or reading performance effect was observed for anti-saccade accuracy in horizontally oriented saccades. However, for vertically oriented anti-saccades only, the effects of reading performance (F(1, 668) = 3.964, $p = 0.047$, $\eta_p^2 = 0.006$) and age (F(5, 668) = 2.709, $p = 0.020$, $\eta_p^2 = 0.020$) were revealed to be statistically significant. Children without reading difficulties were more accurate than children with reading difficulties (Bonferroni post hoc: $p = 0.047$), and 10-year-olds were less accurate than 12-year-old children (Bonferroni post hoc: $p = 0.047$).

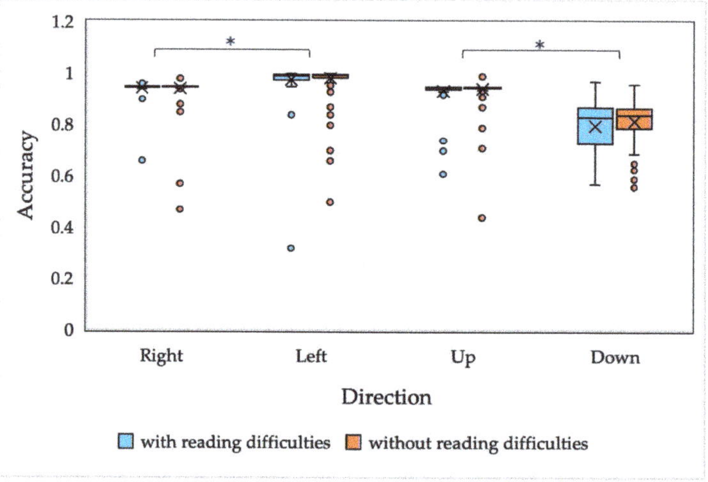

Figure 6. Anti-saccade accuracy in children with and without reading difficulties. Data outliers significantly different from the rest of the data are marked as dots. * Statistically significantly difference (mixed ANOVA, Bonferroni post hoc analysis $p < 0.05$).

3.2. Smooth Pursuit Eye Movements

The results of the mixed repeated measures ANOVA, which evaluated the effect of reading performance, age, and stimulus movement direction on smooth pursuit gain measurements, revealed significant age (F(5; 1334) = 8.606, $p < 0.001$, $\eta_p^2 = 0.031$) and stimulus movement direction (F(3; 1334) = 188.354, $p < 0.001$, $\eta_p^2 = 0.298$) effects (Figure 7). However, the effect of reading performance was not demonstrated to be statistically significant (F(1; 1334) = 2.960, $p = 0.086$, $\eta_p^2 = 0.002$). Furthermore, no significant interaction effects were observed among reading performance, age, and stimulus movement direction ($p > 0.05$).

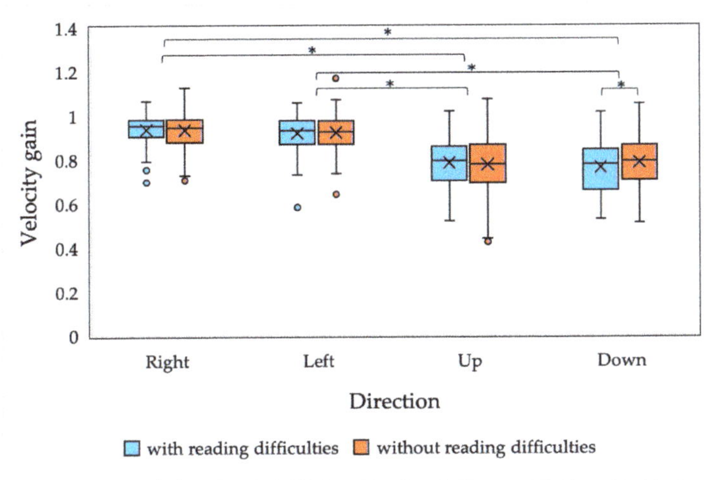

Figure 7. Velocity gain in children with and without reading difficulties. Data outliers significantly different from the rest of the data are marked as dots. * Statistically significantly difference (mixed ANOVA, Bonferroni post hoc analysis $p < 0.05$).

Post hoc pairwise comparisons with Bonferroni adjustments revealed significant differences in velocity gain between specific age groups: 7 vs. 8 years: $p = 0.012$, 8 vs. 11 years (Bonferroni post hoc: $p < 0.001$), 8 vs. 12 years (Bonferroni post hoc: $p < 0.001$), 9 vs. 12 years (Bonferroni post hoc: $p = 0.025$), 10 vs. 11 years (Bonferroni post hoc: $p = 0.032$), and 10 vs. 12 years (Bonferroni post hoc: $p = 0.006$). Horizontal movements (to the left and right) had higher velocity compared to the vertical movements (upwards and downwards): downward vs. leftward (Bonferroni post hoc: $p < 0.001$) downward vs. rightward: (Bonferroni post hoc: $p < 0.001$), upward vs. leftward (Bonferroni post hoc: $p < 0.001$), upward vs rightward (Bonferroni post hoc: $p < 0.001$). The analysis of velocity gain with respect to age, reading performance, and stimulus movement direction revealed a significant difference between children with and without reading difficulties, but only for 7-year-old children during downward stimulus movements (Bonferroni post hoc: $p = 0.007$). No significant main effects or interactions were found for the other groups or movement directions.

3.3. Fixation Stability

The results of the one-way ANOVA did not reveal significant effects of age ($F(5, 267) = 0.212$, $p = 0.957$, $\eta_p2 = 0.004$) or reading performance ($F(1, 267) = 1.242$, $p = 0.266$, $\eta_p2 = 0.005$) on the fixation stability measurements, clearly indicating that fixation stability on a non-reading task was not different in children with and without reading difficulties (see Figure 8).

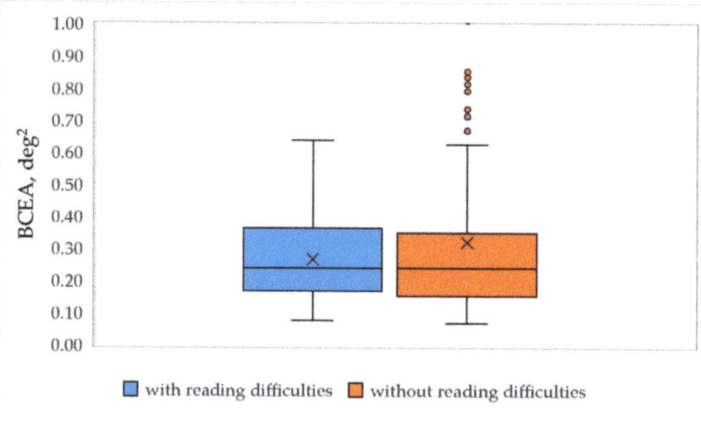

Figure 8. Binocular fixation stability in children with and without reading difficulties. Box plots represent interquartile range; crosses indicate the mean values, and lines indicate the medians.

4. Discussion

The results of the current study demonstrate that there are no significant differences in saccadic and smooth pursuit parameters, nor in fixation stability, in children with and without reading difficulties. A slight difference was observed in the accuracy of the vertically oriented anti-saccades of all children and the smooth pursuit gain of 7-year-old children with and without reading difficulties; however, the main differences were observed only in the downwards-oriented anti-saccades, which, as presented in the current study and in other studies, have been demonstrated to have higher variability [53]. In general, the current findings are consistent with the results demonstrated by Black et al. [39] and Vinuela-Navarro [38], who did not observe significant differences in saccadic latency, accuracy, or velocity parameters between dyslexic and non-dyslexic children. While different studies [29,54,55] have indicated lower performance in anti-saccade tasks in children with dyslexia, the differences are mainly observed in the error rates and correction rates, not related to the dynamics of saccadic eye movements.

Fukushima et al. [56] reported that children with different learning difficulties had not only higher error rates, but also significantly higher latencies in anti-saccade tasks compared to the control group. While the differences observed in the current study and the study by Fukushima et al. [56] could be attributed to differences in research methodologies (e.g., eye tracking equipment, as well as different methods applied in selecting the research participants), the results of the current study are in accordance with other previous studies, indicating that the oculomotor control of saccadic eye movements in children with reading difficulties is intact. And the previously demonstrated differences in the anti-saccade error rates are, as reported by Wilcockson et al. [57], attributed to the impairment of inhibitory control.

Edelman et al. [58] demonstrated that the peak velocity of reflexive saccades is higher than in anti-saccade tasks; however, the current results indicate an opposite tendency. The extent to which the peak velocities in reflexive saccades are lower is still under debate. While the study by Bucci and Seassau [59] observed a tendency of peak velocity in anti-saccade tasks to be slower than in pro-saccade ones, this was not confirmed statistically. Furthermore, in the study by Lunn et al. [60], the control group of healthy children had almost identical mean peak velocity results in pro-saccade and anti-saccade tasks. The results of the current study indicate slightly higher average peak velocity results in the anti-saccade tasks compared to reflexive saccades (with or without the data replacement).

Since the observed differences in the mean results were within the standard error, we assume that they were not of a causal nature; however, the observed differences should be explored in further research.

While previous studies have demonstrated that several brain regions involved in reading also play a role in the neural control of smooth pursuit eye movements [57], not all individuals with reading difficulties exhibit smooth pursuit abnormalities [40,61]. This observation is consistent with the findings of the current research—no significant difference in smooth pursuit eye movements in children with and without reading difficulties was observed (except for in 7-year-old children). Furthermore, the results fell within the normal range according to the normative values defined by Sinno et al. [62]. Data analysis revealed a significant age effect on the smooth pursuit performance which was independent of reading performance, likely due to improvements in binocular coordination, saccade latency, and coordination between saccadic and vergence movements, which continue to develop until 12 to 15 years of age [63].

More recent studies by Fukushima et al. [56] and Caldani et al. [32] have demonstrated that dyslexic children generally exhibit decreased smooth pursuit gain, especially in the left-to-right direction [30]. They also tend to make more catch-up saccades [42]. So what causes the inconsistency with the results of the current research, as well as the results presented by Black et al. [39] and Ygge et al. [61]? While all of the previously mentioned studies have explored smooth pursuit performance in children with dyslexia, a different methodology of participant selection can be observed. The criteria for diagnosing dyslexia vary widely, with some studies using broader definitions or different reading-related tests, which could influence group characteristics and outcomes. The prevalence of developmental dyslexia is estimated around 7% [64], while the prevalence of reading difficulties in the current research was around 25%, including both children with dyslexia and children with reading difficulties caused by other factors (social, environmental, etc.). The results of the current study propose that the smooth pursuit control in children with reading difficulties is intact, and that observed differences in smooth pursuit performance in children with dyslexia might be attributed to magnocellular dysfunction, attention deficits, or other factors previously linked to dyslexia [65].

One of the parameters that characterizes the fixation quality is its stability, i.e., the ability to keep a steady gaze on the fixation target. To assess whether the fixation stability is impaired in children with reading difficulties, fixation stability was assessed in a non-reading task, thereby reducing the involvement of cognitive processes on the eye movement performance. A similar approach of analyzing binocular fixation quality in children with and without reading difficulties was applied by Vinuela-Navarro et al. [38], exploring the number and amplitude of saccadic eye movements during fixation. Both the current study and the study by Vinuela-Navarro et al. [38] did not reveal significant differences in fixation stability in children with and without reading difficulties. The results of our study, however, contradict the findings by Vikesdal et al. [66], demonstrating that children with dyslexia have more unstable fixation in non-reading tasks than children without dyslexia. Although dyslexia is a specific learning disorder of neurobiological origin that primarily affects reading, as mentioned before, the children included in our study had reading difficulties but were not diagnosed with dyslexia, which could explain why the results contradict those of Vikesdal et al. [66].

The dynamics of saccadic and smooth pursuit eye movements are also characterized by other parameters, e.g., latency and initial acceleration; however, latency was not considered in the current study, since prior research and methodological constraints have indicated that latency may not reliably reflect oculomotor performance differences in children with and without reading difficulties. For example, Black et al. [39] and Vinuela-Navarro et al. [38]

found no significant differences in saccadic latency between dyslexic and non-dyslexic children, suggesting that latency might not be a robust marker for reading difficulties. Moreover, latency measurements are often influenced by external factors, such as age-related variability in neural development and task complexity, which could introduce noise rather than providing insight into the specific oculomotor mechanisms linked to reading performance. In this study, the focus was placed on parameters like saccadic accuracy, peak velocity, and fixation stability, as they provide more direct and interpretable measures of oculomotor control in the context of reading difficulties. Additionally, methodological consistency with validated reading tests and eye-tracking protocols ensured that the analysis remained aligned with the study's primary aims.

While the results of the current study do not indicate significant differences in saccadic, smooth pursuit, and fixation parameters in children with and without reading difficulties, the target group (children with reading difficulties) might still include single children that present with eye movement dysfunction. There has been a lengthy discussion on whether children with dyslexia also present with abnormal eye movements in other non-cognitive tasks and whether eye movement abnormalities are the cause or consequence of reading difficulties [67]; however, the studies addressing reading performance in patients with other causes of abnormal eye movements indicate the crucial role of oculomotor system performance in successful reading [68]. The results of the current study do not imply that abnormal eye movements do not cause reading difficulties, but demonstrate that, in general, children with reading difficulties are not prone to oculomotor dysfunction. Furthermore, the absence of general eye movement deficits does not exclude specific oculomotor deficits during reading, since different cognitive processes are involved in oculomotor control during, i.e., attention and lexical processing, which in turn might lead to abnormal eye movements specifically in reading tasks [69,70].

Reading is a multifactorial process that is influenced by psychological, physiological, environmental, sociological, and linguistic factors; therefore, determining the etiology or the exact cause of reading difficulties involves a complex analysis of all previously mentioned factors. The Acadience® Reading test is designed to identify children at risk of developing reading difficulties rather than to diagnose specific reading disorders. In Latvian basic educational schools, reading skills assessment is conducted by speech therapists; however, the diagnostics of specific learning disorders (including dyslexia) requires the involvement of psychologists, neurologists, and other specialists [71]. As the participant selection for the current study was performed only based on the Acadience® Reading test results, the reading difficulties observed among the participants may stem from a variety of underlying causes, including both internal (neurobiological and physical) and external (social environmental) factors [72,73]. Including children with various causes of reading difficulties is beneficial, but complicates the interpretation of results, which is the main limitation of the current study. A clearer distinction between children with dyslexia and those with other causes of reading difficulties would provide deeper insight into the physiology of dyslexia. However, we found that the inclusion of children with different causes of reading difficulties in their natural school environments enhanced the ecological validity of the current findings.

From an optometrist's perspective, reading difficulties may be associated with poor visual acuity, unstable binocular vision, accommodative dysfunction, different ocular diseases, or oculomotor dysfunction [74]. The conventional tests applied to determine the oculomotor performance in reading-like tasks in optometrist practice (e.g., DEM test) are relatively simple and practical; however, these tests are designed for specific eye movement analysis and do not address the overall performance of the oculomotor system, i.e., saccade, smooth pursuit, and fixational tasks. Furthermore, the performance on the DEM test does not correlate with quantitative measurements of eye movements [75]. Other tests for saccadic, fixational, and smooth pursuit system evaluation, like direct clinical observation, are subjected to difficulties in scoring and subjective evaluation of the examiner [76]. The current study suggests that children with reading difficulties should not be directly suspected to have oculomotor dysfunction; however, in the case of specific complaints (difficulties of maintaining fixation or allocating the optimal fixation position in reading), as well as if a child is diagnosed with dyslexia or other developmental disorders, oculomotor performance should be evaluated, if possible, by applying quantitative methods (e.g., eye tracking). Recent evidence demonstrates that short-term visual attention training can improve the oculomotor performance during reading [77].

The current study extends the available information and understanding on oculomotor performance in children with reading difficulties, which, as reported by the OECD Programme for International Student Assessment (PISA) [78], is around 25%, by demonstrating that the oculomotor system in children with reading difficulties (not excluding dyslexia) is intact. The results propose that the observed differences from previous studies exploring eye movements in dyslexic children might be attributed to the participant selection. This implies that, with specific selection of patients diagnosed with developmental dyslexia, differences in the oculomotor performance could still be observed, indicating magnocellular dysfunction, attention deficits, cerebellar functional activation patterns, or other factors previously linked with dyslexia [65,79].

5. Conclusions

Reading is a complex visuomotor task involving precise performance of both the visual and oculomotor systems, guiding the reader throughout the text. Different studies have demonstrated reduced oculomotor performance in children with dyslexia; however, the general take on whether children with reading difficulties of nonspecific cause present with oculomotor abnormalities is still under debate [28–32]. The current study aimed to provide a comprehensive analysis of different eye movements (saccades, smooth pursuit eye movements, and fixation stability) in children with and without reading difficulties who were tested by a validated reading performance test applied in the school environment.

The results regarding saccadic eye movement, smooth pursuit eye movement, and fixation stability in a non-reading task did not reveal a statistically significant difference in oculomotor performance in children with and without reading difficulties. This indicates that the oculomotor system in children with reading difficulties (not excluding dyslexia) is intact. The results propose that the observed differences from previous studies exploring eye movements in dyslexic children might be attributed to the participant selection, and imply that the reduced eye movement performance in children with dyslexia is of neurobiological origin.

Author Contributions: Conceptualization, I.C., E.S., A.S., E.K. and G.K.; methodology, I.C., L.K., E.S., A.S., V.G., L.V., R.T., T.R. and G.K.; software, I.C., E.S., V.G., L.V., R.T. and T.R., validation, E.S., A.S., V.G., S.V., L.V., E.K., T.R. and G.K.; formal analysis, I.C., L.K., E.S., A.S., V.G., S.V. and L.V.; investigation, I.C., L.K., E.S., V.G., S.V., L.V., A.G. and G.K.; resources, I.C., E.S., E.K. and G.K.; data curation, I.C., L.K., E.S., V.G., S.V., L.V., A.G. and G.K.; writing—original draft preparation, I.C., L.K., E.S. and V.G.; writing—review and editing, I.C., L.K., E.S. and A.G.; visualization, I.C., L.K., E.S. and V.G.; supervision, I.C.; project administration, I.C. and G.K.; funding acquisition, I.C. and G.K. All authors have read and agreed to the published version of the manuscript.

Funding: This research was funded by the Latvian Council of Science (project Nr. lzp-2021/1-0219), University of Latvia (project Nr. Y5-AZ77-ZF-N-100) and SIA Mikrotikls and the University of Latvia Foundation (project Nr. 2260).

Institutional Review Board Statement: The study was conducted in accordance with the Declaration of Helsinki, and approved by the UL Life and Medical Sciences Research Ethics Committee (No. 13/06/2022).

Informed Consent Statement: Informed consent was obtained from all legal guardians of the study subjects involved in the study.

Data Availability Statement: The data presented in this study are available on request from the corresponding author. The data are not publicly available due to privacy reasons.

Acknowledgments: We would like to express our gratitude to schools participating in the study (Marupe State Gymnasium, Marupe Elementary School, Riga Cultures Secondary School, and Kuldiga Center Secondary School), as well as speech therapists J. Hanzovska, L. Meiersone, M. Vorza, I. Petuhova, S. Depa, S. Jirgensone.

Conflicts of Interest: The authors declare no conflicts of interest.

References

1. Tindall, E.; Nisbet, D. Exploring the Essential Components of Reading. *J. Adult Educ.* **2010**, *39*, 1–9.
2. Rayner, K. Eye movements in reading and information processing: 20 years of research. *Psychol. Bull.* **1998**, *124*, 372. [CrossRef] [PubMed]
3. Rayner, K.; Pollatsek, A.; Reichle, E.D. Eye movements in reading: Models and data. *Behav. Brain Sci.* **2003**, *26*, 507–526. [CrossRef]
4. Liversedge, S.P.; Findlay, J.M. Saccadic eye movements and cognition. *Trends Cogn. Sci.* **2000**, *4*, 6–14. [CrossRef]
5. Yang, S.N.; McConkie, G.W. Eye movements during reading: A theory of saccade initiation times. *Vis. Res.* **2001**, *41*, 3567–3585. [CrossRef] [PubMed]
6. Reichle, E.D.; Rayner, K.; Pollatsek, A. The EZ Reader model of eye-movement control in reading: Comparisons to other models. *Behav. Brain Sci.* **2003**, *26*, 445–476. [CrossRef]
7. Reichle, E.D.; Sheridan, H. EZ Reader: An overview of the model and two recent applications. In *The Oxford Handbook of Reading*; Oxford Academic: Oxford, UK, 2015; pp. 277–290. [CrossRef]
8. Engbert, R.; Longtin, A.; Kliegl, R. A dynamical model of saccade generation in reading based on spatially distributed lexical processing. *Vis. Res.* **2002**, *42*, 621–636. [CrossRef] [PubMed]
9. Rayner, K.; Reichle, E.D. Models of the reading process. *Wiley Interdiscip. Rev. Cogn. Sci.* **2010**, *1*, 787–799. [CrossRef] [PubMed]
10. Cartwright, K.B. Insights from cognitive neuroscience: The importance of executive function for early reading development and education. *Early Educ. Dev.* **2012**, *23*, 24–36. [CrossRef]
11. Butterfuss, R.; Kendeou, P. The role of executive functions in reading comprehension. *Educ. Psychol. Rev.* **2018**, *30*, 801–826. [CrossRef]
12. Riba, S.S.; Fonseca-Pedrero, E.; Poch-Olivé, M.L. Early development of executive functions: A differential study. *Anal. Psicol.* **2015**, *31*, 607–614. [CrossRef]
13. Blankenship, T.L.; Slough, M.A.; Calkins, S.D.; Deater-Deckard, K.; Kim-Spoon, J.; Bell, M.A. Attention and executive functioning in infancy: Links to childhood executive function and reading achievement. *Dev. Sci.* **2019**, *22*, e12824. [CrossRef] [PubMed]
14. Nouwens, S.; Groen, M.A.; Kleemans, T.; Verhoeven, L. How executive functions contribute to reading comprehension. *Br. J. Educ. Psychol.* **2021**, *91*, 169–192. [CrossRef]
15. Stuart, M.; Coltheart, M. Does reading develop in a sequence of stages? *Cognition* **1988**, *30*, 139–181. [CrossRef]
16. Frey, A. Eye movements in children during reading: A review. In *Actes du Symposium International sur la Litéracie à l'école/International Symposium for Educational Literacy (SILE/ISEL)*; Les Éditions de l'université de Sherbrooke: Québec, CA, USA, 2016; pp. 70–90. [CrossRef]
17. Korneev, A.A.; Matveeva, E.Y.; Akhutina, T.V. What we can learn about reading development from the analysis of eye movements. *Hum. Physiol.* **2018**, *44*, 183–190. [CrossRef]
18. Schroeder, S.; Häikiö, T.; Pagán, A.; Dickins, J.H.; Hyönä, J.; Liversedge, S.P. Eye movements of children and adults reading in three different orthographies. *J. Exp. Psychol. Learn. Mem. Cogn.* **2022**, *48*, 1518. [CrossRef] [PubMed]
19. Nascimento, T.A.; Carvalho, C.A.F.D.; Kida, A.D.S.B.; Ávila, C.R.B.D. Fluency and reading comprehension in students with reading difficulties. *J. Soc. Bras. Fonoaudiol.* **2011**, *23*, 335–343. [CrossRef] [PubMed]
20. Bonifacci, P.; Tobia, V.; Sansavini, A.; Guarini, A. Eye-Movements in a Text Reading Task: A Comparison of Preterm Children, Children with Dyslexia and Typical Readers. *Brain Sci.* **2023**, *13*, 425. [CrossRef]
21. Prabha, A.J.; Bhargavi, R. Predictive model for dyslexia from fixations and saccadic eye movement events. *Comput. Methods Programs Biomed.* **2020**, *195*, 105538. [CrossRef]

22. Lefton, L.A.; Nagle, R.J.; Johnson, G.; Fisher, D.F. Eye movement dynamics of good and poor readers: Then and now. *J. Read. Behav.* **1979**, *11*, 319–328. [CrossRef]
23. Benfatto, M.N.; Seimyr, G.; Ygge, J.; Pansell, T.; Rydberg, A.; Jacobson, C. Screening for dyslexia using eye tracking during reading. *PLoS ONE* **2016**, *11*, e0165508. [CrossRef]
24. Bilbao, C.; Carrera, A.; Otin, S.; Piñero, D.P. Eye Tracking-Based Characterization of Fixations during Reading in Children with Neurodevelopmental Disorders. *Brain Sci.* **2024**, *14*, 750. [CrossRef] [PubMed]
25. Strandberg, A.; Nilsson, M.; Östberg, P.; Seimyr, G.Ö. Eye movements during reading and their relationship to reading assessment outcomes in Swedish elementary school children. *J. Eye Mov. Res.* **2022**, *15*. [CrossRef]
26. Rayner, K.; Chace, K.H.; Slattery, T.J.; Ashby, J. Eye movements as reflections of comprehension processes in reading. *Sci. Stud. Read.* **2006**, *10*, 241–255. [CrossRef]
27. Rezaei, A.; Jeddi, E.M. The contributions of attentional control components, phonological awareness, and working memory to reading ability. *J. Psycholinguist. Res.* **2020**, *49*, 31–40. [CrossRef]
28. Pavlidis, G.T. Do eye movements hold the key to dyslexia? *Neuropsychologia* **1981**, *19*, 57–64. [CrossRef]
29. Biscaldi, M.; Fischer, B.; Hartnegg, K. Voluntary saccadic control in dyslexia. *Perception* **2000**, *29*, 509–521. [CrossRef] [PubMed]
30. Eden, G.F.; Stein, J.F.; Wood, H.M.; Wood, F.B. Differences in eye movements and reading problems in dyslexic and normal children. *Vis. Res.* **1994**, *34*, 1345–1358. [CrossRef]
31. Tiadi, A.; Gérard, C.L.; Peyre, H.; Bui-Quoc, E.; Bucci, M.P. Immaturity of visual fixations in dyslexic children. *Front. Hum. Neurosci.* **2016**, *10*, 58. [CrossRef] [PubMed]
32. Caldani, S.; Gerard, C.L.; Peyre, H.; Bucci, M.P. Pursuit eye movements in dyslexic children: Evidence for an immaturity of brain oculomotor structures? *J. Eye Mov. Res.* **2020**, *13*. [CrossRef] [PubMed]
33. Bilbao, C.; Piñero, D.P. Objective and subjective evaluation of saccadic eye movements in healthy children and children with neurodevelopmental disorders: A pilot study. *Vision* **2021**, *5*, 28. [CrossRef] [PubMed]
34. Protasevica, D.; Kassaliete, E.; Klavinska, A.; Alecka, M.; Berzina, A.; Goliskina, V.; Koleda, M.; Mikelsone, R.; Ozola, E.; Ruza, T.; et al. The Computerized Devel-opmental Eye Movement (DEM) Test: Normative Data for School-Aged Children. *Vision* **2024**, *8*, 47. [CrossRef] [PubMed]
35. Hindmarsh, G.P.; Black, A.A.; White, S.L.; Hopkins, S.; Wood, J.M. Eye movement patterns and reading ability in children. *Ophthalmic Physiol. Opt.* **2021**, *41*, 1134–1143. [CrossRef] [PubMed]
36. Serdjukova, J.; Ekimane, L.; Valeinis, J.; Skilters, J.; Krumina, G. How strong and weak readers perform on the De-velopmental Eye Movement test (DEM): Norms for Latvian school-aged children. *Read. Writ.* **2017**, *30*, 233–252. [CrossRef]
37. Ibrahimi, D.; Aviles, M.; Rodríguez-Reséndiz, J. Oculomotor Patterns in Children with Poor Reading Abilities Meas-ured Using the Development Eye Movement Test. *J. Clin. Med.* **2024**, *13*, 4415. [CrossRef] [PubMed]
38. Vinuela-Navarro, V.; Erichsen, J.T.; Williams, C.; Woodhouse, J.M. Saccades and fixations in children with delayed reading skills. *Ophthalmic Physiol. Opt.* **2017**, *37*, 531–535. [CrossRef] [PubMed]
39. Black, J.L.; Collins, D.W.K.; De Roach, J.N.; Zubrick, S.R. Dyslexia: Saccadic eye movements. *Percept. Mot. Ski.* **1984**, *58*, 903–910. [CrossRef] [PubMed]
40. Black, J.L.; Collins, W.K.; De Roach, J.N.; Zubrick, S.R. Smooth pursuit eye movements in normal and dyslexic children. *Percept. Mot. Ski.* **1984**, *198459*, 91–100. [CrossRef] [PubMed]
41. Good, R.H.; Kaminski, R.A.; Cummings, K.; Dufour-Martel, C.; Petersen, K.; Powell-Smith, K.; Stollar, S.; Wallin, J. *Dynamic Indicators of Basic Early Literacy Skills: DIBELS Next Assessment Manual*; Dynamic Measurement Group: Huston, TX, USA, 2011.
42. Raščevska, M.; Vabale, A.; Griškēviča, I.; Orlovska, M.; Ozola, E.; Koļesovs, A.; Legzdiņš, P.; Mutule, S. *Agrīnās Lasītprasmes Attīstības Rādītāju DIBELS Next Interpretācijas Rokasgrāmata*; Latvijas Universitāte: Riga, Latvia, 2013; p. 184.
43. Goliskina, V.; Ceple, I.; Kassaliete, E.; Serpa, E.; Truksa, R.; Svede, A.; Krauze, L.; Fomins, S.; Ikaunieks, G.; Krumina, G. The effect of stimulus contrast and spatial position on saccadic eye movement parameters. *Vision* **2023**, *7*, 68. [CrossRef]
44. Tobii, AB. *Tobii Pro Lab User Manual (Version 1.152)*; Tobii, AB: Danderyd, Sweden, 2020; p. 168.
45. Faul, F.; Erdfelder, E.; Lang, A.-G.; Buchner, A. G*Power 3: A flexible statistical power analysis program for the social, behavioral, and biomedical sciences. *Behav. Res. Methods* **2007**, *39*, 175–191. [CrossRef]
46. Tabachnick, B.G.; Fidell, L.S. *Experimental designs using ANOVA*; Thomson/Brooks/Cole: Belmont, CA, USA, 2007; pp. 1–724.
47. Scheffer, J. Dealing with missing data. *Res. Lett. Inf. Math. Sci.* **2002**, *3*, 153–160.
48. Zhang, Z. Introduction to machine learning: K-nearest neighbors. *Ann. Transl. Med.* **2016**, *4*, 218. [CrossRef] [PubMed]
49. Niehorster, D.C.; Andersson, R.; Nyström, M. Titta: A toolbox for creating PsychToolbox and Psychopy experiments with Tobii eye trackers. *Behav. Res. Methods* **2020**, *52*, 1970–1979. [CrossRef] [PubMed]
50. Thaler, L.; Schütz, A.C.; Goodale, M.A.; Gegenfurtner, K.R. What is the best fixation target? The effect of target shape on stability of fixational eye movements. *Vis. Res.* **2013**, *76*, 31–42. [CrossRef]
51. Hessels, R.S.; Niehorster, D.C.; Kemner, C.; Hooge, I.T.C. Noise-robust fixation detection in eye movement data: Identification by two-means clustering (I2MC). *Behav. Res. Methods* **2017**, *49*, 1802–1823. [CrossRef] [PubMed]

52. Crossland, M.D.; Rubin, G.S. The use of an infrared eye-tracker to measure fixation stability. *Optom. Vis. Sci.* **2002**, *79*, 735–739. [CrossRef]
53. Abegg, M.; Pianezzi, D.; Barton, J.J. A vertical asymmetry in saccades. *J. Eye Mov. Res.* **2015**, *8*, 1–10. [CrossRef]
54. Lukasova, K.; Silva, I.P.; Macedo, E.C. Impaired oculomotor behavior of children with developmental dyslexia in antisaccades and predictive saccades tasks. *Front. Psychol.* **2016**, *7*, 987. [CrossRef]
55. Wilcockson, T.D.W.; Mardanbegi, D.; Sawyer, P.; Gellersen, H.; Xia, B.; Crawford, T.J. Oculomotor and Inhibitory Control in Dyslexia. *Front. Syst. Neurosci.* **2019**, *12*, 66. [CrossRef]
56. Fukushima, J.; Tanaka, S.; Williams, J.D.; Fukushima, K. Voluntary control of saccadic and smooth-pursuit eye movements in children with learning disorders. *Brain Dev.* **2005**, *27*, 579–588. [CrossRef] [PubMed]
57. Tanabe, J.; Tragellas, J.; Miller, D.; Ross, R.G.; Freedman, R. Brain activation during smooth-pursuit eye movements. *Neuroimage* **2002**, *17*, 1315–1324. [CrossRef]
58. Edelman, J.A.; Valenzuela, N.; Barton, J.J. Antisaccade velocity, but not latency, results from a lack of saccade visual guidance. *Vis. Res.* **2006**, *46*, 1411–1421. [CrossRef] [PubMed]
59. Bucci, M.P.; Seassau, M. Saccadic eye movements in children: A developmental study. *Exp. Brain Res.* **2012**, *222*, 21–30. [CrossRef]
60. Lunn, J.; Donovan, T.; Litchfield, D.; Lewis, C.; Davies, R.; Crawford, T. Saccadic eye movement abnormalities in children with epilepsy. *PLoS ONE* **2016**, *11*, e0160508. [CrossRef] [PubMed]
61. Ygge, J.; Lennerstrand, G.; Rydberg, A.; Wijecoon, S.; Pettersson, B.M. Oculomotor functions in a Swedish population of dyslexic and normally reading children. *Acta Ophthalmol.* **1993**, *71*, 10–21. [CrossRef] [PubMed]
62. Sinno, S.; Najem, F.; Abouchacra, K.S.; Perrin, P.; Dumas, G. Normative Values of Saccades and Smooth Pursuit in Children Aged 5 to 17 Years. *J. Am. Acad. Audio* **2020**, *31*, 384–392. [CrossRef]
63. Yang, Q.; Kapoula, Z. Binocular coordination of saccades at far and at near in children and in adults. *J. Vis.* **2003**, *3*, 554–5561. [CrossRef] [PubMed]
64. Yang, L.; Li, C.; Li, X.; Zhai, M.; An, Q.; Zhang, Y.; Zhao, J.; Weng, X. Prevalence of Developmental Dyslexia in Primary School Children: A Systematic Review and Meta-Analysis. *Brain Sci.* **2022**, *12*, 240. [CrossRef] [PubMed]
65. Kristjansson, A.; Sigurdardottir, H.M. The role of visual factors in dyslexia. *J. Cogn.* **2023**, *6*, 31. [CrossRef]
66. Vikesdal, G.H.; Falkenberg, H.K.; Mon-Williams, M.; Riddell, P.; Langaas, T. Normal saccades but decreased fixation stability in a population of children with dyslexia. *Scand. J. Optom. Vis. Sci.* **2021**, *14*, 1–7. [CrossRef]
67. Stein, J. Reply to: "The Relationship between eye movements and reading difficulties", Blythe, Kirkby & Liversedge. *Brain Sci.* **2018**, *8*, 99. [CrossRef]
68. Rowe, F.; Wright, D.; Brand, D.; Jackson, C.; Price, A.; Walker, L.; Harrison, S.; Eccleston, C.; Maan, T.; Scott, C.; et al. Reading difficulty after stroke: Ocular and non ocular causes. *Int. J. Stroke* **2011**, *6*, 404–411. [CrossRef] [PubMed]
69. Zang, C. New perspectives on serialism and parallelism in oculomotor control during reading: The multi-constituent unit hypothesis. *Vision* **2019**, *3*, 50. [CrossRef]
70. Mancheva, L.; Reichle, E.D.; Lemaire, B.; Valdois, S.; Ecalle, J.; Guérin-Dugué, A. An analysis of reading skill development using E-Z Reader. *J. Cogn. Psychol.* **2015**, *27*, 657–676. [CrossRef]
71. Raščevska, M.; Umbraško, S.; Vabale, A.; Orlovska, M.; Sokola-Nazarenko, M. *Lasītprasmes attīstība 4.-6. klases skolēniem Acadience TM (Dibels Next) mērījumos un intervences efektivitāte*; Valsts izglītības satura centrs: Riga, Latvia, 2019; p. 100.
72. Rohimah, S. Reading difficulties and factors affecting reading difficulties of students of grade 1 elementary school. *J. Lang. Educ.* **2021**, *1*, 189–195. [CrossRef]
73. Taylor, E.K.; Abdurokhmonova, G.; Romeo, R.R. Socioeconomic Status and Reading Development: Moving from "Deficit" to "Adaptation" in Neurobiological Models of Experience-Dependent Learning. *Mind Brain Educ.* **2023**, *17*, 324–333. [CrossRef] [PubMed]
74. Christian, L.W.; Nandakumar, K.; Hrynchak, P.K.; Irving, E.L. Visual and binocular status in elementary school children with a reading problem. *J. Optom.* **2018**, *11*, 160–166. [CrossRef]
75. Ayton, L.N.; Abel, L.A.; Fricke, T.R.; McBrien, N.A. Developmental eye movement test: What is it really measuring? *Optom. Vis. Sci.* **2009**, *86*, 722–730. [CrossRef] [PubMed]
76. Facchin, A. Spotlight on the developmental eye movement (DEM) test. *Clin Optom.* **2021**, *13*, 73–81. [CrossRef]
77. Caldani, S.; Gerard, C.L.; Peyre, H.; Bucci, M.P. Visual attentional training improves reading capabilities in children with dyslexia: An eye tracker study during a reading task. *Brain Sci.* **2020**, *10*, 558. [CrossRef]
78. Pisa, OECD. *2018 Results (Volume I): What Students Know and Can Do*; OECD Publishing: Paris, France, 2019.
79. Stoodley, C.J.; Stein, J.F. Cerebellar function in developmental dyslexia. *Cerebellum* **2013**, *12*, 267–276. [CrossRef]

applied sciences

Article

The Influence of Different Visual Elements of High-Density Urban Observation Decks on the Visual Behavior and Place Identity of Tourists and Residents

Pohsun Wang [1] **and Hao Fu** [1,2,*]

[1] Faculty of Innovation and Design, City University of Macau, Macau 999078, China; phwang@cityu.edu.mo
[2] College of Fine Arts, Shanxi Normal University, Taiyuan 030031, China
* Correspondence: fuhao@sxnu.edu.cn

Abstract: This study focuses on the visual attention of residents and tourists to elements of urban landscapes from above. It screens out elements that attract viewers' attention, assesses their aesthetics, and explores how these aesthetic evaluations affect the psychology of place identity. We tracked data from 30 participants, collected responses from 237 participants, and observed differences in visual behaviors and emotional connections to place identity. The results show that while residents and tourists exhibit similar visual behaviors when viewing the same landscape, they have large differences in their perceptual behaviors for different visual elements and have very different familiarity levels in the place identity dimension. Landmark buildings attract strong visual attention despite their low proportion in the overall landscape. Aesthetic factors such as the color vividness of elements like water, mountains, and landmark buildings significantly affect place identity. This study shows that combining eye-tracking and psychometric analysis can effectively evaluate urban landscape perception and provide valuable insights for visual planning and preservation efforts in historic neighborhoods.

Keywords: visual elements; place identity; landscape; eye-tracking; questionnaires; Historic Centre of Macao

1. Introduction

With the rapid growth of global urbanization, urban population density is increasing dramatically. Landscape views in historic districts are blocked by modern tall buildings, and in the face of declining visual environment quality caused by uncontrolled rapid urban development for urban heritage protection, UNESCO has proposed the theory and methods of urban historical landscape [1]. The "Macao Special Administrative Region Urban Master Plan (2020–2040)" also mentions protecting world heritage as an important goal. In urban planning and design, the sustainable development of the overall urban landscape must be considered [2]. However, perception evaluations of important landscape views remain an urgent problem in the coordinated development and protection of historic cities. Constructing a visual environment aesthetic evaluation system with perception as the core has become a major research topic in the field of urban historical heritage protection and planning in recent years [3]. The use and satisfaction of urban landscape architecture physical spaces mediate the psychological health of the public [4]. Constructing an aesthetic evaluation system of the visual environment with perception as the core has become the core research issue in the fields of urban architecture, landscape architecture, and planning in recent years [5]. Visual perception accounts for approximately 80% of

https://doi.org/10.3390/app15073875

all types of human perception, and the visual system is the most accurate way to obtain external information [6]. The evaluation results of a study on urban streets and buildings provide a corresponding basis for the visual control of urban landscape architecture [7]. In a changing urban landscape, we need to understand which architectural elements can attract viewers' attention from viewing platforms and the psychological impact of their aesthetic evaluation of these visual elements on place identity.

This study focuses on how modern high-rise buildings and key visual elements in urban landscapes affect the beauty and local identity of historical blocks; what methods can be used to construct a visual environmental aesthetic evaluation system based on perception; and how urban landscapes affect the public's mental health. We aim to construct a visual environment aesthetic evaluation system centered on perception to meet the challenges brought about by the uncontrolled urban development of historical blocks and evaluate the impact of modern high-rise buildings on their landscapes. This study explores the effectiveness of urban landscape perception assessment through eye-tracking and psychometric analysis; provides valuable insights for visual planning and protection in historic districts; and fills a research gap in assessments of spatial perception elements of historic urban areas based on place identity emotions.

2. Literature Review

With advancements in ergonomic technology, eye-tracking experiments have emerged as a scientific approach for exploring human emotions through physiological indicators [8]. These experiments play a crucial role in uncovering the characteristics and patterns of visual behavior when people perceive visual elements in public landscapes [9]. In urban park landscape architecture research, fixed time and counting of eye-tracking experimental data have shown that the proportion of architectural elements affects the evaluation of the space [10]. In terms of mountain landscape architecture, artificial visual elements attract more attention than natural elements [11]. In the field of forest research, the visual behavior of the public toward the surrounding elements is different due to the variable spatial elements of forest landscape architecture [12]. In the field of urban public landscapes, eye-tracking technology is combined with online surveys to analyze the correlation between the proportion of different landscape elements and the public's visual attractiveness evaluations [10]. In the field of architecture, this provides a new method for understanding architectural space perception and architectural space interaction. One study used VR eye-tracking technology to deeply analyze the visual perception behavior of subjects in the historical city of Zhangzhou, revealing differences in the perception of the urban atmosphere between individuals from different backgrounds [13]. In the field of street architecture, eye movement experiments demonstrated the visual interaction between pedestrians and buildings on both sides of the street, emphasizing the importance of the visual appeal and social function of street architecture [14].

In summary, most scholars focus on visual perceptions and behavioral preferences in relation to landscape architecture, such as forests, mountains, and urban parks. However, the number of previous studies on the relationship between the perception of physical visual elements in the environment and place identity is very limited [15]. The mechanism by which the objective environment that generates visual perception evaluation affects the viewer's place identity is still unclear. Most research on the public's perceptions of urban landscape architecture focuses on comparing natural and artificial visual elements, ignoring the differences in evaluation objects. In this study, based on the viewers' length of stay and purpose while in historic districts, the public is divided into two main groups: residents and tourists. We explore differences in the visual behavior of different populations toward important landscapes in the same historic district. Therefore, the following hypothesis is proposed:

Hypothesis H1. *The public's perceptions of different visual elements of the same landscape are significantly different.*

Topophilia is the emotional connection between people and the natural environment in which they live [16]. In the 1980s, environmental psychology introduced the concept of "place identity"—the "socialized self in the objective world". The theory of place identity has been empirically studied in relation to urban living and tourist spaces; in terms of theoretical dimension, studies have confirmed the correlation between place identity and familiarity, sense of belonging, environmental preferences, intimacy, cultural atmosphere, the stimulation of former memories, etc. Rather than existing on the same temporal plane, identification processes drive other affective and conative elements that underlie people's attachments to physical environments [17]. People's aesthetic evaluations of the visual environment are important links between them and the environment. Most of the research on place identity and emotion around the environment uses the interview method to evaluate emotional tendencies when people visit environments [18]. In a crowd looking at urban landscape architecture, residents and tourists engage in social interactions, and their perceptions of the architecture and its culture significantly impact their urban place identity [19].

Since the 1980s, the evaluation of visual quality has mainly focused on aesthetic evaluations of visual perception. The evaluation object has changed from a single nature evaluation system to a pluralistic human–nature emotional relationship evaluation system. In recent years, street view images have been used to evaluate visual quality through research methods such as machine learning and deep learning [20], and the experimental approach has been used to assess the specific role of the city's visual and auditory environment in influencing human emotions [21]. However, the impact mechanism between visual aesthetic evaluation and place identity is still unclear. We explore the differences and impact effects between the aesthetic evaluation and place identity of different populations toward important landscapes in the same historic district. Therefore, the following hypotheses are proposed (Figure 1):

Hypothesis H2. *There are differences in the place identity emotions of residents and tourists toward important landscapes in the same historic district.*

Hypothesis H3. *There is a significant impact relationship between the perception evaluation of landscape views in historic districts and place identity emotions.*

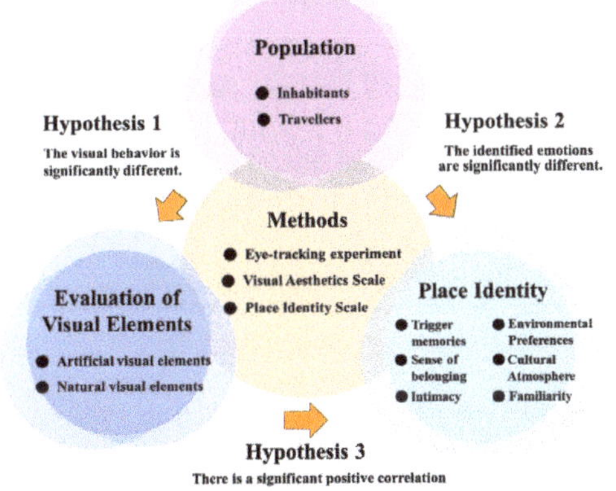

Figure 1. Conceptual framework of the viewer, visual element evaluation, and place identity.

3. Method

3.1. Study Scope and Stimulus Images

Located on the southeastern coast of China, Macao has the highest population density in the world [22] and an urban history of more than 400 years. The Historic Centre of Macau and its buffer zone are located in the city's core, accounting for 12.9% of the area of the Macau Peninsula. The city's urban form has a unique and super-compact style, especially in the Historic Centre [23].

This study aims to define the landscape views of the city based on the overall urban planning of the Macau Special Administrative Region [24]. Viewing points were selected using planning vision—public vision in on-site research—and corridor distribution. Based on observation object categories, the study is divided into 4 themes, with 12 important landscape views (Table 1).

Table 1. Viewpoints and main visual elements viewed from observation decks in Macao.

Theme and Quantity	No.	Sight Direction	Viewing Objects	The Main Visual Elements of the Environment
A. Landscapes of architectural heritage and sites in historic districts	A-01	From the Fortress to the Ruins of St. Paul's	Architectural heritages	Landmarks, low-rise buildings, high-rise buildings, skyscraper buildings, green plants, water, mountains, roads, etc.
	A-02	From the Fortress to White Pigeon Nest Park	Urban greenings	Landmarks, low-rise buildings, high-rise buildings, skyscraper buildings, green plants, water, mountains, etc.
	A-03	From Guia Fort and Lighthouse to Tap Seac Square	Architectural heritages	Landmarks, low-rise buildings, high-rise buildings, skyscraper buildings, green plants, water, mountains, roads, etc.
B. Landscapes of the inner harbor, outer harbor, and water systems	B-01	From Bishop's Hill Observation Deck to the Inner Harbor Waterway	The inner harbor	Landmarks, low-rise buildings, high-rise buildings, skyscraper buildings, green plants, water, mountains, etc.
	B-02	From Bishop's Hill Church to Sai Van Lake	Sai Van Lake	Landmarks, low-rise buildings, high-rise buildings, skyscraper buildings, green plants, water, mountains, roads, bridges, etc.
	B-03	From Guia Fort and Lighthouse to the Outer Harbor	The outer harbor	Landmarks, high-rise buildings, skyscraper buildings, green plants, water, mountains, roads, bridges, etc.
C. Landscapes of the historic district style and urban skyline	C-01	From Guiyang Mountain to Nam Van Lake	The city district and skyline	Landmarks, low-rise buildings, high-rise buildings, skyscraper buildings, green plants, water, mountains, roads, etc.
	C-02	From Fortress to Inner Harbor City	The cityscape of the port area	Landmarks, low-rise buildings, high-rise buildings, skyscraper buildings, green plants, water, etc.
	C-03	From Fortress to Guia Fortress	The landscapes between urban highlands	Landmarks, low-rise buildings, high-rise buildings, skyscraper buildings, green plants, etc.
D. Landscapes of high-point landmark heritage in the historic district	D-01	From the Nam Van Lake to the direction of the Church of Our Lady of Penha	The Church of Our Lady of Penha	Landmarks, low-rise buildings, high-rise buildings, skyscraper buildings, green plants, mountains, water, etc.
	D-02	From the Sai Van Lake to the direction of the Church of Our Lady of Penha	The Church of Our Lady of Penha	Landmark buildings, low-rise buildings, high-rise buildings, skyscraper buildings, green plants, mountains, water, etc.
	D-03	From Lotus Square to the direction of the Guia Lighthouse	Guia Fort and Lighthouse	Landmark buildings, low-rise buildings, high-rise buildings, roads, green plants, mountains, etc.

We categorized the main visual environmental elements seen from each observation deck. Buildings were classified based on height according to the latest version of China's Unified Standard for Civil Building Design [25]. Buildings with heights not exceeding 27 m are low-rise buildings; those ranging from 27 to 100 m are high-rise buildings; and those taller than 100 m are super high-rise buildings. The landscape architecture views were classified based on the viewing targets, and the key environmental visual elements were summarized (Figure 2).

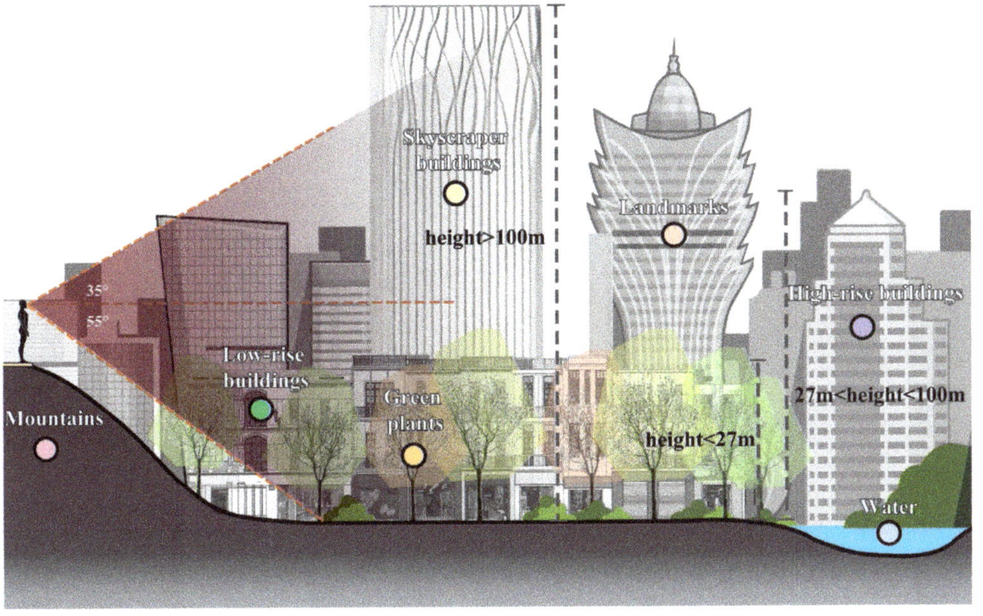

Figure 2. Schematic diagram of the division of sight lines in the landscape's architecture.

Landscape photographs are often used for eye-tracking experiments and can effectively replace visual perceptions of the real environment. Moreover, the use of photographs as stimuli for eye-tracking experiments can eliminate the influence of any external conditions, such as light intensity and seasonal changes [26]. Therefore, experimental images were taken in similar weather conditions using a Canon RP camera with a 16 mm lens; the height of the photographer's eyes was approximately 1.4 m. A total of 168 horizontal landscape architecture images were taken from observation decks in Macao, and five experts were invited to classify them. The experts selected certain images that were then encoded.

The specific practices are as follows: (1) image feature annotation: preliminary observation of 168 images, extraction of keywords of landscape elements (e.g., landmarks, vegetation, water bodies, etc.), and labeling of core features of each image. (2) Free classification: experts group images with similar features according to core landscape features. (3) Integration of grouping: the 168 images were preliminarily categorized into 40 groups by combining obviously overlapping or duplicated categories. (4) Logical correction: cross-category images were adjusted to determine the leader–follower relationship through expert team discussion, including government-defined viewpoints of the landscape, selecting 3–4 images for each viewpoint, retaining 45 landscape images by reducing them to 12 groups. (5) Representativeness verification: This ensured that the most representative image was retained for each group. In total, 12 landscape images of core viewpoints in the Historic Centre of Macao were selected, forming a structured image library that could support the oculomotor experimental study. Finally, the selected images were used for material selection and coding.

The images were normalized with a 16:9 aspect ratio, and the resolution was 6000×3374 dpi. As weather is an uncontrollable factor in the environment, a study of the perception of visual sky elements was outside the scope of this research, so the sky in the images was uniformly replaced with white areas (Figure 3).

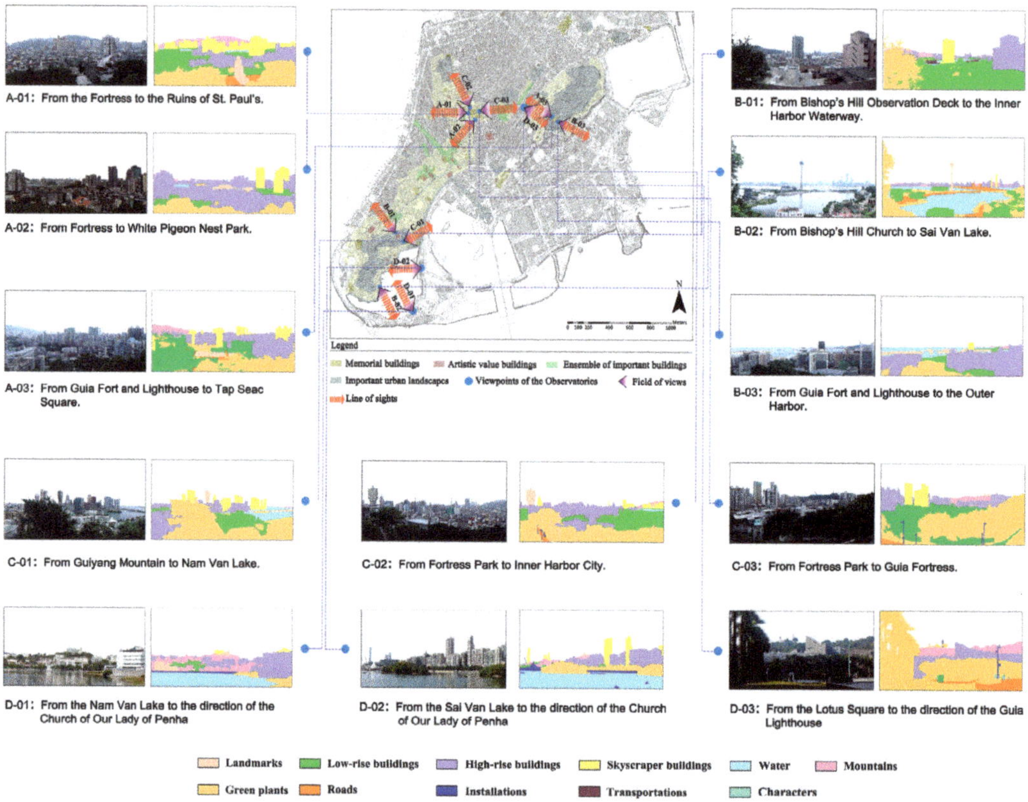

Figure 3. Distribution of important sightlines in the Historic District of Macao and AOI images.

3.2. Eye-Tracking Experiments

3.2.1. Experimental Participants

During the study, 30 members of the public who had visited observation decks in Macao were selected and divided into two groups, residents (n = 15) and tourists (n = 15), according to their purpose in visiting and the length of their stay. Two eye-tracking experiments were conducted at the lab. The participants were exposed to the relevant image samples for the first time; had good visual acuity, with naked eye or corrected visual acuity of 1.0 and above; and had no color weakness or color blindness. They were between 20 and 40 years old and had diverse professional backgrounds, with 12 participants from architecture-related majors and 18 from non-architecture-related majors, covering fields such as industrial design, physics, chemistry, computer science, and fine arts. In the viewing scene section, we allowed participants to view the scene freely, as free viewing is the most natural and common way to view landscape architecture, maximizing the simulation of daily viewing conditions [27].

3.2.2. Experimental Protocol

Considering the interference effect of light on eye-tracking experiments, the testing was conducted in a laboratory. Curtains were used throughout the experiment to block out outdoor light sources, and the laboratory light environment was adjusted and controlled through internal lighting. The eye-tracking experiments were conducted with Tobii Pro Fusion, manufactured by a Swiss company, and images of the experimental material were displayed on a 23.8-inch screen (1920 × 1080 dpi, 600 Hz) in the lab.

To begin, participants were placed in a laboratory with a capacity of more than 40 people, and the experimenter explained the purpose, procedures, and requirements to the participants individually before the start of each eye movement experiment, with no conversation between participants. Once participants understood the experiment, they were invited to sit in a comfortable chair and were assisted by the experimenter in calibrating the eye tracker to make sure that it was calibrated accurately. Then, nine sample images of the study site were displayed. Each image sample was displayed for 20 s with a two-second gap between images; this was intended to alleviate eyestrain and ensure visual focus adjustment.

Each session lasted approximately 30 min, with approximately 4 min of introduction and preparation before the experiment, approximately 6 min for the eye-tracking experiment, and approximately 20 min for the questionnaire content (Figure 4).

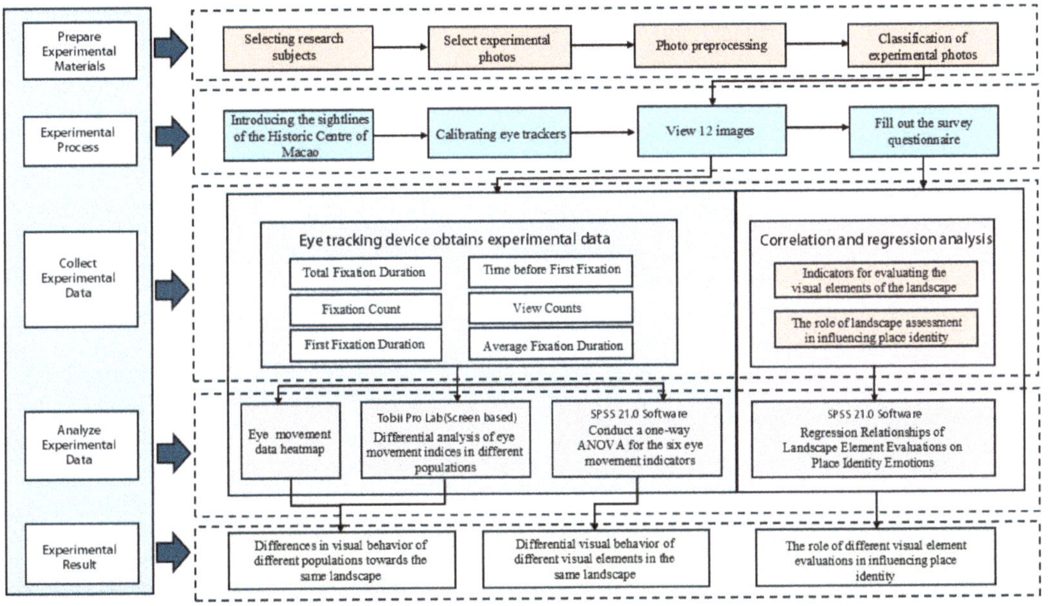

Figure 4. Step-by-step diagram of the eye-tracking experiment.

3.2.3. Drawing AOIs

The relationship between visual behavior and the visual environment was explored using areas of interest (AOIs). The AOIs in the samples were determined according to the main fixed elements of the images to analyze the spatial and visual elements of the samples. The visual elements in the environment around the observation decks in Macao were divided into features such as landmarks, low-rise buildings, high-rise buildings, skyscraper buildings, green plants, water, mountains, roads, and so on.

3.2.4. Data Index Selection

To explore the visual behavior characteristics of the residents and tourists in relation to AOIs in the images, several eye-tracking indicators were analyzed. Six were selected—total fixation time (TFD), fixation count (FC), time before first fixation (TFF), time to first fixation

(FFD), average fixation duration (AFD), and view count (VC) [28] (Table 2). These metrics were used to identify the characteristics of viewer visual behavior toward visual elements of the urban overlook landscape.

Table 2. Eye-tracking indexes for different landscape elements.

Metrics	Indicator	Abbreviation	Basic Significance
Fixation metrics	Total fixation duration/(s)	TFD	The longer the TFD, the more attention participants paid to the area and the more difficulty they had in processing relevant information.
	Fixation count/(no. of visits)	FC	Fixation count is the total number of gaze points produced by the subject's eyes during a specific experimental task or observation. This can reflect the degree of attention to different areas or stimuli and the subject's processing of information.
	Time to first fixation (s)	TFF	It takes time for the eyes to move to an AOI. If the TFF was short, it was easier to notice an element. TFF was used to measure visual saliency.
	First fixation duration/(s)	FFD	FFD for the AOI. As the FFD lengthens, respondents either have greater difficulty in recognizing the AOI or perceive it as more appealing.
	Average fixation duration (s)	AFD	Mean gaze duration is the average of the subjects' gaze duration for each gaze point in the eye movement experiment.
	Visit count/(no. of visits)	VC	The visit count of an AOI represents the number of times a respondent views that region. A higher count generally implies the AOI's greater attractiveness to the respondent.

3.3. *Visual Aesthetics and Place Identity Scale*

3.3.1. Questionnaire Design

After the experiment and the retrospective interviews, the same A1–D3, 12 landscape pictures from the eye-tracking experiment were used to evaluate the aesthetics of the different visual elements. The color labels of the visual elements were divided by the AOIs; combined with the dimension of place identity, the degree of identification of the tourists and residents was psychologically measured. The questionnaire took the residents and tourists as the subjects and preliminarily screened the visual elements according to the distribution of eye movement. Moreover, we invited six experts in the field of architecture to classify and describe the visual elements of the overlooked landscapes. We selected eight evaluation indicators—the proportion of coordination, form richness, neatness and aesthetics, color vividness, change difference, and so on—to evaluate the visual aesthetics of the landscape architecture viewed from urban observation decks. This was combined with the five-point Psychological Scale of Place Identity Likert scale (Table 3), which was completed by the residents and tourists [29].

3.3.2. Questionnaire Participants

A total of 250 participants were selected at observation decks in Macao, including 30 participants from the eye-tracking experiment. In total, 250 questionnaires were distributed, 225 of which were online. The questionnaire was collected from 1 November 2023 to 30 December 2023. A total of 237 valid questionnaires were collected, with a response rate of 94.8%.

Table 3. Visual aesthetic and place identity evaluation scale.

Evaluation Index	Specific Dimensions	Online Questionnaire Items
Visual aesthetics evaluation	Proportional harmonization	What do you think is the proportion and coordination of visual elements (the larger the number, the more coordinated)?
	Morphological richness	What do you think is the morphological richness of visual elements (the larger the number, the richer)?
	Neatness and beauty	What do you think of the neatness and beauty of visual elements (the larger the number, the more beautiful)?
	Vividness of color	What do you think the color of the visual element is (the larger the number, the more vivid)?
	Variation difference	What do you think is the difference in changes in visual elements (the larger the number, the greater the difference)?
	Quantitative intensity	What do you think is the numerical density of visual elements (the larger the number, the denser)?
	Significance of features	What do you think is the prominent style of the visual elements (the larger the number, the more salient)?
	Decorative sophistication	What do you think of the decorative sophistication of visual elements (the larger the number, the more refined)?
Place identity evaluation	Trigger memories	I have a lot of fond memories of this place
	Sense of belonging	I feel this place is a part of me
	Familiarity	I feel very familiar with this place
	Intimacy	I feel intimate with this place
	Environmental preferences	This place is very special to me
	Cultural identity	I have a strong cultural identity with this place
	Identity responsibility	Living here says a lot about who I am

3.3.3. Data Analysis

The data were analyzed using SPSS 21.0 (IBM Corp., Armonk, NY, USA) and plotted using Origin 2022 (OriginLab Corp., Northampton, MA, USA) and Adobe Photoshop CS6 (Adobe Inc., San Jose, CA, USA). The procedure was as follows.

(1) One-way analysis of variance (ANOVA) was used to analyze the differences in visual behavior indicators and to determine the differences between groups of people at the same viewing platform for the same image and different visual element eye movement indicators. An independent sample *t*-test and univariate analysis were performed, and H1 was verified.
(2) A visual aesthetic evaluation and psychometric measurement of different scenes were carried out using the questionnaire data. Reliability and validity analyses and an independent sample *t*-test were carried out to determine the psychological differences between different groups of people in relation to places with the same overall landscape architecture. The aim was to verify H2.
(3) Correlation analysis and multiple regression equations were used to determine the correlation between the aesthetic evaluation indexes of different visual elements and place identity, and H3 was verified.

4. Results and Analysis

4.1. Eye-Tracking Heat Map Distribution and Analysis

The eye-tracking data collected using Tobii Pro Lab (v1.194) were processed for analysis to create an eye movement heat map of the images. Moreover, through notation, the most crucial visual elements, as perceived by both residents and tourists, were analyzed and filtered out. Taking the thermal distribution in Figure A-01 as an illustration, visual elements such as landmark buildings, high-rise buildings, super high-rise buildings, and water systems within the figure draw a high level of attention. These elements are represented by dark red and orange-red colors and the "•" symbol. Lower buildings and mountains are also considered, but their prominence is less evident. They are depicted with light green spots and the "◎" symbol. Green plants and roads attract relatively little attention. There is no heat distribution for them, and no light green spots are present, with the "○" symbol being used. As for Figure A-02′s heat distribution, there are no visual elements of landmark buildings in the figure, and the "×" symbol is used. The identification situation for the other heat maps is the same, and the visual elements corresponding to the visual hot spots in the image are identified similarly (Table 4).

Table 4. Eye-tracking heat distribution of different types of viewing platform landscape images.

NO.	A-01				A-02				A-03			
Resident eye-tracking experiment heat map	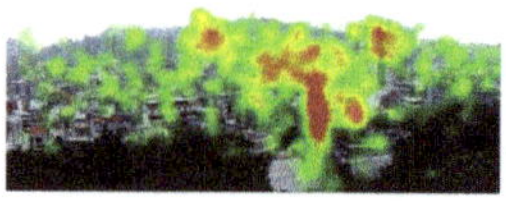								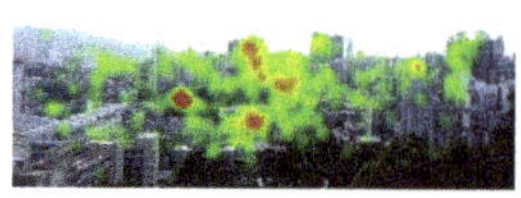			
Tourist eye-tracking experiment heat map	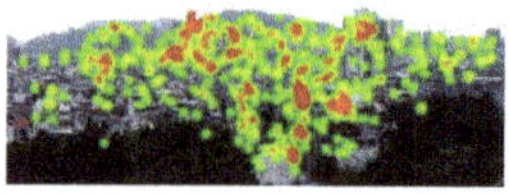				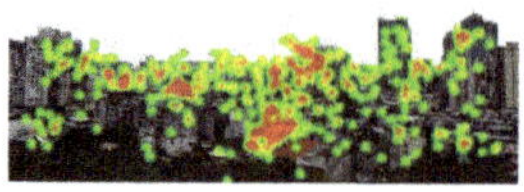							
Hot spots corresponding to visual features	landmarks	•	mountains	•	landmarks	•	mountains	•	landmarks	•	mountains	◎
	low-rise buildings	•	water	×	low-rise buildings	•	water	×	low-rise buildings	◎	water	○
	high-rise buildings	•	green plants	○	high-rise buildings	•	green plants	◎	high-rise buildings	◎	green plants	◎
	skyscraper buildings	•	roads	○	skyscraper buildings	•	roads	×	skyscraper buildings	•	roads	◎
NO.	B-01				B-02				B-03			
Resident eye-tracking experiment heat map					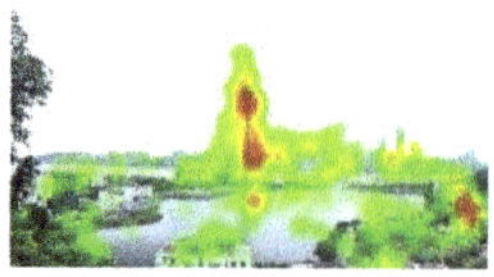				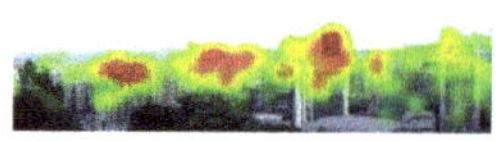			
Tourist eye-tracking experiment heat map												
Hot spots corresponding to visual features	landmarks	×	mountains	◎	landmarks	•	mountains	○	landmarks	•	mountains	○
	low-rise buildings	◎	water	◎	low-rise buildings	•	water	◎	low-rise buildings	○	water	•
	high-rise buildings	•	green plants	◎	high-rise buildings	◎	green plants	◎	high-rise buildings	•	green plants	◎
	skyscraper buildings	•	roads	×	skyscraper buildings	•	roads	○	skyscraper buildings	•	roads	•
NO.	C-01				C-02				C-03			
Resident eye-tracking experiment heat map	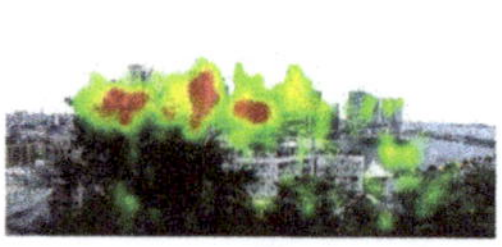								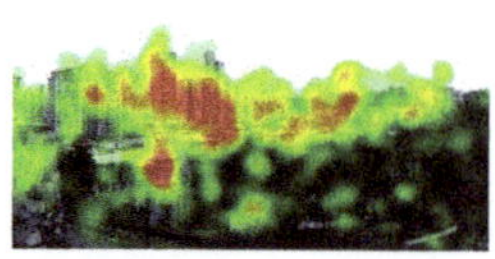			

Table 4. *Cont.*

NO.	C-01				C-02				C-03			
Tourist eye-tracking experiment heat map	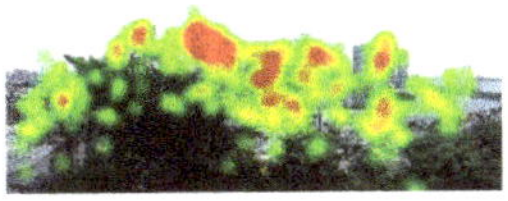								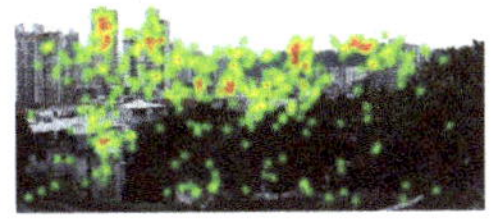			
Hot spots corresponding to visual features	landmarks	•	mountains	○	landmarks	•	mountains	○	landmarks	•	mountains	○
	low-rise buildings	◎	water	◎	low-rise buildings	•	water	○	low-rise buildings	•	water	×
	high-rise buildings	•	green plants	◎	high-rise buildings	◎	green plants	○	high-rise buildings	◎	green plants	◎
	skyscraper buildings	•	roads	○	skyscraper buildings	•	roads	○	skyscraper buildings	•	roads	○
NO.	D-01				D-02				D-03			
Resident eye-tracking experiment heat map												
Tourist eye-tracking experiment heat map												
Hot spots corresponding to visual features	landmarks	•	mountains	•	landmarks	•	mountains	•	landmarks	•	mountains	•
	low-rise buildings	•	water	•	low-rise buildings	•	water	◎	low-rise buildings	×	water	×
	high-rise buildings	•	green plants	◎	high-rise buildings	◎	green plants	•	high-rise buildings	•	green plants	•
	skyscraper buildings	•	roads	×	skyscraper buildings	×	roads	×	skyscraper buildings	×	roads	○

• This element is of significant concern. ◎ This element attracts attention but is not significant. ○ The image has this element, but it does not receive attention. × The feature does not appear in the image.

Comparing the eye movement heat distribution maps of residents and tourists, the visual elements that residents pay attention to are more concentrated; the focus is on the visual center of the image and the visual elements of the buildings with a regional identity. Meanwhile, the visual elements that tourists pay attention to are more scattered, but the attention paid to different visual elements tends to be consistent. Looking at the heat distribution map of the landscape architecture image eye movement experiment performed on the observation decks in Macao, the visual elements concentrated on are landmarks, skyscraper buildings, high-rise buildings, water, and other visual elements, and the colors are dark red and orange.

Using the eye-tracking heat map and the visual element markers corresponding to the hot spots, a preliminary screening of visual elements was carried out. The visual element of roads was marked with an "○" symbol and an "×" symbol in the nine images, indicating that the residents and tourists paid less attention to roads. We screened out landmark buildings, low-rise buildings, high-rise buildings, super high-rise buildings, water systems, and mountains. Seven visual elements, including green plants, were used to support the ANOVA and the results of the perceptual identity questionnaires.

4.2. Perceptual Behavior of Visual Elements

4.2.1. Visual Behaviors of Different Groups of Viewers

The eye movement index of one image was randomly selected from the landscape architecture images viewed by each group on the observation decks, and six indicators of the residents (n = 15) and tourists (n = 15) were analyzed. Finally, the eye-tracking experiment data for the three landscape architecture images of A-01, B-02, and C-02 were selected as the data sources. Based on the experimental results for the different groups of participants on the same observation deck, SPSS 21.0 (IBM Corp.) was used to perform an independent sample *t*-test (Figure 5, Supplementary Materials Table S1), and the significance of the six visual behaviors in the two groups was compared.

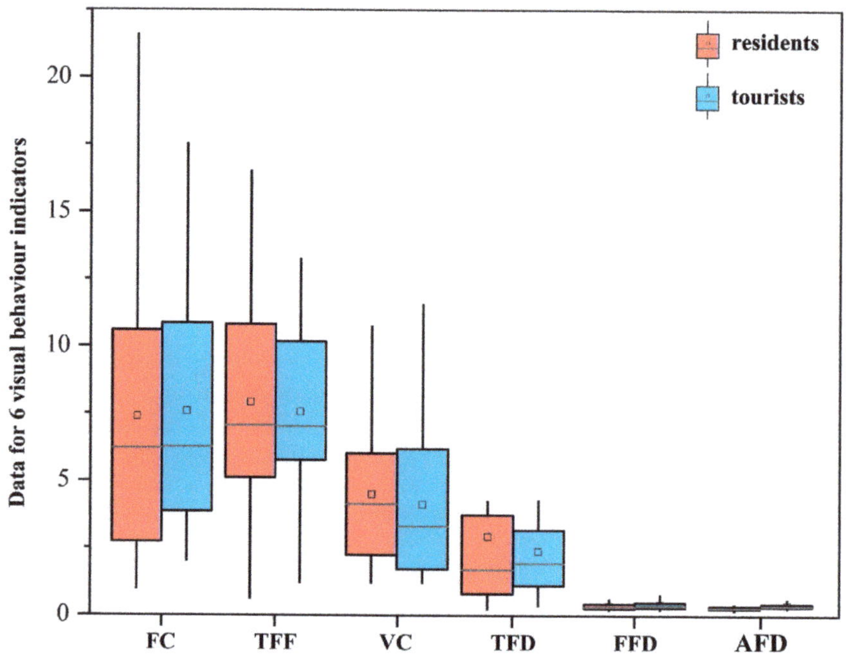

Figure 5. Viewers' visual behavior data for different landscape architecture elements.

The data in Table S1 show that the F-value (i.e., the Levene F-statistic) has a *p*-value > 0.05 for six variables, and these six eye-tracking data all show the same variance. In the two-tailed test, the *p*-value is also >0.05; combined with the F-statistic, this suggests that the visual behavior of the two groups toward the same landscape architecture elements has no significant difference, and the visual behavior characteristics of the same view are maintained.

4.2.2. Eye Movement Indicators of Different Visual Elements

A multivariate model with the proportion of visual elements as a covariate was used to analyze the influence of visual element types on visual behavior and to determine the differences. Again, we selected the eye-tracking experimental data for A-01, B-02, and C-02 as the data sources and used SPSS 21.0 (IBM Corp.) to conduct a one-way ANOVA for six eye movement indicators of seven visual elements in the landscape architecture image: landmark buildings, low-rise buildings, high-rise buildings, super high-rise buildings, water systems, mountains, and green plants. We then conducted a least significant difference multiple comparison test and obtained statistical data results (Figure 6, Supplementary Materials Table S2).

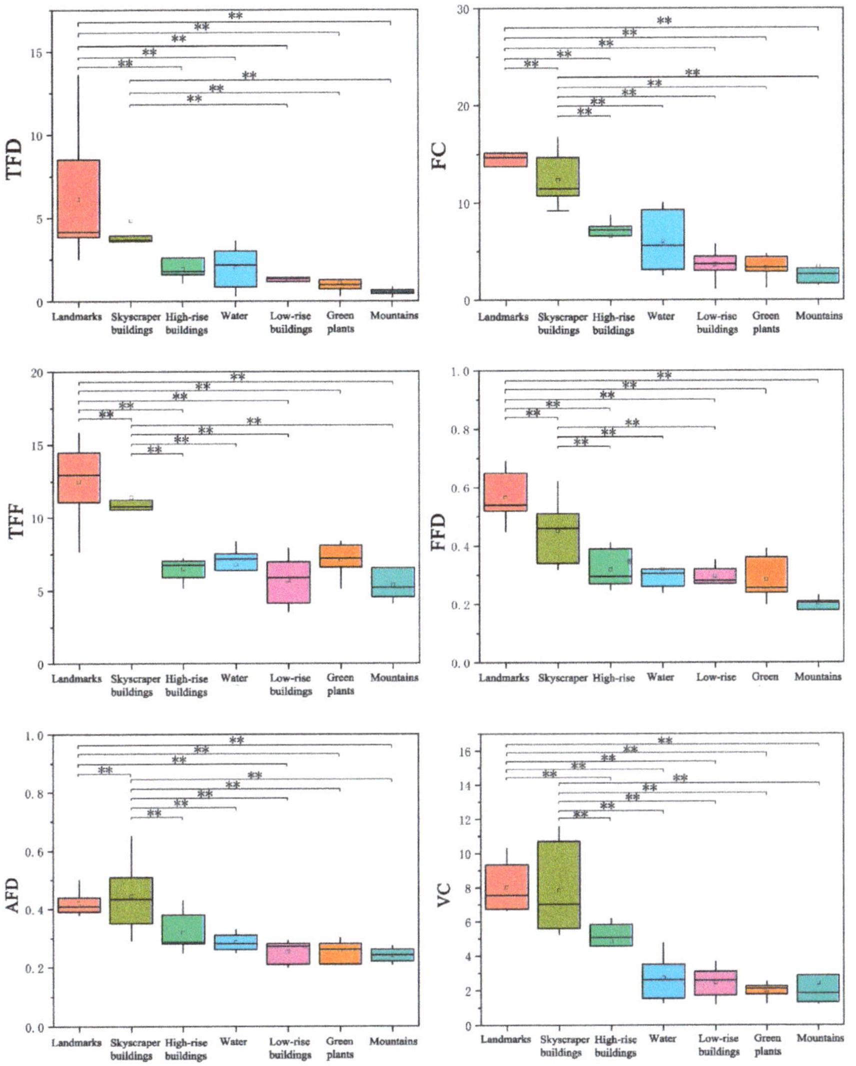

Figure 6. Viewers' visual behavior data for different landscape architecture elements (** Represents $p < 0.01$, which is significantly correlated at the 0.01 level).

From the data in Figure 6, after examining the eye movement indicators of visual elements in various landscape architecture images viewed from observation decks, it becomes evident that a significant difference exists for TFD and AFD at the 0.01 significance level ($p < 0.01$); the FC, FFD, and VC indicators display a significant difference at the 0.05 significance level ($p < 0.05$). However, no significant difference can be observed in TFF ($p > 0.05$).

Among the eye movement indicators for different visual elements, those of super high-rise buildings and landmark buildings showed the greatest significance. In terms of the AFD and FC eye movement indicators, the average values of the visual elements of super high-rise buildings and landmark buildings were relatively high. This implies that participants faced the most difficulty in extracting data from super high-rise and landmark buildings, as this demanded a larger cognitive load. Generally, the trends for the FFD, TFF, AFD, and VC eye movement indicators were largely in line with that of TFD. They also corresponded to the distribution outcomes of the eye movement heat map.

In summary, the eye-tracking indicators revealed differences in the TFD, FC, FFD, AFD, and VC data between the visual behavior indicators of the viewers when looking at urban landscape architecture ($p < 0.05$). Different groups of people have consistent perceptions of the same landscape, but their perceptions of different visual elements vary. Therefore, we can verify that H1 is valid.

4.3. Questionnaire Data Results and Analysis

4.3.1. Questionnaire Validity Test

The data from 237 questionnaires were imported into SPSS 21.0 for reliability and validity testing. The reliability of the four dimensions of the questionnaire was monitored, and a Cronbach's value of >0.7 was found, indicating that the overall reliability of the questionnaire was good. A Kaiser–Meyer–Olkin (KMO) test revealed a KMO value = 0.978 > 0.8 and a *p*-value (Sig.) = 0.000 < 0.05, with high significance. The questionnaire passed the BART spherical test of the KMO checksum.

4.3.2. Psychological Measurement of Place Identity in Different Groups of Viewers

A total of 237 questionnaires were used as data sources to divide the participants into residents (n = 126) and tourists (n = 111), and an independent sample *t*-test compared the significance of the identity psychology between the two groups when visiting the same landscape architecture and place (Figure 7).

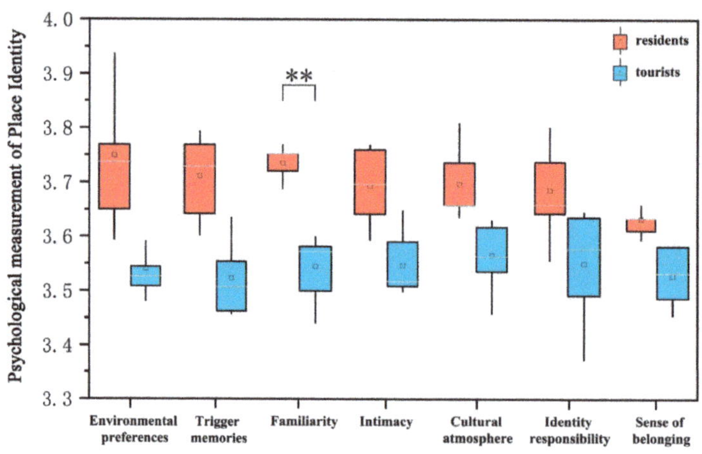

Figure 7. Psychometric results for participants at the view scene (** Represents $p < 0.01$, which is significantly correlated at the 0.01 level).

The data in Figure 7 show that among the seven measurement indicators of place identity emotion for nine spots in the landscape architecture, the mean value of the resident group is greater than that of the tourist group. This indicates that the average value of residents' place identity emotion relating to landscape architecture is higher than that of tourists. However, the independent samples *t*-test shows that the six place identity sentiment indicators have *p*-values > 0.05. That said, the familiarity index has a *p*-value of < 0.05, so there was a significant difference in the familiarity index in the place identity between residents and tourists in the process of viewing the same landscape, and there was no significant difference in other indicators. Although Hypothesis 2 is established, and residents have a higher sense of place identity than tourists, there is a significant difference only in the familiarity of place identity dimension.

4.3.3. Psychological Correlation with Place Identity

To verify the validity of H3, a two-variable Pearson correlation analysis was carried out between the aesthetic evaluation scores of the different visual elements of the landscape architecture seen from the observation decks and the psychological dimensions of place identity. Moreover, the aesthetic evaluation indexes and data for visual elements with a direct correlation with the psychology of place identity were screened out (Table 5).

Table 5. Biased correlation analysis of aesthetic evaluation of visual elements and place identity.

Landscape Visual Elements	Evaluation Factor	Trigger Memories	Sense of Belonging	Familiarity	Environmental Preferences	Cultural Atmosphere	Intimacy	Place Identity
landmarks	cleanliness	0.464	0.622 *	0.479	0.556 *	0.464	0. 643 **	0.530 *
	vividness of color	0.698 **	0.667 **	0.589 **	0.603 **	0.620 **	0.644 **	0.698 **
low-rise buildings	proportional harmonization	0.532 *	0.428	0.539 *	0.401	0.401	0.583 *	0.420
	cleanliness	0.572 *	0.646 **	0.745 **	0.510 *	0.552 *	0.541 *	0.521 *
	vividness of color	0.487	0.452	0.344	0.490	0.534 *	0.509 *	0.561 *
high-rise buildings	cleanliness	0.707 **	0.580 *	0.712 **	0.467	0.509 *	0.672 **	0.534 *
	vividness of color	0.454	0.681 **	0.372	0.656 **	0.589 *	0.733 **	0.645 **
skyscraper buildings	morphological richness	0.377	0.482	0.551 *	0.445	0.390	0.429	0.396
	cleanliness	0.471	0.500 *	0.462	0.343	0.528 *	0.463	0.493
	vividness of color	0.548 *	0.616 **	0.409	0.638 **	0.577 *	0.730 **	0.648 **
	quantitative intensity	0.426	0.460	0.513 *	0.460	0.431	0.505 *	0.514 *
water	proportional harmonization	0.584 *	0.633 **	0.637 **	0.581 *	0.593 *	0.580 *	0.587 *
	morphological richness	0.550 *	0.629 **	0.667 **	0.594 *	0.479	0.706 **	0.510 *
	cleanliness	0.411	0.382	0.611 **	0.505 *	0.407	0.554 *	0.504 *
	vividness of color	0.569	0.770 **	0.700 **	0.737 **	0.665 **	0.837 **	0.706 **
mountains	proportional harmonization	0.660 **	0.678 **	0.742 **	0.580 *	0.574 *	0.765 **	0.590 *
	cleanliness	0.262	0.441	0.439	0.359	0.290	0.594 *	0.445
	vividness of color	0.540 *	0.670 **	0.366	0.713 **	0.628 **	0.704 **	0.690 **
green plants	proportional harmonization	0.313	0.423	0.249	0.376	0.334	0.533 *	0.444
	morphological richness	0.650 **	0.655 **	0.726 **	0.565 *	0.516 *	0.733 **	0.579 *
	cleanliness	0.369	0.449	0.536 *	0.383	0.371	0.568 *	0.394
	vividness of color	0.666 **	0.779 **	0.513 *	0.702 **	0.680 **	0.834 **	0.754 **

** Represents $p < 0.01$, which is significantly correlated at the 0.01 level. * Represents $p < 0.05$, which is significantly correlated at the 0.05 level.

From the results in Table 5, we can draw the following conclusions:

(1) There were 10 visual aesthetic evaluation indexes with a 0.05 level of significance for place identity psychology and six visual evaluation factors with a 0.01 level of significance for place identity psychology.
(2) From the perspective of visual element categories, the aesthetic perception of green plant visual elements had the strongest psychological correlation with place identity (0.754); the intimacy correlation coefficient between the color vividness of green plants and the psychological dimension of place identity was the highest (0.834); and the number of evaluation indicators with a correlation between water visual elements and place identity psychology was the highest. There is a significant positive correlation between natural visual elements and place identity.
(3) Regarding the evaluation factor categories, the visual aesthetic evaluation index of decorative sophistication and the psychological correlation coefficient of place identity exhibited a negative relationship. Therefore, the visual aesthetic evaluation of decorative sophistication had a negative correlation with place identity psychology. The correlation coefficients between the other significance evaluation factors and place identity psychology were positive; therefore, the seven types of visual aesthetic evaluation indexes (e.g., proportion coordination, form richness, neatness, and color vividness) had a positive correlation with place identity psychology.
(4) Among the 16 aesthetic evaluation indexes of visual elements that correlated with the psychology of place identity, the correlation coefficients of the color vividness of green plants and the color vividness of water were >0.7, indicating that there was a strong positive correlation between these two items. The correlation coefficients between the remaining 14 visual element aesthetic evaluation indexes and place identity psychology were within 0.3–0.7, indicating that these 14 aesthetic evaluation indicators strongly correlated with the psychological notion of identity.

4.3.4. Multiple Linear Regression Results and Analysis

Multiple linear regression was used to verify common visual elements affecting identity psychology in landscape architectures seen from different urban observation decks. The dependent variable was "city identity sense", while the independent variables were the data of eight visual elements and fifty-six corresponding visual aesthetic evaluation indicators. Model summary, ANOVA results, and coefficient results (Table 6) were obtained, yielding the multiple regression equation.

$$Y = 0.573 \times x_1 + 0.316 \times x_2 + 0.515 \times x_3 + 0.169 \times x_4 + 0.617 \times x_5 + 0.223 \times x_6 + 0.799 \times x_7 + 0.094 \times x_8 + 1.021 \times x_9 - 3.10$$

Based on the data from the overall regression model presented in the table, the following results can be derived.

(1) F = 14.251 and p-value = 0.009 < 0.05, which is less than 0.05. This indicates that the model is significant at the 0.05 level. It passes the F-test and has a linear correlation with the dependent variable. Hence, the model is validated.
(2) The adjusted R^2 considers the number of predictors used to forecast the target variable, helping to decide if adding new variables improves the model fit. Here, the adjusted R^2 value was 0.726. This implies that the nine influencing factors could account for 72.6% of the variation in identity psychology. In other words, 72.6% of the attachment emotion resulted from these nine factors, suggesting that the model had a moderate fit.
(3) The Durbin–Watson (D–W) value = 2.006; since this value is close to 2, it implies that there is no autocorrelation present. As a result, the data used in the model are independent.

Table 6. Multiple linear regression model data table.

Y	X			Non-Normalized B	Non-Normalized Standard Error	Standard Coefficient Beta	t	p-Value
Place identity	(constant)			−3.10	0.159		20.401	0.009
	landmarks	vividness of color	X_1	0.573	0.037	0.425	15.663	0.000
	high-rise buildings	vividness of color	X_2	0.316	0.025	0.271	12.608	0.000
	skyscraper buildings	vividness of color	X_3	0.515	0.040	0.391	12.750	0.000
	water	proportional harmonization	X_4	0.169	0.028	0.162	5.991	0.001
		vividness of color	X_5	0.617	0.036	0.470	17.340	0.000
	mountains	proportional harmonization	X_6	0.223	0.029	0.245	7.806	0.000
		vividness of color	X_7	0.799	0.024	0.628	33.048	0.000
	green plants	morphological richness	X_8	0.094	0.024	0.100	4.010	0.007
		vividness of color	X_9	1.021	0.028	0.978	37.009	0.000

Based on the significance and correlation coefficients of the independent and dependent variables, the following conclusions can be reached.

(1) Among the independent variables, the color brightness of green plants ($p = 0.000 < 0.01$, B = 1.021), the color vividness of mountains ($p = 0.000 < 0.01$, B = 0.799), the color vividness of water ($p = 0.000 < 0.01$, B = 0.617), the color vividness of landmarks ($p = 0.000 < 0.01$, B = 0.573), the color brightness of skyscraper buildings ($p = 0.000 < 0.01$, B = 0.515), the color brightness of high-rise buildings ($p = 0.001 < 0.05$, B = 0.316), the proportional coordination of mountains ($p = 0.001 < 0.01$, B = 0.223), and the morphological richness of green plants ($p = 0.007 < 0.01$, B = 0.094) were prominent. There was a significant positive correlation between the evaluation factors of these nine visual elements and identity psychology.

(2) The correlation coefficient of the decorative delicacy evaluation is less than 0, but the p-value is more than 0.05, and the correlation coefficient is less than 0.3; therefore, although there is a negative correlation between decorative delicacy and the identity of place of each visual element, the influence is small. Overall, the multiple linear regression results are consistent with the correlation analysis, and Hypothesis 3 is verified.

5. Discussion

5.1. Visual Aesthetic Evaluation and Psychological Verification of Place Identity

Our results show that people's visual perception of different landscape elements has significant differences; the aesthetic evaluation of landscape elements significantly influences the psychology of place identity; and the influence of natural landscape elements is higher than that of artificial landscape elements. In our analysis of the eye-tracking visual behavior data, the public showed a strong interest in and visual preference for artificial elements, particularly recognizable landmark buildings and skyscraper buildings, providing support for similar results obtained in research on visual preference behavior relating to mountain landscape architecture [30]. In previous studies, the perceptions of a landscape supporting autonomy, relatedness, and competence were associated with identification, dependence, and emotional connection with that landscape [31]. The proportion of artificial visual elements in the urban environment of the historic urban area is much higher than that of natural visual elements. However, the visual aesthetic perception of natural elements such as water, mountains, and green plants, has a more significant impact on the psychological perception of place identity, which is also supported by a study of visual perception among different groups of tourists [32]. Studies of visual perception in different populations have also shown that people experience a more positive psychological impact from natural landscape elements [33]. From visual behaviors to emotional cognition, viewers display great subjectivity in expressing different emotions. Combined with the experimental results of this study, this indicates that there are differences in people's perceptual behaviors toward different visual elements of the landscape.

After previous studies showed that the most characteristic feature of the historic urban landscape of Macao is the diversity of colors, eye-tracking experiments and data collection were used to determine the most common colors in urban and landscape architecture in 60 districts of the city [34]. The results of that study also showed the significant influence of color vividness on the psychological impact of identification with different visual elements. Further research found that landmarks and super-tall buildings are more visually appealing to the public in urban landscape environments.

For the landscape of Macao's historic district, although eye movement experiments found that different groups have significant visual behaviors toward visually attractive landmarks, the proportion of visual elements in the landscape images revealed that important landscapes in Macao's historic district are affected by super high-rise buildings and that the proportion of landmarks in these images is relatively small, less than 1%. The lighthouse landmark can only be seen from a wide perspective, and the former outer harbor cannot be seen from the Guia Lighthouse. In 2008, the Macau government promulgated the first visual landscape control regulation after the Historic Centre of Macao was deemed a World Heritage Site, "Setting the Maximum Allowable Altitude for Buildings to be Constructed in the Area Around the Guia Lighthouse". This set the height limit for buildings around the Guia Lighthouse at 60 m, which was relaxed to 90 m in 2015 for a particular planned building. However, UNESCO warned against this, and the developer cooperated with the government in reducing the height of this building [22]. In this study, our analysis of visual behavior and measurement of place identity provides data support for controlling building height within the important line of sight in the historic urban area.

5.2. Limitations and Future Work

This study reveals that residents and tourists do not have significant differences in their perceived behaviors regarding the same urban landscape. However, there are significant differences in people's perceptions of different landscape elements, and the aesthetic perceptions of different landscape elements have a significant effect on place identity, providing a new perspective for enhancing the sense of identity of the city. That said, there are some limitations. To enhance the universality and representativeness of these findings, it is essential to significantly expand the number of subjects. This should involve recruiting participants from diverse cultural backgrounds, various social and economic classes, and different geographical regions [35].

Future work will need to explore the differences in participants' affective attitudes related to their own identity features in the environment in more depth. Previous research has found that men and women differ in their overall and localized focus on the environment, with women being more concerned with the overall image and men being more interested in specific localized areas [36]. In addition to eye-tracking, physiological sensing devices, GPS location devices, and human behavior recognition are all useful methods for capturing behavioral data. These behavioral metrics data can be processed by artificial neural networks, and visual quality models can be built to support decision-making. Therefore, future research will require deep intersections with other disciplines to provide data to inform landscape architecture theory development and guide urban overlook planning and management.

6. Conclusions

Synthesizing the discussion of this research and previous studies, we aimed to investigate whether there are differences in the visual perception behaviors of different groups of people toward the same historic urban landscape and different landscape elements. By analyzing the experimental data, we determined that there was no significant difference in the visual behavior of residents and tourists viewing the same landscape images. However, in the analysis of the behavioral data for different visual elements in the landscape architecture, TFD, AFD, FC, FFD, and VC showed significant differences ($p < 0.05$). Eye movement behavior related to the visual elements of landmark buildings and super high-rise buildings was the most significant, indicating that H1 is valid.

This study also examines the variability of place identity sentiments toward different groups of people in the same landscape. The questionnaire data show that the average level of place identity for residents is higher than that of tourists, and there is a significant difference in familiarity indexes ($p < 0.05$), indicating that H2 holds true.

Regarding the influence of the aesthetic perception of different visual elements on place identity, we combined the results of the visual aesthetic evaluation and the psychological measurement of identity through correlation analysis and multiple linear regression analysis. We concluded that different landscape visual elements in Macao have correlations with place identity, such as the color brightness of green plants, the vividness of the color of mountains, the vividness of the color of water bodies, the vividness of the color of landmark buildings, the vividness of skyscrapers, the vividness of high-rise buildings, the proportionality and harmony of mountains, and the morphological richness of green plants. They also have a significant positive effect on place identity. Regarding the visual evaluation factors, the delicacy of decoration is negatively correlated with identity psychology. However, the effect is not significant, indicating that H3 holds.

This study is significant for improving viewers' perceptions and evaluations of urban landscapes. Using an eye-tracking experiment, the visual elements of building facades and street styles that were the most important to the public were extracted. Combined with an evaluation of the visual elements, an aesthetic evaluation of landscape architecture viewed from various observation decks was performed. The ability of the visual landscape architecture to stimulate viewers' sense of urban identity was explored using a place identity psychological scale, and the results can serve as a basis for urban landscape architecture planning and tourism management. In future research on the visual environment of architecture, eye-tracking experiments can be applied to virtual reality scenes to compare viewers' evaluations of the visual environment of landscape architecture before and after a change.

Supplementary Materials: The following supporting information can be downloaded at: https://www.mdpi.com/article/10.3390/app15073875/s1, Table S1: Average of resident and visitor eye movement indicator data. Table S2: One-way ANOVA of eye movement indexes for different visual elements. Table S3: Coefficient of correlation between visual perception and place identity.

Author Contributions: Conceptualization, P.W. and H.F.; methodology, H.F.; software, H.F.; validation, H.F.; formal analysis, H.F.; investigation, H.F.; resources, H.F.; data curation, H.F.; writing—original draft preparation, H.F.; writing—review and editing, P.W.; visualization, H.F.; supervision, P.W.; project administration, P.W.; funding acquisition, P.W. All authors have read and agreed to the published version of the manuscript.

Funding: This research was funded by the Science and Technology Development Fund of the Macao Special Administrative Region, grant number 0036/2022/A, Philosophy and Social Science Planning Projects in Shanxi Province, grant number 2023YY103.

Institutional Review Board Statement: This study was conducted in accordance with the Declaration of Helsinki and approved by the Institutional Ethics Committee of the City University of Macau (protocol code 2024-RE-24, 10 April 2024).

Informed Consent Statement: Informed consent was obtained from all subjects involved in this study. Written informed consent was obtained from the participants to publish this paper.

Data Availability Statement: The original contributions presented in this study are included in the article/Supplementary Materials. Further inquiries can be directed to the corresponding author.

Conflicts of Interest: The authors declare no conflicts of interest.

References

1. UNESCO World Heritage Convention. 44 COM 7B.141—Decision. Available online: https://whc.unesco.org/en/decisions/7856/ (accessed on 21 January 2025).
2. Macaomagazine. City of the Future. 6 April 2022. Available online: https://macaomagazine.net/city-of-the-future/ (accessed on 26 May 2024).
3. Liu, M.; Nijhuis, S. Mapping landscape spaces: Methods for understanding spatial-visual characteristics in landscape design. *Environ. Impact Assess. Rev.* **2020**, *82*, 106376. [CrossRef]
4. Hadavi, S. Direct and Indirect Effects of the Physical Aspects of the Environment on Mental Well-Being. *Environ. Behav.* **2017**, *49*, 1071–1104. [CrossRef]
5. Liu, Q.; Zhu, Z.; Zeng, X.; Zhuo, Z.; Ye, B.; Fang, L.; Huang, Q.; Lai, P. The impact of landscape complexity on preference ratings and eye fixation of various urban green space settings. *Urban For. Urban Green.* **2021**, *66*, 127411. [CrossRef]
6. Gholami, Y.; Taghvaei, S.H.; Norouzian-Maleki, S.; Mansouri Sepehr, R. Identifying the stimulus of visual perception based on Eye-tracking in Urban Parks: Case Study of Mellat Park in Tehran. *J. For. Res.* **2021**, *26*, 91–100. [CrossRef]
7. Fu, H.; Wang, P.; Zhou, J.; Zhang, S.; Li, Y. Investigating Influence of Visual Elements of Arcade Buildings and Streetscapes on Place Identity Using Eye-Tracking and Semantic Differential Methods. *Buildings* **2023**, *13*, 1580. [CrossRef]
8. Scott, N.; Zhang, R.; Le, D.; Moyle, B. A review of eye-tracking research in tourism. *Curr. Issues Tour.* **2017**, *22*, 1244–1261. [CrossRef]
9. Amati, M.; Ghanbari Parmehr, E.; McCarthy, C.; Sita, J. How eye-catching are natural features when walking through a park? Eye-tracking responses to videos of walks. *Urban For. Urban Green.* **2018**, *31*, 67–78. [CrossRef]
10. Li, J.; Zhang, Z.; Jing, F.; Gao, J.; Ma, J.; Shao, G.; Noel, S. An evaluation of urban green space in Shanghai, China, using eye tracking. *Urban For. Urban Green.* **2020**, *56*, 126903. [CrossRef]
11. Zhou, S.; Gao, Y.; Zhang, Z.; Zhang, W.; Meng, H.; Zhang, T. Visual Behaviour and Cognitive Preferences of Users for Constituent Elements in Forest Landscape Spaces. *Forests* **2022**, *13*, 47. [CrossRef]
12. Guo, S.; Sun, W.; Chen, W.; Zhang, J.; Liu, P. Impact of Artificial Elements on Mountain Landscape Perception: An Eye-Tracking Study. *Land* **2021**, *10*, 1102. [CrossRef]
13. Zhang, L.-M.; Zhang, R.-X.; Jeng, T.-S.; Zeng, Z.-Y. Cityscape Protection Using VR and Eye Tracking Technology. *J. Vis. Commun. Image Represent.* **2019**, *64*, 102639. [CrossRef]
14. Simpson, J.; Freeth, M.; Simpson, K.J.; Thwaites, K. Visual Engagement with Urban Street Edges: Insights Using Mobile Eye-Tracking. *J. Urban. Int. Res. Placemaking Urban Sustain.* **2019**, *12*, 259–278. [CrossRef]
15. Justulin Zacarias, E.F.; Gasparetto Higuchi, M.I. Overview of studies on place identity. *Novos Cad. Naea* **2021**, *24*, 57–72.
16. Brian, F. Blake Topophilia: A Study of Environmental Perception, Attitudes and Values. By Yi-Fu Tuan. *J. Leis. Res.* **1974**, *6*, 323–325. [CrossRef]
17. Kyle, G.T.; Jun, J.; Absher, J.D. Repositioning Identity in Conceptualizations of Human–Place Bonding. *Environ. Behav.* **2014**, *46*, 1018–1043. [CrossRef]
18. Tenzer, M.; Schofield, J. People and places: Towards an understanding and categorisation of reasons for place attachment—Case studies from the north of England. *Landsc. Res.* **2023**, *49*, 340–358. [CrossRef]
19. Lin, Z.; Chen, Y.; Filieri, R. Resident-tourist value co-creation: The role of residents' perceived tourism impacts and life satisfaction. *Tour. Manag.* **2017**, *61*, 436–442. [CrossRef]
20. Zhao, X.; Lu, Y.; Lin, G. An integrated deep learning approach for assessing the visual qualities of built environments utilizing street view images. *Eng. Appl. Artif. Intell.* **2024**, *130*, 107805. [CrossRef]

21. Jiang, B.; Xu, W.; Ji, W.; Kim, G.; Pryor, M.; Sullivan, W.C. Impacts of nature and built acoustic-visual environments on human's multidimensional mood states: A cross-continent experiment. *J. Environ. Psychol.* **2021**, *77*, 101659. [CrossRef]
22. Government of Macao Special Administrative Region Statistics and Census Service. Statistics—Statistics and Census Service. Available online: https://www.dsec.gov.mo/en-US/Statistic?id=101 (accessed on 17 February 2024).
23. Zhang, H.; Lai, P.Y.; Jim, C.Y. Species diversity and spatial pattern of old and precious trees in Macau. *Landsc. Urban Plan.* **2017**, *162*, 56–67. [CrossRef]
24. Zheng, J. *The Cultural Construction and Conservation Dilemma of Architectural Cultural Heritage in the Later World Heritage Era of Macao World Architecture*; The Cultural Affairs Bureau of the Macao Special Administrative Region Government: Macao, China, 2019; Volume 11, pp. 44–49. Available online: https://www.culturalheritage.mo/Survey/sgchm2017/cn/?AspxAutoDetectCookieSupport=1 (accessed on 13 December 2024).
25. Announcement of the Ministry of Housing and Urban-Rural Development on the Issuance of the National Standard "Unified Standard for Civil Building Design". Available online: https://www.mohurd.gov.cn/gongkai/zc/wjk/art/2024/art_17339_240715.html (accessed on 9 January 2024).
26. Gao, Y.; Sun, X.M.; Zhang, T.; Zhan, H. Analysis on the characteristics of eye movement and the evaluation of psychological perception for forest waterscape space. *IOP Conf. Ser. Earth Environ. Sci.* **2021**, *626*, 012007. [CrossRef]
27. Van Loo, E.J.; Caputo, V.; Nayga, R.M.; Seo, H.-S.; Zhang, B.; Verbeke, W. Sustainability labels on coffee: Consumer preferences, willingness-to-pay and visual attention to attributes. *Ecol. Econ.* **2015**, *118*, 215–225. [CrossRef]
28. Li, N.; Zhang, S.; Xia, L.; Wu, Y. Investigating the Visual Behavior Characteristics of Architectural Heritage Using Eye-Tracking. *Buildings* **2022**, *12*, 1058. [CrossRef]
29. Boley, B.B.; Strzelecka, M.; Yeager, E.P.; Ribeiro, M.A.; Aleshinloye, K.D.; Woosnam, K.M.; Mimbs, B.P. Measuring place attachment with the Abbreviated Place Attachment Scale (APAS). *J. Environ. Psychol.* **2021**, *74*, 101577. [CrossRef]
30. Li, Z.; Sun, X.; Zhao, S.; Zuo, H. Integrating eye-movement analysis and the semantic differential method to analyze the visual effect of a traditional commercial block in Hefei, China. *Front. Archit. Res.* **2021**, *10*, 317–331. [CrossRef]
31. Landon, A.C.; Woosnam, K.M.; Kyle, G.T.; Keith, S.J. Psychological Needs Satisfaction and Attachment to Natural Landscapes. *Environ. Behav.* **2021**, *53*, 661–683. [CrossRef]
32. Huang, J.; Liang, J.; Yang, M.; Li, Y. Visual Preference Analysis and Planning Responses Based on Street View Images: A Case Study of Gulangyu Island, China. *Land* **2022**, *12*, 129. [CrossRef]
33. Deng, L.; Li, X.; Luo, H.; Fu, E.-K.; Ma, J.; Sun, L.-X.; Huang, Z.; Cai, S.-Z.; Jia, Y. Empirical study of landscape types, landscape elements and landscape components of the urban park promoting physiological and psychological restoration. *Urban For. Urban Green.* **2020**, *48*, 126488. [CrossRef]
34. Wang, P.; Song, W.; Zhou, J.; Tan, Y.; Wang, H. AI-Based Environmental Color System in Achieving Sustainable Urban Development. *Systems* **2023**, *11*, 135. [CrossRef]
35. Liu, W.; Hu, Z.; Fei, Y.; Chen, J.; Yu, C. Eye Tracking and Semantic Evaluation for Ceramic Teapot Product Modeling. *Appl. Sci.* **2025**, *15*, 46. [CrossRef]
36. Wang, P.; Yang, W.; Wang, D.; He, Y. Insights into Public Visual Behaviors through Eye-Tracking Tests: A Study Based on National Park System Pilot Area Landscapes. *Land* **2021**, *10*, 497. [CrossRef]

applied sciences

MDPI

Article

Combining Eye-Tracking Technology and Subjective Evaluation to Determine Building Facade Color Combinations and Visual Quality

Zhanzhu Wang [1], Maoting Shen [1] and Yongming Huang [2,*]

[1] College of Communication and Art Design, University of Shanghai for Science and Technology, Shanghai 200093, China; zhuzi4880@126.com (Z.W.); 223512986@st.usst.edu.cn (M.S.)
[2] Architecture and Environmental Design Field, Major of Art and Environment, Arts Studies, Graduate School, Kyoto University of The Arts, 2-116 Kitashirakawa Uryuzancho, Sakyo Ward, Kyoto 606-8271, Japan
* Correspondence: y-huang2166@st.kyoto-art.ac.jp

Featured Application: Exploring the visual quality of architectural color combinations through eye physiological feedback using eye-tracking technology.

Abstract: Architectural colors significantly influence urban culture, city imagery, regional vitality, and residential experiences. Previous studies have demonstrated that appropriate architectural colors can enhance urban vitality, but research on multicolored buildings remains limited. This study examines the relationship between hue, color variations, and visual quality by cross-verifying eye-tracking physiological indicators with subjective assessments. Using digital models of old residential buildings in Shanghai's Yangpu District, different color combinations were applied to explore real-world architectural color impacts. Results showed that blue and green combinations reduced visual pressure and created a calming space, while purple combinations were rated highly in both visual perception and subjective evaluations. Brightness differences notably influenced visual quality more than hue differences. However, larger hue variations, when paired with suitable brightness and saturation contrasts, also achieved better visual evaluations. This study fills a research gap by providing mathematical support for color combinations in architectural design, improving visual comfort and enhancing urban vitality.

Keywords: visual perception technology; architectural color; color combination; color harmony; visual perception evaluation system; visual quality

1. Introduction

1.1. Research Status of Architectural Color

Urban color plays a crucial role in urban spaces, significantly influencing a city's image, attractiveness, and addressing specific economic and practical issues [1]. Additionally, it conveys materiality, spatial context, and cultural significance [2]. As a key component of urban color, architectural color holds significant value in creating visually appealing cities due to its distinct visual characteristics [3]. With the continuous evolution of society, the demand for aesthetic sensitivity in the appearance of residential buildings has further increased [4]. Architectural color is highly flexible in its expressiveness, and the color choices for individual buildings are based on a comprehensive consideration of external factors. Beyond their ornamental and decorative functions, unified, harmonious, and stable color combinations play a decisive role in shaping the surrounding environment and influencing people's emotions. The harmony and diversity of building facade colors significantly affect people's perceptions and preferences regarding the urban environment [3].

Architectural color also functions as a form of cultural capital, offering innovative approaches to the sustainable development of culture [5]. Collette and Nguyen revealed how appropriate architectural color plays a fundamental role in maintaining the identity of specific environments in the face of cultural homogenization driven by globalization [6]. The symbolic value embedded in architectural color enhances the identity and cultural recognition of the built environment through the "color–emotion association" mechanism [7]. Beyond its cultural significance, well-applied architectural color can improve consumer preferences, thereby increasing its economic value [8]. Consequently, the visual quality of architectural color plays a crucial role in regional development.

Currently, scholars are researching architectural color from various perspectives. J. Wang et al. [4] analyzed the quantity and proportion of different color attributes at various evaluation levels by incorporating subjective evaluation factors, systematically identifying the characteristics of the three elements of residential building color. Gou [9] researched the spatial attributes of color configuration, highlighting the importance of color layout and visual hierarchy in architectural design. Hogg et al. [10], Shi et al. [11], Z. Wang [12] and Zarghami et al. [13] from a humanistic perspective, explored the relationship between the three elements of color and visual preferences, providing theoretical support for architectural color renewal. Existing research on architectural color primarily focuses on color selection, color proportion, and subjective analysis of color elements [14]. Various studies have shown that warm color schemes can achieve better visual effects in residential buildings [4], the evaluation of residential buildings decreases with lower brightness [15], and high-saturation colors can cause visual discomfort [11]. In China, architectural color reflects spatial configuration through its hierarchical structure [9].

In addition to the color elements of building facades, colors themselves possess various attributes and planning principles [16]. For instance, the harmony, diversity, and comfort of color combinations are closely linked to visual quality [17]. However, current research on the harmony of color combinations requires further exploration. Therefore, systematic studies of color combinations and their impact on the visual quality of multicolored buildings could offer new research potential [18].

1.2. Visual Quality

Visual quality is defined as environmental characteristics that are conducive to visual comfort [19]. It is a critical aspect of urban planning [20], playing a vital role in shaping a city's image and significantly enhancing the overall livability of urban areas [21]. Visual comfort serves as a key indicator for measuring visual quality and is one of the quantifiable metrics that directly reflect it. The visual quality of a city greatly influences public appreciation of the urban environment [22].

In natural landscapes, Mundher et al. [23] described and measured visual quality by considering both the visual characteristics and comfort of urban natural landscapes, integrating objective and subjective factors. This research provided urban planners with theoretical support for preserving urban natural landscapes. Ma et al. [24] explored the relationship between plant mechanisms, color, and visual aesthetic quality by studying the colors of vegetation in urban spaces, aiming to enhance the visual quality of urban streets. Z. Zhang et al. [25] examined the relationship between the color composition elements of forest landscapes and the estimated scenic beauty values from a macro-perspective, establishing an evaluation framework for assessing the visual quality of forest landscapes.

In architectural landscapes, numerous studies have focused on the visual quality of various building types, including historic preservation buildings, commercial buildings, and residential structures. These studies examine the relationship between the physical design elements of building facades and visual quality, aiming to enhance the overall visual comfort of urban streets. Bu et al. [26] investigated the balance of visual comfort between new high-rise buildings and historic structures, with the goal of better protecting historic buildings amidst rapid urban development. Zanon et al. [19] conducted simulations of light comfort in indoor built environments to explore the relationship between lighting conditions and the visual quality of indoor spaces. Other studies, focusing on aspects such as the visual comfort of advertising signs, plant colors and combinations, and lighting environments [22,24,26,27], propose more targeted strategies—based on the user perspective—to improve regional visual quality.

Mundher et al. [23] explicitly proposed defining and interpreting urban forest aesthetics based on the classification of visual characteristics of urban forests. The correlation between the visual quality of forest colors and the scenic beauty of forests has already been confirmed in previous studies [25]. Additionally, Ma et al. [24] found that street plant landscapes with red hues tend to achieve higher visual landscape quality.

Previous studies have systematically examined the visual quality of various visual elements. However, old residential buildings, as a significant part of the urban visual landscape [4], have rarely been the focus of color-related visual quality research. Therefore, this study focuses on the colors of aging residential buildings in urgent need of renovation, systematically analyzing the visual quality of their colors.

1.3. Evolution of Research Techniques

Objective and quantitative research on visual quality is becoming increasingly important in urban planning [28] (Appendix A). Previous studies have relied on user evaluations within specific geographical areas to assess visual perception quality. However, subjective evaluations are influenced by constantly changing psychological preferences and experiences [29]. The lack of robust generalizability in empirical surveys based on subjective evaluations makes them unsuitable for supporting more rigorous visual quality research.

The widespread application of computer technology has created new data landscapes, providing quantitative tools for detailed studies of color and visual quality. Compared to traditional urban color research methods, these tools can efficiently process larger volumes of data than manual methods, thereby expanding the scope of research [14]. Long and Liu [30] utilized street view images to explore the relationship between the urban green view index and visual comfort, aiming to guide future street greening planning. Cheng et al. [20] employed street view technology combined with metrics, such as regional saturation, visual entropy, green view index, sky index, and sky openness index, to measure the visual quality of urban landscapes. Hu et al. [31] used machine learning and human–machine adversarial models to assess residents' perceptions of urban neighborhood visual quality. Verma et al. [32] collected visual and audio files, employing deep learning and standard algorithms to measure perceptions of audiovisual environments within urban areas.

Previous studies that have integrated various metrics for urban visual quality have validated the feasibility of using street view data and machine learning in visual quality research. While these methods offer a reliable approach for comprehensive quantitative analysis in urban color studies, macro-level color research based on big data often lacks insight into the psychological cognition of individual users regarding spatial colors. This presents potential limitations for research on visual perception.

1.4. Advantages of Eye-Tracking Technology

With advancements in physiological and psychological research, methods for obtaining emotion feedback based on physiological indicators have emerged as a new quantitative research approach [33]. In recent years, eye-tracking has become a common tool in psychological experiments, used to better understand the psychological drivers behind behavior [34]. Eye-tracking provides objective measurements of users' cognitive processes through non-invasive methods [35], offering insights into problem-solving, behavioral reasoning, mental imagery, and search strategies [35]. By recording the objective movements of the eyes, eye-tracking technology precisely captures the rapid cognitive processing of the environment and provides quantitative data for analyzing this process [36,37]. This allows researchers to intuitively understand visual cognitive processing, from spatial elements to behavioral decision-making, thereby enhancing the objectivity and effectiveness of cognitive psychological analysis. The extensive use of eye-tracking technology in consumer behavior research has further substantiated its connections with psychology and behavioral science [38,39]. Consequently, in recent years, eye-tracking technology has been increasingly applied in areas such as physiological evaluation metrics [40] and environmental experience [41].

In urban studies, eye-tracking technology provides quantifiable metrics for visual behavior in urban environments. Gao et al. [42], Hasse & Weber [43], J. Li et al. [44] and N. Li et al. [45] used eye-tracking technology to explore the relationship between urban visual elements and visual perception by analyzing the visual behavior characteristics of subjects. Z. Li et al. [46] and Lv et al. [47] combined eye-tracking technology with the semantic differential method to investigate the multidimensional relationship between urban landscapes and visual quality. These studies have shown that well-considered architectural colors, facade designs, landscape richness, transparency, and layering significantly enhance the visual quality of urban environments [42–44,46,47].

These studies have confirmed the feasibility of using eye-tracking technology as a cognitive tool for assessing the built environment. By exploring subjective physiological indicators in depth, they have helped to explain the psychological principles and mechanisms underlying the appeal of urban landscapes. Additionally, they have clarified the correlation between eye movement behavior and psychological perception assessments [42], which holds innovative significance for interdisciplinary research on urban color and visual perception. The heterogeneity of landscapes in urban environments further supports the idea that significant differences impact attention capture during decision-making, which in turn reflects the degree of visual preference [48].

1.5. Research Questions and Objectives

In real-world architecture, residential buildings, with which residents interact frequently and for extended periods, often lack proper color planning [49]. This deficiency negatively impacts residents' psychological well-being in their daily lives and commutes [4], reducing the overall living experience. Recently, an increasing number of scholars have focused on the impact of architectural color on users' emotions, leading to more in-depth research in this area. In China, the study of architectural color is still in an exploratory phase, with most research concentrating on monochromatic buildings. These studies typically involve organizing architectural colors in various ways and collecting user feedback. However, the prevalence of buildings with poorly planned combined color schemes is starting to disrupt the overall aesthetic tone of cities [4], diminishing residents' quality of life. There is an urgent need for systematic theoretical support to guide the reorganization and renovation of color combinations in urban architecture.

This study involved on-site visits to old residential buildings in the Yangpu District of Shanghai, analyzing the relationship between facade color elements, eye-tracking metrics, and subjective evaluations. The aim was to explore the potential application of eye-tracking metrics in studying architectural color combinations, with the objective of developing an analytical framework for the visual perception of multicolored residential buildings by integrating eye-tracking data with subjective evaluation metrics. Using two-tone buildings as experimental subjects, this study quantitatively analyzes the relationship between color element differences and visual quality. Additionally, through eye-tracking experiments, the study explores how differences in color elements impact visual quality, expressing these effects more precisely through mathematical models. The study raises the following questions:

(1) Can eye-tracking metrics serve as new indicators for evaluating architectural color combinations?
(2) How can the relationship between architectural color combinations and visual quality be quantitatively analyzed? What is the relationship between the two?
(3) What common characteristics are shared by color combinations that receive high visual quality evaluations?

2. Materials and Methods

2.1. Research Subject

This study, based on color differences in architectural color combinations (total color difference, brightness difference, chroma difference, and hue difference), integrated eye-tracking metrics with subjective color preferences to introduce new dimensions for measuring color visual quality in existing research. Focusing on residential buildings, we systematically explain the relationship between eye-tracking physiological indicators, color selection, and color differences. The findings aim to provide insights for urban and architectural color planning (Figure 1). In selecting color combinations, to avoid overlooking those that might yield heterogeneous conclusions, all possible color combinations were included in the experiment. To ensure the generalizability and objectivity of the conclusions, we also assessed participants' visual comfort and preferences. This approach allowed us to maintain the diversity of color combinations while gaining insights into subjective opinions, which will be further analyzed alongside objective eye-tracking metrics.

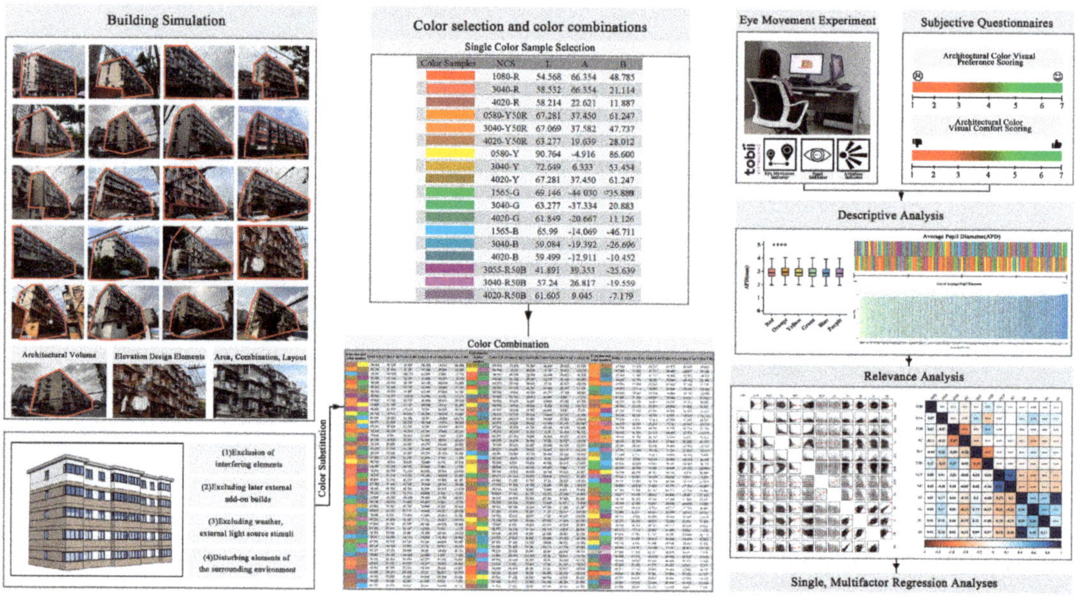

Figure 1. Research framework.

2.2. Color Selection

The color selection method used in [50] study has been shown to comprehensively reflect the color preferences of subjects, and we adopted this method for our research. We utilized the Natural Color System (NCS) and selected six hues (red, orange, yellow, green, blue, and purple). Based on the Munsell color system, 18 different monochromatic samples were chosen at specific intervals, referencing 3 key color attributes: hue, brightness, and chroma. These 18 monochromatic samples were displayed in the CIELAB color space using color values (L, a, and b values; Figure 2). The 18 color samples were then combined to create 153 different color combinations, which were applied to building surfaces for experimentation (Appendix B) [50]. To avoid missing any combinations that might yield heterogeneous results, all possible color combinations were included in the experiment. Additionally, to objectively compare the visual perception and subjective evaluation of different color combinations, we calculated the total color difference (Equation (1)), brightness difference (Equation (2)), saturation difference (Equation (3)), and hue difference (Equation (4)) between the two colors in the simulated building images using the CIELAB color difference formula, based on different color elements [50]. In this formula, L represents lightness, a represents the color range from green to red, and b represents the color range from blue to yellow:

$$\Delta E_{ab}^{*} = \sqrt{\left(L_1^{*} - L_2^{*}\right)^2 + \left(a_1^{*} - a_2^{*}\right)^2 + \left(b_1^{*} - b_2^{*}\right)^2} \tag{1}$$

$$\Delta L_{ab}^{*} = L_1^{*} - L_2^{*} \tag{2}$$

$$\Delta C_{ab}^{*} = \sqrt{\left(a_1^{*}\right)^2 + \left(b_1^{*}\right)^2} - \sqrt{\left(a_2^{*}\right)^2 + \left(b_2^{*}\right)^2}\Delta h_{ab}^{*} \tag{3}$$

$$\Delta h_{ab}^{*} = \sqrt{\left(\Delta E_{ab}^{*}\right)^2 + \left(\Delta L_{ab}^{*}\right)^2 - \left(\Delta C_{ab}^{*}\right)^2} \tag{4}$$

ΔE_{ab}^{*} : The total color difference between Color 1 and Color 2.
ΔL_{ab}^{*} : The brightness difference between Color 1 and Color 2.
ΔC_{ab}^{*} : The saturation difference between Color 1 and Color 2.
Δh_{ab}^{*} : The hue difference between Color 1 and Color 2.

Color Samples	NCS	L	A	B
	1080-R	54.568	66.354	48.785
	3040-R	58.532	66.354	21.114
	4020-R	58.214	22.621	11.887
	0580-Y50R	67.281	37.450	61.247
	3040-Y50R	67.069	37.582	47.737
	4020-Y50R	63.277	19.639	28.012
	0580-Y	90.764	−4.916	86.600
	3040-Y	72.649	6.333	53.454
	4020-Y	65.771	4.125	33.285
	1565-G	69.146	−44.03	35.883
	3040-G	63.277	−37.334	20.883
	4020-G	61.849	−20.667	11.126
	1565-B	65.99	−14.069	−46.711
	3040-B	59.084	−19.392	−26.696
	4020-B	59.499	−12.911	−10.452
	3055-R50B	41.891	39.353	−25.639
	3040-R50B	57.24	26.817	−19.559
	4020-R50B	61.605	9.045	−7.179

Figure 2. Color selection: monochrome color samples and color values.

2.3. Simulation of Experimental Subjects

Most existing old residential buildings in China were constructed between the 1970s and 1990s and have been in use for over 40 years. As a result, the overall community environment has deteriorated, functions have become outdated, pipelines have aged, and the quality of living has declined. Through on-site visits to various residential communities in the Yangpu District of Shanghai, key elements, such as building volume, architectural forms, facade design elements, and the area and design methods of different components, were extracted. Based on these observations, a building measuring 24,000 mm × 11,000 mm × 15,000 mm was designed, reflecting the most prevalent Soviet-style row housing of that era, resulting in a five-story row housing unit.

Simulated images can create virtual experimental scenarios, allowing for controlled testing conditions and providing low-cost color combination options for individual buildings. This approach offers both convenience and cost-effectiveness for the overall experimental design. Therefore, SketchUp 2020 software was used for modeling, and Photoshop 2020 was employed to apply 153 different color combinations to the building unit (Figure 3).

Figure 3. Overview of experimental subjects.

2.4. Subjects

The aesthetic preferences of college students generally align with those of the broader public [46,51,52]. For this experiment, 52 students from the University of Shanghai for Science and Technology were randomly selected as participants. Their professional backgrounds included architecture, urban planning, landscape architecture, management, economics, and environmental engineering. Of the participants, 35.4% were male and 64.6% were female, with an age range of 18 to 30 years. Specifically, 95.8% were aged 18 to 25, and 4.2% were aged 25 to 30. Regional and cultural diversity were also considered, as most participants were from various cities, such as Shanghai (47.9%), Jiangsu (14.5%), Jiangxi (14.5%), and Henan (10.4%). All participants had normal color vision and were seeing the experimental images for the first time. Four students were excluded from the study due to a sampling rate below 70%, which affected the reliability of their data.

2.5. Experimental Environment, Experimental Equipment, and Experimental Procedures

The experiment was conducted in a laboratory at the Shanghai Institute of Technology. To avoid interference with eye-tracking data collection, natural light was controlled, and indoor lighting served as the sole light source (Figure 4a). The Tobii-X120 eye tracker, manufactured by Tobii in Switzerland, was used to collect eye-tracking data from participants. Experimental materials were displayed on a secondary monitor with dimensions of 17 inches (1024 × 768 dpi). The experimenters observed and performed calibration and data collection using Tobii-x2-60 software on the main monitor (Figure 4b).

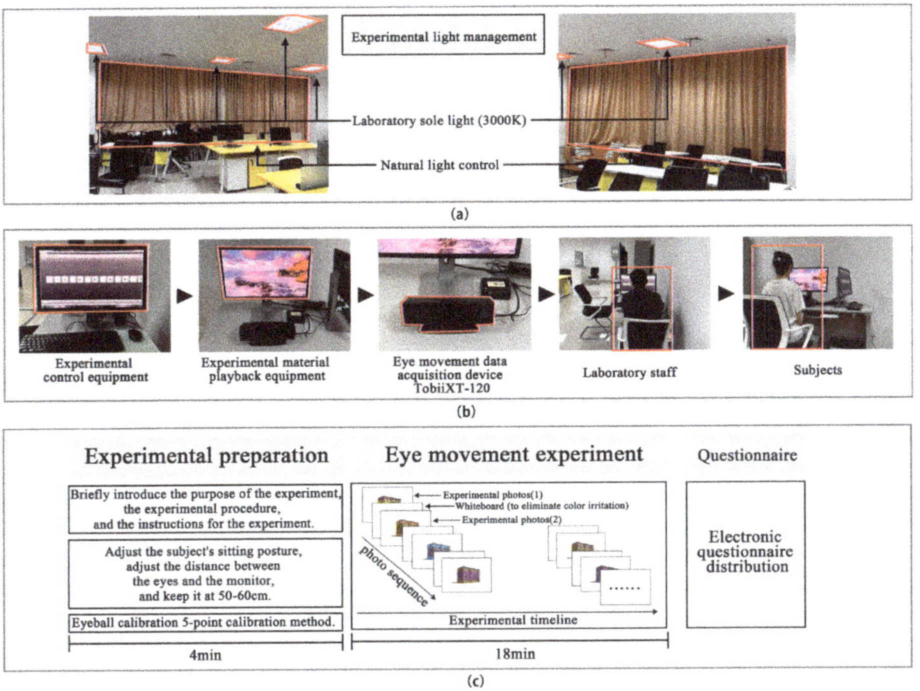

Figure 4. Laboratory environment, equipment, and personnel: (**a**) laboratory ambient light control, (**b**) experimental equipment, and (**c**) experimental procedure.

After familiarizing participants with the experimental setup, the staff verbally explained the experiment's purpose, requirements, and overall procedure. Once participants understood these aspects, initial eye calibration procedures were conducted before starting the experiment. During the experiment, the secondary monitor automatically displayed 153 color images, each for 5 s. Between each image, a blank screen was shown for 2 s to minimize color visual stimuli and reduce eye fatigue among participants. After the eye-tracking experiment concluded, a questionnaire was distributed to gather participants' subjective evaluations of the building colors (Figure 4c).

2.6. Selection of Eye-Tracking Data Metrics

During scene viewing, eye movements can be divided into two relatively independent stages: fixation and saccade [45]. In this experiment, six eye movement behavior characteristics were selected to observe architectural color combinations (Table 1). TFF reflects the visual attractiveness and alertness of the participants. The combination of APD and ASA measures the visual stimulus level, indicating the comfort of the color combinations for the participants [53]. FFD, FC, and TFD track annotation behavior, reflecting the participants' preferences for architectural color combinations [54].

Table 1. Meaning of eye movement indicators.

Eye Movement Index	Abbreviations	Meaning
First Fixation Duration	FFD (s)	The duration of the first fixation within the area of interest, representing the ease or difficulty of recognizing or finding attractiveness of the experimental object.

Table 1. *Cont.*

Eye Movement Index	Abbreviations	Meaning
Fixation Count	FC (count)	The sum of all fixation durations within the area of interest, representing the level of interest of the participants toward the experimental object.
Time to First Fixation	TFF (s)	The time taken by the participants to first fixate on an area of interest, i.e., the duration until the first entry into the area of interest. It represents the visual saliency of the experimental material.
Total Fixation Duration	TFD (s)	The sum of the durations of all fixations within the area of interest. It represents the level of interest the participants have in the experimental stimulus.
Average Pupil Diameter	APD (mm)	It directly reflects the level of visual stimulation perceived by the participants toward the stimuli.
Average Saccadic Amplitude	ASA (degrees)	The average distance between saccades, typically measured in degrees of visual angle. Eye movements, facilitated by involuntary saccades, help alleviate visual fatigue and reflect the ease of information processing by the participants in the experiment.

2.7. Subjective Data Collection

This study gathered participants' subjective color perceptions of the experimental objects using a rating scale. Simultaneously, participants filled out an electronic questionnaire, where the order of building colors matched the sequence in the eye-tracking experiment. Participants rated the visual comfort and visual preference of each color combination based on their subjective perceptions. The questionnaire employed a 7-point Likert scale, with scores ranging from 1 to 7 (Appendix C). A score of 1 indicated "very uncomfortable" or "strongly dislike", while a score of 7 indicated "very comfortable" or "strongly like" (Table 2).

Table 2. Questionnaire for subjective evaluation of color comfort on building facades.

Evaluation Indicators	Abbreviations	Subjective Rating						
Architectural Color Preferences	ACP	1	2	3	4	5	6	7
Visual Comfort	VC	1	2	3	4	5	6	7

2.8. Data Processing and Analysis

Data preprocessing, management, and analysis were conducted using R v4.3.2 software. Inter-group differences for each experimental color group (assuming normal distribution) were tested using ANOVA. To examine the correlation between eye-tracking and subjective evaluation metrics with ΔE, ΔL, ΔC, and Δh, correlation analysis was performed. For regression analysis, eye-tracking and subjective evaluation metrics were treated as dependent variables, while color element differences were treated as independent variables. Based on the trend of the dependent variables relative to the independent variables, linear regression models were established for pairs of variables that exhibited a linear relationship. Multi-factor regression models were developed for each eye-tracking and subjective evaluation metric with the image's ΔE, ΔL, ΔC, and Δh. Two-sided tests were used in this study, with $p \leq 0.05$ indicating statistically significant differences.

The experiment collected eye-tracking indices and subjective evaluation data from 48 participants for 153 sets of images, resulting in a total of 58,752 datasets. The ANOVA *p*-values for the different color hue groups of the participants' APD, ASA, FFD, FC, TFF, TFD, ACP, and VC were all less than 0.05, indicating significant changes in these eye-tracking metrics as the building color differences in ΔE, ΔL, ΔC, and Δh varied.

3. Results

3.1. Descriptive Analysis

By categorizing each experimental color combination according to hue and plotting box plots, comparisons of various metrics across different hues were observed (Figure 5a). Additionally, the 153 color combinations were ranked for each experimental metric based on their values (Figure 5b), providing a more intuitive representation of the relationship between color element differences and the various metrics. As shown in Figure 5, differences in APD, ASA, FFD, FC, TFF, TFD, ACP, and VC among participants were evident across different hues.

In the warm color group, the orange color combinations exhibited higher mean values for APD, ASA, and VC (2.97 mm, 2.29 mm, and 3.01, respectively), while the mean value for FFD was lower (0.24 s). The values for FC and VC showed significant fluctuations, whereas TFD and ACP displayed smaller fluctuations. In the yellow color group, the color combinations had higher mean values for FFD, TFD, and VC (0.25 s, 3.13 s, and 3.01, respectively). The values for FC and VC showed significant fluctuations, while TFF, TFD, and ACP exhibited smaller fluctuations.

In the intermediate color group, the purple color combinations exhibited higher mean values for FC and VC (9.1 and 3.01, respectively) and lower mean values for ASA and FFD (2.19 mm and 0.24 s, respectively). The values for TFF, ACP, and VC showed significant fluctuations, while FFD, TFD, and FC displayed smaller fluctuations.

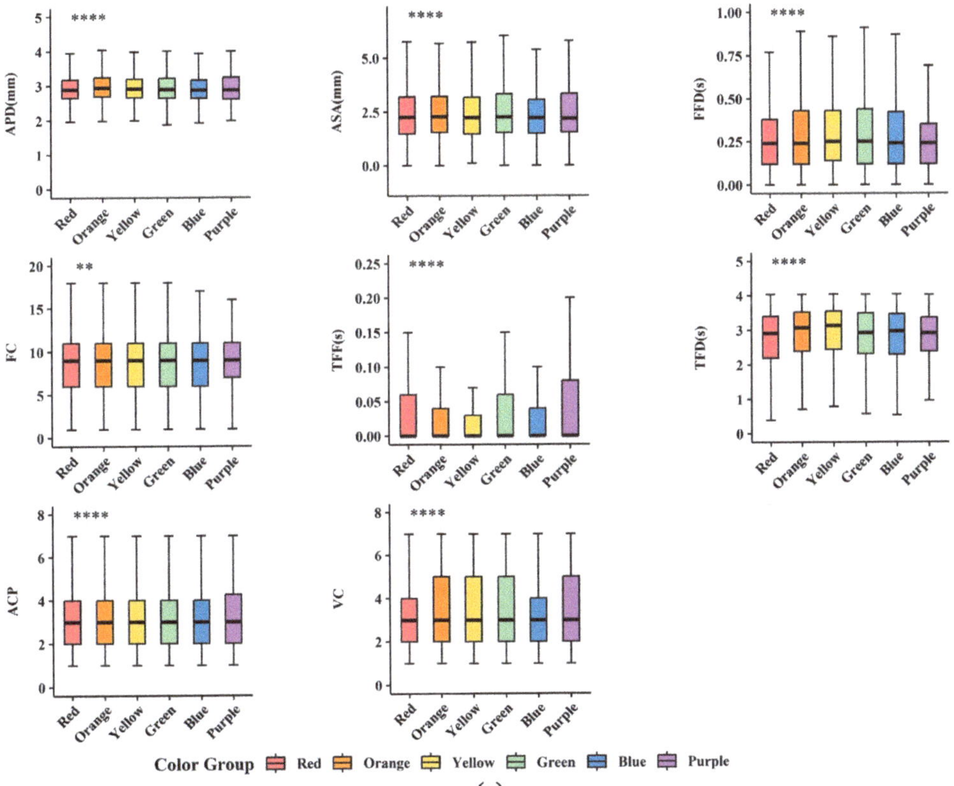

(a)

Figure 5. *Cont.*

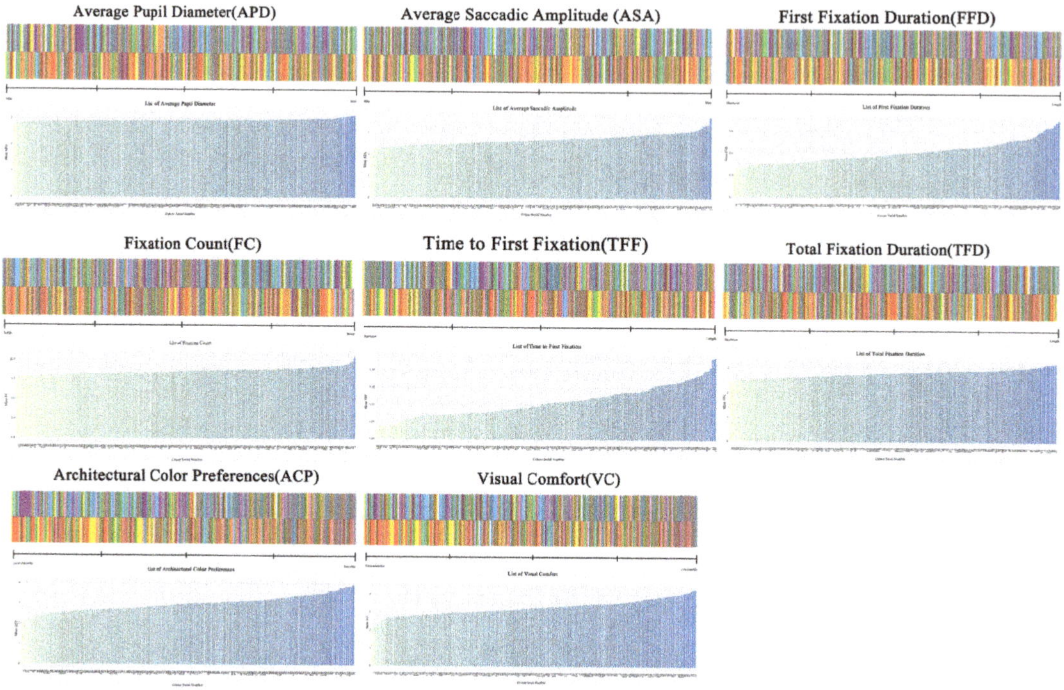

(b)

Figure 5. (**a**) Box plots of participants' APD, ASA, FFD, FC, TFF, TFD, ACP, and VC across different hue groups (the middle line represents the median, the upper and lower bounds of the box represent the 25th and 75th percentiles, respectively, and the upper and lower ends of the whiskers represent the 5th and 95th percentiles, respectively). **: $p \leq 0.01$; ****: $p \leq 0.0001$. (**b**) The colour combination is sorted by different indicators and their corresponding values.

In the cool color group, the blue color combinations exhibited lower mean values for APD and ASA (2.90 mm and 2.19 mm, respectively). TFD values showed significant fluctuations, while APD, ASA, TFF, ACP, and VC displayed smaller fluctuations. In the green color group, the color combinations had a higher mean value for FFD (0.25 s). The values for TFF, TFD, and VC showed significant fluctuations, while ACP showed smaller fluctuations.

3.2. Correlation Analysis

The color element differences in ΔE, ΔL, ΔC, and Δh showed a positive linear correlation with APD, ASA, FFD, and TFF ($p < 0.05$) and a negative linear correlation with FC, TFD, and ACP ($p < 0.05$; Figure 6a).

To develop an evaluation model for comfortable colors, this study analyzed the correlation between eye-tracking and subjective evaluation metrics with ΔE, ΔL, ΔC, and Δh (Figure 6b). The calculations indicated that APD had a relatively low overall correlation with ΔE, ΔL, ΔC, and Δh (0.01, 0.02, 0.01, and 0.01). ASA showed a strong correlation with ΔL (0.31) and a relatively strong correlation with ΔE, ΔC, and Δh (0.17, 0.18, and 0.16). FFD demonstrated a moderate correlation with ΔL and ΔC (0.06 and 0.04), with weaker correlations for the other variables. FC showed a strong correlation with ΔL (−0.33) and a relatively strong correlation with ΔE, ΔC, and Δh (−0.15, −0.19, and −0.13). TFF had a relatively strong correlation with ΔE, ΔL, ΔC, and Δh (0.1, 0.19, 0.11, and 0.09). TFD exhibited a moderate correlation with ΔE, ΔL, ΔC, and Δh (−0.08, −0.15, −0.08, and

−0.07). ACP showed a strong correlation with ΔE, ΔL, ΔC, and Δh (−0.27, −0.31, −0.19, and −0.24). VC also showed a strong correlation with ΔE, ΔL, ΔC, and Δh (−0.3, −0.33, −0.21, and −0.26).

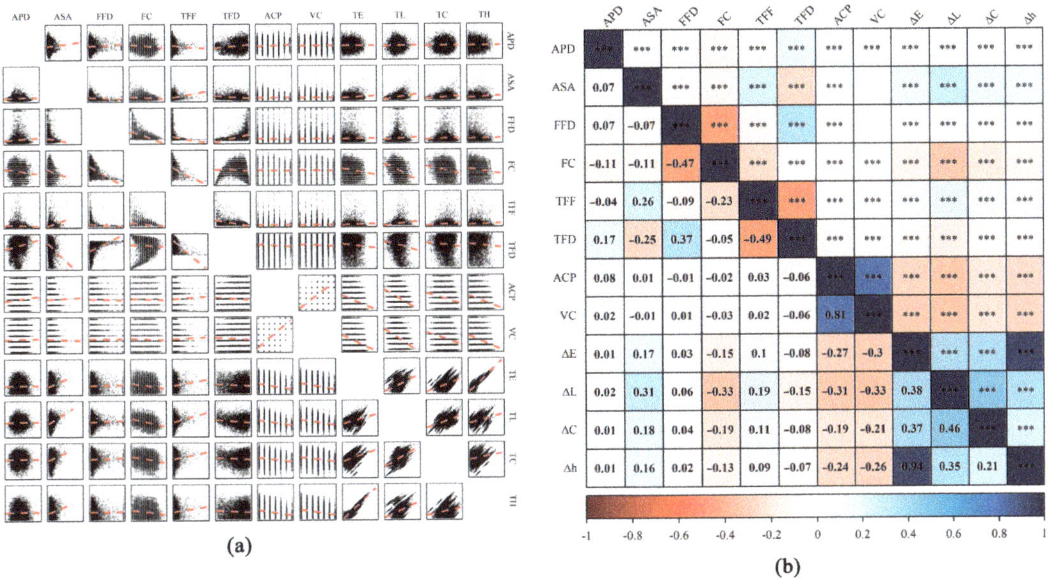

Figure 6. The correlation analysis between ΔE, ΔL, ΔC, and Δh and eye-tracking metrics and subjective comfort metrics APD, ASA, FFD, FC, TFF, TFD, ACP, and VC is depicted in the figures: (**a**) scatter plot matrix and (**b**) correlation analysis chart, with Pearson correlation coefficients. ***: $p < 0.001$.

It is important to note that ΔE, ΔL, ΔC, and Δh showed a strong correlation with the subjective evaluation metrics ACP and FC, and a relatively strong correlation with the eye-tracking metrics ASA, FC, TFF, and TFD. They exhibited a moderate correlation with FFD, while APD showed a notable correlation only with ΔL.

3.3. Regression Analysis

Each eye-tracking and subjective evaluation metric was treated as the dependent variable, with ΔE, ΔL, ΔC, and Δh as the independent variables. Based on the trends observed in Figure 6, linear regression models were established for pairs of variables that exhibited linear changes. First, single-factor regression models were created. When the regression coefficients of these models showed statistical significance, multi-factor regression models were subsequently developed (Table 3).

In the single-factor linear regression models established between the color element differences and the eye-tracking metrics and subjective comfort, all regression coefficients were statistically significant (Table 3). To develop a mathematical model for the relationship between ΔE, ΔL, ΔC, Δh, and user visual comfort, multi-factor linear regression analysis models were created, resulting in fitted equations. These equations revealed the trends in variations between the color metrics and the eye-tracking metrics, clarifying the impact of different colors on visual comfort.

This study established multi-factor linear regression models to predict APD, VD, FC, TFD, VC, and ACP based on differences in color elements. The R^2 values and regression coefficients of these models were tested and found to be statistically significant ($p < 0.05$; Table 4).

Table 3. Single-factor regression models of color elements with eye-tracking metrics and subjective comfort.

Variable	Factors	Unstandardized Coefficients		Standardized Coefficients	t	Sig.
		β	SE	β		
APD	ΔE	<0.001	<0.001	0.013	3.207	0.001
	ΔL	0.001	<0.001	0.022	5.398	<0.001
	ΔC	<0.001	<0.001	0.014	3.329	0.001
	Δh	<0.001	<0.001	0.011	2.765	0.006
ASA	ΔE	0.012	<0.001	0.167	40.988	<0.001
	ΔL	0.038	<0.001	0.306	77.756	<0.001
	ΔC	0.013	<0.001	0.176	43.219	<0.001
	Δh	0.012	<0.001	0.157	38.334	<0.001
FFD	ΔE	0.001	<0.001	0.028	6.721	<0.001
	ΔL	0.002	<0.001	0.06	14.425	<0.001
	ΔC	0.001	<0.001	0.045	10.794	<0.001
	Δh	0.001	<0.001	0.023	5.49	<0.001
FC	ΔE	−0.019	0.001	−0.148	−36.253	<0.001
	ΔL	−0.069	0.001	−0.326	−83.298	<0.001
	ΔC	−0.023	0.001	−0.189	−46.572	<0.001
	Δh	−0.016	0.001	−0.127	−31.068	<0.001
TFF	ΔE	0.001	<0.001	0.096	23.199	<0.001
	ΔL	0.004	<0.001	0.189	46.492	<0.001
	ΔC	0.001	<0.001	0.111	27.051	<0.001
	Δh	0.001	<0.001	0.086	20.855	<0.001
TFD	ΔE	−0.003	<0.001	−0.078	−18.837	<0.001
	ΔL	−0.008	<0.001	−0.15	−36.749	<0.001
	ΔC	−0.003	<0.001	−0.079	−19.115	<0.001
	Δh	−0.002	<0.001	−0.07	−17.001	<0.001
ACP	ΔE	−0.015	<0.001	−0.268	−67.297	<0.001
	ΔL	−0.029	<0.001	−0.311	−79.059	<0.001
	ΔC	−0.011	<0.001	−0.19	−46.815	<0.001
	Δh	−0.014	<0.001	−0.24	−59.762	<0.001
VC	ΔE	−0.017	<0.001	−0.297	−75.171	<0.001
	ΔL	−0.032	<0.001	−0.326	−83.474	<0.001
	ΔC	−0.012	<0.001	−0.207	−51.088	<0.001
	Δh	−0.016	<0.001	−0.264	−66.089	<0.001

In the regression model for APD, the influence of ΔL was positive; when ΔL increased by 1 unit, APD decreased by less than 0.001 units. In the regression model for ASA, the influences of ΔL, ΔC, and Δh were positive. When these variables increased by 1 unit, the ASA value increased by 0.034, 0.003, and 0.002 units, respectively. In the regression model for FFD, the influences of ΔL and ΔC were positive; when ΔL and ΔC increased by 1 unit, the FFD value increased by 0.002 and less than 0.001 units, respectively. In the regression model for FC, the influences of ΔE, ΔL, and ΔC were negative; when these variables increased by 1 unit, the FC value decreased by 0.011, 0.064, and 0.004 units, respectively. Conversely, the influence of Δh was positive; when Δh increased by 1 unit, the FC value increased by 0.009 units. In the regression model for TFF, the influences of ΔE, ΔL, and ΔC were positive. When these variables increased by 1 unit, the TFF value increased by less than 0.001, 0.003, and 0.001 units, respectively. In the regression model for TFD, the influences of ΔE and ΔL were negative; when these variables increased by 1 unit, the TFD value decreased by 0.002 and 0.008 units, respectively. In the regression model for ACP, the influences of ΔE and ΔL were negative; when these variables increased by 1 unit, the ACP value decreased by 0.016 and 0.023 units, respectively. Conversely, the influence of Δh was positive; when Δh increased by 1 unit, the ACP value increased by 0.006 units. In the regression model for VC, the influences of ΔE and ΔL were negative; when these variables increased by 1 unit, the VC value decreased by 0.019 and 0.025 units, respectively.

Conversely, the influence of Δh on VC was positive; when Δh increased by 1 unit, the VC value increased by 0.008 units.

Table 4. Multi-factor regression models of color element differences with eye-tracking metrics and subjective comfort.

		Unstandardized Coefficients		Standardized Coefficients		
Variable	Factors	β	SE	β	t	Sig.
APD	Intercept	2.944	0.004		687.156	<0.001
	ΔE	<0.001	<0.001	0.012	0.831	0.406
	ΔL	<0.001	<0.001	0.02	3.971	<0.001
	ΔC	<0.001	<0.001	0.002	0.335	0.738
	Δh	>−0.001	<0.001	−0.007	−0.503	0.615
ASA	Intercept	3.293	0.022		153.04	<0.001
	ΔE	0.002	0.001	0.022	1.657	0.097
	ΔL	0.034	0.001	0.27	57.769	<0.001
	ΔC	0.003	<0.001	0.036	6.904	<0.001
	Δh	0.002	0.001	0.033	2.547	0.011
FFD	Intercept	0.451	0.007		67.425	<0.001
	ΔE	<0.001	<0.001	0.008	0.55	0.582
	ΔL	0.002	<0.001	0.05	10.114	<0.001
	ΔC	<0.001	<0.001	0.02	3.708	<0.001
	Δh	>−0.001	<0.001	−0.006	−0.476	0.634
FC	Intercept	7.379	0.036		203.959	<0.001
	ΔE	−0.011	0.002	−0.089	−6.71	<0.001
	ΔL	−0.064	0.001	−0.303	−65.304	<0.001
	ΔC	−0.004	0.001	−0.03	−5.925	<0.001
	Δh	0.009	0.002	0.07	5.48	<0.001
TFF	Intercept	0.172	0.004		47.659	<0.001
	ΔE	<0.001	<0.001	0.032	2.311	0.021
	ΔL	0.003	<0.001	0.17	35.145	<0.001
	ΔC	<0.001	<0.001	0.023	4.234	<0.001
	Δh	<0.001	<0.001	−0.009	−0.667	0.505
TFD	Intercept	2.648	0.01		262.649	<0.001
	ΔE	−0.002	<0.001	−0.048	−3.453	0.001
	ΔL	−0.008	<0.001	−0.141	−28.945	<0.001
	ΔC	<0.001	<0.001	−0.001	−0.193	0.847
	Δh	0.001	<0.001	0.025	1.86	0.063
ACP	Intercept	2.912	0.016		181.793	<0.001
	ΔE	−0.016	0.001	−0.28	−21.385	<0.001
	ΔL	−0.023	<0.001	−0.247	−53.705	<0.001
	ΔC	<0.001	<0.001	0.006	1.111	0.267
	Δh	0.006	0.001	0.109	8.645	<0.001
VC	Intercept	3.251	0.016		198.37	<0.001
	ΔE	−0.019	0.001	−0.331	−25.594	<0.001
	ΔL	−0.025	<0.001	−0.252	−55.491	<0.001
	ΔC	<0.001	<0.001	0.005	0.993	0.321
	Δh	0.008	0.001	0.135	10.867	<0.001

3.4. Summary

3.4.1. The Relationship between Hue and Visual Perception Evaluation

In the warm color spectrum, building color combinations involving red had significantly lower VC and ACP values compared to other color combinations (Figure 6). Red is associated with high arousal levels and is generally linked to active emotions such as anger, excitement, and passion [55]. These characteristics contrast with the intended architectural attributes of residential buildings and can negatively impact visual quality by intensifying emotional responses in users. In contrast, the orange and yellow groups exhibited significantly higher FFD, TFD, and VC values compared to the red group, indicating that orange and yellow tones are more stable and are generally associated with positive emotions, such as pleasure and happiness [55]. As a result, visual preference for color combinations in the orange and yellow hues was significantly higher than that for the red group.

As hues transition from warm to cool, users' emotions shift from high activity to low activity. Purple, as a transitional color between warm and cool tones, exhibited the highest FC value. Participants showed a greater preference for building color combinations in purple tones, with visual quality rated higher than other hue combinations. The ACP mean was also the highest among the different color groups, indicating a stronger overall preference for purple color combinations. However, it is important to note that while the ACP mean for purple color combinations was higher than that of other color groups, the variability was significant, suggesting that participants may provide polarized visual quality evaluations for purple combinations.

In the natural color spectrum, the green and blue groups exhibited higher average visual quality indicators than the warm color groups. The green group showed higher FFD and VC values compared to the blue group, indicating that the visual quality of the green group was superior. More participants preferred building color combinations that included green. Conversely, the blue group had lower mean values for APD and ASA compared to the other color groups.

3.4.2. The Relationship between ΔE, ΔL, ΔC, Δh, and Visual Perception Evaluation

In the correlation analysis, ΔE showed a strong correlation with ACP and VC, and a moderate correlation with ASA, FC, and TFF. This suggests that excessive total color difference significantly reduced users' subjective color evaluations and visual preferences, leading to visual fatigue.

ΔL showed a strong correlation with ASA, FC, ACP, and VC, and a moderate correlation with TFF and TFD. This indicates that excessive brightness differences caused significant visual stimulation, which can substantially lower users' visual perception evaluations when viewing buildings.

ΔC showed a moderate correlation with ASA, FC, TFF, ACP, and VC. This suggests that excessive chroma differences not only reduced the subjective color evaluation of residential buildings but also increased visual fatigue and cognitive load, thereby lowering the overall visual quality of the buildings.

Δh showed a strong correlation with ACP and VC, and a moderate correlation with ASA, FC, TFF, and TFD. This indicates that excessive hue differences significantly impacted users' subjective evaluation of color combinations and, to some extent, reduced the visual quality of residential buildings, leading to visual fatigue.

In the regression analysis of different models, ΔE strongly influenced certain eye-tracking metrics and subjective color evaluations. ΔL affected all eye-tracking metrics and subjective evaluations in every model. Additionally, in some models, the extreme values of the regression coefficients for eye-tracking metrics were higher for ΔL than for ΔE. ΔC affected eye-tracking metrics only in specific regression models, with overall regression coefficients remaining relatively low. Notably, Δh showed a positive influence on eye-tracking metrics related to visual preference and subjective color evaluation, with relatively high regression coefficients. This suggests that appropriately increasing Δh may improve the visual quality of residential buildings and enhance the living experience.

4. Discussion

4.1. Insights into Color Selection Related to Hue Factors

In traditional studies, the warm color groups of orange and yellow have been considered the optimal choice for residential building colors due to their high visual quality [4,56,57]. However, in this study, among the six hues examined, the purple group had significantly higher mean values for FC, ACP, and VC compared to the orange and yellow groups. This suggests that the color combinations in the purple group provided superior visual stimulation and subjective color preference. In the subjective metric ACP, the purple group exhibited a large range of variation. Therefore, when using purple in building color combinations, it is essential to consider users' acceptance of the color, as well as the local cultural context and its tolerance for purple.

Furthermore, factors shaping urban color environments include natural and climatic characteristics, as well as the interplay between color and form [16]. The green and blue spaces, represented by "ecological spaces" and "environmental spaces", have been shown to influence the associations between physiological and psychological health outcomes [58] and positively impact residents' sense of well-being [17]. Additionally, the sky view factor and the visibility of natural greenery have been shown to significantly affect neighborhood vitality.

In this study, for the first time, blue-colored groups, as revealed through eye-tracking physiological indicators, demonstrated lower average values and smaller numerical fluctuations in response to visual stimuli compared to other color hues. This suggests that blue, as an architectural decorative tone, can create a calming atmosphere that promotes relaxation, inner security, and confidence [59]. In urban color planning, for certain individuals or specific residential locations, selecting color combinations from the blue group can help reduce visual pressure, lower cognitive burden, and emphasize the differentiation of neighborhood characteristics.

4.2. Experimental Insights into Color Element Differences

Through regression analysis of ΔE, ΔL, ΔC, and Δh using eye-tracking data and subjective questionnaires, we found the following:

(1) Residential colors with medium to high brightness and medium to low saturation often receive higher color evaluations [4,50,60]. Further studies have explored architectural color combinations, finding that higher brightness contrast increases visual fatigue [58]. When accent and auxiliary colors fall within the dark range (low ΔL), the public may develop negative emotions toward these combinations [61], which can affect the vitality of streets in the area. This experiment indicated that ΔL had a more significant impact compared to other color difference factors (ΔE, ΔC, and Δh) on both user visual stimulation and subjective visual evaluations. Regression models further revealed that, in various models related to visual and subjective preferences, the regression coefficient of ΔL was higher than that of other color difference factors, indicating a stronger and more precise connection between ΔL and subjective visual quality.

(2) In fields such as art appreciation, consumer decision-making, and targeted information retrieval, numerous studies have shown that a larger Δh tends to increase user visual attention, enhance visual perception, and boost consumer engagement [38,62,63]. However, this study found that Δh had a relatively smaller impact on visual perception compared to other color difference elements, based on eye-tracking data and visual perception evaluations. Additionally, its influence on subjective ratings was not particularly significant. Thus, it can be observed that in the field of architectural colors, Δh does not play as significant a role in visual perception ratings as it does in other domains. Conversely, when the ΔL and ΔC values of architectural color combinations are kept close, combinations with a larger Δh can still improve the visual quality of architectural color compositions, thereby enhancing the vitality of the streetscape.

(3) In experiments exploring harmonious complementary landscapes, vibrant plant colors, such as oranges and yellows, can play a more significant role, even when combined with cooler tones [64]. This study's regression results similarly indicated that in architectural color research, a certain degree of Δh can enhance subjective evaluations of buildings. This may be because moderate color contrast can elevate users' EEG activity levels [65]. Previous studies have demonstrated a strong correlation between EEG activity and visual comfort [66]. Therefore, it can be inferred that controlling a certain degree of Δh in architectural color combinations can effectively enhance the visual quality of buildings and improve the residential experience, creating better spatial dynamics.

4.3. Strengths and Limitations of This Study

This study focused on dual-tone architectural color combinations, exploring differences in perceptual quality under various combinations by examining the relationships between color element differences, eye-tracking metrics, and subjective evaluations. Current architectural color experiments often emphasize single colors, which diverge from the reality of residential color schemes in China. By focusing on dual-tone architectural colors, this study breaks away from the limitations of previous research on color experimental samples. Through regression analysis, we precisely delineated and explained the specific impact levels of color element differences, providing accurate data support for research on dual-tone architectural color combinations.

Previous architectural color studies often used open-ended scales to gather subjective evaluations from participants. However, subjective expressions of color preferences can be influenced by factors such as individual education, geographic location, and cultural background [6], leading to inconsistencies between physiological responses and subjective evaluations. This study introduced eye-tracking as a physiological indicator to bypass the subjective processing of color perception, directly collecting users' visual feedback on color stimuli to validate subjective evaluations. The aim was to enhance the accuracy and representativeness of experimental data. It should be acknowledged that, although eye-tracking experiments have improved the accuracy of color perception to some extent, human visual perception and subjective preferences cannot be fully captured through a single eye-tracking technique [15]. Future research should incorporate a broader range of physiological indicators, such as EEG, skin conductance, and blood oxygen levels, to conduct a more comprehensive quantitative analysis of color perception.

Additionally, this study preliminarily validated that eye-tracking metrics can serve as evaluation indicators for measuring color visual quality. However, relying on a single experiment may not accurately capture individual preferences. Previous eye-tracking studies have shown that FC and FFD can reflect participants' preferences to some extent regarding experimental stimuli [46,67]. Despite this, color combinations with excessive Δh and ΔL tend to elicit heightened physiological attention [38,68], which is not necessarily driven by preference [65]. Therefore, future eye-tracking experiments in the field of color should focus on identifying the specific causes behind eye movement indicators. Comprehensive data metrics are crucial for color-related visual quality studies to fully understand individual visual experiences.

During the experiment, the focus was on investigating the relationship between color combinations and visual quality, so only architectural units were selected as the experimental subjects. Factors such as the surrounding ancillary buildings, community environment, sky view factor, and green view index were not considered in evaluating the visual quality of individual buildings. In future research on the overall community environment, it will be necessary to adopt a more comprehensive approach, considering various influencing factors to assess their impact on the visual quality of architectural color combinations.

This experiment also has certain regional limitations. The architectural styles used as subjects were based on a survey of old residential buildings in the Shanghai area. It is undeniable that the architectural forms of old residential buildings vary across different regions of China. Therefore, the generalizability of the experimental conclusions requires further validation through color studies conducted in other regions. Additionally, the extent to which cultural background influences individual color perception remains under-researched and lacks quantifiable studies. Incorporating color perception data from users with diverse cultural backgrounds is essential to enhance the generalizability of the study's conclusions.

5. Conclusions

In this study, through the combination of eye-tracking technology and subjective color evaluation, two-tone architectural monoliths were used as research objects. Mathematical models were established to examine the relationships between two-tone color combinations, visual quality, and subjective evaluation using single-factor and multi-factor regression models. By collecting various eye-tracking data and color evaluation scores from participants, the study explored the impact of two-tone color combinations on visual quality and subjective evaluations. This provides mathematical support and multidimensional perspectives for the updating of architectural colors, addressing a research gap in existing studies that primarily focus on monochromatic colors. The findings offer more practical and targeted advice for color planning in street areas, contributing to the creation of a more vibrant and pleasant neighborhood environment.

Additionally, through a mathematical model analyzing color element differences in relation to visual quality and subjective evaluation, this study revealed the specific impact of color element differences on architectural colors, providing more scientific references for the future selection of architectural color combinations. Compared to previous methods that explored architectural color evaluation through open-ended questions, this study more accurately captured participants' subconscious reactions to color combinations by integrating eye-tracking experiments. This approach reduced the influence of subjective bias on experimental results, thereby improving the reliability of the overall data feedback [69].

Moreover, new measurement indicators were introduced for studying the visual quality of multicolored buildings, exploring the potential of incorporating eye-tracking metrics into the rating system for the color visual quality of multicolored residential buildings. The visual quality evaluation framework for multicolored buildings was initially constructed by combining physiological indicators with subjective evaluations. A humanistic perspective was also introduced, providing a new measurement dimension for studying the relationship between color and neighborhood vitality, while enhancing the research framework for street vitality.

Author Contributions: Conceptualization, M.S.; methodology, M.S. and Y.H.; software, M.S. and Y.H.; validation, M.S. and Y.H.; formal analysis, M.S. and Y.H.; investigation, M.S.; resources, M.S.; data curation, M.S.; writing—original draft preparation, M.S. and Y.H.; writing—review and editing, M.S. and Y.H.; visualization, M.S. and Y.H.; supervision, Y.H.; project administration, Y.H.; funding acquisition, Z.W. All authors have read and agreed to the published version of the manuscript.

Funding: The authors received no specific funding for this work.

Institutional Review Board Statement: Not applicable.

Informed Consent Statement: Not applicable.

Data Availability Statement: The original contributions presented in the study are included in the article, further inquiries can be directed to the corresponding author.

Conflicts of Interest: The authors declare that they have no known competing financial interests or personal relationships that could have appeared to influence the work reported in this paper.

Appendix A. Methods and Research Variables

Theme	Use Case	Methods	Data Analysis	
Building color	Colors of old residential facades in Shanghai, China	/	AHP hierarchical analysis	[4]
	Color geography, the relationship between color and architectural form	Literature search	/	[9]
	Color and visual perception	Questionnaire, fNIRS extraction of hemodynamic responses	Descriptive analysis	[11]

Theme	Use Case	Methods	Data Analysis	
Visual quality	Xuzhou historical building color plan	On-site measurement, color card comparison	/	[56]
	Color planning for Shanghai Historic Landscape District	Expert scoring, questionnaires	Descriptive analysis	[12]
	Tall building height and color and perceived psychological repair	Questionnaires, measurement of psychological variables of recovery	Descriptive analysis, Pearson's correlation, multiple regression, and analysis of variance (ANOVA)	[13]
	Evaluation of architectural color combinations	Questionnaire	Descriptive analysis, regression analysis	[50]
	Visual quality of natural landscapes within cities	Photographic model simulation of streetscapes, questionnaires	Scenic beauty estimation (SBE), Wilcoxon signed-rank test	[24]
	Relationship between forest color and landscape aesthetics estimates	ArcGIS and FRAGSTATS extraction of spatial indices of color plates	One-way analysis of variance (ANOVA) and partial correlation analysis were used	[25]
	Visual quality of historic buildings and surrounding buildings	Questionnaire	One-way ANOVA, multiple linear regression analyses	[26]
	Visual quality of indoor light environments in buildings	Simulation of indoor models	Relative weight calculation	[19]
	Visual quality of street outdoor billboards	Photo survey	R.G.B bivariate histogram calculation	[22]
	Visual comfort and preference in specific indoor scenes	Forced-choice approach to psychophysical experiments	/	[27]
	Quality of visual perception of urban streets	Street view imaging technology	Salience detection algorithm combining simple priorities (SDSP)	[20]
	Perception of visual and auditory landscapes in urban environments	Customized deep learning (DL) models	/	[32]
Eye movement technique	Evaluation of visual behavioral characteristics and psychological perception of forest water feature spaces	Eye-tracking techniques and psycho-perceptual questionnaires	Wilcoxon rank sum test, T test, correlation analysis	[42]
	Balance and visual aesthetics of building facade compositions	Eye-tracking technology (building facade elements)	Descriptive analysis	[43]
	Proportion of landscape elements and residents' evaluation of the attractiveness of urban complexes	Eye-tracking techniques (landscape element composition), online questionnaires	Descriptive analysis, correlation analysis	[44]
	Visual behavioral characteristics of built heritage and what factors influence visual perception	Eye movement techniques (visual search paths, gaze ratios)	Descriptive analysis, correlation analysis	[45]
	Influences on the visual characteristics of spatial elements in traditional Chinese commercial neighborhoods	Eye-tracking techniques (streetscape elements), semantic differentiation	Descriptive analysis, correlation analysis	[46]
	Key factors and landscape elements of concern in urban parks	Eye movement technology (green vision)	Descriptive analysis, correlation analysis	[67]
	Color elements and visual comfort in metro space	Eye-tracking techniques (color in indoor environments), questionnaires	Correlation analysis, regression analysis	[54]

Appendix B. Color Combination and Color Value Correspondence

Experimental color number	Color 1 (L)	Color 1 (a)	Color 1 (b)	Color 2 (L)	Color 2 (a)	Color 2 (b)
	58.214	22.621	11.887	90.764	−4.916	86.600
	67.281	37.450	61.247	67.281	37.450	61.247
	65.99	−14.069	−46.711	61.605	9.045	−7.179
	67.281	37.450	61.247	41.891	39.353	−25.639
	54.568	66.354	48.785	69.146	−44.03	35.883
	67.281	37.450	61.247	61.849	−20.667	11.126
	58.214	22.621	11.887	67.281	37.450	61.247
	41.891	39.353	−25.639	61.605	9.045	−7.179
	67.069	37.582	47.737	41.891	39.353	−25.639
	67.069	37.582	47.737	90.764	−4.916	86.600
	41.891	39.353	−25.639	57.24	26.817	−19.559
	90.764	−4.916	86.600	69.146	−44.03	35.883
	61.849	−20.667	11.126	65.99	−14.069	−46.711
	63.277	19.639	28.012	90.764	−4.916	86.600
	67.281	37.450	61.247	65.99	−14.069	−46.711
	54.568	66.354	48.785	67.281	37.450	61.247
	72.649	6.333	53.454	61.605	9.045	−7.179
	69.146	−44.03	35.883	41.891	39.353	−25.639
	58.214	22.621	11.887	63.277	19.639	28.012
	67.281	37.450	61.247	59.499	−12.911	−10.452
	58.214	22.621	11.887	63.277	−37.334	20.883
	67.281	37.450	61.247	61.605	9.045	−7.179
	90.764	−4.916	86.600	72.649	6.333	53.454
	54.568	66.354	48.785	61.605	9.045	−7.179
	58.532	66.354	21.114	41.891	39.353	−25.639
	58.532	66.354	21.114	69.146	−44.03	35.883
	67.069	37.582	47.737	67.281	37.450	61.247
	67.281	37.450	61.247	59.084	−19.392	−26.696
	54.568	66.354	48.785	41.891	39.353	−25.639
	58.532	66.354	21.114	61.605	9.045	−7.179
	59.084	−19.392	−26.696	59.499	−12.911	−10.452
	69.146	−44.03	35.883	61.605	9.045	−7.179
	67.069	37.582	47.737	72.649	6.333	53.454
	67.069	37.582	47.737	59.084	−19.392	−26.696
	72.649	6.333	53.454	61.849	−20.667	11.126
	67.281	37.450	61.247	57.24	26.817	−19.559
	67.069	37.582	47.737	69.146	−44.03	35.883
	67.281	37.450	61.247	61.605	9.045	−7.179
	58.214	22.621	11.887	59.499	−12.911	−10.452
	65.99	−14.069	−46.711	59.084	−19.392	−26.696
	67.281	37.450	61.247	69.146	−44.03	35.883
	69.146	−44.03	35.883	57.24	26.817	−19.559
	63.277	−37.334	20.883	65.99	−14.069	−46.711
	59.084	−19.392	−26.696	61.605	9.045	−7.179
	58.532	66.354	21.114	63.277	19.639	28.012
	67.069	37.582	47.737	63.277	19.639	28.012
	58.532	66.354	21.114	58.214	22.621	11.887
	61.849	−20.667	11.126	57.24	26.817	−19.559
	58.214	22.621	11.887	59.084	−19.392	−26.696
	58.214	22.621	11.887	61.605	9.045	−7.179
	63.277	19.639	28.012	65.99	−14.069	−46.711

Experimental color number	Color 1 (L)	Color 1 (a)	Color 1 (b)	Color 2 (L)	Color 2 (a)	Color 2 (b)
	67.281	37.450	61.247	61.849	−20.667	11.126
	90.764	−4.916	86.600	57.24	26.817	−19.559
	58.532	66.354	21.114	61.849	−20.667	11.126
	54.568	66.354	48.785	63.277	−37.334	20.883
	63.277	−37.334	20.883	57.24	26.817	−19.559
	54.568	66.354	48.785	61.849	−20.667	11.126
	90.764	−4.916	86.600	63.277	−37.334	20.883
	59.499	−12.911	−10.452	61.605	9.045	−7.179
	58.214	22.621	11.887	67.069	37.582	47.737
	90.764	−4.916	86.600	59.499	−12.911	−10.452
	63.277	19.639	28.012	61.605	9.045	−7.179
	58.532	66.354	21.114	59.084	−19.392	−26.696
	69.146	−44.03	35.883	59.499	−12.911	−10.452
	63.277	19.639	28.012	67.281	37.450	61.247
	67.281	37.450	61.247	69.146	−44.03	35.883
	58.532	66.354	21.114	65.99	−14.069	−46.711
	63.277	−37.334	20.883	41.891	39.353	−25.639
	63.277	19.639	28.012	57.24	26.817	−19.559
	58.214	22.621	11.887	41.891	39.353	−25.639
	63.277	19.639	28.012	41.891	39.353	−25.639
	65.99	−14.069	−46.711	59.499	−12.911	−10.452
	90.764	−4.916	86.600	65.99	−14.069	−46.711
	72.649	6.333	53.454	41.891	39.353	−25.639
	90.764	−4.916	86.600	61.605	9.045	−7.179
	90.764	−4.916	86.600	58.214	22.621	11.887
	63.277	−37.334	20.883	61.849	−20.667	11.126
	67.069	37.582	47.737	57.24	26.817	−19.559
	67.281	37.450	61.247	59.499	−12.911	−10.452
	58.214	22.621	11.887	72.649	6.333	53.454
	67.069	37.582	47.737	63.277	−37.334	20.883
	58.532	66.354	21.114	67.281	37.450	61.247
	58.532	66.354	21.114	72.649	6.333	53.454
	58.532	66.354	21.114	67.281	37.450	61.247
	67.281	37.450	61.247	57.24	26.817	−19.559
	58.214	22.621	11.887	57.24	26.817	−19.559
	90.764	−4.916	86.600	67.281	37.450	61.247
	72.649	6.333	53.454	63.277	−37.334	20.883
	58.532	66.354	21.114	90.764	−4.916	86.600
	69.146	−44.03	35.883	61.849	−20.667	11.126
	63.277	19.639	28.012	59.084	−19.392	−26.696
	67.281	37.450	61.247	59.084	−19.392	−26.696
	67.281	37.450	61.247	63.277	−37.334	20.883
	58.214	22.621	11.887	65.99	−14.069	−46.711
	59.499	−12.911	−10.452	57.24	26.817	−19.559
	67.281	37.450	61.247	65.99	−14.069	−46.711
	72.649	6.333	53.454	57.24	26.817	−19.559
	54.568	66.354	48.785	57.24	26.817	−19.559
	54.568	66.354	48.785	59.499	−12.911	−10.452
	67.281	37.450	61.247	90.764	−4.916	86.600
	72.649	6.333	53.454	69.146	−44.03	35.883
	58.214	22.621	11.887	61.849	−20.667	11.126

Experimental color number	Color 1 (L)	Color 1 (a)	Color 1 (b)	Color 2 (L)	Color 2 (a)	Color 2 (b)
	67.281	37.450	61.247	63.277	19.639	28.012
	67.069	37.582	47.737	61.849	−20.667	11.126
	67.069	37.582	47.737	59.499	−12.911	−10.452
	54.568	66.354	48.785	59.084	−19.392	−26.696
	67.281	37.450	61.247	67.069	37.582	47.737
	63.277	−37.334	20.883	59.084	−19.392	−26.696
	61.849	−20.667	11.126	59.084	−19.392	−26.696
	67.281	37.450	61.247	72.649	6.333	53.454
	63.277	19.639	28.012	63.277	−37.334	20.883
	58.214	22.621	11.887	67.281	37.450	61.247
	67.281	37.450	61.247	41.891	39.353	−25.639
	54.568	66.354	48.785	65.99	−14.069	−46.711
	72.649	6.333	53.454	67.281	37.450	61.247
	58.532	66.354	21.114	57.24	26.817	−19.559
	54.568	66.354	48.785	58.532	66.354	21.114
	58.532	66.354	21.114	59.499	−12.911	−10.452
	54.568	66.354	48.785	72.649	6.333	53.454
	58.532	66.354	21.114	67.069	37.582	47.737
	61.849	−20.667	11.126	59.499	−12.911	−10.452
	65.99	−14.069	−46.711	57.24	26.817	−19.559
	72.649	6.333	53.454	59.084	−19.392	−26.696
	90.764	−4.916	86.600	59.084	−19.392	−26.696
	69.146	−44.03	35.883	65.99	−14.069	−46.711
	69.146	−44.03	35.883	59.084	−19.392	−26.696
	54.568	66.354	48.785	90.764	−4.916	86.600
	57.24	26.817	−19.559	61.605	9.045	−7.179
	65.99	−14.069	−46.711	41.891	39.353	−25.639
	69.146	−44.03	35.883	63.277	−37.334	20.883
	67.281	37.450	61.247	63.277	−37.334	20.883
	63.277	19.639	28.012	61.849	−20.667	11.126
	54.568	66.354	48.785	67.281	37.450	61.247
	58.214	22.621	11.887	69.146	−44.03	35.883
	58.532	66.354	21.114	63.277	−37.334	20.883
	72.649	6.333	53.454	59.499	−12.911	−10.452
	90.764	−4.916	86.600	41.891	39.353	−25.639
	63.277	19.639	28.012	72.649	6.333	53.454
	54.568	66.354	48.785	63.277	19.639	28.012
	63.277	−37.334	20.883	61.605	9.045	−7.179
	67.069	37.582	47.737	65.99	−14.069	−46.711
	63.277	19.639	28.012	69.146	−44.03	35.883
	90.764	−4.916	86.600	61.849	−20.667	11.126
	59.084	−19.392	−26.696	57.24	26.817	−19.559
	67.281	37.450	61.247	61.605	9.045	−7.179
	59.499	−12.911	−10.452	41.891	39.353	−25.639
	61.849	−20.667	11.126	41.891	39.353	−25.639
	67.069	37.582	47.737	61.605	9.045	−7.179
	63.277	−37.334	20.883	59.499	−12.911	−10.452
	54.568	66.354	48.785	67.069	37.582	47.737
	63.277	19.639	28.012	59.499	−12.911	−10.452
	59.084	−19.392	−26.696	41.891	39.353	−25.639
	61.849	−20.667	11.126	61.605	9.045	−7.179

Appendix C. Questionnaire

Architectural Color Visual Comfort Score	(1) very uncomfortable
	(2) less comfortable
	(3) slightly uncomfortable
	(4) general
	(5) slightly comfortable
	(6) more comfortable
	(7) very comfortable
Architectural color visual preference scoring	(1) very much disliked
	(2) less favorable
	(3) slightly disliked
	(4) general
	(5) slightly preferred
	(6) prefer
	(7) favorite

References

1. Anter, K.F. Forming Spaces with Colour and Light: Trends in Architectural Practice and Swedish Colour Research. *Colour Des. Creat.* **2008**, *2*, 1–10.
2. Jaglarz, A. Perception of Color in Architecture and Urban Space. *Buildings* **2023**, *13*, 2000. [CrossRef]
3. Askari, A.H.; Soltani, S. Contribution of Building Façades to Attractive Streetscapes: Study of Two Main Streets in Kuala Lumpur City. *JDBE* **2018**, *18*, 29–40. [CrossRef]
4. Wang, J.; Zhang, L.; Gou, A. Study of the Color Characteristics of Residential Buildings in Shanghai. *Color Res. Appl.* **2021**, *46*, 240–257. [CrossRef]
5. Xu, Z.; Zheng, X. Roadmap of Moving Urban Colour toward Cultural Sustainability in China. *Color Res. Appl.* **2021**, *46*, 222–239. [CrossRef]
6. Odetti, J.V. The Study of Color as a Component of Urban Cultural Construction: A Theoretical Reflection and Methodological Proposal. *Color Res. Appl.* **2023**, *48*, 453–467. [CrossRef]
7. Manav, B. Color-emotion Associations, Designing Color Schemes for Urban Environment-architectural Settings. *Color Res. Appl.* **2017**, *42*, 631–640. [CrossRef]
8. Chen, K.; Lin, H.; Chen, Y.-J.; Xu, Y.; Ding, S.; Guo, Y.; You, S. Do Consumers Have Colour Aesthetic Preferences for the Facade Materials of Condominium Buildings? *Buildings* **2024**, *14*, 557. [CrossRef]
9. Gou, A. Method of Urban Color Plan Based on Spatial Configuration. *Color Res. Appl.* **2013**, *38*, 65–72. [CrossRef]
10. Hogg, J.; Goodman, S.; Porter, T.; Mikellides, B.; Preddy, D.E. Dimensions and Determinants of Judgements of Colour Samples and a Simulated Interior Space by Architects and Non-Architects. *Br. J. Psychol.* **1979**, *70*, 231–242. [CrossRef]
11. Shi, Y.; Tu, Y.; Wang, L.; Zhu, N.; Zhang, D. How Visual Discomfort Is Affected by Colour Saturation: A fNIRS Study. *IEEE Photonics J.* **2022**, *14*, 1–7. [CrossRef]
12. Wang, Z.; Sun, H.; Li, J. Research on Architectural Color and Visual Comfort in Historic Landscape Areas. *Buildings* **2023**, *13*, 1004. [CrossRef]
13. Zarghami, E.; Ghanbaran, A.; Karimimoshaver, M.; Saadativaghar, P. Effect of Height and Color of Tall Buildings on the Perceived Psychological Restoration of Residents. *J. Archit. Plan. Res.* **2019**, *36*, 321–342.
14. Zhai, Y.; Gong, R.; Huo, J.; Fan, B. Building Façade Color Distribution, Color Harmony and Diversity in Relation to Street Functions: Using Street View Images and Deep Learning. *ISPRS Int. J. Geo-Inf.* **2023**, *12*, 224. [CrossRef]
15. Wang, Z.; Shen, M.; Huang, Y. Exploring the Impact of Facade Color Elements on Visual Comfort in Old Residential Buildings in Shanghai: Insights from Eye-Tracking Technology. *Buildings* **2024**, *14*, 1758. [CrossRef]
16. Vasiljevic-Tomic, D.; Maric, I. Colour in the City: Principles of Nature-Climate Characteristics. *Facta Univ. Archit. Civ. Eng.* **2011**, *9*, 315–323. [CrossRef]
17. Wan, J.; Zhou, Y.; Li, Y.; Su, Y.; Cao, Y.; Zhang, L.; Ying, L.; Deng, W. Research on Color Space Perceptions and Restorative Effects of Blue Space Based on Color Psychology: Examination of the Yijie District of Dujiangyan City as an Example. *Int. J. Environ. Res. Public Health* **2020**, *17*, 3137. [CrossRef]
18. Serra, J.; García, Á.; Torres, A.; Llopis, J. Color Composition Features in Modern Architecture. *Color Res. Appl.* **2012**, *37*, 126–133. [CrossRef]
19. Zanon, S.; Callegaro, N.; Albatici, R. A Novel Approach for the Definition of an Integrated Visual Quality Index for Residential Buildings. *Appl. Sci.* **2019**, *9*, 1579. [CrossRef]

20. Cheng, L.; Chu, S.; Zong, W.; Li, S.; Wu, J.; Li, M. Use of Tencent Street View Imagery for Visual Perception of Streets. *ISPRS Int. J. Geo-Inf.* **2017**, *6*, 265. [CrossRef]
21. Bharmoria, R.; Sharma, V. Analyzing the Urban Sustainable Development Impact of Degraded Visual Quality on Streetscape Causing Visual Variation: A Case of Dharamshala, Himachal Pradesh. *Environ. Dev. Sustain.* **2024**. [CrossRef]
22. Adam, M.; Al-Sharaa, A.; Ab Ghafar, N.; Mundher, R.; Abu Bakar, S.; Alhasan, A. The Effects of Colour Content and Cumulative Area of Outdoor Advertisement Billboards on the Visual Quality of Urban Streets. *ISPRS Int. J. Geo-Inf.* **2022**, *11*, 630. [CrossRef]
23. Mundher, R.; Abu Bakar, S.; Al-Helli, M.; Gao, H.; Al-Sharaa, A.; Mohd Yusof, M.J.; Maulan, S.; Aziz, A. Visual Aesthetic Quality Assessment of Urban Forests: A Conceptual Framework. *Urban Sci.* **2022**, *6*, 79. [CrossRef]
24. Ma, B.; Hauer, R.J.; Xu, C. Effects of Design Proportion and Distribution of Color in Urban and Suburban Green Space Planning to Visual Aesthetics Quality. *Forests* **2020**, *11*, 278. [CrossRef]
25. Zhang, Z.; Qie, G.; Wang, C.; Jiang, S.; Li, X.; Li, M. Relationship between Forest Color Characteristics and Scenic Beauty: Case Study Analyzing Pictures of Mountainous Forests at Sloped Positions in Jiuzhai Valley, China. *Forests* **2017**, *8*, 63. [CrossRef]
26. Bu, X.; Chen, X.; Wang, S.; Yuan, Y.; Han, C. The Influence of Newly Built High-Rise Buildings on Visual Impact Assessment of Historic Urban Landscapes: A Case Study of Xi'an Bell Tower. *J. Asian Archit. Build. Eng.* **2022**, *21*, 1304–1319. [CrossRef]
27. Wang, Q.; Xu, H.; Zhang, F.; Wang, Z. Influence of Color Temperature on Comfort and Preference for LED Indoor Lighting. *Optik* **2017**, *129*, 21–29. [CrossRef]
28. Markopoulos, E.; Goonetilleke, R.S.; Ho, A.G.; Luximon, Y. (Eds.) *Advances in Creativity, Innovation, Entrepreneurship and Communication of Design: Proceedings of the AHFE 2021 Virtual Conferences on Creativity, Innovation and Entrepreneurship, and Human Factors in Communication of Design, USA, 25–29 July 2021*; Lecture Notes in Networks and Systems; Springer International Publishing: Cham, Switzerland, 2021; Volume 276, ISBN 978-3-030-80093-2.
29. Sarasso, P.; Neppi-Modona, M.; Sacco, K.; Ronga, I. "Stopping for Knowledge": The Sense of Beauty in the Perception-Action Cycle. *Neurosci. Biobehav. Rev.* **2020**, *118*, 723–738. [CrossRef]
30. Long, Y.; Liu, L. How Green Are the Streets? An Analysis for Central Areas of Chinese Cities Using Tencent Street View. *PLoS ONE* **2017**, *12*, e0171110. [CrossRef] [PubMed]
31. Hu, K.; Xu, Z.; Wang, X.; Wang, Y.; Li, H.; Zhang, Y. Research on Street Color Environment Perception Based on CEP-KASS Framework. *Buildings* **2023**, *13*, 2649. [CrossRef]
32. Verma, D.; Jana, A.; Ramamritham, K. Predicting Human Perception of the Urban Environment in a Spatiotemporal Urban Setting Using Locally Acquired Street View Images and Audio Clips. *Build. Environ.* **2020**, *186*, 107340. [CrossRef]
33. Csikszentmihalyi, M.; Larson, R. The Ecology of Adolescent Activity and Experience. *J. Youth Adolesc.* **1977**, *6*, 281–294. [CrossRef] [PubMed]
34. Worthy, D.A. An Examination of the Effects of Eye-Tracking on Behavior in Psychology Experiments. *Behav. Res. Methods* **2024**, *56*, 6812–6825. [CrossRef] [PubMed]
35. Mele, M.L.; Federici, S. Gaze and Eye-Tracking Solutions for Psychological Research. *Cogn. Process.* **2012**, *13*, 261–265. [CrossRef] [PubMed]
36. Hamamé, C.M.; Vidal, J.R.; Perrone-Bertolotti, M.; Ossandón, T.; Jerbi, K.; Kahane, P.; Bertrand, O.; Lachaux, J.-P. Functional Selectivity in the Human Occipitotemporal Cortex during Natural Vision: Evidence from Combined Intracranial EEG and Eye-Tracking. *NeuroImage* **2014**, *95*, 276–286. [CrossRef]
37. Yarbus, A.L. *Eye Movements and Vision*; Springer US: Boston, MA, USA, 1967; ISBN 978-1-4899-5381-0.
38. Bix, L.; Seo, W.; Sundar, R.P. The Effect of Colour Contrast on Consumers' Attentive Behaviours and Perception of Fresh Produce. *Packag. Technol. Sci.* **2013**, *26*, 96–104. [CrossRef]
39. Siegrist, M. Consumers' Food Selection Behaviors in Three-Dimensional (3D) Virtual Reality. *Food Res. Int.* **2019**, *117*, 50–59. [CrossRef]
40. Shoval, N.; Schvimer, Y.; Tamir, M. Real-Time Measurement of Tourists' Objective and Subjective Emotions in Time and Space. *J. Travel Res.* **2018**, *57*, 3–16. [CrossRef]
41. Scott, N.; Zhang, R.; Le, D.; Moyle, B. A Review of Eye-Tracking Research in Tourism. *Curr. Issues Tour.* **2019**, *22*, 1244–1261. [CrossRef]
42. Gao, Y.; Sun, X.M.; Zhang, T.; Zhan, H. Analysis on the Characteristics of Eye Movement and the Evaluation of Psychological Perception for Forest Waterscape Space. *IOP Conf. Ser. Earth Environ. Sci.* **2021**, *626*, 012007. [CrossRef]
43. Hasse, C.; Weber, R. Eye Movements on Facades: The Subjective Perception of Balance in Architecture and Its Link to Aesthetic Judgment. *Empir. Stud. Arts* **2012**, *30*, 7–22. [CrossRef]
44. Li, J.; Zhang, Z.; Jing, F.; Gao, J.; Ma, J.; Shao, G.; Noel, S. An Evaluation of Urban Green Space in Shanghai, China, Using Eye Tracking. *Urban For. Urban Green.* **2020**, *56*, 126903. [CrossRef]
45. Li, N.; Zhang, S.; Xia, L.; Wu, Y. Investigating the Visual Behavior Characteristics of Architectural Heritage Using Eye-Tracking. *Buildings* **2022**, *12*, 1058. [CrossRef]
46. Li, Z.; Sun, X.; Zhao, S.; Zuo, H. Integrating Eye-Movement Analysis and the Semantic Differential Method to Analyze the Visual Effect of a Traditional Commercial Block in Hefei, China. *Front. Archit. Res.* **2021**, *10*, 317–331. [CrossRef]
47. Lv, J.; Xu, X.; Ding, N. Research on the Quantitative Method of Cognitive Loading in a Virtual Reality System. *Information* **2019**, *10*, 170. [CrossRef]

48. Orquin, J.L.; Loose, S.M. Attention and Choice: A Review on Eye Movements in Decision Making. *Acta Psychol.* **2013**, *144*, 190–206. [CrossRef] [PubMed]
49. Guo, H.Y.; Cai, Y.N. China's Urban Architecture and Environmental Color Practicein in Ten Years. *AMM* **2012**, *174–177*, 2512–2515. [CrossRef]
50. Li, K.; Zheng, Z.; Wang, P.; Yan, W. Research on the Colour Preference and Harmony of the TWO-COLOUR Combination Buildings. *Color Res. Appl.* **2022**, *47*, 980–991. [CrossRef]
51. De La Fuente Suárez, L.A. Subjective Experience and Visual Attention to a Historic Building: A Real-World Eye-Tracking Study. *Front. Archit. Res.* **2020**, *9*, 774–804. [CrossRef]
52. Kiefer, P.; Giannopoulos, I.; Raubal, M.; Duchowski, A. Eye Tracking for Spatial Research: Cognition, Computation, Challenges. *Spat. Cogn. Comput.* **2017**, *17*, 1–19. [CrossRef]
53. Wang, H.; Tang, J.; Xu, P.; Chen, R.; Yao, H. Research on the Influence Mechanism of Street Vitality in Mountainous Cities Based on a Bayesian Network: A Case Study of the Main Urban Area of Chongqing. *Land* **2022**, *11*, 728. [CrossRef]
54. Zhang, L.; Li, X.; Li, C.; Zhang, T. Research on Visual Comfort of Color Environment Based on the Eye-Tracking Method in Subway Space. *J. Build. Eng.* **2022**, *59*, 105138. [CrossRef]
55. Hanada, M. Correspondence Analysis of Color–Emotion Associations. *Color Res. Appl.* **2018**, *43*, 224–237. [CrossRef]
56. Hong, X.; Ji, X.; Wu, Z. Architectural Colour Planning Strategy and Planning Implementation Evaluation of Historical and Cultural Cities Based on Different Urban Zones in Xuzhou (China). *Color Res. Appl.* **2022**, *47*, 424–453. [CrossRef]
57. Wang, J.; Zhang, L.; Gou, A. Study on the Preference of City Color Image Selection Based on the Logistic Model: A Case Study of Shanghai. *Color Res. Appl.* **2020**, *45*, 542–557. [CrossRef]
58. Labib, S.M.; Lindley, S.; Huck, J.J. Spatial Dimensions of the Influence of Urban Green-Blue Spaces on Human Health: A Systematic Review. *Environ. Res.* **2020**, *180*, 108869. [CrossRef]
59. Kaya, N.; Crosby, M. Color Associations with Different Building Types: An Experimental Study on American College Students. *Color Res. Appl.* **2006**, *31*, 67–71. [CrossRef]
60. Gou, A.; Wang, J. Research on the Location Characters of Urban Color Plan in China. *Color Res. Appl.* **2008**, *33*, 68–76. [CrossRef]
61. He, F. Gender Differences in Color Perceptions and Preferences of Urban Façades Based on a Virtual Comparison. *Build. Environ.* **2023**, *245*, 110907. [CrossRef]
62. Brychtova, A.; Coltekin, A. An Empirical User Study for Measuring the Influence of Colour Distance and Font Size in Map Reading Using Eye Tracking. *Cartogr. J.* **2016**, *53*, 202–212. [CrossRef]
63. Massaro, D.; Savazzi, F.; Di Dio, C.; Freedberg, D.; Gallese, V.; Gilli, G.; Marchetti, A. When Art Moves the Eyes: A Behavioral and Eye-Tracking Study. *PLoS ONE* **2012**, *7*, e37285. [CrossRef]
64. Zhang, L.; Dempsey, N.; Cameron, R. 'Blossom Buddies'—How Do Flower Colour Combinations Affect Emotional Response and Influence Therapeutic Landscape Design? *Landsc. Urban Plan.* **2024**, *248*, 105099. [CrossRef]
65. Abromavičius, V.; Serackis, A. Eye and EEG Activity Markers for Visual Comfort Level of Images. *Biocybern. Biomed. Eng.* **2018**, *38*, 810–818. [CrossRef]
66. Lu, M.; Hu, S.; Mao, Z.; Liang, P.; Xin, S.; Guan, H. Research on Work Efficiency and Light Comfort Based on EEG Evaluation Method. *Build. Environ.* **2020**, *183*, 107122. [CrossRef]
67. Gholami, Y.; Taghvaei, S.H.; Norouzian-Maleki, S.; Mansouri Sepehr, R. Identifying the Stimulus of Visual Perception Based on Eye-Tracking in Urban Parks: Case Study of Mellat Park in Tehran. *J. For. Res.* **2021**, *26*, 91–100. [CrossRef]
68. Forbes, S.H.; Plunkett, K. Colour Perception Changes with Basic Colour Word Comprehension. *Dev. Sci.* **2023**, *26*, e13406. [CrossRef]
69. Babiker, A.; Faye, I.; Prehn, K.; Malik, A. Machine Learning to Differentiate Between Positive and Negative Emotions Using Pupil Diameter. *Front. Psychol.* **2015**, *6*, 1921. [CrossRef]

applied sciences

Article

The Influence of Monochromatic Illustrations on the Attention to Public Health Messages: An Eye-Tracking Study

Marina Milošević, Dorotea Kovačević * and Maja Brozović

Faculty of Graphic Arts, University of Zagreb, 10000 Zagreb, Croatia; mmilosevic1@grf.hr (M.M.); maja.brozovic@grf.unizg.hr (M.B.)
* Correspondence: dorotea.kovacevic@grf.unizg.hr

Abstract: Past research suggests that images can be a useful tool in attracting people's attention and improving public health communication. This is especially important for the effective transmission of health-related messages to the youth, who should be warned about risks of alcohol and tobacco use. This study explores the application of eye-tracking methodology to investigate the influence of monochromatic illustrations on youths' visual attention to digital ads, with a particular emphasis on those presenting harmful effects of alcohol and tobacco consumption. Four digital ads were designed for the purpose of the experiment. Two ad topics (alcohol versus tobacco) and two ad contents (text-only versus text with a monochromatic illustration) were used as variables in the digital ad design. Analyses of eye fixations and viewing time revealed that more students noticed the digital ads with the illustration than the text-only ads. Furthermore, we found that the ads with illustrations had a longer viewing time and a higher number of fixations, regardless of the ad topic. The findings highlight the effectiveness of eye-tracking technology in examining the role of illustrations in digital ads, focusing on a better perception of addiction prevention campaigns whose message is targeted primarily toward young people.

Keywords: attention; graphic design; warnings; youth; illustration

1. Introduction

In the current digital era, online content plays a major role in the optimization of visual advertising, especially for young target audiences. As new media continue to evolve, the creative use of digital images has expanded. Previous research suggests that images can be a useful tool in attracting people's attention and improving marketing communication. Images serve more than just a visual embellishment; sometimes, they are key components that can significantly enhance the effectiveness of marketing messages, particularly in the domain of public health communication. For campaigns that require seriousness and directness, such as addiction prevention, the use of a black monochromatic palette can amplify the emotional effect of the advertised content. Additionally, monochromatic imagery can stand out from the typical colorful marketing content, thereby emphasizing its uniqueness in a mass of information and attracting more attention and the target group's awareness. With the development of sophisticated technologies for the analysis of graphic content such as eye tracking, researchers are equipped to directly measure how digital ads attract attention and how successfully they convey visual messages. As a result, they are able to conduct precise image analyses and adjust its visual parameters to maximize viewer engagement.

For example, a recent study [1] examined visual attention to ad features and found that images of people attracted greater attention than other ad elements. Focusing on the ad format, another research [2] revealed that a native advertisement attracted more attention than a banner advertisement. More recently, an eye-tracking experiment was conducted by Chen-Sankey et al. [3], who analyzed visual attention metrics in the context of e-cigarette advertisement and showed that the participants directed attention differently to various ad features. All of these studies have predominantly focused on digital advertising and photos, while monochromatic illustrations conveying public health warnings have not been sufficiently evaluated in terms of objective direct measurements. Digital ads and corresponding photos can offer defined parameters for effectiveness analyses, making them popular for both academic and commercial studies. On the other hand, illustrations frequently present more challenges for quantitative assessment due to their specific structure. The illustration format may exhibit different potential for attracting attention, calling for a more in-depth investigation. Therefore, in this study, eye tracking was employed to examine the effects of monochromatic illustrations on the visual processing of public health warnings. This is especially important for the effective transmission of health-related messages to the youth, who should be warned about risks of alcohol and tobacco use since both are harmful and potentially addictive [4].

2. Problem Statement

According to the 2030 Agenda for Sustainable Development [5], one of the Sustainable Development Goals is to ensure health and promote wellbeing for people of all ages (Goal 3). In that context, the youth require careful consideration because supporting their wellbeing sets the foundations for a generation that can drive progress of society, economic stability, and sustainable development. Under the goal of wellbeing promotion, the Agenda for Sustainable Development addresses prevention of substance abuse—including harmful consumption of alcohol (Goal 3.5) and strengthening the capacity of all countries for the early warning and management of global health risks (Goal 3.d). Within that framework, public health campaigns can achieve a great amount in addressing the importance of the youth's healthy lifestyle on the national level. Thus, this study seeks to explore whether digital advertising with a specific, yet simple, type of illustration can be effective in promoting anti-alcohol and anti-smoking behavior among young Croatian people.

In Croatia, alcohol use is a public health concern, so national strategy aims to reduce alcohol-related harm by concentrating on consumption among youth [6], with a special focus on prevention [7] and changes in drinking culture by promoting non-drinking as a socially accepted behavior [8]. Particularly alarming are the findings of a study on the role of social marketing in the prevention of alcohol abuse [9], which indicated that alcohol consumption is a very common lifestyle among Croatian students. The concern is not limited to drinking only, but extends to smoking. A recent study on students in southern Croatia [10] revealed that there is a lack of knowledge about the harmful effects of tobacco, which emphasizes the need for better prevention in this area also. Therefore, our study covers both anti-alcohol and anti-smoking messages as a component of the support to the public health authorities in tailored initiatives aimed at the young population and their healthy lifestyle [11,12].

Previous studies investigated whether specific forms of warning messages regarding tobacco harm can be effectively communicated through packaging [13–16]. These studies examined the perception of warning messages in conditions where a harmful product is already packed and offered to a potential consumer. In an effort to emphasize the importance of prevention and warning about product harm in earlier stages (before a young person is in direct contact with the product), the present study is focused on ads as a medium for transmitting preventive messages. Nowadays, digital advertising is the most rapidly expanding type of promotion [17]. It can quite successfully target message recipients, ranging from a general to a more specific audience. In current advertising-related targeting, more focus is on the sentiment of target groups instead of just their demographic data [18], and this is also the case for persuasive communication in community-based public health campaigns. For example, a study on public health advertisements confirmed that emotion-based advertisements increased willingness to donate more than rational-based ones [19]. Even in situations in which people search for health information of a more sensitive nature, such as topics on psychotic disorders, digital ads can be an effective method for encouraging help seeking in the mental health community [20].

Still, there are studies that suggest that people sometimes pay less attention to digital ads than other content elements regularly incorporated into a webpage design. In a study [21], Sørum reported that the participants noticed the ads but their attention was not maintained on the ads for a long time. This implies that, in order to understand the complexities of people's visual attention in a digital environment, it is essential to consider the visual perception of an ad as a process unfolded in two primary stages. The first stage refers to the ability of the ad to attract attention, and the subsequent stage refers to its ability to maintain the attention of viewers. A powerful objective measure of both attention stages is viewers' eye movements. The effectiveness of eye movement analyses for understanding aspects of viewer behavior is recognized in a wide range of research disciplines [22], and marketing is not an exception [23–27]. Therefore, to explore the potential of the deliberated selection of visual elements in public health marketing, this study employed eye-tracking technology to investigate viewers' visual attention to digital ads that convey health-related messages. More specifically, our objectives were to examine whether digital ads with illustrations promoting anti-alcohol and anti-smoking behavior can attract the attention of Croatian students, and to understand which types of ad designs are most likely to maintain their attention.

Hypotheses

Building upon the defined research objectives, this chapter introduces a set of hypotheses that served as framework that guided a systematic design of the ads in our study, testing their attention-grabbing properties.

The foundation of our work lies in the acknowledgment that design quality is an important aspect of visual advertising that influences visual perception. Past research provides strong evidence for this. In a study on health advertisements [28], the authors selected the ads that varied in three levels of design quality (i.e., low, mid, high) and examined whether they gain different levels of participants' attention. They found that advertisements of low levels of design quality were viewed for a shorter amount of time than those of higher levels of design quality. A study on print versions of ads [29] showed that highly rated advertisements attracted more attention compared to those that were poorly rated. Judging the quality of design often involves considering the attractiveness of its elements. In a study on content elements of online banner ads, it was found that the image was the most attractive element [30]. The positive effect of images is supported by results of past research that examined the ability of visual ads to capture visual attention [31,32]. The superiority of pictorial content was also demonstrated in a study in the field of health-related advertising. O'Malley et al. [33] used eye-tracking technology to measure the attention toward text-only, image-only, and image–text ads promoting osteoporosis prevention. They found that the combination of text and images was more effective at attracting the participants' visual attention than the image-only and text-only ad formats. Past research suggests that graphic images included in cigarette warning labels on advertising improved recall of the warning and associated health risks [34]. They did so by attracting and maintaining the visual attention of smokers. Images used as an addition to text warnings on cigarette packaging may be particularly effective for youth populations, as suggested by Casado-Aranda et al. [35].

Therefore, we set three hypotheses as follows. The first hypothesis refers to the effect of images (particularly illustrations) on the ad's ability to attract youth visual attention, while the second hypothesis refers to attention maintenance. Finally, the third hypothesis refers to the effect of illustrations on the ad's general noticeability. All the hypotheses focus on the visual attention variables because the attention is necessary for further processing [36] and greater visual attention may increase the possibility that young people will comprehend and remember the viewed information more easily.

H1: *A digital ad with text and an illustration attracts the attention of young viewers faster than a digital ad with no illustration.*

H2: *A digital ad with text and an illustration maintains the attention of young viewers for longer than a digital ad with no illustration.*

H3: *Young participants notice a digital ad with text and an illustration more than a digital ad with no illustration.*

To test the hypotheses, we employed an eye-tracking experiment. Eye tracking is considered as a suitable tool for consumer neuroscience research [37]. It allows a researcher to obtain quantitative data by recording viewers' eye movements during advertising exposition, and provides an insight into the viewers' underlying cognitive processes [35]. Hence, understanding eye movement behavior can be particularly useful in improving the persuasive quality of digital ads in public health promotion [27], especially of those that target young generations and gain their attention in today's media-saturated environment.

3. Materials and Methods

3.1. Stimuli Design

We designed four digital ads that had different topics (alcohol vs. smoking) and different content elements (text only vs. text with an illustration). Other design elements and visual attributes were consistent in all the ads, such as size, color, layout, and alignment. Different messages were used for different ad topics, but both consisted of two words that were presented with the same typeface and in the same type size. Translated in English, the text "Alcohol harms." was used for the anti-alcohol message, and the text "Stop smoking." was used for the anti-smoking message. Both ads contained a brief description of health-related issues that were directly associated with the topic. The brief description did not exceed more than three lines of text, similar to the conditions in previous studies [33]. The number of characters in the brief description ranged from 91 to 100, and the word count ranged from 18 to 19 words. Each ad took up 12.65% of the webpage area. The webpage was designed as a static page with a neutral color palette [38] and generic layout to reduce the effects of the color saturation and surrounding on the perception of the ad. Since the target group in this study was students, all textual content for the webpage was taken from the website of the University of Zagreb.

The results from human–computer interaction research suggest that banner blindness is strongest for digital banners placed on the right side of the screen [39]. A study on teenagers' perception of advertisements [40] showed that ads in the left column have a positive effect on visual attention. Therefore, the ads in our study were positioned on the left part of the webpage.

The illustrations used in those digital ads that contained the combination of text and illustration were originally created by one of the authors (M.M.), who made them particularly for the purpose of the study. The anti-alcohol ad contained an illustration with a person in a bottle while the anti-smoking ad contained an illustration with a person in an ashtray. In creating the illustrations, the goal was to present dramatic, almost shocking visual representations of negative consequences arising from alcohol abuse and cigarette smoking. The designer sought to apply the same graphic style to both illustrations and to make their level of visual complexity as similar as possible. Both illustrations were horizontally oriented. Since previous research [30] suggests that attention to digital ads may vary depending on the viewers' familiarity with the advertised brand, the digital ads in our study contained an imaginary brand presented through a simple generic logo.

3.2. Participants

There were 58 participants in the study, of which 47 were females. Given that the study deals with messages that target a young population, age was a criterion for participant selection. Their age ranged from 20 to 25 years (M = 22.5, SD = 1.71). All had normal or corrected vision and reported no visual or neurological disorders. All signed informed consent.

3.3. Equipment

The experiment took place in a controlled, quiet testing lab at the Faculty of Graphic Arts, University of Zagreb, designed to minimize external distractions and reduce background noise. Distinct light sources were removed and the test materials were presented directly in front of the participants to focus attention. The participants' eye movements were recorded by Tobii Eye Tracker X60 (Tobii Technology, Stockholm, Sweden), with a sampling rate of 60 Hz and an accuracy of 0.5 degrees. The digital ads were presented on a computer display (model LEN L1900pA, Lenovo, Morrisville, NC, USA) with a screen resolution of 1280 × 1024 pixels. Tobii Studio 3.2.1 software was used for the presentation of stimuli.

3.4. Procedure

Each student participated in an eye-tracking test individually. After a successful 5-point calibration, the experiment started. Each trial started by displaying a central fixation dot on the computer screen. Then, a webpage with one of the ads was presented for 10 s, and then the same procedure was carried out for a webpage with another ad. Each student was exposed to just one version of the ad for each topic. The participants were pre-informed of the time constraints for viewing the webpage. They were instructed to view the page freely, without a specific task, since a recent study showed that attention to digital ads may be influenced by a goal-oriented process [41]. Order of stimuli presentation is also noted to have an impact on people's responses [25]. Therefore, the order of ad presentations was randomized by selecting the Counterbalance mode in the presentation software (Tobii Studio 3.2.1), which was enabled for all materials in the session. The software arranged the order according to a Latin square method.

3.5. Dependent Measures

One of the benefits of the eye-tracking software is that it allows the experimenter to predefine areas (or regions) of interest [42]. The eye-tracking data in our experiment were calculated for the following areas of interest: area of the digital ad, area of the illustration on the ad, and area of the slogan. An example of one digital ad can be seen in Figure 1. For the digital ads with no illustration, only two areas of interest were analyzed (i.e., area of the digital ad and area of the slogan). We analyzed the measures of visual attention that were calculated in similar studies, which employed vision-based tracking systems [43–46]. In order to test the hypotheses, we categorized these measures into two groups, with one related to attention attraction and one related to attention maintenance.

For the first one (attention attraction), two variables were calculated: time to first fixation (TFF) and fixations before (FB). TFF is the time spent from the stimulus onset until the particular area of interest was fixated on for the first time [47]. A shorter time usually indicates that the visual element is more effective in attracting the attention. FB refers to the number of fixations that the participant directed toward other visual elements on the webpage before noticing an area of interest.

For the second group of measures (attention maintenance), three variables were calculated: total fixation duration (TFD), fixation count (FC), and total visit duration (TVD). TFD is the total time that the participant spent fixating on visual content. FC is the number of times that the participant fixated on the area of interest. TVD is the sum of visit durations that the participant spent looking at the area of interest.

For the third hypothesis, which refers to a general ad's noticeability, a binary variable was used, also known as hit ratio [48]. If a participant fixated on the ad at least once, the value was 1. If a participant did not fixate on the ad at all, the value was 0.

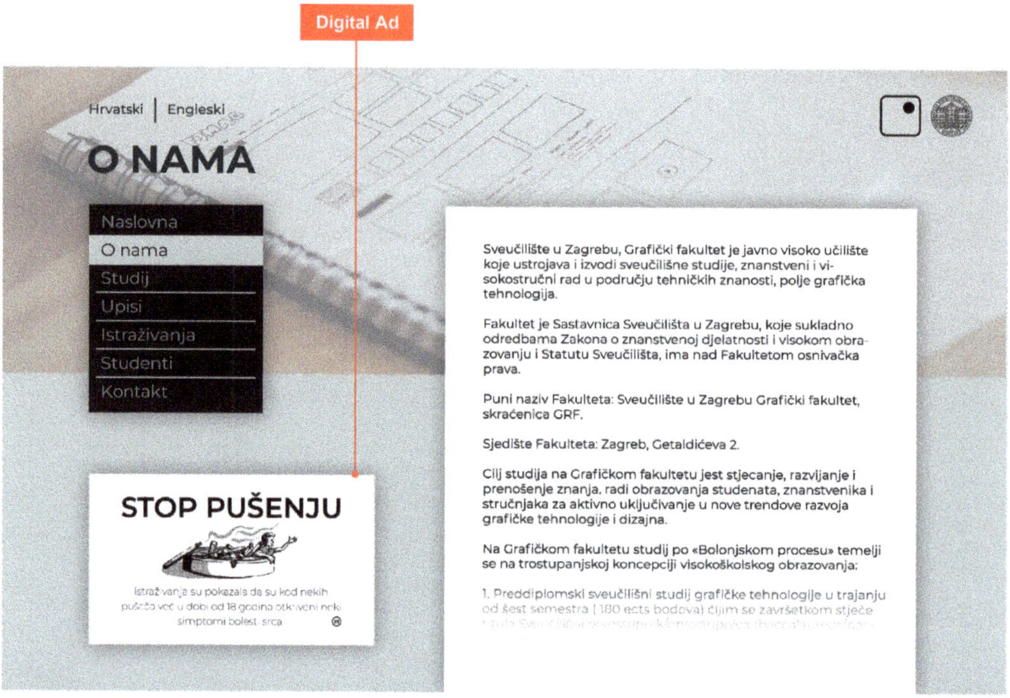

Figure 1. An example of a stimulus containing one of the digital ads displayed on the webpage used in the experiment (ad topic: smoking, ad content: text with illustration).

4. Results

The participants' fixations were first analyzed by checking the attention maps generated by the eye-tracking software (Figure 2). Attention maps in the form of heatmaps are a common way to present a visualization of the density of viewers' eye-tracking data. Different color hues represent different amounts of participants' fixations. Longer fixations recorded on the stimulus are presented by a red color, followed by yellow and green colors. By comparing the regions of the webpage marked in red (Figure 2), a higher prevalence of the red color can be noticed in the ads containing an illustration than on the ads with no illustration, indicating a longer average fixation duration on the illustrated ads.

The eye-tracking data were then statistically analyzed with IBM SPSS 23. A two-factor repeated measures analysis of variance (ANOVA) was carried out to examine whether ad topic (alcohol vs. smoking) and ad content (text only vs. text with an illustration) significantly influenced the participants' eye-tracking measures. The results showed that ad topic had no influence on any of the measures of visual attention (for both attention attraction and attention maintenance), and this applies to all the examined areas of interest. Ad content had no statistically significant effect on the attention attraction variables (TFF, FB), $p > 0.05$. However, ad content significantly influenced attention maintenance variables. The ads with an illustration had a longer TFD (M = 2.19, SD = 0.18) than the ads with no illustration ((M = 1.78, SD = 1.18), F(1.57) = 4.69, $p < 0.0$). Furthermore, the illustrated ads had a significantly higher FC (M = 10.36, SD = 0.78) than the ads with no illustration ((M = 8.53, SD = 0.84), F(1.57) = 5.92, $p < 0.05$), and TVD was also longer for the illustrated digital ads (M = 2.76, SD = 0.23) compared to the ads with no illustration ((M = 2.13, SD = 0.22), F(1.57) = 6.59, $p < 0.05$). Figure 3 shows the mean values for each measure.

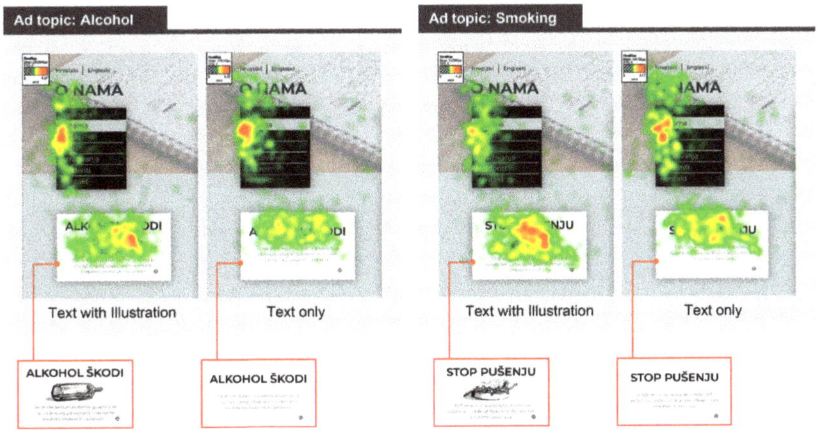

Figure 2. Heatmaps showing the duration of fixations by all participants (note: only the left area of the webpage is shown to optimize data presentation).

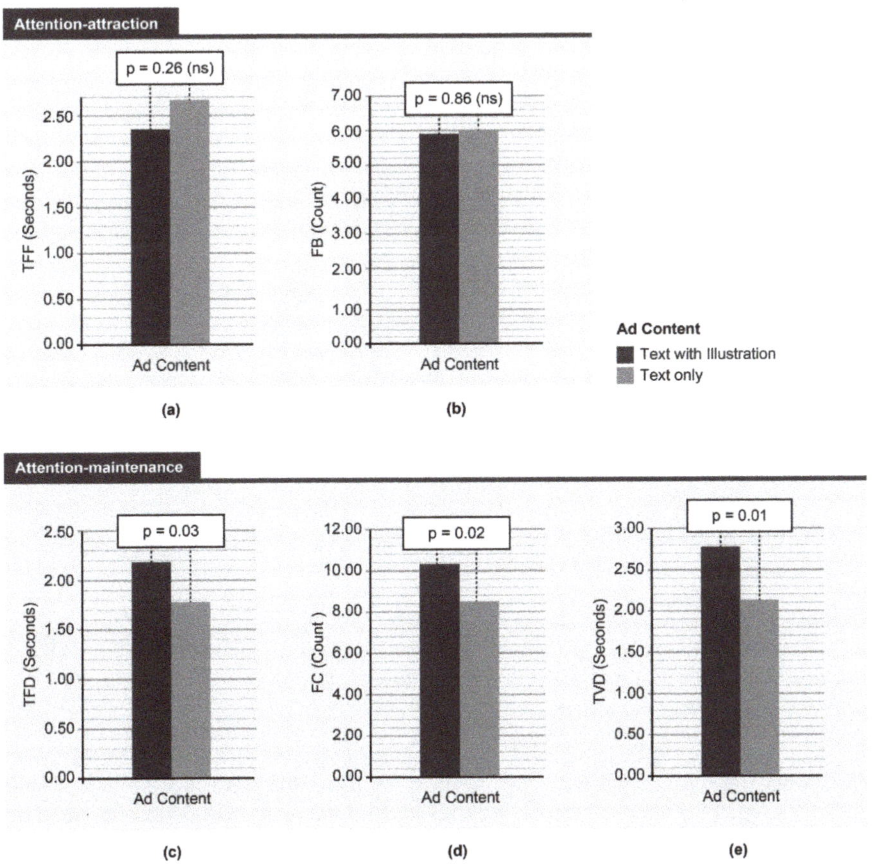

Figure 3. Mean values for eye-tracking data on the digital ads: (**a**) time to first fixation; (**b**) fixations before; (**c**) total fixation duration; (**d**) fixation count; (**e**) total visit duration.

By analyzing the eye-tracking data recorded only in the area of the illustration (Table 1), we found no significant influence of the illustration's visual representation (man in the bottle vs. man in the ashtray) on any of the variables (all $ps > 0.05$).

Table 1. Descriptive statistics and comparisons between two ad topics for all dependent variables (note: area of interest was the illustration).

Ad Topic	Statistic	TFF	FB	TFD	FC	TVD
Alcohol	Mean (SD)	4.03 (0.45)	9.71 (1.21)	1.20 (0.15)	5.43 (0.61)	1.35 (0.17)
Smoking	Mean (SD)	3.98 (0.44)	7.68 (0.82)	1.21 (0.16)	5.28 (0.63)	1.41 (0.18)
ANOVA results	F-value	0.12	2.14	1.74	0.43	0.72
	p-value	0.91	0.15	0.19	0.084	0.79

McNemar's tests were used to examine the influence of ad topic and ad content on the general noticeability. Figure 4 shows the distribution of participants depending on whether they noticed the ad. The only statistically significant difference was found between the ads containing an illustration and those without, indicating that more participants noticed the ads with text and an illustration (96.6%) than the ads with text only (89.7%); $p < 0.05$.

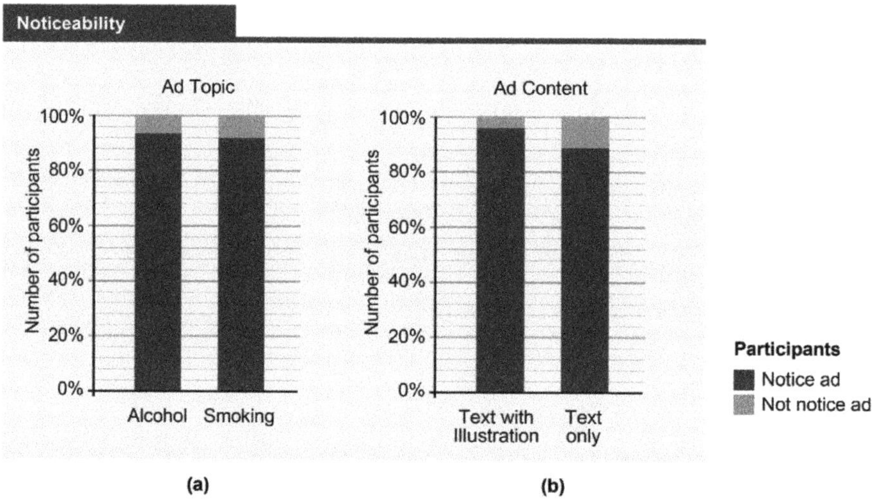

Figure 4. Participant distribution across conditions based on whether they noticed the digital ad; (**a**) ad topic as factor; (**b**) ad content as factor.

5. Discussion

The goal of this study was to determine whether adding an illustration in digital ads that promote anti-alcohol and anti-smoking behavior can make the ads more attention grabbing. Thus, we compared the viewers' eye-tracking data for the ads with an illustration and those with text only. The main finding was that the inclusion of the illustration in the ads had a beneficial influence on the variables that measured how long the ads maintain the participants' attention for. This confirms hypothesis H2 and suggests that adding pictorial content to health-related advertising has the potential to hold the youth's focus in a digital environment. This is not surprising since it is reasonable to expect that the visual richness of an ad maintains longer visual attention [49]. The result is consistent with prior research [32], which demonstrated the effect of ad content on participants' perception and showed that a higher level of attention was recorded among the participants exposed to pictorial ads than those exposed to text ads. This is also in line with the results reported by O'Malley et al. [33], who found that an ad with an image and text was more effective at attracting participants' attention than an ad with text only. This could be of particular importance for the effectiveness of public health promotions, which should not only be easily noticed by young people but also remembered well. As reported in a study on banner advertisement exposure [50], a longer duration of fixation on an ad can lead to better memory performance. Thus, our results contribute to a more general perspective provided by prior works [51] that showed that presentation style in ads can greatly influence whether the message will be seen and remembered. High values for two measures recorded on illustrated ads in our study (i.e., TFD and FC) could also be considered as an indicator of the ad's visual interest. According to findings of earlier studies in advertising, total fixation time can be a significant predictor of ad clicking [41], and longer fixations are also associated with an increased depth of ad processing [23]. Additionally, more fixations correlate with high memory [52,53]. Taken together, these results suggest that illustrations in digital ads can assist marketing practitioners in developing a digital ad that is more noticeable and, consequently, more memorable. Still, this does not necessarily apply to all formats of advertising, such as television commercials. A neuroscience-based research study on online advertising [54] showed that television commercials that had a longer visit duration were less recalled. Since our results refer to a static digital ad (as well as to a static webpage), this suggests that the level of content dynamics likely influences attention and recall performances. Therefore, using an illustration for improving the attention-grabbing properties of an ad does not guarantee good memorability for all types of advertising.

Additionally, we made an analysis of the ads' noticeability on a more general level, in which we compared the frequency of the people who detected the ad and those who did not. More participants noticed the ads with text and an illustration than the ads with no illustration, which supports hypothesis H3. This is in line with previous findings on the strong effect of the pictorial content in ads on attention [55]. Still, caution is needed when generalizing these results to diverse ad categories. For example, results from the field of tourism advertising indicated that ads containing text without an image attracted the participants' attention more effectively [56]. This implies that relying just on imagery in digital ads is not enough. Specific contextual characteristics of the promotional message, such as field of advertising and target audience, should also be taken into account.

Regarding the ability of a digital ad to attract attention, our results showed no effect of ad content on the variables that measured how fast the ads attracted the participants' attention, which refutes hypothesis H1. This outcome deviates from other studies upon which we based our assumptions, so it could be ascribed to the dominance of the textual component. As Gómez-Carmona et al. [57] state, ads with negative text can activate the viewer's emotional system faster. Thus, it seems that the verbal aspect of the digital ads in our study surpassed the pictorial aspect.

Although they were not included in the hypotheses, additional conclusions regarding the ad's pictorial content can be drawn from our results. The students in our experiment were exposed to two different illustrations (one for each topic), and the variations in the illustration did not influence their visual attention. This could be explained by the position of the illustration, which was centrally located in both versions of the ad. Since the central part of the composition has been identified as the most effective location in terms of attracting the viewers' interest in previous research [58], it is possible that placing the illustrations in the ad center had a more profound influence on attention than their motifs. However, a previous study on online health messages [59] indicated that the total fixation duration was influenced by different illustration types. Also, a study on a tourist hotel's marketing images [60] showed that different images led to different levels of customers' attention, which suggests that the lack of an image content effect should not be expected in all types of marketing materials.

Although our results may not be representative enough to apply these findings to broader demographic groups, they still indicate how a simple form of an ad image, such as a monochromatic illustration, can be effective in the context of digital advertising. Still, some of the discrepancies with other neuromarketing studies highlight the need for further investigation of more specific illustration attributes to comprehensively understand how illustrated content enhances the visual interest of a digital ad. In a larger context of public health promotion, the results of this study may contribute to the ongoing efforts to promote wellbeing for the youth and strengthen the prevention of harmful use of alcohol and tobacco among Croatian students.

6. Conclusions

This study explored how a combination of text and image can be effective in promoting anti-alcohol and anti-smoking behavior among young people through digital media. The objective was to obtain direct insight into the visual attention of Croatian students when they view digital ads signaling alcohol and tobacco harm and to compare the amount of attention attracted by the text-only and the text–illustration ad format. Based on the findings, it can be concluded that the presence of an illustration in the ads improved the level of visual interest to some extent. Although it did not reduce the time needed for ad detection, it did increase the amount of attention that students dedicated to the ad. It also increased the possibility of spotting the ad on the webpage. This study contributes to the existing body of knowledge on the visual perception of digital ads by giving further evidence toward the influence of images on people's visual processing.

The findings may be of interest for those engaged in public health marketing and those who are developing efficient marketing approaches for the promotion of youth wellbeing, supporting Goal 3 of the 2030 Agenda for Sustainable Development. Regarding theoretical implications, this study contributes to research-based knowledge on youths' visual stimulation, suggesting that monochromatic illustrations can be particularly effective in retaining their attention on important messages. The findings could also contribute to the development of decision-making authorities' guidelines or regulations that promote the use of specific visual techniques to enhance public health initiatives.

However, this study still has some limitations. Our sample was not representative enough to analyze the potential influence of gender on the eye-tracking outcomes. Furthermore, certain aspects of the ad effectiveness, such as comprehension or recall, were not addressed in our investigation. Investigating further hypotheses on the long-term effects on memory from ad exposure could provide a more comprehensive understanding of the effectiveness of the monochromatic illustrations. Future studies would benefit from the inclusion of the variable of whether the participants themselves have a precedent of alcohol and tobacco intake. A broader potential of illustrated anti-alcohol and anti-smoking messages in current non-commercial marketing practices, such as the power to encourage changes in youth behavior, is also worthy of a detailed examination. This calls for teams of researchers and specialists from different domains, like graphic design, marketing, and neurosciences. It is highly likely that cross-disciplinary collaboration among these fields could lead to better persuasive performances of digital ads in promoting a healthy lifestyle for the youth.

Author Contributions: Conceptualization, M.M. and D.K.; methodology, D.K.; software, D.K.; validation, D.K. and M.B.; formal analysis, D.K.; investigation, M.M.; resources, D.K. and M.M.; data curation, M.M.; writing—original draft preparation, M.M.; writing—review and editing, D.K. and M.B.; visualization, M.M.; supervision, M.B.; project administration, D.K. and M.B.; funding acquisition, D.K. and M.B. All authors have read and agreed to the published version of the manuscript.

Funding: This work was supported by the University of Zagreb.

Institutional Review Board Statement: Not applicable.

Informed Consent Statement: Informed consent was obtained from all subjects involved in the study.

Data Availability Statement: The data presented in this study are available upon reasonable request from the corresponding author. The data are not publicly available due to privacy.

Conflicts of Interest: The authors declare no conflicts of interest. The funders had no role in the design of the study; in the collection, analyses, or interpretation of data; in the writing of the manuscript; or in the decision to publish the results.

References

1. Zahmati, M.; Azimzadeh, S.M.; Sotoodeh, M.S.; Asgari, O. An eye-tracking study on how the popularity and gender of the endorsers affected the audience's attention on the advertisement. *Electron. Commer. Res.* **2023**, *23*, 1665–1676. [CrossRef]
2. De Keyzer, F.; Dens, N.; De Pelsmacker, P. The processing of native advertising compared to banner advertising: An eye-tracking experiment. *Electron. Commer. Res.* **2023**, *23*, 1921–1940. [CrossRef]
3. Chen-Sankey, J.; Weiger, C.; La Capria, K. Using Eye tracking to examine young adults' visual attention to e-cigarette advertising features and associated positive e-cigarette perceptions. *Ann. Behav. Med.* **2024**, *58*, 445–456. [CrossRef] [PubMed]
4. Morin, C. The Neurophysiological Effect of Emotional Ads on the Brains of Late Adolescents and Young Adults. Ph.D. Thesis, Fielding Graduate University, Santa Barbara, CA, USA, 2014.
5. United Nations. *Transforming Our World: The 2030 Agenda for Sustainable Development*; A/RES/70/1; United Nations: New York, NY, USA, 2015.
6. World Health Organization, Regional Office for Europe. *Alcohol as a Public Health Issue in Croatia: Situation Analysis and Challenges*; World Health Organization: Geneva, Switzerland, 2017.
7. Zoričić, Z.; Gašpar, V.; Ilić, S. Problems associated with drinking in young people in Croatia. *Alcoholism* **2008**, *44*, 29–35.
8. Mišević, Ž.; Bogdan, A.; Mišević, M.; Ružić, T. Cultural patterns of drinking and alcoholism in north and south of Croatia. *Interdiscip. Descr. Complex Syst.* **2020**, *18*, 36–56. [CrossRef]
9. Leko Šimić, M.; Turjak, S. Croatian Students and Alcohol Consumption—Implications for Social Marketing. *Ekon. Misao Praksa* **2018**, *27*, 77–101.
10. Istenic, D.; Gavic, L.; Tadin, A. Prevalence of use and knowledge about tobacco products and their harmful effects among university students in southern Croatia. *Healthcare* **2023**, *11*, 771. [CrossRef] [PubMed]
11. Dragun, R.; Veček, N.N.; Marendić, M.; Pribisalić, A.; Đivić, G.; Cena, H.; Polašek, O.; Kolčić, I. Have lifestyle habits and psychological well-being changed among adolescents and medical students due to COVID-19 lockdown in Croatia? *Nutrients* **2020**, *13*, 97. [CrossRef] [PubMed]
12. Modica, E. Evaluation of Perception to Different Stimuli by Cognitive and Emotional Reaction for Neuromarketing Application. Ph.D. Thesis, Sapienza University of Rome, Rome, Italy, 2019.
13. Wakefield, M.; Germain, D.; Durkin, S.; Hammond, D.; Goldberg, M.; Borland, R. Do larger pictorial health warnings diminish the need for plain packaging of cigarettes? *Addiction* **2012**, *107*, 1159–1167. [CrossRef]
14. Nilsson, T. Legibility and visual effectiveness of some proposed and current health warnings on cigarette packages. In *Report of Research Done under Contract to the Bureau of Tobacco Control of Health Canada*; Government of Canada: Ottawa, ON, Canada, 1999; pp. 1–17.
15. Byrne, S.; Kalaji, M.; Niederdeppe, J. Does visual attention to graphic warning labels on cigarette packs predict key outcomes among youth and low-income smokers? *Tob. Regul. Sci.* **2018**, *4*, 18–37. [CrossRef]
16. Süssenbach, P.; Niemeier, S.; Glock, S. Effects of and attention to graphic warning labels on cigarette packages. *Psychol. Health* **2013**, *28*, 1192–1206. [CrossRef]
17. Huebner, C. Digital Advertising and the Marketing Mix: A review of current literature and implications for Higher Education Marketeers. *J. Mark. Commun. High. Educ.* **2019**, *1*, 1–25. [CrossRef]
18. Raj Rocque, S.; Singh Sisodia, N. Digital advertising and its impact on the economy in the future. *Cent. Asian J. Math. Theory Comput. Sci.* **2023**, *4*, 29–34. [CrossRef]
19. Harris, J.M.; Ciorciari, J.; Gountas, J. Consumer neuroscience and digital/social media health/social cause advertisement effectiveness. *Behav. Sci.* **2019**, *9*, 42. [CrossRef] [PubMed]
20. Birnbaum, M.L.; Garrett, C.; Baumel, A.; Scovel, M.; Rizvi, A.F.; Muscat, W.; Kane, J.M. Using digital media advertising in early psychosis intervention. *Psychiatr. Serv.* **2017**, *68*, 1144–1149. [CrossRef] [PubMed]
21. Sørum, H. Visual attention and recall in website advertisements: An eye tracking study. In *Advances in Intelligent Systems and Computing*; Springer: Cham, Switzerland, 2018; Volume 607, pp. 205–216, ISBN 9783319604916.
22. Tatler, B.W.; Kirtley, C.; Macdonald, R.G.; Mitchell, K.M.A.; Savage, S.W. The active eye: Perspectives on eye movement research. In *Current Trends in Eye Tracking Research*; Horsley, M., Eliot, M., Knight, B., Reilly, R., Eds.; Springer International Publishing: Cham, Switzerland, 2014; pp. 3–16, ISBN 9783319028682.
23. Venkatraman, V.; Dimoka, A.; Vo, K.; Pavlou, P.A. Relative effectiveness of print and digital advertising: A memory perspective. *J. Mark. Res.* **2021**, *58*, 827–844. [CrossRef]
24. Myers, S.D.; Deitz, G.D.; Huhmann, B.A.; Jha, S.; Tatara, J.H. An eye-tracking study of attention to brand-identifying content and recall of taboo advertising. *J. Bus. Res.* **2020**, *111*, 176–186. [CrossRef]
25. Gómez-Carmona, D.; Cruces-Montes, S.; Marín-Dueñas, P.P.; Serrano-Domínguez, C.; Paramio, A.; García, A.Z. Do you see it clearly? The effect of packaging and label format on google ads. *J. Theor. Appl. Electron. Commer. Res.* **2021**, *16*, 1648–1666. [CrossRef]

26. Segijn, C.M.; Voorveld, H.A.M.; Vakeel, K.A. The role of ad sequence and privacy concerns in personalized advertising: An eye-tracking study into synced advertising effects. *J. Advert.* **2021**, *50*, 320–329. [CrossRef]
27. Alsharif, A.H.; Salleh, N.Z.M.; Al-Zahrani, S.A.; Khraiwish, A. Consumer behaviour to be considered in advertising: A systematic analysis and future agenda. *Behav. Sci.* **2022**, *12*, 472. [CrossRef]
28. Champlin, S.; Lazard, A.; Mackert, M.; Pasch, K.E. Perceptions of design quality: An eye tracking study of attention and appeal in health advertisements. *J. Commun. Healthc.* **2014**, *7*, 285–294. [CrossRef]
29. Watson, J.; Ye, H.; Zhang, J.; Topoglu, Y.; Suri, R.; Ayaz, H. The face of bad advertising: Assessing the effects of human face images in advertisement design using eye-tracking. In *Advances in Neuroergonomics and Cognitive Engineering*; Ayaz, H., Asgher, U., Paletta, L., Eds.; Springer: Cham, Switzerland, 2021; pp. 143–148, ISBN 978-3-030-80285-1.
30. Peker, S.; Dalveren, G.G.M.; İnal, Y. The effects of the content elements of online banner ads on visual attention: Evidence from an-eye-tracking study. *Future Internet* **2021**, *13*, 18. [CrossRef]
31. Pieters, R.; Wedel, M. Attention Capture and Transfer in Advertising: Brand, Pictorial, and Text-Size Effects. *J. Mark.* **2004**, *68*, 36–50. [CrossRef]
32. Goodrich, K. Anarchy of effects? Exploring attention to online advertising and multiple outcomes. *Psychol. Mark.* **2011**, *28*, 417–440. [CrossRef]
33. O'Malley, D.; Latimer, A.; Berenbaum, E. Using eye tracking technology to determine the most effective viewing format and content for osteoporosis prevention print advertisements. *J. Appl. Biobehav. Res.* **2011**, *16*, 167–186. [CrossRef]
34. Strasser, A.A.; Tang, K.Z.; Romer, D.; Jepson, C.; Cappella, J.N. Graphic warning labels in cigarette advertisements. *Am. J. Prev. Med.* **2012**, *43*, 41–47. [CrossRef] [PubMed]
35. Casado-Aranda, L.-A.; Sánchez-Fernández, J.; Ibáñez-Zapata, J.-Á. Evaluating communication effectiveness through eye tracking: Benefits, state of the art, and unresolved questions. *Int. J. Bus. Commun.* **2023**, *60*, 24–61. [CrossRef]
36. McGuire, W.J. Attitudes and opinions. *Annu. Rev. Psychol.* **1966**, *17*, 475–514. [CrossRef]
37. Harris, J.M.; Ciorciari, J.; Gountas, J. Consumer neuroscience for marketing researchers. *J. Consum. Behav.* **2018**, *17*, 239–252. [CrossRef]
38. Lewandowska, A.; Olejnik-Krugly, A. Do background colors have an impact on preferences and catch the attention of users? *Appl. Sci.* **2022**, *12*, 225. [CrossRef]
39. Resnick, M.; Albert, W. The impact of advertising location and user task on the emergence of banner ad blindness: An eye-tracking study. *Int. J. Hum. Comput. Interact.* **2014**, *30*, 206–219. [CrossRef]
40. Gidlöf, K.; Holmberg, N.; Sandberg, H. The use of eye-tracking and retrospective interviews to study teenagers' exposure to online advertising. *Vis. Commun.* **2012**, *11*, 329–345. [CrossRef]
41. Simonetti, A.; Bigne, E. Does banner advertising still capture attention? An eye-tracking study. *Spanish J. Mark. ESIC* **2023**, *28*, 3–20. [CrossRef]
42. Carter, B.T.; Luke, S.G. Best practices in eye tracking research. *Int. J. Psychophysiol.* **2020**, *155*, 49–62. [CrossRef]
43. Murphy, G.; Corcoran, C.; Tatlow-Golden, M.; Boyland, E.; Rooney, B. See, like, share, remember: Adolescents' responses to unhealthy-, healthy- and non-food advertising in social media. *Int. J. Environ. Res. Public Health* **2020**, *17*, 2181. [CrossRef]
44. Im, H.; Ju, H.W.; Johnson, K.K.P. Beyond visual clutter: The interplay among products, advertisements, and the overall webpage. *J. Res. Interact. Mark.* **2021**, *15*, 804–821. [CrossRef]
45. Bočaj, N.; Ahtik, J. Effects of visual complexity of banner ads on website users' perceptions. *Appl. Sci.* **2023**, *13*, 13317. [CrossRef]
46. Kang, D.; Kwon, J.; Nam, S. Research on effective advertising types in virtual environment. *Appl. Sci.* **2023**, *13*, 7063. [CrossRef]
47. Lai, M.L.; Tsai, M.J.; Yang, F.Y.; Hsu, C.Y.; Liu, T.C.; Lee, S.W.Y.; Lee, M.H.; Chiou, G.L.; Liang, J.C.; Tsai, C.C. A review of using eye-tracking technology in exploring learning from 2000 to 2012. *Educ. Res. Rev.* **2013**, *10*, 90–115. [CrossRef]
48. Andrychowicz-Trojanowska, A. Basic terminology of eye-tracking research. *Appl. Linguist. Pap.* **2018**, *25*, 123–132. [CrossRef]
49. Filiopoulou, D.; Rigou, M.; Faliagka, E. Display ads effectiveness: An eye tracking investigation. In *Advances in E-Business Research Series*; IGI Global: Hershey, PA, USA, 2019; pp. 205–230, ISBN 9781522572626.
50. Lee, J.W.; Ahn, J.H. Attention to banner ads and their effectiveness: An eye-tracking approach. *Int. J. Electron. Commer.* **2012**, *17*, 119–137. [CrossRef]
51. Krugman, D.M.; Fox, R.J.; Fletcher, J.E.; Rojas, T.H. Do adolescents attend to warnings in cigarette advertising? An eye-tracking approach. *J. Advert. Res.* **1994**, *34*, 39–52.
52. Lu, Y.; Qi, W.; Qin, J. Research on the memory of online advertising based on eye-tracking technology. In Proceedings of the 2018 IEEE 15th International Conference on e-Business Engineering (ICEBE), Xi'an, China, 12–14 October 2018; IEEE: Piscataway, NJ, USA; pp. 324–328.
53. Kovačević, D.; Brozović, M. Noticeability and recall of visual elements on packaging. In Proceedings of the 9th International Symposium on Graphic Engineering and Design, Novi Sad, Serbia, 8–10 November 2018; Kašiković, N., Ed.; Faculty of Technical Sciences: Novi Sad, Serbia, 2018; pp. 261–266.
54. Guixeres, J.; Bigné, E.; Azofra, J.M.A.; Raya, M.A.; Granero, A.C.; Hurtado, F.F.; Ornedo, V.N. Consumer neuroscience-based metrics predict recall, liking and viewing rates in online advertising. *Front. Psychol.* **2017**, *8*, 1808. [CrossRef] [PubMed]
55. Pieters, R.; Wedel, M.; Zhang, J. Optimal feature advertising design under competitive clutter. *Manag. Sci.* **2007**, *53*, 1815–1828. [CrossRef]

56. Kong, S.; Huang, Z.; Scott, N.; Zhang, Z.; Shen, Z. Web advertisement effectiveness evaluation: Attention and memory. *J. Vacat. Mark.* **2019**, *25*, 130–146. [CrossRef]
57. Gómez-Carmona, D.; Muñoz-Leiva, F.; Liébana-Cabanillas, F.; Nieto-Ruiz, A.; Martínez-Fiestas, M.; Campoy, C. The effect of consumer concern for the environment, self-regulatory focus and message framing on green advertising effectiveness: An eye tracking study. *Environ. Commun.* **2021**, *15*, 813–841. [CrossRef]
58. Sałabun, W.; Karczmarczyk, A.; Mejsner, P. Experimental study of color contrast influence in Internet advertisements with eye tracker usage. In *Neuroeconomic and Behavioral Aspects of Decision Making*; Springer Proceedings in Business and Economics; Nermend, K., Łatuszynska, M., Eds.; Springer: Cham, Switzerland, 2017; pp. 365–375, ISBN 9783319629377.
59. Bol, N.; Van Weert, J.C.M.; Loos, E.F.; Romano Bergstrom, J.C.; Bolle, S.; Smets, E.M.A. How are online health messages processed? Using eye tracking to predict recall of information in younger and older adults. *J. Health Commun.* **2016**, *21*, 387–396. [CrossRef]
60. Wang, T.C.; Tsai, C.L.; Tang, T.W. Exploring advertising effectiveness of tourist hotels' marketing images containing nature and performing arts: An eye-tracking analysis. *Sustainability* **2018**, *10*, 3038. [CrossRef]

applied sciences

MDPI

Article

The Design of Informational and Promotional Messages by Cooperative Banks and Their Perception Among Young Consumers—An Eye-Tracking Analysis Versus Conscious Identification Based on Empirical Research

Przemysław Pluskota [1,*], Kamila Słupińska [2], Agata Wawrzyniak [3] and Barbara Wąsikowska [3]

1 Institute of Spatial Management and Socio-Economic Geography, University of Szczecin, 71-101 Szczecin, Poland
2 Institute of Management, University of Szczecin, 71-004 Szczecin, Poland; kamila.slupinska@usz.edu.pl
3 Institute of Economics and Finance, University of Szczecin, 71-101 Szczecin, Poland; agata.wawrzyniak@usz.edu.pl (A.W.); barbara.wasikowska@usz.edu.pl (B.W.)
* Correspondence: przemyslaw.pluskota@usz.edu.pl

Featured Application: This research makes a significant contribution to understanding how young people perceive informational and promotional materials published on the websites of small financial institutions. This knowledge helps avoid mistakes in designing websites containing information and promotional content aimed at this group of respondents.

Abstract

The article explores how the design of informational and promotional messages from financial institutions influences their reception by young people. The study combined eye tracking, individual in-depth interviews (IDIs), and text mining analysis to examine both visual attention and participants' conscious reactions. The aim was to identify young users' preferences, determine factors influencing content perception, and assess the effectiveness of visual and audiovisual communication strategies. The main hypothesis proposed that minimalistic and visually attractive messages, enhanced with dynamic graphics, more effectively shape attitudes and elicit positive emotions. Specific aspects examined included the role of infographics, color schemes, message dynamics, and references to financial institutions in attracting attention and engagement. The results indicate that young people operate primarily in virtual space and express limited interest in traditional media such as television or print. They favor short, clear, and visually structured messages. Excessive textual content and lack of clarity provoked negative reactions and discouraged further engagement. Elements like infographics, colors, and logos were found to be strongly associated with brand recognition and memorability.

Keywords: eye tracking; young consumers; designing informational and promotional messages; cooperative banks; regional financial institutions

1. Introduction

Generational changes influence the banking services market, shaping its perception and usage. Membership in different age groups affects how banks communicate, what offerings they provide, and their inclination toward adopting new technological solutions [1]. One such modern development is the increasing use of eye tracking by researchers to study consumer behavior [2], including by banks. Initially used by psychologists to study basic mental functions, the technique has gained recognition in other fields. Eye tracking is a method of analyzing visual attention, particularly effective in capturing unconscious cognitive processes. It has gained popularity because of its ability to track cognitive actions in real time, without subjective judgments [3]. In consumer behavior research, eye tracking makes it possible to answer the question of how visual stimuli influence purchasing and transactional behavior. By recording participants' eye movements, instead of asking direct questions, it is possible to stimulate actual purchasing scenarios, including financial services. This makes it possible to identify visual elements, such as color, texture, appearance, and the form of the message, which influence consumer decisions at the point of purchase [4]. Eye movements are now widely used to study cognitive processes during reading, scene perception, and visual searches [5].

One of the key factors significantly impacting individuals' use of banking services is age, which is reflected in the diversified range of products offered by financial institutions. There are products available for school-aged children (up to 13 years), teenagers (aged 13–18), and young adults (18–26 years) [6]. This diversification arises from the differing financial needs and capabilities of consumers at various life stages. Age-specific services and products are widely applied in the financial sector, particularly in banking. Both age and other household characteristics determine financial needs and attitudes toward modern solutions, including those used in banking and strategies for reaching target groups.

The priority target groups of the National Financial Education Strategy for Poland, in relation to financial education programs, include

- Low-income households;
- Consumers at risk of over-indebtedness or already over-indebted;
- Working individuals, especially those nearing retirement age;
- Children, youth, parents, and teachers;
- Migrants and refugees (particularly from Ukraine) [7] (p. 33).

This categorization relates both to financial situations and age. Age influences a group's acceptance of technology [8], allowing us to distinguish several groups and sub-groups that vary in their acceptance of banking service methods [6].

1. Older generations:

- Baby Boomers—primarily driven by financial considerations and service quality, generally less receptive to mobile technology, and less inclined to change existing habits;
- Generation X—tends to find mobile banking complex, is cost-sensitive regarding mobile services, and prefers traditional payment methods such as cash;

2. Younger generations:

- Generation Y (Millennials)—favors quick payments and values a positive bank image, frequently uses online payment systems, and prioritizes convenience and transaction speed;
- Generation Z—expects personalized services accessible through mobile apps, often uses phone transfers and POS payments with QR codes, and favors innovative, user-friendly technology.

These generations also differ in their behavior in the economy. Their characteristics generate different needs in today's labor market, which is reflected in the way of communicating and creating information messages [9,10]. It is also said that Generation C has emerged—one that bridges the generational divide by being classified based on the digital environment they inhabit and use for communication [11,12]. They stand out with unique behaviors, values, attitudes, and a digital lifestyle. Members of this generation are constantly connected to the web, actively use technology, and navigate seamlessly within digital environments [13]. They are highly tech-savvy and active on social media, creating their own digital worlds through the content they post and view as attractive.

Research on the financial inclusion of Poles confirms that the percentage of people with bank accounts [14] is increasing year by year [15]. An age group analysis reveals the following account ownership rates:

- 18–24 years—93.3%;
- 25–39 years—98.6%;
- 40–54 years—95.0%;
- 55–64 years—91.0%;
- 65 years and older—80.2% [16].

This shows the special role that the younger generation plays as potential bank customers. Consequently, financial institutions are shifting their communication towards more intuitive, image-based messages and mobile applications. These changes are driven by the preferences and characteristics of various age groups as well as the younger generation's evolving view of reality. Different information and marketing messages are needed for each generation [8,17]. Generation Y, for instance, is more visually oriented than text-focused and is at or has already reached an age when major decisions—such as those related to careers or finances—must be made [18] (p. 44). Young people appreciate modern technological solutions, favoring mobile payments and apps. With easy access to the Internet and mobile devices, cashless payments are popular among this group because they are perceived as convenient and fast. Young users also value intuitive and user-friendly solutions and being aware of potential risks, place trust in modern technologies, including biometric security measures [19].

Modern technologies adopted by financial institutions, including banks—such as those based on digitization of activities, artificial intelligence—are shaping how banks operate, reach their target audiences, and sell their services.

Research indicates that for community banks, geography matters, as local conditions strongly influence the activities of small banks [20], and banks have a stronger impact on the development of their customers, becoming key players in existing SME financing mechanisms, important institutions in regional financial ecosystems [21]. Replacing bank branches with remote forms of access increases the bank's competitiveness. To maintain customer confidence and provide effective service in an increasingly digitized environment, cooperative banks are being forced to use digital technology [22]. A study of German cooperative and savings banks diagnosed the barriers to the digitalization of their operations. These were mainly related to technological and regulatory complexity and a lack of employee involvement in the digitization process [23]. The development of remote distribution channels in exchange for the expansion of traditional branches has a positive impact on the bank's operating costs, availability of products, including loans, positively influencing local economic development [24]. The implementation and development of e-banking by banks support their competitive advantage. A study of Spanish banks on customer adoption of e-banking indicates that positive customer attitudes, ease of use, and usability significantly influence customer attitudes [25]. Therefore, simplifying the message and the interfaces, supporting customers, pointing out the strengths of using this form of access, and encouraging customers to co-create services are tasks for banks, including cooperative banks, which, in order to compete with the larger, commercial ones, must constantly improve their services. To this end, a survey of the cooperative bank's message was conducted among young customers who rarely visit bank branches, using mainly the bank via the Internet or an app.

Taking into account also the fact that personal finances play an important role in the economy, as well as their growing role in everyday life and the impact on financial decisions, J. K. Solarz notes the transformation of personal finances and the role of generations in shaping the financial needs of a generation, referring to this phenomenon as financial socialization [26]. This process begins in early childhood and continues into early old age, involving parents, grandparents, friends, teachers, and social media in shaping an individual's financial identity.

Financial socialization involves not only acquiring theoretical knowledge but also developing attitudes that form habits and impulses toward certain behaviors. People who place excessive trust in their financial competence, despite a lack of knowledge, are vulnerable to the temptations of abuse and mis-selling [27]. According to J. K. Solarz, generational changes are shaping the modern banking system. For the younger generation, key factors in choosing a bank and using its services include the following:

- Technology: Young people value modern technologies such as mobile applications and contactless payments, which make managing finances easier.
- Convenience: They prefer fast, straightforward payment methods that enable instant transactions.
- Security: They focus on the safety of their data and transactions, opting for proven and secure payment methods.
- Financial education: Acquiring knowledge about financial management is essential for making informed decisions.
- Self-control of expenses: They appreciate tools that help monitor and control spending, enabling them to manage their budgets more effectively.

Young consumers tend to favor cashless payment methods, avoid complex banking procedures, and seek simple, intuitive solutions. As a result of banking activity diversification, products aimed at younger customers should meet specific criteria. The most important of these is to offer personalization, i.e., creating solutions tailored to the unique needs and preferences of the young generation [6]. For younger consumers, advertising messages should be adapted using language and interactive communication channels that appeal to them, primarily through social media.

The characteristics of the young generation not only shape their shopping behaviors but also require financial institutions to adopt a more tailored approach. In their view, marketing messages should be specially prepared. Young people expect a personalized offer, aligned to their specific needs in terms of dynamism, language, communication channels, and the form of the presented content.

These factors should be incorporated into the marketing strategies of banks that aim to reach this group of specific, but also prospective, customers.

2. Literature Review

2.1. Perception of Informational and Promotional Content Among Young People (Young Adults)

Young adults from ages 18 to 26, belonging to the so-called Generation Z (and partially the youngest Millennials), have grown up in a highly digitalized environment. Their way of processing informational and promotional content differs significantly from that of previous generations, both in terms of preferred communication formats and cognitive mechanisms related to attention and memory.

Characteristic features of message reception among young adults:

- Preference for visual content—young recipients respond more quickly to images, infographics, emojis, and short videos than to continuous text. Aesthetic quality and graphic consistency play a crucial role in capturing their attention.

- Demand for short, concise messages (snackable content)—brief, clear messages that are easy to scan (headings, bullet points, highlights) are favored. Long, unstructured texts are often ignored.
- Low tolerance for information overload—excessive data or unclear formatting leads to frustration and message rejection.
- Expectation of immediate value—messages that do not immediately convey a benefit for the recipient are quickly dismissed.
- Importance of authenticity and credibility—young people value transparency and quickly recognize artificial or manipulative tones. Brand consistency and values matter.
- Digital and mobile-first communication—most content is consumed on smartphones, mainly through social media platforms. Therefore, messages must be responsive and optimized for mobile formats.

Characteristics of the 18–26 age group:

- Digital natives—raised with the Internet and mobile technology, they are fluent in using digital tools.
- Multitasking and selective attention—they process multiple stimuli simultaneously but have a short attention span.
- High visual awareness—they are sensitive to visual quality and graphic details.
- Experience-driven—they expect personalized, interactive, and enjoyable user experiences.
- Social and emotional sensitivity—they respond positively to messages that appeal to emotions, values, and a sense of community.

Table 1 presents a summary of the literature concerning various aspects of Generation Z.

Table 1. A literature review on various aspects of Generation Z.

	Generation Z Characteristics
Age range definitions	Most dictionaries, like the *Merriam-Webster Online Dictionary*, the *Oxford Dictionary*, and *Encyclopedia Britannica*, define Generation Z as the generation of people born in the late 1990s and early 2000s [28–30]. Researchers and popular media use the mid-to-late 1990s as starting birth years and the early 2010s as ending birth years, with the generation loosely being defined as people born around 1997 to 2012 (Twenge 2017) [31] (Dimock 2019) [32]. Generation Z is the generation reaching adulthood in the second decade of the 21st century, and is perceived as being familiar with the Internet from a very young age [33,34].
Terminology	Generation Z follows the millennial generation (or Millennials), sometimes called Generation Y, which followed Generation X. Reaching the end of the standard Latin alphabet, Generation Z is succeeded by Generation Alpha, the first generation to be assigned a Greek letter. Generation Z (often shortened to Gen Z or Gen Zers) is also known as Zoomers. Other proposed names for the generation included: Homeland Generation, Homelanders, Centennials, iGeneration, Post-millennials, Pluralist Generation, Internet Generation, Net Gen, Digital Natives, and Neo-Digital Natives.

Table 1. *Cont.*

	Generation Z Characteristics
General description	Members of Gen Z are true digital natives; from a very young age, they have been exposed to the Internet, social networking sites, e-commerce, and mobile systems. Members of Generation Z have been born or raised around the time of the 9/11 terrorist attacks and have been undeniably shaped by the recession of 2007–2009 and the COVID-19 pandemic. The *Economist* has described Generation Z as being more educated, well-behaved, stressed, depressed, and exam-obsessed in comparison to previous generations [35]. While Generation Z proved to be less resilient than older cohorts, their fundamental values did not change, and they remained open to change, such as the transition towards hybrid school and remote work. They prefer to get their news from social media networks such as Instagram and TikTok and the video-sharing site YouTube rather than more traditional media, such as radio or television. Generation Z members spend much leisure time watching streaming, Internet scrolling, social networking, watching YouTube videos, and playing games on smartphones. Generation Z has gradually become the main force of market consumption [36]. As consumers, Generation Z members typically rely on the Internet to research their options and place orders. They are generally skeptical and avoid companies whose practices and values are contradictory [37,38]. Their purchases are heavily influenced by trends promoted by "influencers" on social media [39–41], as well as fear of missing out (FOMO) and peer pressure [42]. The need to be "trendy" is a key motivator. Due to their relatively high income, members of Generation Z have higher spending habits. According to new research, they rely on social media to make purchasing decisions, with health and beauty products being the most frequently purchased category on these platforms [43]. Generation Z shows greater concern for the environment and an eco-friendly lifestyle, which may influence consumer and social trends [44]. Generation Z expects flexibility, development opportunities, the use of new technologies, and transparency from employers.
Use of information and communications technology (ICT)	Generation Z is made up of individuals who grew up in a fully digitalized society. Information and communications technology (ICT), including the Internet and mobile devices, is a natural and integral part of their lives. They are characterized by high proficiency in using digital tools and a preference for online communication. They grew up in a world where the Internet and mobile technologies are commonplace, unlike older generations who had to learn them [45]. They are often referred to as digital natives [46]. Generation Z is accustomed to processing information quickly and is capable of multitasking. Generation Z lives in an era of widespread access to social media and consistently integrates it into their daily lives, using it not only to communicate with friends and family but also to interact with people they would otherwise never meet in person. The vast majority use social media and spend a lot of time online [47]. They prefer to communicate via social media, instant messaging, and other online platforms. They actively create online content, not just consume it like older generations [48]. They are typically proficient in using various devices and software, often without the need to read manuals. Generation Z is characterized by openness to new technologies and work methods. In summary, Generation Z is a generation deeply connected to ICT, shaping their way of thinking, communicating, and lifestyle. Understanding their needs and preferences is crucial for businesses, education, and society as a whole.

Table 1. *Cont.*

	Generation Z Characteristics
Marketing for Generation Z	Marketing to Generation Z requires understanding their unique characteristics and preferences. These individuals have been raised in the digital age and are fluent in technology and social media. Key principles of marketing to this generation include authenticity, speed, visuality, interactivity, social and ecological engagement, and personalization. Generation Z demands new forms of communication and marketing. Companies must adapt their communication strategies to their preferences, prioritizing authenticity, storytelling, and social media engagement. **Gen Zers have short attention spans, requiring brands to create concise, visual, and engaging content** [49,50]. Gen Z values authenticity and credibility in brands. They expect honesty and transparency in communication, as well as a commitment to important social and environmental issues. Gen Z is the generation with responsible behavior regarding environmental protection and a real promoter of Green Marketing culture [44]. Zoomers enjoy interactive content that allows for engagement and personalization, such as contests, quizzes, and the ability to create their own content. Gen Z is highly engaged with social media and follows influencers, making influencer marketing an effective way to reach this group. Gen Z spends most of its time on mobile devices, so content should be optimized for smartphones and tablets.
Website design	Website design for Generation Z should prioritize minimalism, speed, intuitiveness, and engaging visuals. Gen Z values authenticity, inclusivity, and seamless mobile experiences, favoring websites that are fast, easy to navigate, and visually appealing with interactive elements [51]. Gen Z prefers websites that are easy to scan and understand, with ample white space and clear typography. They prioritize simple and straightforward navigation that allows users to find information quickly [52]. Since Gen Z is highly mobile-centric, websites have to be fully responsive and optimized for mobile devices. Gen Z enjoys visually stimulating designs with strong colors and dynamic elements. Interactive elements, like animations, micro-interactions, and scroll-triggered effects, make websites more engaging [53]. Videos, GIFs, and other multimedia content can enhance the user experience and capture Gen Z's attention [54]. Gen Z expects websites to load quickly. Designers should incorporate visuals and content that reflect a range of backgrounds and perspectives to showcase inclusivity. Websites have to be accessible to users with disabilities and offer personalized experiences based on user preferences [55]. By focusing on these key elements, website designers can create engaging and effective online experiences that resonate with Gen Z users.

Source: own research.

2.2. Application of Eye Tracking in Economic Studies

Eye-tracking technology is an affordable and relatively simple tool widely used in experimental research on information perception and processing. Numerous studies in experimental economics and finance confirm that the way information is presented, received, and processed significantly influences decision-making processes. Eye tracking allows researchers to collect data that complement traditional observations of choice variables and has gained considerable popularity among scholars in behavioral economics and finance.

The use of eye-tracking technology in experimental economics and finance research is a relatively recent development, although similar topics have previously been addressed in literature related to marketing, organizational studies, and accounting. Table 2 presents examples of eye-tracking applications in the broadly understood field of economics.

Table 2. Examples of eye-tracking applications in economic research (including consumer behavior).

Authors	Title	Year
K. Balcombe, I. Fraser, L. Williams, E. McSorley [56]	Examining the relationship between visual attention and stated preferences: A discrete choice experiment using eye-tracking	2017
P. Ballco, T. De-Magistris, V. Caputo [57]	Consumer preferences for nutritional claims: An exploration of attention and choice based on an eye-tracking choice experiment	2019
P. Bansal, E. J. Kim, S. Ozdemir [58]	Discrete choice experiments with eye-tracking: How far we have come and ways forward	2024
M. Bartels, K. Tillack, C. T. Jordan Lin [59]	Communicating nutrition information at the point of purchase: An eye-tracking study of shoppers at two grocery stores in the United States	2018
N. Bault, P. Wydoodt, G. Coricelli [60]	Different attentional patterns for regret and disappointment: An eye-tracking study	2016
M. G. Ceravolo, V. Farina, L. Fattobene, L. Leonelli, and G. Raggetti [61]	Presentational format and financial consumers' behaviour: An eye-tracking study	2019
M. G. Ceravolo, V. Farina, L. Fattobene, L. Leonelli, G. Raggetti [62]	Anchoring effect in visual information processing during financial decisions: An eye-tracking study	2022
M. Chen, Y. Chen, R. Qi, J. H. W. Hsiao, W. W. T. Lam, Q. Liao [63]	Testing the effects of health-benefit, environmental-benefit and co-benefit priming for promoting sustainable food choice and their psychological mechanisms: A randomized controlled trial combined with eye tracking	2024
Y. Chen, J. Jermias, T. Panggabean [64]	The role of visual attention in the managerial judgment of balanced-scorecard performance evaluation: Insights from using an eye-tracking device	2016
G. Devetag, S. Di Guida, L. Polonio [65]	An eye-tracking study of feature-based choice in one-shot games	2016
J. Ding, J. G. Yu, P. Bansal [66]	Preferences for electric vehicles under uncertain charging prices: An eye-tracking study	2025
R. Duclos [67]	The psychology of investment behavior: (De)biasing financial decision making one graph at a time	2015
S. Fiedler et al. [68]	Social value orientation and information search in social dilemmas: An eye-tracking analysis	2013
B. Garcia-Carrion, S. Del Barrio-Garcia, F. Munoz-Leiva, L. Porcu [69]	Exploring destination positioning and message congruence in tourism management: An eye-tracking and fMRI study	2025
L.P. Hilbert, M.K. Noordewier, L. Seck et al. [70]	Financial scarcity and financial avoidance: an eye-tracking and behavioral experiment	2024
A. Hüsser, W. Wirth [71]	Do investors show an attentional bias toward past performance? An eye-tracking experiment on visual attention to mutual fund disclosures in simplified fund prospectuses	2014
J. Jin, A. Wang, C. Wang, Q. Ma [72]	How do consumers perceive and process online overall vs. individual text-based reviews? Behavioral and eye-tracking evidence	2023
N. Krucien, M. Ryan, F. Hermens [73]	Visual attention in multi-attributes choices: What can eye-tracking tell us?	2017
J. N. Lahey, D. Oxley [74]	The power of eye tracking in economics experiments	2016

Table 2. *Cont.*

Authors	Title	Year
K. E. Lewis, C. Grebitus, R. M. Nayga [75]	The impact of brand and attention on consumers' willingness to pay: Evidence from an eye tracking experiment	2016
N. Modi, J. Singh [76]	Understanding online consumer behavior at e-commerce portals using eye-gaze tracking	2023
J. Mou, D. Shin [77]	Effects of social popularity and time scarcity on online consumer behaviour regarding smart healthcare products: An eye-tracking approach	2018
C. Rasch, J. J. Louviere, T. Teichert [78]	Using facial EMG and eye tracking to study integral affect in discrete choice experiments	2015
Y. Ren, Q. Liu, G. Wu, J. P. Loy [79]	Consumer preferences for sugar-sweetened beverages: Evidence from online surveys and laboratory eye-tracking choice experiments	2025
E. Reutskaja, R. Nagel, C. F. Camerer, A. Rangel [80]	Search dynamics in consumer choice under time pressure: An eye-tracking study	2011
T. Ruppenthal [81]	Eye-tracking studies on sustainable food consumption: A systematic literature review	2023
M. Ryan, N. Krucien, F. Hermens [82]	The eyes have it: Using eye tracking to inform information processing strategies in multi-attributes choices	2018
T. Shavit et al. [83]	Using an eye tracker to examine behavioral biases in investment tasks: An experimental study	2010
M. Sielicka-Różyńska, E. Jerzyk, N. Gluza [84]	Consumer perception of packaging: An eye-tracking study of gluten-free cookies	2021
A. Simonetti, E. Bigne [85]	Customer journey-based smart technology of new brands: a self-reported and eye-tracking study	2025
L. P. Sirois, J. Bédard, P. Bera [86]	The informational value of key audit matters in the auditor's report: Evidence from an eye-tracking study	2018
F. M. Toma, C. O. Cepoi, M. N. Kubinschi et al. [87]	Gazing through the bubble: an experimental investigation into financial risk-taking using eye-tracking	2023
J. Vila, Y. Gomez [88]	Extracting business information from graphs: An eye tracking experiment	2016
Q. Wang, D. Ma, H. Chen, X. Ye, Q. Xu [89]	Effects of background complexity on consumer visual processing: An eye-tracking study	2020
E. Wästlund et al. [90]	Heuristics and resource depletion: Eye-tracking customers' in situ gaze behavior in the field	2015
S. Zhen, X. Xia, L. Huang, Y. Cao, H. Fu, Y. Ren [91]	Does risk preference matter to consumers' willingness to pay for functional food: Evidence from lab experiments using the eye-tracking technology	2024

Source: own research.

Making individual financial decisions—such as those related to spending, saving, mortgages, retirement planning, asset distribution, or insurance—typically requires choosing between alternatives under constraints like limited time, information, or cognitive capacity. Eye-tracking studies in this area have examined a range of economic theories and have been applied across fields including behavioral economics, finance, accounting, and organizational economics. The power of eye-tracking technology lies in the possibility of capturing the visual attention of individuals for different types of economic behavior.

3. Materials and Methods

Eye-tracking surveys are used to analyze the advertising messages of consumer goods manufacturers, advertising messages [92], and, less frequently, to study the transparency of messages and transaction services by financial institutions, particularly cooperative banks. It can therefore be concluded that this study, although a pilot study, is pioneering in this group of financial institutions.

Combining the eye-tracking method with in-depth face-to-face interviews (IDIs) adds value to the results, as it allows for the additional analysis of the conscious and unconscious aspects of the participants. In-depth interviews are often used alongside methods such as eye tracking to provide contextual understanding and deeper insights into the cognitive and emotional processes underlying the observed behavior [93].

Combining eye tracking with qualitative techniques, like in-depth interviews, allows researchers to move beyond 'where people look' to explore 'why they look,' thereby enriching interpretation and generating more holistic insights. Therefore, the authors conducted their research using a mixed method [94].

The research involved a group of 48 young people (volunteers), aged from 18 to 26. Among the 48 participants, 23 were high school students from two regional secondary schools, and the remaining 25 were university students from the largest university in the region. In total, the study included 27 men and 21 women. The study received approval from the appropriate bioethics committee. It was carried out in two stages. Initially, an eye-tracking session was conducted (with the usage of a Tobii Pro Fusion 60 Hz eye tracker to capture eye movements), followed by direct interviews with the participants.

The aim of the study was to identify the cognitive, behavioral, and emotional reactions of young users when exposed to the promotional and informational content of a financial institution represented by a cooperative bank. For this purpose, a mixed approach was used. Quantitative data from eye-tracker measurements (identification of gaze focus points, fixation time, visual regression) and qualitative data obtained in in-depth face-to-face interviews were used.

The emotional reception of the content was analyzed using text data mining, including frequency analysis, i.e., which words and phrases most often describe the analyzed messages and keywords indicating attitudes towards the messages. Special attention was paid to the comparison of emotional reactions in the context of audiovisual messages (commercials) and static content (information texts, advertising leaflets).

Analysis of the emotions that arose when viewing the promotional materials and reading the content was carried out in two ways:

1. Visual data analysis (eye tracking)—this allowed for precise tracking of the visual engagement of young audiences and also made it possible to identify moments when participants focused their attention on the most salient elements of the advertising materials. This made it possible to identify which parts of the ad triggered stronger emotional responses (e.g., curiosity, surprise, or joy).
2. Emotion analysis in texts (text mining) during IDIs—by using sentiment analysis and word frequency analysis, it was possible to identify positive and negative emotions associated with different elements of the communications. Young people reacted emotionally to different aspects of communication—infographics, colors, or images of famous people.

4. Results of Eye-Tracking Study and In-Depth Interviews (IDIs)

In the first part of the eye-tracking study, the participants' task was to watch a promotional spot for cooperative banks and answer the question "Is this advertisement for you?" When analyzing the recordings in Tobii Pro Lab, we examined whether the participants focused their attention on the logo shown in the spot and on other information about the cooperative banks—such as the website address or the company's mission (i.e., a declaration that clearly specifies which customer needs the banks intend to satisfy and which values they represent) (see Figure 1).

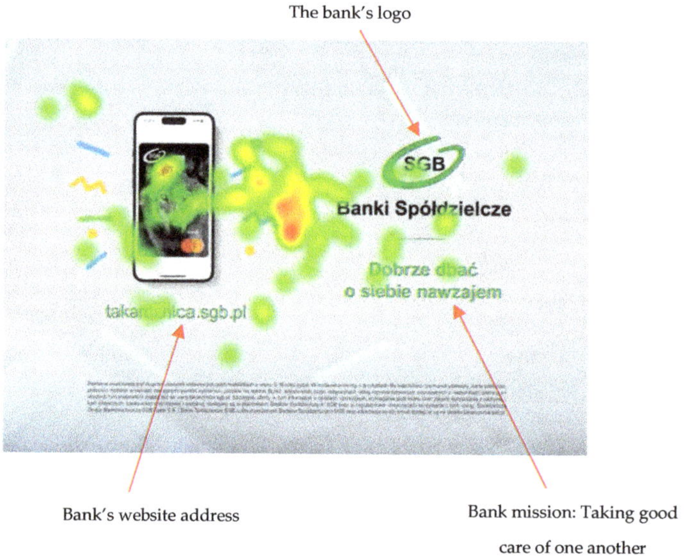

Figure 1. Example of heat map concentrated on the logo, website address, and mission of the cooperative banks. Source: own research.

The metrics calculated in the Tobii Pro Lab software (Full edition 25.7.1400 (x64)) are presented in Appendix A, with examples provided in Table A1. The four most relevant ones in this study are listed below:

1. **The time to the first fixation inside an AOI during an interval (TTFF):** This metric measures how long it takes for a participant's gaze to land on a specific area of interest (AOI) for the first time. A shorter TTFF indicates that the visual features of the element are more successful in drawing immediate attention. It is especially useful for determining which items attract the gaze most quickly.
2. **The total duration of the fixation inside an AOI during an interval:** This indicator reflects the total time and the mean position of a sequence of fixations within a defined AOI. It may consist of several fixations and short saccades between them. The sequence is considered finished once the gaze shifts outside the AOI.
3. **The number of fixations occurring in an AOI during an interval:** This measure shows how often an AOI is fixated upon. A higher number of fixations suggests that the element is perceived as more important or relevant. Closely related to fixation duration, it helps evaluate the overall level of attention devoted to different stimuli across tasks of varying length.
4. **The last AOI fixated during an interval:** This identifies the final area of interest (AOI) where the participant's gaze remained within a defined time window of the study.

As part of statements during IDIs, in the context of audiovisual messages, dynamism appeared to be a key factor that attracted the attention of young audiences. Although some respondents indicated that the ad was intended for a younger age group, it was noted that the dynamism of the message—involving rapid visual changes, intense colors, and movement—had a strong impact on their perception. Young people reacted positively to the commercial and enjoyed watching it (Figure 2).

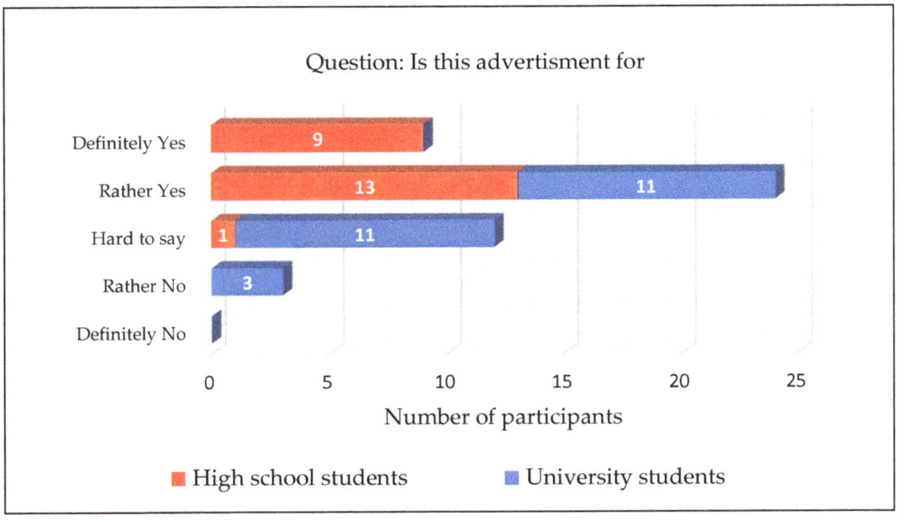

Figure 2. Distribution of respondents' answers to the question: "Is this advertisement for you?". Source: own research.

Interview participants were asked what aspects could increase their involvement, and the participants indicated that the scenes presented should be more focused on situations relevant to their generation. This may mean that active emotional and visual involvement during the viewing of such material is an important factor for them in determining the attractiveness of the message, with increased involvement being additionally achieved by making the message more personalized to their age group.

In interpreting the keywords of the reception of what they saw, it can be pointed out that young people prefer messages that offer dynamic visual scenarios. The slogan dynamism and dynamic were the main summaries of what they saw while expressing a positive reaction to it. Although the advertising message itself may have been dedicated to a younger age group, participants perceived the positive aspects of this dynamism, which promoted a better assimilation of the content, even if the advertising itself was not directly aimed at them.

The analysis of eye-tracking data related to the presented commercial spot, together with the responses provided by the participants during interviews (IDIs), made it possible to develop guidelines and recommendations for the cooperative bank. These could significantly improve the informational message so that it appeals to both groups of young people, namely, high school students and university students.

In the second part of the study, traditional advertising materials in the form of leaflets were presented. For the purposes of the eye-tracking experiment, they were converted into electronic form, and some of them were additionally discussed during the IDI. The participants' task was to review selected bank leaflets. The aim of this part of the study was to examine whether the information contained in the leaflet, as well as the way it was arranged, was understandable and clear to young people.

4.1. Importance of Readability in Written Messages

When analyzing and interpreting written messages (e.g., information materials and leaflets of the cooperative bank), the readability of the content played a key role in the reception. The study revealed that excessive text (see Figure 3) and a lack of clear structure generated negative emotions in the form of discouragement, which was noticeable in participants' textual responses and sentiment analysis. They often emphasized the need for the information presented to be straightforward, avoiding overburdening content that did not capture their attention. The colored dots in Figure 3 represent fixation points and scan paths, illustrating how individual high school students processed the information presented in the leaflet. A higher density of fixation points observed at the beginning of the leaflet suggests a more detailed and attentive reading of the initial sections. In contrast, the decreasing number of fixations in later parts implies that cognitive fatigue and information overload may have diminished participants' engagement with the subsequent content.

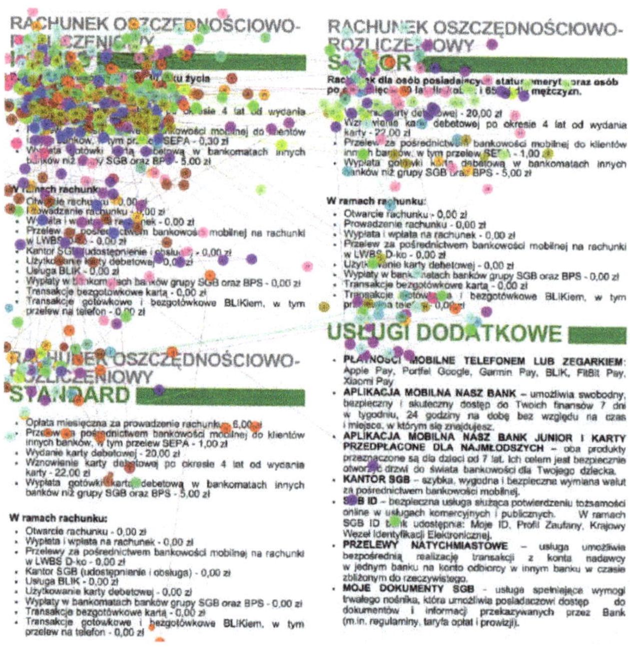

Figure 3. Bank leaflet with excessive text discouraging young people from reading. Source: own research.

Interview participants emphasized the need for the information presented to be straightforward, avoiding overburdening content that did not capture their attention. In terms of emotion, the readability of the content was considered an important factor influencing the level of satisfaction with communication. During individual in-depth interviews (IDIs), respondents indicated that too much text or complicated messages, full of industry jargon, made the message unattractive. Simplicity and clarity of information appeared to be an important factor in the ease of reception of the material. Young people preferred short, concise messages that were easy for them to understand and remember. From the analysis of the results, it appears that an excess of information resulted in a decrease in engagement and, therefore, a decreased effectiveness of communication. Participants expected information to be presented in a minimalist way, without unnecessary detail, and to be clearly organized in the form of headlines, lists, or infographics.

4.2. Importance of Visual Elements—Infographics, Logos, Colors

Very important elements that young people identify with financial institution brands are infographics, colors, and logos. The IDIs showed that they regard these elements as key brand identifiers that enable them to quickly recognize the institution. In this context, participants paid particular attention to the visual consistency of the materials, as well as the aesthetic appeal. Infographics were seen as a particularly useful tool that allowed for quick understanding of complex content, which is important in the case of offers regarding banking services.

The sample bank information leaflet presented in Figure 4 contains very short and easy-to-remember information for young people regarding payment card security, its expiration date, card number, PIN, and CVV code. Accompanied by an illustration of the card, it uses arrows to indicate the location of each element.

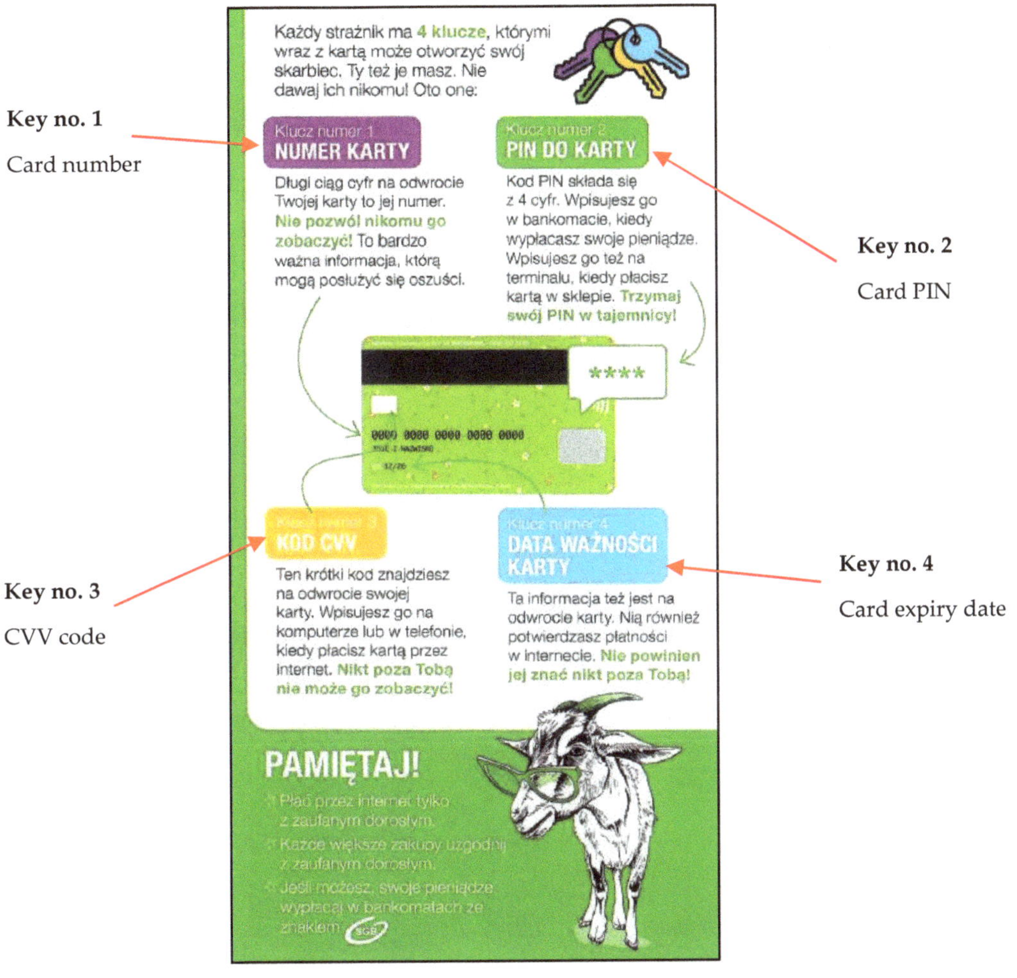

Figure 4. Example of a bank leaflet that graphically (using arrows) and with short notes indicates the location of individual elements of a payment card and how they should be protected. Source: own research.

Additionally, with the use of text highlighted in green font, guidelines were provided to help young people understand how to protect their payment card information, for example:

- Card number—"Do not let anyone see it";
- Card PIN—"Keep it secret";
- CVV code—"No one but you can see it";
- Card expiry date—"No one but you should know it".

4.3. *Role of Celebrity Image*

Another important element that significantly influenced young people's perception was the images of celebrities in the advertising material (see Figure 5).

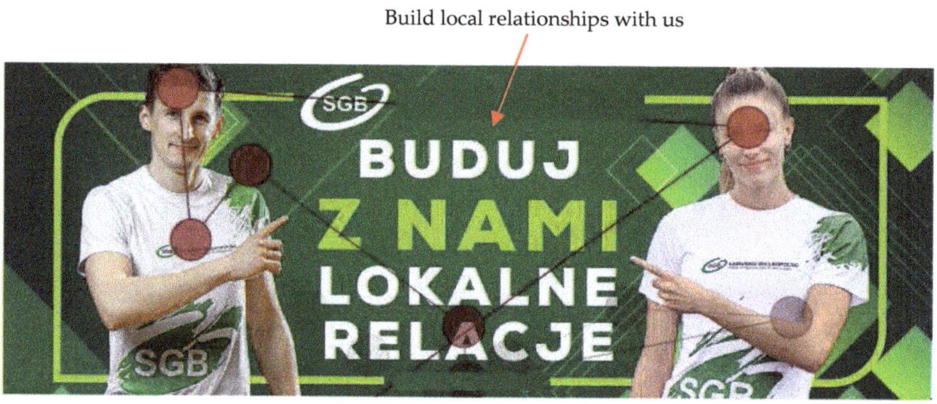

Figure 5. An example of an advertisement featuring images of athletes well-known to the local community. Source: own research.

Figure 5 presents an example of an advertisement placed on the homepage of a cooperative bank, featuring images of athletes well-known to the local community. The caption indicated by the athletes encourages potential customers of the bank to build local relationships together.

It was observed that participants reacted positively to the use of celebrity images, but their perception depended on the degree of familiarity with this person and the link to their personal context. As their statements during interviews indicated, if the celebrity was close to them, e.g., came from their region or was related to their interests, the effect of positive perception was clearly stronger. In contrast, in the case of less familiar characters, the effectiveness of advertising featuring them declined, indicating the importance of recognizability and emotional connection with the depicted character.

4.4. *Interaction with Online Advertising*

The results of the study indicated that young people are mainly involved in the virtual world, where they encounter a variety of advertising and information messages on a daily basis. They occasionally associate forms outside of it. They openly declare a lack of interest in traditional media such as television or the press. Nonetheless, as the study participants declared during the IDI, young people's attention is selective, and they treat most advertisements and messages as background or ignore them, especially when these are received on social media platforms. Young people rarely pay particular attention to adverts unless they are specifically tailored to their interests or offer content that they find interesting and engaging.

In conclusion, it should be stated that the key element is the simplicity of the message and the use of graphic forms that support the cognitive process based on identifying keywords. The study revealed that the way young people and students browse pages resembles social media habits—rapid, fragmented reading, even in contexts requiring concentration. This indicates the necessity of designing content with consideration for new models of perception, based on minimalism and visual clarity.

5. Discussion and Future Work

In the context of the study, an important element that emerged among young participants was the need for personalization of the offer. Young people indicated that advertising and informational messages must be tailored to their individual needs and preferences, as well as to their associations with the financial institution, in order to be more attractive and effective. In particular, it was noted that messages corresponding to their interests or needs were better received and evoked positive emotions, such as interest in the mentioned material.

A positive reception among the younger generation was observed in the case of leaflets with colorful characters, even though the association with the financial institution was weaker. These leaflets were more willingly picked up. The way in which the informational content was positioned was also important; when it required more analysis or effort, participants reacted with irritation. This was the case with a leaflet presenting a colorful circle with descriptions of the functionalities offered by the cooperative bank (see Figure 6). During interviews, participants emphasized that it should not be necessary to turn one's head in order to read the information.

Figure 6. A graphic element from a leaflet promoting various bank services ("We Have It All Here" Campaign). Source: own research.

Students also responded positively to colorful materials but paid more attention to the alignment of the leaflet with its content. Both groups were more likely to choose materials that clearly identified with the bank, which makes this form more preferable for them. The balance between graphic design and content must therefore be adjusted in favor of less text and more slogans or highlights.

Students were interested in accounts offering added value; therefore, one option worth considering is the introduction of deposit offers on favorable terms, which could persuade them to consider the bank. Moreover, since they tend to monitor their expenses more closely, the possibility of analyzing them through a transparent graphical cost planner would meet their stated need.

High school students, on the other hand, are much less inclined to change their current accounts but remain open to exploring new opportunities offered by the bank. Therefore, to build their openness to the brand and encourage them to consider it in the future, it would be worthwhile to introduce workshops or classes where they could simultaneously learn something new while engaging with the brand. This could also reshape their perception of the bank, which they often associate with people outside their age group.

The emotional reception of the content was analyzed using text data mining, including frequency analysis, i.e., which words and phrases most often describe the analyzed messages and keywords indicating attitudes towards the messages. Special attention was paid to the comparison of emotional reactions in the context of audiovisual messages (commercials) and static content (information texts, advertising leaflets).

The combination of the results of visual analysis (eye tracking) and interviews with text mining analysis shows that young people's emotions were strongly dependent on the dynamism of the visual message and the simplicity of the content. Infographics, colors and logos were perceived as positive elements that built an emotional bond with the brand, while an excess of text or complexity of the message resulted in negative emotional reactions, such as fatigue.

The study thus made it possible to accurately measure the emotional reactions of young people in the context of different forms of promotional materials, as well as to identify factors that determine the positive or negative reception of communication. The results point to the need to proactively tailor content to the emotional preferences of young consumers in order to increase their engagement and the effectiveness of communications.

The authenticity of the communication was also an important element. Young people preferred adverts that have a more "human touch", rather than just to general marketing standards. They associated advertisements in the form of storytelling in a positive way and were much more likely to indicate that they associated such advertisements that they had watched and enjoyed.

Future research (concerning the same set of audiovisual advertisement materials from cooperative banks) will involve a larger cohort of participants, focusing on young adults aged between 18 and 26 years. This will allow for the identification of potential differences in the behavior of young people representing high school students and college students. The study will incorporate a deeper eye-tracking analysis of fixation and saccade patterns.

Additionally, different approaches to designing informational and promotional messages across various banks will be compared by creating a controlled set of stimuli, which will enable drawing clear conclusions based on predefined hypotheses.

6. Limitations

The above section presents the results of a study aimed at analyzing the behavior of young customers seeking information in a regional bank. The study was of a pilot nature, and the sample size (N = 48) limits the possibility of generalizing the findings to the entire population. The participants were students and young adults from a single region of one country. The analysis focused mainly on the indicators available in the Tobii Pro Lab software (e.g., number and duration of fixations), while other indicators that could potentially deepen the interpretation of the results (such as regressions, etc.) will be the subject of future analyses planned by the authors as part of subsequent in-depth studies.

Despite these limitations, the research and conclusions presented in the article offer valuable insights for practitioners, particularly in the design of advertising leaflets tailored to young customers. These findings can be valuable for a wide range of businesses, including banks and cooperative banks, as well as companies that target young consumers.

Author Contributions: Conceptualization, P.P., K.S., A.W. and B.W.; methodology, P.P., K.S., A.W. and B.W.; software, A.W. and B.W.; formal analysis, P.P., K.S., A.W. and B.W.; investigation, P.P., K.S., A.W. and B.W.; resources, P.P., K.S., A.W. and B.W.; data curation, A.W. and B.W.; writing—original draft preparation, P.P., K.S., A.W. and B.W.; writing—review and editing, P.P., K.S., A.W. and B.W.; funding acquisition, P.P. and K.S. All authors have read and agreed to the published version of the manuscript.

Funding: This publication is co-financed by the Minister of Science under the "Regional Excellence Initiative".

Institutional Review Board Statement: Approval of the Bioethics Committee at the Regional Medical Chamber (OIL) in Szczecin for a period of 5 years—Resolution No. 14/2022 of 8 December 2022.

Informed Consent Statement: Informed consent was obtained from all subjects involved in the study.

Data Availability Statement: The data presented in this study are available upon request from the corresponding author due to the information they contain regarding a specific financial institution (e.g., its name and address).

Conflicts of Interest: The authors declare no conflicts of interest. The funders had no role in the design of the study; in the collection, analyses, or interpretation of data; in the writing of the manuscript; or in the decision to publish the results.

Appendix A

Data analysis in Tobii Pro Lab enabled the identification of the following eye-tracking indicators (see Table A1):

1. **Interval metrics:**

- The duration of an interval.
- The start time of an interval.

2. **Area of Interest (AOI) fixation metrics:**

- The total duration of the fixation inside an AOI during an interval.
- The average duration of the fixation inside an AOI during an interval.
- The duration of the shortest fixation inside an AOI during an interval.
- The duration of the longest fixation inside an AOI during an interval.
- The number of fixations occurring in an AOI during an interval.
- The time to the first fixation inside an AOI during an interval.
- The duration of the first fixation inside an AOI during an interval.
- The last AOI fixated during an interval.
- The AOI fixated at the end of an interval.
- The average pupil diameter of all fixation samples in an AOI in an interval. Calculated using the resulting pupil diameter after applying a pupil diameter filter.

3. **AOI glance metrics:**

- The total duration of the glances inside an AOI during an interval.
- The average duration of the glances inside an AOI during an interval.

- The duration of the shortest glances inside an AOI during an interval.
- The duration of the longest glances inside an AOI during an interval.
- The number of glances occurring in an AOI during an interval.
- The time to the first glance inside an AOI during an interval.
- The duration of the first glances inside an AOI during an interval.

4. **AOI saccade metrics:**

- The number of saccades occurring in an AOI during an interval.
- The duration until the start of the first saccade that ends in an AOI during an interval.
- The duration until the start of the first saccade that exists in an AOI during an interval.
- The peak velocity of the first saccade that ends in an AOI during an interval.
- The peak velocity of the first saccade that exists in an AOI during an interval.

5. **Event properties:**

- The type of event of the current row (fixation, saccade, blink).
- The validity of the event of the row, either whole or partial.
- Represents the order of the events in the current TOI interval.
- The start time, counted from the current TOI interval start.
- The start time, counted from the current TOI interval stop.
- The duration of the event.

6. **Fixation properties:**

- The name of the AOI that the current fixation hits.
- The proportion of the fixation that occurs within the AOI.
- The normalized horizontal coordinate of the fixation point.
- The normalized vertical coordinate of the fixation point.

7. **Saccade properties:**

- The angle of the straight line between the preceding fixation and succeeding fixation. This can only be applied to whole saccades.
- The average velocity across all samples belonging to the saccade, even outside the interval.
- The maximum velocity across all samples belonging to the saccade, even outside the interval.
- The amplitude for whole saccades.
- The name of the AOI, in which the current saccades started, if any, as determined by the preceding fixation position.
- The name of the AOI, in which the current saccades landed, if any, as determined by the succeeding fixation position.
- The normalized horizontal coordinate of the current saccade's start position, as determined by the preceding fixation.
- The normalized vertical coordinate of the current saccade's start position, as determined by the preceding fixation.
- The normalized horizontal coordinate of the current saccade's landing position, as determined by the succeeding fixation.
- The normalized vertical coordinate of the current saccade's landing position, as determined by the succeeding fixation.

Table A1. Part of the table with metrics for selected participants.

Participant	The Start Time of an Interval	The Average Duration of the Fixation Inside an AOI During an Interval	The Number of Fixations Occurring in an AOI During an Interval	The Time to the First Fixation Inside an AOI During an Interval	The Average Duration of the Fixations Inside an AOI During an Interval	The Number of Fixations Occurring in an AOI During an Interval	The Average Duration of the Visits Inside an AOI During an Interval	The Number of Visits Occurring in an AOI During an Interval	Time to First Visit Inside an AOI During an Interval	The Average Duration of the Glances Inside an AOI During an Interval	The Duration of the First Glances Inside an AOI During an Interval	The Number of Saccades Occurring in an AOI During an Interval	The Duration Until the Start of the First Saccade that Ends in an AOI During an Interval
k01	604699	183	2	29037	183	2	383	1	29037	417	417	1	29004
k01	604699	119	3	8578	158	2	425	1	8578	425	425	2	
k01	604699	190	5	19936	200	4	1151	1	19936	1151	1151	3	
k01	604699	133	2	27887	133	2	300	1	27887	350	350	1	27837
k01	604699	203	6	370	203	6	317	4	370	375	283	2	320
k01	604699	112	4	29504	128	3	630	1	29504	713	713	1	29421
k01	604699	158	2	4286	158	2	350	1	4286	367	367	1	
k01	604699	733	1	28271	733	1	733	1	28271	750	750	0	
k01	604699	340	5	21537	408	4	1798	1	21537	1865	1865	4	21470
k03	598355	124	3	8550	150	1	440	1	8550	440	440	2	
k03	598355	319	3	19924	275	2	1022	1	19924	1056	1056	2	19890
k03	598355	133	1	27907	133	1	133	1	27907	150	150	0	27891
k03	598355	261	3	29141	261	3	867	1	29141	900	900	2	29107
k03	598355	254	2	4090	333	1	542	1	4090	575	575	1	4056
k03	598355	250	3	28107	250	3	783	1	28107	850	850	2	28041
k03	598355	288	4	21724	288	4	600	2	21724	675	350	2	22290

Table A1. *Cont.*

Participant	The Start Time of an Interval	The Average Duration of the Fixation Inside an AOI During an Interval	The Number of Fixations Occurring in an AOI During an Interval	The Time to the First Fixation Inside an AOI During an Interval	The Average Duration of the Fixations Inside an AOI During an Interval	The Number of Fixations Occurring in an AOI During an Interval	The Average Duration of the Visits Inside an AOI During an Interval	The Number of Visits Occurring in an AOI During an Interval	Time to First Visit Inside an AOI During an Interval	The Average Duration of the Glances Inside an AOI During an Interval	The Duration of the First Glances Inside an AOI During an Interval	The Number of Saccades Occurring in an AOI During an Interval	The Duration Until the Start of the First Saccade that Ends in an AOI During an Interval
k04	594612	106	3	8561	142	2	434	1	8561	451	451	2	8544
k04	594612	522	2	19916		0	1094	1	19916	1094	1094	1	
k04	594612	1200	1	28111	1200	1	1200	1	28111	1233	1233	0	28078
k04	594612	517	1	4611	517	1	517	1	4611	567	567	0	4561
k04	594612	192	2	3944	192	2	433	1	3944	500	500	1	
k04	594612	739	1	29378		0	739	1	29378	806	806	0	29312
k04	594612	820	1	22497		0	820	1	22497	820	820	0	
k05	620186	202	2	8544		0	438	1	8544	438	438	1	
k05	620186	211	5	19909	278	3	561	2	19909	577	1049	3	21324
k05	620186	197	5	27953	242	4	534	2	27953	559	17	3	27991
k05	620186	156	3	790	156	3	242	2	790	292	383	1	723
k05	620186	198	3	29124	217	2	305	2	29124	355	183	1	29091
k05	620186	51	1	4590		0	51	1	4590	68	68	0	
k05	620186	183	1	29341	183	1	183	1	29341	250	250	0	29274
k05	620186	309	4	21974	256	3	1338	1	21974	1438	1438	2	21874

Source: own research.

References

1. Król, P. Dopasowanie ofert kont instytucji finansowych do zmieniających się trendów w zachowaniu pokolenia Z. *Przedsiębiorczość Eduk.* **2024**, *20*, 31–44. [CrossRef]
2. Andreou, G.; Gkantaki, M. An Eye-Tracking Study on Text Comprehension While Listening to Music: Preliminary Results. *Appl. Sci.* **2025**, *15*, 3939. [CrossRef]
3. Malheiros, B.A.; Spers, E.E.; Contreras Castillo, C.J.; Aroeira, C.N.; de Lima, L.M. The Role of Visual Attention and Quality Cues in Consumer Purchase Decisions for Fresh and Cooked Beef: An Eye-Tracking Study. *Appl. Sci.* **2025**, *15*, 7360. [CrossRef]
4. Chae, H.; Lee, Y.; Kim, S. The influence of visual attention on consumer choice in online grocery shopping: An eye-tracking study. *Sustainability* **2019**, *11*, 6347.
5. Rayner, K. Eye movements and attention in reading, scene perception, and visual search. *Q. J. Exp. Psychol.* **2009**, *62*, 1457–1506. [CrossRef]
6. Kaczmarek, M. Przynależność Do Kohorty Pokoleniowej Jako Determinanta Korzystania z BLIK-a. *Bank I Kredyt* **2023**, *54*, 221–238.
7. Krajowa Strategia Edukacji Finansowej Polityka Na Rzecz Rozwijania Kompetencji Finansowych, Przyjęto uchwałą Rady Ministrów nr 71 z dnia 25 czerwca 2024 r. w sprawie przyjęcia polityki publicznej „Krajowa Strategia Edukacji Finansowej. Polityka na rzecz rozwijania kompetencji finansowych". 2024. Available online: https://www.gov.pl/web/finanse/krajowa-strategia-edukacji-finansowej (accessed on 15 July 2025).
8. Camilleri, M.A. Higher Education Marketing: Opportunities and Challenges in the Digital Era. In *SSRN Scholarly Paper*; Social Science Research Network: Rochester, NY, USA, 2019. Available online: https://papers.ssrn.com/abstract=3484035 (accessed on 15 June 2025).
9. Bielińska-Dusza, E. The Motivation of Generations: What Drives Generation X, Y, Z? *J. Hum. Resour. Manag. Res.* **2022**, *2022*, 637177. [CrossRef]
10. Urbanovičová, P.; Mikulàšková, J.; Čambàl, M.; Edl, M. How Millennials Affect Corporate Culture. *Res. Pap. Fac. Mater. Sci. Technol. Slovak Univ. Technol.* **2019**, *27*, 76–83. [CrossRef]
11. Hardey, M. Generation C: Content, Creation, Connections and Choice. Marketing Communications. Available online: https://www.academia.edu/8472490/Generation_C_content_creation_connections_and_choice_MARKETING_COMMUNICATIONS (accessed on 29 April 2025).
12. Reports Insights State of the Media:, U.S. Digital Consumer Report, Nielsen. Available online: https://www.nielsen.com/insights/2012/us-digital-consumer-report/ (accessed on 29 April 2025).
13. Capitals Coalition. The Rise of Generation C & Implications for the World of 2020. 9 February 2011. Available online: https://capitalscoalition.org/the-rise-of-generation-c-implications-for-the-world-of-2020/ (accessed on 15 June 2025).
14. Stola, E. Ubankowienie społeczeństwa a rozwój sektora bankowego w Polsce. *J. Insur. Financ. Mark. Consum. Prot.* **2018**, *4*, 81–95.
15. Maison, D. Postawy Polaków Wobec Obrotu Bezgotówkowego–Raport z Badania 2021 i Analiza Porównawcza z Danymi z 2009, 2013 i 2016 Roku, NBP. 2021. Available online: https://nbp.pl/system-platniczy/dane-i-analizy/analizy-i-opracowania/zwyczaje-platniczepolakow (accessed on 1 June 2025).
16. Zwyczaje Płatnicze w Polsce w 2023 r., NBP. 2024. Available online: https://static.nbp.pl/systemy/platniczy/Zwyczaje-platnicze-w-Polsce-2023.pdf (accessed on 1 June 2025).
17. Valtere, L. Perspective on Generations: Their Impact on Higher Education Marketing. *Changes Soc. Bus. Environ.* **2013**, 213–219. Available online: https://www.ceeol.com/search/article-detail?id=531507 (accessed on 1 June 2025).
18. Pluskota, P.; Rzempała, J.; Skweres-Kuchta, M. *Lifelong Learning—A Short Version. Children's University–Regional View of Competence Development of Young People*; Wydawnictwo Naukowe Uniwersytetu Szczecińskiego: Szczecin, Poland, 2024. [CrossRef]
19. Banaszewska, M.; Iwańczuk-Kaliska, A.; Marchewka-Bartkowiak, K.; Warchlewska, A. *Zachowania Płatnicze Młodzieży W Polsce W Warunkach Izolacji Pandemicznej Raport z Projektu*; Uniwersytet Ekonomiczny w Poznaniu, Instytut Finansów: Poznań, Poland, 2022.
20. Aiello, F.; Bonanno, G. Multilevel empirics for small banks in local markets. *Pap. Reg. Sci.* **2018**, *97*, 1017–1038. [CrossRef]
21. Bezzon, B.; Labrouche, G.; Levy, R. Regional cooperative banks, ecosystems and small and medium-sized enterprise financing: The importance of cognitive, social and geographic proximities. *J. Small Bus. Enterp. Dev.* **2024**, *31*, 810–828. [CrossRef]
22. Colamartino, C.; Barone, M. Can cooperative banks maintain local engagement amidst digital transformation? Evidence from Italy. In *Digit Finance*; Springer: Berlin/Heidelberg, Germany, 2025. [CrossRef]
23. Ulrich-Diener, F.; Dvouletý, O.; Špaček, M. The future of banking: What are the actual barriers to bank digitalization? *Bus. Res. Q.* **2025**, *28*, 491–513. [CrossRef]
24. Bernini, C.; Brighi, P. Bank branches expansion, efficiency and local economic growth. *Reg. Stud.* **2017**, *52*, 1332–1345. [CrossRef]
25. Carranza, R.; Díaz, E.; Sánchez-Camacho, C.; Martín-Consuegra, D. e-Banking Adoption: An Opportunity for Customer Value Co-creation. *Front. Psychol.* **2021**, *11*, 621248. [CrossRef] [PubMed]
26. Solarz, J.K. Metapersonalizacja finansów osobistych w Polsce. In *Finanse Wobec Wyzwań Nowej Gospodarki*, 1st ed.; Marchewka-Bartkowiak, K., Waliszewski, K., Eds.; Wydawnictwo Uniwersytetu Ekonomicznego w Poznaniu: Poznań, Poland, 2024; pp. 115–125. [CrossRef]

27. Ameer, R.; Khan, R. Financial Socialization, Financial Literacy, and Financial Behavior of Adults in New Zealand. *J. Financ. Couns. Plan.* **2020**, *31*, 313–329. [CrossRef]
28. Definition of Generation, Z. Merriam-Webster. Available online: https://www.merriam-webster.com/dictionary/Generation%20Z (accessed on 17 July 2025).
29. Definition of Generation Z noun from the Oxford Advanced Learner's Dictionary. Oxford Learner's Dictionaries. Available online: https://www.oxfordlearnersdictionaries.com/us/definition/english/generation-z?q=generation+z (accessed on 17 July 2025).
30. Gen Z–Years, Age Range, Meaning, & Characteristics–Britannica. Available online: https://www.britannica.com/topic/Generation-Z (accessed on 17 July 2025).
31. Twenge, J.M. Have Smartphones Destroyed a Generation? Atlantic Monthly, September 2017. Available online: https://bob.cs.ucdavis.edu/assets/ecs188/smartphone.pdf (accessed on 25 August 2025).
32. Dimock, M. Defining generations: Where Millennials end and Generation Z begins. *Pew Res. Cent.* **2019**, *17*, 1–7.
33. Shapiro, J. *The New Childhood: Raising Kids to Thrive in a Connected World*; Little, Brown Spark: New York, NY, USA, 2018.
34. Lexico. *Generation Z*; Oxford University Press: Oxford, UK, 2017; Available online: https://web.archive.org/web/20190817140603/https://www.lexico.com/en/definition/generation_z (accessed on 17 July 2025).
35. Generation Z Is Stressed, Depressed and Exam-Obsessed. The Economist. 27 February 2019. Available online: https://www.economist.com/graphic-detail/2019/02/27/generation-z-is-stressed-depressed-and-exam-obsessed (accessed on 17 July 2025).
36. Grigoreva, E.A.; Garifova, L.F.; Polovkina, E.A. Consumer behavior in the information economy: Generation Z. *Int. J. Financ. Res.* **2021**, *12*, 164–171. [CrossRef]
37. Djafarova, E.; Foots, S. Exploring ethical consumption of generation Z: Theory of planned behaviour. *Young Consum.* **2022**, *23*, 413–431. [CrossRef]
38. Ribeiro, M.A.; Seyfi, S.; Elhoushy, S.; Woosnam, K.M.; Patwardhan, V. Determinants of generation Z pro-environmental travel behaviour: The moderating role of green consumption values. *J. Sustain. Tour.* **2025**, *33*, 1079–1099. [CrossRef]
39. Sun, Y.; Wang, R.; Cao, D.; Lee, R. Who are social media influencers for luxury fashion consumption of the Chinese Gen Z? Categorisation and empirical examination. *J. Fash. Mark. Manag. Int. J.* **2022**, *26*, 603–621. [CrossRef]
40. How to Sell to the Young. The Economist. 19 January 2023. Available online: https://archive.ph/20230122104027/https://www.economist.com/leaders/2023/01/19/how-to-sell-to-the-young (accessed on 18 July 2025).
41. How the Young Spend Their Money. The Economist. 16 January 2023. Available online: https://archive.ph/20230128084943/https://www.economist.com/business/2023/01/16/how-the-young-spend-their-money (accessed on 18 July 2025).
42. Kale, S. Out of Style: Will Gen Z Ever Give Up Its Dangerous Love of Fast Fashion? The Guardian. 6 October 2021. Available online: https://www.theguardian.com/fashion/2021/oct/06/out-of-style-will-gen-z-ever-give-up-its-dangerous-love-of-fast-fashion (accessed on 18 July 2025).
43. Roman, E. Gen Z Is Projected to Become the Largest, Wealthiest Generation, According to Report. Yahoo Finance. 15 July 2024. Available online: https://finance.yahoo.com/news/gen-z-is-projected-to-become-the-largest-wealthiest-generation-according-to-report-185059231.html?guccounter=1&guce_referrer=aHR0cHM6Ly9lbi53aWtpcGVkaWEub3JnLw&guce_referrer_sig=AQAAABddqOX_sE0_dvQSaZsFEJBILkcyFd4SsUOj3v8LX01RL0JrSiQY2Y6IM2DiudpmaxF5dXepEld5p7LmVdO5-2RmiPxaIRDFF0k2Xsbg4S29k2d25vuuME546XEb595npriwHSF0KBNazXIG0w0Uqc0m5nZkJBChfa3tfAls3cR0 (accessed on 16 July 2025).
44. Dragolea, L.L.; Butnaru, G.I.; Kot, S.; Zamfir, C.G.; Nuţă, A.C.; Nuţă, F.M.; Ştefănică, M. Determining factors in shaping the sustainable behavior of the generation Z consumer. *Front. Environ. Sci.* **2023**, *11*, 1096183. [CrossRef]
45. Turner, A. Generation Z: Technology And Social Interest. *J. Individ. Psychol.* **2015**, *71*, 103–113. [CrossRef]
46. Bourke, B. Connecting with Generation Z through social media. In *Preparing the Higher Education Space for Gen Z*; IGI Global: Hershey, PA, USA, 2019; pp. 124–147.
47. Elkatmış, M. Examination of social media usage habits of generation Z. *Front. Psychol.* **2024**, *15*, 1370823. [CrossRef]
48. Shetu, S.N. Do user-generated content and micro-celebrity posts encourage generation Z users to search online shopping behavior on social networking sites—The moderating role of sponsored ads. *Future Bus. J.* **2023**, *9*, 100. [CrossRef]
49. Insights Consumers. *Attention Spans*; Microsoft Canada, Spring: Toronto, ON, Canada, 2015.
50. Diz, M.R. Gen Z and Millennials in the Workplace: How Are Leaders Adapting to Their Short Attention Span and How Will They Keep Them from Leaving a Qualitative Study. 2021. Available online: https://www.ucas.com/connect/blogs/how-to-capture-the-eight-second-attention-span-of-gen-z (accessed on 18 July 2025).
51. Raghavan, R. Web Design for Generation Z: Understanding and Capturing The Digital Natives. acodez.in. 2024. Available online: https://acodez.in/web-design-for-generation-z/ (accessed on 18 July 2025).
52. Volfova, H.; Janků, A.; Landa, M.; Rausova, N. The Website through Gen Z's Eyes: Key Insights and Important Elements for Effective Online Pr Promotion of Universities. *Mark. Sci. Inspir.* **2024**, *19*, 14–23. [CrossRef]
53. Banu, A.R.; Tresha, T.K.; Chowdhury, S.S.; Srabonty, S.N. Online Library Interfaces: A User-Centered Study on Design and Functionality Preferences of Gen-Z Users. *J. Creat. Writ.* **2024**, *8*, 40–57. [CrossRef]

54. Arizal, N.; Nofrizal; Listihana, W.D.; Hadiyati. Gen Z Customer Loyalty in Online Shopping: An integrated model of trust, website design, and security. *J. Internet Commer.* **2024**, *23*, 121–143. [CrossRef]
55. Manca, M.; Palumbo, V.; Paternò, F.; Santoro, C. The transparency of automatic web accessibility evaluation tools: Design criteria, state of the art, and user perception. *ACM Trans. Access. Comput.* **2023**, *16*, 3. [CrossRef]
56. Balcombe, K.; Fraser, I.; Williams, L.; McSorley, E. Examining the relationship between visual attention and stated preferences: A discrete choice experiment using eye-tracking. *J. Econ. Behav. Organ.* **2017**, *144*, 238–257. [CrossRef]
57. Ballco, P.; De-Magistris, T.; Caputo, V. Consumer preferences for nutritional claims: An exploration of attention and choice based on an eye-tracking choice experiment. *Food Res. Int.* **2019**, *116*, 37–48. [CrossRef]
58. Bansal, P.; Kim, E.J.; Ozdemir, S. Discrete choice experiments with eye-tracking: How far we have come and ways forward. *J. Choice Model.* **2024**, *51*, 100478. [CrossRef]
59. Bartels, M.; Tillack, K.; Jordan Lin, C.T. Communicating nutrition information at the point of purchase: An eye--tracking study of shoppers at two grocery stores in the United States. *Int. J. Consum. Stud.* **2018**, *42*, 557–565. [CrossRef]
60. Bault, N.; Wydoodt, P.; Coricelli, G. Different attentional patterns for regret and disappointment: An eye-tracking study. *J. Behav. Decis. Mak.* **2016**, *29*, 194–205. [CrossRef] [PubMed]
61. Ceravolo, M.G.; Farina, V.; Fattobene, L.; Leonelli, L.; Raggetti, G. Presentational format and financial consumers' behaviour: An eye-tracking study. *Int. J. Bank Mark.* **2019**, *37*, 821–837. [CrossRef]
62. Ceravolo, M.G.; Farina, V.; Fattobene, L.; Leonelli, L.; Raggetti, G. Anchoring effect in visual information processing during financial decisions: An eye-tracking study. *J. Neurosci. Psychol. Econ.* **2022**, *15*, 19–30. [CrossRef]
63. Chen, M.; Chen, Y.; Qi, R.; Hsiao, J.H.W.; Lam, W.W.T.; Liao, Q. Testing the effects of health-benefit, environmental-benefit and co-benefit priming for promoting sustainable food choice and their psychological mechanisms: A randomized controlled trial combined with eye tracking. *J. Environ. Psychol.* **2024**, *100*, 102485. [CrossRef]
64. Chen, Y.; Jermias, J.; Panggabean, T. The Role of Visual Attention in the Managerial Judgment of Balanced-Scorecard Performance Evaluation: Insights from Using an Eye-Tracking Device. *J. Account. Res.* **2016**, *54*, 113–145. [CrossRef]
65. Devetag, G.; Di Guida, S.; Polonio, L. An eye-tracking study of feature-based choice in one-shot games. *Exp. Econ.* **2016**, *19*, 177–201. [CrossRef]
66. Ding, J.; Yu, J.G.; Bansal, P. Preferences for electric vehicles under uncertain charging prices: An eye-tracking study. *Transp. Res. Part D: Transp. Environ.* **2025**, *140*, 104608. [CrossRef]
67. Duclos, R. The Psychology of Investment Behavior: (De)biasing Financial Decision making One Graph at a Time. *J. Consum. Psychol.* **2015**, *25*, 317–325. [CrossRef]
68. Fiedler, S.; Glöckner, A.; Nicklisch, A.; Dickert, S. Social Value Orientation and Information Search in Social Dilemmas: An Eye-Tracking Analysis. *Organ. Behav. Hum. Decis. Process.* **2013**, *120*, 272–284. [CrossRef]
69. Garcia-Carrion, B.; Del Barrio-Garcia, S.; Munoz-Leiva, F.; Porcu, L. Exploring destination positioning and message congruence in tourism management: An eye-tracking and fMRI study. *Tour. Manag.* **2025**, *108*, 105111. [CrossRef]
70. Hilbert, L.P.; Noordewier, M.K.; Seck, L.; van Dijk, W.W. Financial scarcity and financial avoidance: An eye-tracking and behavioral experiment. *Psychol. Res.* **2024**, *88*, 2211–2220. [CrossRef]
71. Hüsser, A.; Wirth, W. Do Investors Show an Attentional Bias toward Past Performance? An Eye-Tracking Experiment on Visual Attention to Mutual Fund Disclosures in Simplified Fund Prospectuses. *J. Financ. Serv. Mark.* **2014**, *19*, 169–185. [CrossRef]
72. Jin, J.; Wang, A.; Wang, C.; Ma, Q. How do consumers perceive and process online overall vs. individual text-based reviews? Behavioral and eye-tracking evidence. *Inf. Manag.* **2023**, *60*, 103795. [CrossRef]
73. Krucien, N.; Ryan, M.; Hermens, F. Visual attention in multi-attributes choices: What can eye-tracking tell us? *J. Econ. Behav. Organ.* **2017**, *135*, 251–267. [CrossRef]
74. Lahey, J.N.; Oxley, D. The Power of Eye Tracking in Economics Experiments. *Am. Econ. Rev.* **2016**, *106*, 309–313. [CrossRef]
75. Lewis, K.E.; Grebitus, C.; Nayga, R.M., Jr. The impact of brand and attention on consumers' willingness to pay: Evidence from an eye tracking experiment. *Can. J. Agric. Econ. Rev. Can. D'agroeconomie* **2016**, *64*, 753–777. [CrossRef]
76. Modi, N.; Singh, J. Understanding online consumer behavior at e-commerce portals using eye-gaze tracking. *Int. J. Hum. –Comput. Interact.* **2023**, *39*, 721–742. [CrossRef]
77. Mou, J.; Shin, D. Effects of social popularity and time scarcity on online consumer behaviour regarding smart healthcare products: An eye-tracking approach. *Comput. Hum. Behav.* **2018**, *78*, 74–89. [CrossRef]
78. Rasch, C.; Louviere, J.J.; Teichert, T. Using facial EMG and eye tracking to study integral affect in discrete choice experiments. *J. Choice Model.* **2015**, *14*, 32–47. [CrossRef]
79. Ren, Y.; Liu, Q.; Wu, G.; Loy, J.P. Consumer preferences for sugar-sweetened beverages: Evidence from online surveys and laboratory eye-tracking choice experiments. *Food Policy* **2025**, *130*, 102791. [CrossRef]
80. Reutskaja, E.; Nagel, R.; Camerer, C.F.; Rangel, A. Search dynamics in consumer choice under time pressure: An eye-tracking study. *Am. Econ. Rev.* **2011**, *101*, 900–926. [CrossRef]

81. Ruppenthal, T. Eye-tracking studies on sustainable food consumption: A systematic literature review. *Sustainability* **2023**, *15*, 16434. [CrossRef]
82. Ryan, M.; Krucien, N.; Hermens, F. The eyes have it: Using eye tracking to inform information processing strategies in multi-attributes choices. *Health Econ.* **2018**, *27*, 709–721. [CrossRef]
83. Shavit, T.; Giorgetta, C.; Shani, Y.; Ferlazzo, F. Using an Eye Tracker to Examine Behavioral Biases in Investment Tasks: An Experimental Study. *J. Behav. Financ.* **2010**, *11*, 185–194. [CrossRef]
84. Sielicka-Różyńska, M.; Jerzyk, E.; Gluza, N. Consumer perception of packaging: An eye—tracking study of gluten—free cookies. *Int. J. Consum. Stud.* **2021**, *45*, 14–27. [CrossRef]
85. Simonetti, A.; Bigne, E. Customer journey-based smart technology of new brands: A self-reported and eye-tracking study. *J. Consum. Mark.* **2025**. [CrossRef]
86. Sirois, L.P.; Bédard, J.; Bera, P. The informational value of key audit matters in the auditor's report: Evidence from an eye-tracking study. *Account. Horiz.* **2018**, *32*, 141–162. [CrossRef]
87. Toma, F.M.; Cepoi, C.O.; Kubinschi, M.N.; Miyakoshi, M. Gazing through the bubble: An experimental investigation into financial risk-taking using eye-tracking. *Financ. Innov.* **2023**, *9*, 28. [CrossRef]
88. Vila, J.; Gomez, Y. Extracting business information from graphs: An eye tracking experiment. *J. Bus. Res.* **2016**, *69*, 1741–1746. [CrossRef]
89. Wang, Q.; Ma, D.; Chen, H.; Ye, X.; Xu, Q. Effects of background complexity on consumer visual processing: An eye-tracking study. *J. Bus. Res.* **2020**, *111*, 270–280. [CrossRef]
90. Wästlund, E.; Otterbring, T.; Gustafsson, A.; Shams, P. Heuristics and Resource Depletion: Eye-Tracking Customers' in Situ Gaze Behavior in the Field. *J. Bus. Res.* **2015**, *68*, 95–101. [CrossRef]
91. Zhen, S.; Xia, X.; Huang, L.; Cao, Y.; Fu, H.; Ren, Y. Does risk preference matter to consumers' willingness to pay for functional food: Evidence from lab experiments using the eye-tracking technology. *Food Qual. Prefer.* **2024**, *119*, 105197. [CrossRef]
92. Mancini, M.; Cherubino, P.; Martinez, A.; Vozzi, A.; Menicocci, S.; Ferrara, S.; Giorgi, A.; Aricò, P.; Trettel, A.; Babiloni, F. What Is behind In-Stream Advertising on YouTube? A Remote Neuromarketing Study employing Eye-Tracking and Facial Coding techniques. *Brain Sci.* **2023**, *13*, 1481. [CrossRef]
93. Bojko, A. *Eye Tracking the User Experience: A Practical Guide to Research*; Rosenfeld Media: New York, NY, USA, 2013.
94. Duchowski, A.T. *Eye Tracking Methodology: Theory and Practice*; Springer International Publishing AG: Cham, Switzerland, 2017.

applied sciences MDPI

Article

The Role of Visual Attention and Quality Cues in Consumer Purchase Decisions for Fresh and Cooked Beef: An Eye-Tracking Study

Bruna Alves Malheiros [1], Eduardo Eugênio Spers [2,*], Carmen Josefina Contreras Castillo [3], Carolina Naves Aroeira [4] and Lilian Maluf de Lima [5]

1 Symrise, Descalvado 13690-000, SP, Brazil; brumalheiros@outlook.com
2 Department Economics, Administration and Sociology (LES), Luiz de Queiroz College of Agriculture (ESALQ), University of São Paulo (USP), Piracicaba 13418-900, SP, Brazil
3 Department of Food Science and Technology, Luiz de Queiroz College of Agriculture (ESALQ), University of São Paulo (USP), Piracicaba 13418-900, SP, Brazil; ccastill@usp.br
4 AgroProx, Avenida Limeira, Areião, Piracicaba 13414-018, SP, Brazil; cnaroeira@gmail.com
5 Pecege Institute, Piracicaba 13414-157, SP, Brazil; lmlima05@hotmail.com
* Correspondence: edespers@usp.br

Abstract

This study analyzes Brazilian consumer behavior regarding quality and visual cues in fresh red meat and cooked beef. Using eye tracking to collect visual attention metrics and psychological scales to assess food behavior, the research examines how visual attention to beef attributes impacts product choice. A discrete choice method combined nine hypothetical products with varied attributes. Results showed that consumers display different visual behaviors toward cues, influencing their probability of choosing a product. For fresh beef, color was the most significant factor, especially bright red and brown hues. Color influenced both the time to first fixation and total fixation time, while breed also affected total fixation time. Dark-red color and unspecified breed information increased the purchase probability, while Nellore breed and brown color decreased it. Total fixation numbers were significantly impacted by color, breed, marbling, and price. In cooked beef, tenderness, price, and flavor were key visual cues. Tenderness and flavor influenced the time to first fixation, whereas price and flavor impacted the number of fixations. This research contributes to understanding visual cues and attention in food choices, suggesting strategies for enhancing beef labeling and communication to better inform Brazilian consumers.

Keywords: eye tracking; visual attention; visual cues; meat; consumer behavior

1. Introduction

Eye tracking is a powerful and precise method for analyzing visual attention, particularly effective in capturing non-conscious cognitive processes. Traditionally used by psychologists to investigate fundamental mental functions, this technique has gained recognition in other fields such as education and consumer behavior research due to its capacity to track real-time cognitive actions without relying on self-reported data [1].

In consumer research, eye tracking offers an objective means to assess how visual stimuli influence purchasing behavior. By recording participants' eye movements, rather than asking direct questions, researchers can closely simulate real-world shopping scenarios. This allows for the identification of visual elements—such as color, texture, and packaging—that drive consumer decisions at the point of purchase [2].

Beef products provide a compelling case for studying the role of visual attention, as consumers often rely on visual cues to assess quality. For fresh beef, color is the primary indicator, while packaging plays a similar role for cooked beef. Appearance is not only the first sensory attribute to be noticed but is also a key driver of appetite and product acceptability [3]. Prior research has shown that visual features significantly influence expectations related to tenderness, flavor, and overall eating quality—even before tasting the product [4,5].

Despite these insights, consumers often face challenges in accurately predicting beef quality based on visual inspection alone, creating a major challenge for the meat industry. Understanding how consumers interpret visual cues—especially in the context of both raw and cooked forms of the same product—is essential for aligning marketing strategies and product development with consumer expectations.

The existing literature confirms the relevance of visual attention in food choice [6,7], yet few studies have specifically focused on Brazilian consumers. Recent findings indicate that Brazilian consumers associate bright red coloration in fresh beef with higher quality, while a tough texture in cooked beef negatively affects choice and willingness to pay [8].

Therefore, this study has two main objectives: (1) to evaluate how visual attention to different visual and quality cues affects the probability of purchase for both fresh and cooked beef among Brazilian consumers; and (2) to introduce a novel statistical approach to analyze eye tracking data in the context of food choice.

By incorporating both raw and prepared forms of beef, this research captures a more complete understanding of how visual cues shape consumer decision-making throughout the purchasing and consumption cycle.

2. Literature Review

2.1. Visual Attention for Food

Vision is the most used sense during the process of buying food [9,10] because consumers need to visualize information before they can use it in the selection process [11]. Visual attention broadly represents the cognitive process of understanding specific visual information and demarcating these into areas of interest (AOI) [12–15]. Visual attention is understood to be the first stage of a series of processes that influence consumer choice [16].

The attention given by consumers to a stimulus is selective and can be presented in the following two ways: bottom-up, in which attention is directed to stimuli, and top-down, which is related to the objectives of the observer [17,18]. The bottom-up situation occurs when a stimulus stands out in the visual field, that is, attention is guided by visual characteristics, such as the size, color, and shape of a stimulus or internal characteristics of stimuli that automatically capture the attention of individuals [19–22]. Researchers have explored how the appearance of cereal bars influences the consumer purchasing decisions across different brands and flavors [13]. They found that brand and variety were key determinants in driving consumer attention, suggesting that greater attention given to a brand increases the likelihood of its selection. Another study evaluated the effect of visual attention on product labels on chocolates, yoghurt, and tomato with the organic symbol in different sizes [23]. They found that the use of bottom-up effects to improve the design and visibility of this information increased attention capture as well as the probability of product choice.

Conversely, top-down attention occurs when consumers focus on stimuli that are in line with their goals, expectations, and even their emotions [19,24–26]. Research has observed that consumers perceive the healthiness of mayonnaise, bread, and yoghurt, as well as their willingness to buy, and found that consumers directed visual attention to specific information such as ingredients and nutritional information [27]. The effect of beef fat content on visual attention and purchase choice has also been examined, with researchers finding that consumers pay more attention and more often choose beef with a lower fat content due to the perception of healthiness [28]. Specific objectives can focus on visual attention and food choice [29].

When observing a stimulus, the consumer's eye movements can be measured with eye-tracking equipment, which allows accurate recording of periods and fixation sites. The literature suggests that consumers process AOI information with fixations on specific attributes. Alternations between these fixations, e.g., color and size, are called saccades. Fixations are the periods of attention given to an area of interest, characterized by periods of discrete immobility of the eye while the information is visualized and processed before looking for another AOI. Although it is understood that the view is suppressed during the saccades, they are useful to reveal elements of the consumer's visual search, such as trade-offs and the order seen in the AOI [18,30,31]. In the present study, the AOIs contemplate information on quality and visual cues based on the color, marbling, breed, flavor, tenderness, and aroma of beef.

The main parameters of visual attention contemplated in the present study include the following: (1) time of the first fixation—reflects the time it takes the consumer to process information of an AOI for the first time. In general, information that is more complex for the consumer takes longer to be visualized, so the shorter the time for the first fixation, the greater the ability of information to capture visual attention. (2) The total fixation time—indicates the total time of fixing on an AOI throughout the stimulus. Long fixations may indicate that the consumer has difficulty in interpreting the information presented. (3) The number of fixations in an AOI (number of looks)—refers to the total number of fixations on a specific attribute or area of interest. The more a consumer fixes on an attribute of interest, the more importance the consumer places on this, e.g., color [1,32,33].

The information contained in the food itself is an important communication channel that directs how visual attention and consumer choices will be affected [10,16]. It is important to understand which elements can effectively capture the consumer's attention to explore their behavior and increase the probability of choosing a food [9,13,27]. Thus, eye tracking is used to measure which stimulus information has the shortest time for the first fixation, the path taken by the eye, total fixation time, the percentage of areas fixated on, and the quantity of fixations [16,18,34,35].

Studies focusing on the visual attention of Brazilian consumers have previously been conducted [16,36,37]; however, they focused on other products than beef. As such, the current study aims to contribute novel data that can assist private and government institutions to explore which quality and visual cues are more responsible for the visual attention of consumers. Using this information, it is possible to design strategies to increase the probability of choice and gain insights for future research.

2.2. *Quality and Visual Cues for Foods*

The quality of a product may have an impact on the food choice process, but only if it is perceived [38]. The visualization of quality cues is necessary for understanding and used at the time of choosing a food product since the first sensory contact of the consumer with food is through the eyes [39]. Thus, before food is effectively chosen by the consumer, visual attention is directed to visual cues, which provide information about the characteristics of food through words, images, symbols, or the appearance of the product [39,40].

Consumers make purchasing decisions based on intrinsic and extrinsic visual cues that they correlate to specific attributes based on prior experience and credibility [41]. Intrinsic visual cues are inherent to the food itself and cannot be manipulated without affecting its physical properties—these include appearance, taste, and texture. Extrinsic visual cues are related to the product and include labelling, packaging, marketing information or situational contexts [42]. Regarding beef, the main intrinsic visual cues reported are cut, color, fat content, flavor, tenderness, and taste [41–48]. Extrinsic visual cues represent all the other product characteristics, such as brand, price, packaging, origin, and information regarding animal production [5,43,49].

Exposure to visual cues can promote variations in consumption, acceptance, visual attention, and the creation of expectations by the consumer [39], which can have a halo effect, modifying the perception of flavor and acceptability of food. In this sense, visual properties are critical for holding visual attention and swaying the choosing of products sold by their own appearance, and not by the packaging design. According to previous studies, foods marketed by their own appearance stand out by visual cues of color, texture, and the perception of flavor stimulated by color [40]. Regarding beef, studies have shown that the acceptance of beef with information highlighting a dark red color is higher compared to those with a bright red color (cherry red) [50,51]. Additionally, it was found that intrinsic and extrinsic visual cues significantly influence consumer emotions, hedonic perception, and even food purchase intention, and the provision of additional information, such as names or information on the packaging, can increase the choice and probability of purchase [52].

From a Brazilian perspective, the main visual cues of beef presented to consumers at points of sale can be divided into two groups: (a) regulated: mandatory labelling information by governmental agencies, such as beef cut name, manufacturer name, expiration date, etc.; and (b) strategic: explored in the form of claims or seals by private initiatives.

When buying fresh beef in Brazil, consumers can purchase from one of three situations: meat available in butcher shops and packed according to the consumer's request, chilled meat cuts, and vacuum-packed meat [53]. When purchasing from a butchery or as chilled meat, the consumer has little extrinsic information available about the product—there is a focus mainly on price and visual cues such as appearance. On the other hand, vacuum-packed meats are labelled with regulated visual cues for packaged products, namely: weight, origin, federal inspection seal, product preservation, brand, batch identification, manufacture date, expiration date, and registration number [54]; and strategic visual cues, such as breed, premium quality seals, livestock system, and age, among others.

When purchasing cooked beef (in a restaurant), two main options are available: a la carte dishes prepared with specific beef cuts or all-you-can-eat buffets with various types of meats and beef cuts [55]. In both cases, consumers are provided with specific information on price, regulated visual cues, and strategic visual cues, such as animal breed. Usually, additional information can be obtained from the waiter or even from the appearance and aroma of the meat.

Visual cues of food are determinants in the food choice process; however, the way foods are presented to consumers is not standardized. Previous studies on visual attention attempted to use such data to estimate the amount of consumption, eating behavior, reward process, stimulation of healthy consumption and reducing obesity, differences in dietary perceptions between generations, and effect of expanding areas of exposure to increase sales [39,56–62]. However, few studies have focused on the contribution of visual attention to the probability of choice [52], and studies focusing on beef are scarce.

2.3. The Relationship Between Eye Tracking and Packaging in Consumer Research

Visual attention plays a crucial role in consumer decision-making processes, particularly in environments characterized by high information density such as retail shelves. Eye-tracking technology has emerged as a powerful methodological tool to objectively measure visual attention and explore how packaging elements influence consumer behavior.

Research has consistently shown that packaging design elements—such as color, typography, shape, and visual claims—capture attention and can influence product evaluation and purchase intention [63]. Eye-tracking metrics such as time to first fixation, fixation duration, and number of fixations on areas of interest (AOIs) are commonly used to evaluate which package attributes attract consumer attention and for how long.

It has been demonstrated that visual saliency and branding elements on packaging significantly affect gaze patterns and ultimately influence brand choice. Findings indicate that consumers tend to fixate first and longer on visually salient packages, and that such visual prominence increases the likelihood of product selection [64]. Likewise, strong design features on packages have been shown to not only increase attention but also enhance memory and preference, particularly when integrated into coherent branding strategies [65].

Moreover, eye-tracking studies have revealed that consumers do not process all available visual information on packaging but tend to rely on heuristic cues, often guided by top-down expectations [66]. This behavior underscores the importance of strategically placing key product attributes—such as sustainability claims or nutritional information—in visually dominant positions on the package.

In the context of sustainable packaging, eye-tracking techniques have been used to investigate how eco-labels on food packaging affect consumer attention and choices. Findings suggest that eco-labels can influence purchasing decisions, but only when they are visually accessible and aligned with consumer values [67].

Overall, the integration of eye tracking into packaging research enables a nuanced understanding of the visual mechanisms that drive product perception and choice. These insights are invaluable for marketers and designers aiming to create more effective, attention-capturing packaging in competitive marketplaces.

3. Materials and Methods

3.1. Research Data

The study was conducted in the city of Piracicaba (SP, Brazil) in December 2020, with 23 participants. The participants were randomly selected from the researchers' contact network and through recommendations from other participants. All individuals who participated in the research reported consuming and buying beef at least once a week. The study was conducted in person, with a scheduled time to avoid large groups, in an open space and respecting the World Health Organization (WHO) and local authorities' protocols to prevent COVID-19 infection. Data collection addressed all the requirements of Resolution 510/16 of the National Commission for Research Ethics [68] regarding the protection of research participants and was approved by the Ethics Committee on Research with Human Beings of ESALQ/USP, the researchers' institution. In the experimental space, the participants were given eye-tracking equipment and directed towards an online questionnaire that aimed to collect socioeconomic and consumer behavior data.

3.2. Visual Attention—Eye Tracking

The study presented consumers with nine hypothetical products to choose from according to the protocol [8]. Two purchase situations were used for fresh and cooked beef, as well as a combination of information on visual cues including color, marbling, breed, tenderness, flavor, and aroma, in order to evaluate whether visual attention interferes with the probability of consumer choice.

Two different visual stimuli were created to represent hypothetical products to consumers, a vacuum-packed striploin representing fresh beef and a dish with striploin steak representing cooked beef. The attributes were inserted directly into the images with the same spacing, font size, and image, mitigating the risks of interference in the readability of the information. Each product was compared by occupying the right and left positions to avoid favoring visual field position, and a total of 18 visual stimuli were presented in each purchasing situation. After visualizing the stimulus, the consumer had to indicate which product was their choice ("product on the right", "product on the left", or "none of the products").

All visual stimuli were presented to participants on Microsoft® PowerPoint® with standard slide (4:3) maximized and used eye-tracking equipment (Tobii T120, Edge Eye, Stockholm, Sweden) with an integrated monitor, on a 17″ screen with a resolution of 1280 × 935 pixels and full-screen adjustment. Before starting the presentation of the images to consumers, the equipment was calibrated for each participant, considering equipment distance and height, to ensure better capture of the field of vision. Calibration was performed according to the manufacturer's recommendations.

The images were displayed in sequence and the transition time was 10 s. Amongst the transitions, there was a black background image, with the aim of "cleaning the image" and preparing the consumer for comparison with the following products. Before being presented to consumers, the areas of interest for capturing visual attention in the images were marked around the quality cues studied, with the geometric shape of identical areas for all attributes, according to Figure 1.

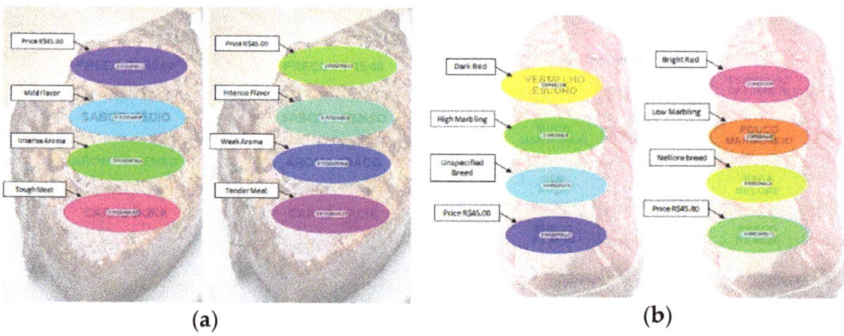

Figure 1. Areas of interest demarcated: (**a**) cooked beef; (**b**) fresh beef.

Eye tracking metrics used were the tree [18]:

- Time to First Fixation: This refers to the elapsed time until the user's gaze first fixates on a specific area of interest. A shorter time indicates that the visual characteristics of the area are more effective in capturing attention. This metric is particularly valuable when analyzing attention to a specific target [1].
- Fixation Duration: This metric captures the total time and average location of a sequence of fixations within a designated area of interest. It may include multiple fixations and short saccades between them. The fixation sequence is considered complete when the gaze moves outside the area of interest [1].
- Number of Fixations (Visits) on an Area of Interest: A higher number of fixations suggests that the area holds greater significance for the viewer. This measure is closely related to fixation duration and helps assess the total number of fixations in tasks of varying lengths. The number of times an element is fixated on reflects its perceived importance [69,70].

A total of 414 observations were collected from the 23 participating consumers (23 participants × 18 questions = 414 observations) for each situation. These data were initially used as variables of a Logit regression model, making it possible to find their effects on the choice of the visualized product.

3.3. Data Analysis—Logit Model

Logistic or logit regression is used as a nonlinear model designed specifically for binary dependent variables [8]. This regression allows the estimation of the probability associated with the occurrence of a given event, considering a set of explanatory variables [71]; this type of analysis also aims to understand the differences between groups and to estimate the probability that an individual belongs to a particular category, based on a set of explanatory variables [72].

Two logit models were estimated, one for fresh meat and another for cooked meat. The explanatory variables of each model, as well as the details of the logistic equations, are presented in Appendix A. The data related to the variables of the two models (fresh and cooked beef) were considered significant when $p < 0.05$. The analyzes of the models were based on the evaluation of the marginal effect provided by the coefficients of the explanatory variables on the dependent variable. Other models were estimated considering the presence of other variables, using a stepwise method and with the aid of the lowest value for the Akaike Information Criterion (AIC), and the most robust model (best statistical adjustment) was defined for each stage. The lower the AIC, the better the fit of the model [73,74]. A way to evaluate the adjustment of logistic regression to data is to use Pseudo R-Squared. They play a similar role to R^2 in linear regression, by representing the proportion of variation that occurred in the dependent variable that is explained by the model [74]. These values can also be used to compare the performance of competing models, between two equally valid logistic equations. For this, the one with the highest Pseudo-R^2 should be chosen.

4. Results

4.1. Participant's Profile

The participant's profile was obtained by data collection via online surveys (Table 1).

Table 1. Socioeconomic profile of participants.

Information	Description	%	N
Frequency of weekly consumption	Once a week	13%	3
	2 or 3 times a week	61%	14
	4 or 7 times a week	26%	6
Frequency of month purchase	Once a month	9%	2
	2 or 3 times a month	52%	12
	4 or 5 times a month	22%	5
	6 or more times a month	17%	4
Gender	Male	57%	13
	Female	43%	10
Age range (years)	18–30	74%	17
	31–40	26%	6
Education	High School completed	13%	3
	University completed	57%	13
	Postgraduate	30%	7
Average income (month)	USD 200.00–600.00	61%	14
	USD 601.00–1000.00	26%	6
	USD 1001.00–3000.00	13%	3
Location	São Paulo	96%	22
	Others	4%	1

The most prominent participants included men aged 24 to 30 years, with a college degree, a minimum frequency of purchase of 2 times a month, weekly consumption of beef

2 to 3 times per week, living in the state of São Paulo, Brazil, and a salary ranging from USD 200.00–600.00, which corresponded to approximately 22% of the research participants.

4.2. Visual Attention—Fresh Beef

The results obtained from the Logit model for fresh and cooked beef are presented in Tables 2 and 3, respectively. When comparing the predicted values obtained with the estimated models for fresh beef to the observed values of the sample, an accuracy index of 95.19% was obtained, which is considered a good fit [75].

Table 2. Estimates of Logit model coefficients for fresh beef and respective values of marginal effects.

Variables	Coefficients	Standard Deviation	p-Value	MgE #
Intercepto	17.359	4.2607	0.0000 ***	-
FRESH.PRICE	0.0952	0.0370	0.0101 **	0.0318
CBROWNCO	−3.4719	1.1442	0.0024 ***	−0.1116
FRESH.FBRIGHTCO	−0.5713	0.2405	0.0175 **	−0.0050
FRESH.FBROWNCO	−0.7133	0.3780	0.0592 *	−0.0062
FRESH.TDARKCO	3.7014	1.9604	0.0590 *	0.0325
FRESH.TNELLOREB	−3.3003	1.6080	0.0401 **	−0.0289
FRESH.TWITHOUTB	13.2524	4.4611	0.0029 ***	0.1163
FRESH.TBROWNCO	−2.8524	1.3094	0.0293 **	−0.0250
FRESH.VDARKCO	−1.7553	0.6874	0.0106 **	−0.0154
FRESH.VSMALLMAR	−2.4407	0.7195	0.0007 ***	−0.0214
FRESH.VMODERATEMAR	−0.9295	0.41418	0.0248 **	−0.0081
FRESH.VABUNDANTMAR	−1.6202	0.4790	0.0007 ***	−0.0142
FRESH.VPRICE380	−1.1141	0.5596	0.0465 **	−0.0097
FRESH.VWITHOUTB	−5.4550	1.5222	0.0004 ***	−0.0479
FRESHQCONS_C	5.3391	2.0705	0.0099 ***	0.0265
FRESHQPURCHASE_D	−7.3969	2.0462	0.0004 ***	−0.7977
MED_COMP	−1.4765	0.5006	0.0031 ***	−0.0129
n				229
AIC				92.213
Mc Fadden (Pseudo-R2)				0.74
Cox–Snell (Pseudo-R2)				0.51
NagelKerke (Pseudo-R2)				0.83

MgE: marginal effect; * p value < 0.10; ** p value < 0.05; *** p value < 0.01.

Appendix B describes the guidelines that were given to consumers regarding breed, color, and marbling. For the item color and marbling, photos were provided with guidelines.

First Fixation (Table 2)

Considering the first fixation, two visual cues related to color were significant: FRE.FBRIGHTCO and FRE.FBROWNCO. In this study, visual cues for fresh beef including color, breed, marbling, and price were considered; however, the information about color was the only one found to be significant for first fixation. Consumers who first fixated on color information had a 0.50% lower purchase probability after looking at visual cue information on FRE.FBRIGHTCO and 0.62% lower probability after looking at visual cue information on FRE.FBROWNCO.

Table 3. Estimates of Logit model coefficients for cooked beef and respective values of marginal effects.

Variables	Coefficients	Standard Deviation	*p*-Value	MgE #
Intercepto	1.14007	1.62154	0.4820 ND	-
COOK.PRICE	0.16837	0.06606	0.0108 **	0.0050
COO.TOUGHMC	−2.21684	0.86441	0.0103 **	−0.1660
COO.FTOUGH	−0.88719	0.20065	0.0000 ***	−0.0266
COO.FWEAKFLAVOUR	−0.36660	0.15630	0.0190 **	−0.0110
COO.VPRICE580	−0.99133	0.32722	0.0024 ***	−0.0297
COO.VPRICE900	−2.28778	0.65485	0.0005 ***	−0.0686
COO.VINTENSEFLAVOUR	−0.61987	0.29383	0.03489 **	−0.0186
COO.FREQCONSU_B	2.09243	0.84142	0.0128 **	0.0867
COO.FREQPURCHASE_D	−1.74068	0.87018	0.0454 **	−0.0944
N				226
AIC				85.109
Mc Fadden (Pseudo-R2)				0.68
Cox–Snell (Pseudo-R2)				0.46
NagelKerke (Pseudo-R2)				0.77

MgE: marginal effect; ** *p* value < 0.05; *** *p* value < 0.01; ND: not significant (*p* value > 0.10).

Total Fixation (Table 2)

Considering the total time (seconds) that participants fixated on specific attributes, visual cues including color and breed were significant: FRE.TDARKCO, FRE.TBROWNCO, FRE.TNELLOREB, and FRE.TWITHOUTB. The information on visual cues FRE.TDARKCO and FRE.TWITHOUTB increased the probability of choice by ~3% and ~12%, respectively. However, FRE.TNELLOREB and FRE.TBROWNCO reduced the probability of choice by ~3% and ~2.5%, respectively. Fixation on color, before the presentation of the products in the equipment, was studied; consumers were asked to visualize a color pattern, guided by the authors.

Total Number of Fixations (Table 2)

Considering the number of times that each participant fixated on an AOI (visit), color, marbling, price, and breed were significant, namely FRE.VDARKCO, FRE.VSMALLMAR, FRE.VMODERATEMAR, FRE.VABUNDANTMAR, FRE.VPRICE380, and FRE.VWITHOUTB.

4.3. Visual Attention—Cooked Beef

Table 3 presents the results of the final Logit model estimated for cooked beef. Regarding the predicted values obtained with the estimated model for cooked beef, when comparing them with the observed values of the sample, an accuracy index of ~96.7% concordance was established, which indicates a good fit of the data [75].

First Fixation (Table 3)

Considering the first fixation, tenderness and flavor were significant: COO.FTOUGH and COO.FWEAKFLAVOUR. Consumers who first fixated on tenderness information reduced the purchase probability by 2.6% by looking at visual cue information COO.FTOUGH and by 1.1% after looking at visual cue information COO.FWEAKFLAVOUR.

Total Fixation (Table 3)

The best-fitting model did not present any significant variables related to "total fixation" metrics for explaining the choice of cooked meat.

Total Number of Fixations (Table 3)

Considering the number of times that individuals visited an AOI, the information regarding price and taste was significant: COO.VPRICE580, COO.VPRICE900, and COO.VINTENSEFLAVOUR. Only two levels of price and one level of flavor had a high frequency of visits.

Regarding COO.VPRICE580 and COO.VPRICE900, each time the consumer revisited the AOI, there was a reduction in the probability of choice by 2.97% and 6.86%, respectively. To a lesser extent, on the COO.VINTENSEFLAVOUR: each time the consumer revisited the AOI, there was a reduction in the probability of choice by 1.86%.

5. Discussion

Knowledge about the socio-economic profile of participants is very important in research such as this. For this, firstly, the possible participants were indicated by people that knew what the focus would be. Then, these participants were randomly chosen, resulting in a total of 23 individuals. The indications were to better select a priori the education of the participants, having, at least, graduated from high school. Almost half belong to the same gender, with 57% being male and 43% being female (Table 1). Differences were found among gender about specific preferences and attitudes towards meat color; for instance, female participants more commonly held skeptical attitudes towards red meat and meat-eating [76].

Parameters such as frequency of consumption and average income are very important to measure, mainly given that beef is an expensive component of a meal. The consumer is motivated by both internal and external influences, at the time of purchase, and the external part is determined by social class, culture, family, and the groups that they are included [77]. In this study, beef was typically consumed by the participants two or three times a week (61%), and at least once a week (Table 1).

Visual attention was measured both in fresh and cooked beef. First, observations regarding the fresh meat will be discussed, as the results of both were analyzed separately. For fresh beef, as mentioned above, the accuracy index was very representative with 95.19%, comparing predicted values with the estimated models.

The results of the first fixation, the aspect that the participants first pay attention to, were significant, regarding fresh beef, for the FRE.FBRIGHTCO and FRE.FBROWNCO, stimulus of beef with bright red color and brown color, respectively. This was the parameter that they perceived more compared to breed, marbling, and price. This was corroborated with other studies [39,40]. These results indicate that, for the sample of consumers studied, the color of fresh beef is the visual cue that attracts their visual attention of during the first fixation.

Changes between the bright red color (oxymyoglobin) and brown color (metmyoglobin) of fresh beef can be considered as "extreme" changes [78]. Brazilian consumers typically find dark-colored beef (deoxymyoglobin) at the point of sale [79]. It is suggested that the first fixation is concentrated on the color. This can be due to the difference from what is typically expected, which can cause distrust [9]. Alternatively, a possible justification for why other quality cues were not significant when first fixating may be the lack of understanding of the consumer about those parameters or not being able to determine other attributes for the beef choice [13].

Regarding the time to first fixation, which is the first stimulus displayed, the color was the main parameter that was associated with it, as discussed above. However, regarding the total fixation, which is the duration of all fixations within an AOI, or within all AOIs belonging to an AOI group [80], it was indicated that the respondents focused significantly on color and breed attributes (FRE.TDARKCO, FRE.TBROWNCO, FRE.TNELLOREB, and FRE.TWITHOUTB), regarding fresh beef. The dark red color and any breed information increased the purchase chance while the information regarding brown color and Nellore breed reduced it.

Regarding FRE.TDARKCO, the increased purchase probability as a consequence of visual attention increase may be motivated by the consumer's familiarity with that information [10,81]. For FRE.TBROWNCO, the reduction in the probability of choice may be due to doubts in the interpretation of this information, because it is known that the color of beef can change according to the type of packaging—typically, vacuum-packed beef is brown in color [82,83]. In addition, microbiological, physicochemical, or temperature changes may also give beef a brown color [84,85]. Thus, it can be posited that the consumer's preference may be influenced by their knowledge about color variations in beef, but also due to their concerns with quality, which can be correct.

The increase in the probability of choice when viewing the information on FRE.TWITHOUTB (total time of fixation in the AOI related to the stimulus of beef without any breed information) may indicate that the consumers participating understand that less information about beef is more attractive and facilitates their choice; thus, visual cues of the beef itself can be a sufficient source of information, as reported by [40,86]. There was a reduction in the probability of sale caused by the information on FRE.TNELLOREB, which may be due to previous negative experiences associated with beef from the Nellore breed, lack of knowledge about the contribution of the breed to the quality of beef, or even the habit to consume beef without the identification of the breed [41,87,88].

It is also possible that some consumers have preferences for specific breeds, such as Angus (Bos taurus taurus) than Nellore (Bos taurus indicus), as it has been proven by many authors [89,90] that they present more tender meat. Meat from Bos indicus crossbred cattle is less tender than meat from Bos taurus cattle, regardless of marbling score [91]. This can lead to decreased acceptance of Nellore meat by consumers, especially since such research is more feasible today. Moreover, advertising for the Brazilian Angus, the main Bos taurus breed in Brazil, uses tenderness as a key selling point.

The correlation found between total fixation time and probability of choice is similar to the findings of a study in which an increase in visual attention was found to increase the probability of choice [10]. The total number of fixations, which is the number of times that each participant fixated on an AOI, with respect to fresh beef and considering the parameters color, marbling, price, and breed, were FRE.VDARKCO, FRE.VSMALLMAR, FRE.VMODERATEMAR, FRE.VABUNDANTMAR, FRE.VPRICE380, and FRE.VWITHOUTB.

Analyzing the information on FRE.VDARKCO, it was observed that the probability of the consumer choosing beef with this information was reduced by 1.54% each time the individual revisited fixating on the dark red color. The return of the consumer to the same visual cue indicates importance, doubt, or difficulty in understanding the attribute [18,32]. Additionally, the understanding of information can be facilitated or hindered according to the level of consumer involvement with beef [2,88]. Participants' level of involvement with fresh and cooked beef was not measured.

The information on the visual cues FRE.VSMALLMAR, FRE.VMODERATEMAR, and FRE.VABUNDANTMAR reduced consumer choice by 2%, 0.8%, and 1.4% at each return fixation, respectively. This study indicated that all levels of marbling were important, and they reduced the probability of purchase. These results differ partially from the findings of [92], who identified a higher preference for beef with lower fat content, and in this case, lower marbling lead to more rejection. These authors found that beef variation in fat content can produce different patterns in terms of visual attention and product choice; the present study found the same pattern for all levels of marbling. The differences found may be due to different cultural habits since the study [87] was conducted with Portuguese consumers. It is important to consider the level of knowledge about the contribution of marbling to flavor, succulence, and tenderness of beef [92]. Consumers who make health-based food choices are more likely to seek information that helps them assess the healthiness of products but may encounter difficulties related to knowledge and quality cues of the product [93,94].

Regarding FRE.VPRICE380, the high number of visits indicates the importance of this visual cue for consumers, which may represent a reduction in the probability of choice due to mistrust and difficulty in identifying trade-offs. In Brazil, one kg of beef typically costs more than USD 3.80. When consumers find cheaper prices, they tend to distrust the establishment or the characteristics of the product; this is reflected in the reduction in the probability of purchase by 0.97% each time the individual fixates on the price. These results agree with the study [8], who also identified a reduction in the probability of choice for beef with lower prices. Additionally, consumers make decisions based on a variety of quality and visual cues. For beef, a range of products are available, and consumers need to evaluate both regulated and strategic visual cues before deciding which confers greater complexity [27].

Regarding breed, some consumers may understand that the presence of information about the breed can help them distinguish between beef quality. Regarding the information on FRE.VWITHOUTB, the high number of visits to this attribute may indicate distrust or lack of understanding of differences between breeds, as there was a reduction in the probability of choice by 4.79% for each fixation, or that the breed has no influence on their decision. This can be attributed to the fact that in Brazil, the purchase of the meat often occurs directly in the butcher shops that offer a variety of cuts fresh or cooked, without the label, and some information about breed, gender, age, and others.

The dark color of beef, without breed information, showed different behaviors between the metrics. When evaluated for total fixation, they increased the probability of choice; however, when evaluated considering the number of fixations (visits), the probability of choice was reduced.

For both the total fixation time and the number of visits, greater visual attention reduced the probability of choice for all visual cue information. One hypothesis may be that beef consumers are more sensitive to information about visual cues, leading them to reject beef, than they are to information that is of interest. It is likely that the lack of familiarity with the information presented to consumers may justify difficulty in interpreting the information, resulting in a preference for not buying beef with unavailable information.

Considering the first fixation about cooked beef, only tenderness and flavor were significant: COO.FTOUGH and COO.FWEAKFLAVOUR. The participants contemplated visual cues including flavor, aroma, tenderness, and price. The most significant for first fixation was 'weak flavor' and 'tough beef'. First fixation on these attributes suggests that consumers looking for cooked beef primarily considered information on taste and tenderness. Those with weak flavor and tough beef reduced the purchase probability by 2.6% and 1%, respectively, as this is negative information. The findings agree with research [8] that evaluated the same visual cues through online research in cooked beef and identified which ones could increase the probability of choice. These authors found that there is a significant reduction in the probability of choice when the product has weak flavor and a tough consistency. It included negative information on the product description; this is not likely to occur in a real purchase situation, but we found it important to highlight that visual cues that refer to undesirable characteristics should be the target of future studies.

For the metric referring to time of the first fixation, the visual cues that were most significant were tough meat and weak flavor, again demonstrating that, contrary to the behaviors identified for fresh beef, consumers of cooked beef are more interested in avoiding meat beef with negative information about attributes. These results are in accordance with other research [8]. A possible explanation for this may be the repeated experiences of beef consumption with these negative characteristics.

Regarding cooked beef, the number of times that individuals visited an AOI was analyzed, and it was found to be significant for price and taste: COO.VPRICE580, COO.VPRICE900, and COO.VINTENSEFLAVOUR. Prices of USD 5.80 (COO.VPRICE580) and USD 9.00 (COO.VPRICE900) reduced the probability of choice by 2.97% and 6.86%, respectively. The participants were more resistant to choosing more expensive beef. A decrease in the probability of purchase for beef with a price of USD 9.00/kg was also found [8].

Considering COO.VINTENSEFLAVOUR, the high number of returns to this information may indicate consumer preference. It was observed that the probability of the consumer choosing beef reduces by 1.86% each time they revisit the AOI related to intense flavor.

Unlike fresh beef, metrics referring to the number of return fixations of cooked beef were composed of information about attributes that generate difficulty in understanding or differentiation (i.e., breed). For cooked beef, the total time of fixation on AOIs was not significant, so they were not included in the model. A possible justification for this may be the greater understanding and familiarity of consumers with visual qualities of cooked beef compared to fresh beef [9,10]. Interestingly, cooked beef is usually presented with little information available, so the increase in information in the study may have caused confusion and difficulty, because they are desirable quality cues, but not normally reported.

6. Conclusions

Visual attention is the process by which consumers use visual attributes to determine if they are interested in a product. These may be attributes that consumers know, and these can either increase or reduce the probability of choice. Attributes regarding color and breed maintained more attention among the consumers than the others, specifically, dark color and any information about the breed. The other visual attributes studied reduced the probability of consumers making a choice, which may be an indication that the participants of this study focused visual attention on the information that they reject as the brown color; it was also proved that the consumers would like to have more information, such as breed, and that this can be more representative of the purchasing decisions of Brazilian consumers.

Information on marbling and different breeds should be highlighted if it is clear to the consumer what the benefit of these attributes is for the final product. It may therefore be beneficial to include this information as explanatory texts on the packaging. Legislation on strategic visual cues can be implemented, so mandatory additional information should be placed on the packaging, or standardization of specific information should be enforced to highlight relevant health attributes.

This study contributes to a better understanding of how visual attention influences consumer decision-making in the context of beef purchases, integrating eye-tracking technology with experimental design to capture non-conscious responses. By identifying which visual cues—such as color, marbling, breed, price, tenderness, and flavor—drive attention and affect choice probability, the findings confirm that visual information is a key determinant in food selection, particularly in fresh and cooked meat contexts.

These insights have practical implications for product packaging design, retail marketing strategies, and food labeling practices. For instance, the prominence of color in driving attention toward fresh beef suggests that enhancing the visual appeal of product displays can increase purchase likelihood. In the case of cooked beef, the findings emphasize the relevance of flavor and tenderness cues, which can be strategically highlighted in visual communication and promotional materials.

For future research and applications, these findings support the expansion of eye-tracking methods to other food categories and retail environments. Moreover, they encourage the integration of psychological metrics and biometric tools in market analysis, allowing a more comprehensive and precise approach to understanding consumer behavior and optimizing visual merchandising.

Author Contributions: All authors contributed to the study's conception and design. B.A.M., E.E.S. and C.J.C.C. conceived the study. All authors acquired and interpreted the data. C.N.A. drafted the manuscript. L.M.d.L. performed the analyses and interpretation of data and statistical analyses. B.A.M. wrote the original draft and reviewed and edited the manuscript. C.N.A., E.E.S., C.J.C.C. and L.M.d.L. performed critical manuscript revisions. E.E.S. and C.J.C.C. supervised the study. All authors have read and agreed to the published version of the manuscript.

Funding: This research was funded by grant 2017/26667-2, São Paulo Research Foundation (FAPESP).

Institutional Review Board Statement: All procedures and protocols used in this study complied with the Institutional Animal Care and Use Committee Guidelines (protocol 2019/22, 11 December 2020) and were approved by the committees of the Luiz de Queiroz College of Agriculture at the University of São Paulo.

Informed Consent Statement: All participants in the tests provided informed consent, in accordance with the form approved by the ethics committee, under CAAE number 39155320.5.0000.5395.

Data Availability Statement: The original contributions presented in this study are included in the article. Further inquiries can be directed to the corresponding author.

Conflicts of Interest: Author Bruna Alves Malheiros was employed by the company Symrise. Author Carolina Naves Aroeira was employed by the company AgroProx. Author Lilian Maluf de Lima was employed by the company Pecege Institute. The remaining authors declare that the research was conducted in the absence of any commercial or financial relationships that could be construed as a potential conflict of interest.

Appendix A

Detailed description of the models for fresh and cooked meat, according to the explanatory variables.

Thus, for the fresh beef model:

$$Y = \beta_0 + \beta_1 X_1 + \beta_2 X_2 + \ldots + \beta_{17} X_{17} + u_t$$

Y is the binary variable where Y = 1 characterizes the consumer who chose the product (right or left) of the stimulus visualized in fresh beef; Y = 0 characterizes the consumer who chose neither the product on the right nor the product on the left of the slide; β_0, β_1, β_2, ...β_{17} are the parameters of the model to be estimated (coefficients) and it refers to the term of stochastic error. The values assumed by X are described below:

X_1 = "FRESH.PRICE", price that the participant would be willing to pay for the beef visualized in the stimulus, assuming values of USD 3.80, USD 5.80, and USD 9.00;

X_2 = "FRESH.BROWNCO", brown-colored beef, assuming value 1 when the beef visualized by the participant has a brown color and value 0 when the beef has dark red or bright red color;

X_3 = "FRESH.FBRIGHTCO", total time (seconds) of the first fixation in the AOI, related to the stimulus of beef with bright red color;

X_4 = "FRESH.FBROWNCO", total time (seconds) of the first fixation in the AOI related to the stimulus of the beef with brown color;

X_5 = "FRESH.TDARKCO", total time (seconds) of fixation in the AOI related to the stimulus of beef with dark red color;

X_6 = "FRESH.TNELLOREB", total time (seconds) of fixation in the AOI related to the stimulus of beef from animals of the Nellore breed;

X_7 = "FRESH.TWITHOUTB", total time (seconds) of fixation in the AOI related to the stimulus of beef without any breed information;

X_8 = "FRESH.TBROWNCO", total time (seconds) of fixation in the AOI related to the stimulus of beef with brown color;

X_9 = "FRESH.VDARKCO", number of times that the individual fixates on the AOI related to the stimulus of beef with dark red color;

X_{10} = "FRESH.VSMALLMAR", number of times that the individual looks at the AOI related to the stimulus of beef with small marbling;

X_{11} = "FRESH.VMODERATEMAR", number of times that the individual looks at the AOI related to the stimulus of beef with moderate marbling;

X_{12} = "FRESH.VABUNDANTMAR", number of times that the individual looks at the AOI related to the stimulus of beef with abundant marbling;

X_{13} = "FRESH.VPRICE380", number of times that the individual looks at the AOI related to the stimulus of beef price of USD 3.80;

X_{14} = "FRESH.VWITHOUTB", number of times that the individual looks at the AOI related to the stimulus of beef without breed information;

X_{15} = "FREQCONS_C", frequency of beef consumption in the week, assuming value 1 when consumption is 4 to 5 times a week and value 0 when otherwise;

X_{16} = "FREQPURCHASE_D", frequency of purchase of beef in the month, assuming value 1 when the frequency of purchase is 6 or more times per month and value 0 when otherwise;

X_{17} = "MED_COMP", average of the scores assigned by the participants during the product choice experiment. These scores[1] refer to information on comparisons of knowledge of the general population with the interviewed consumer.

For the cooked beef model, the values assumed by X are described below:

X_1 = "COOK.PRICE", price that the consumer would be willing to pay for the beef visualized in the stimulus, considering values of USD 3.80, USD 5.80, and USD 9.00;

X_2 = "COOK.FREQCONS_B", frequency of beef consumption in the week, assuming value 1 when consumption is 2 to 3 times a week and value 0 when otherwise;

X_3 = "COOK.FREQPURCHASE_D", frequency of buying beef in the month, assuming value 1 when the purchase is 6 or more times per month and value 0 when otherwise;

X_4 = "COOK.TOUGHMC", tough beef, assuming value 1 when the beef visualized by the consumer is tough and value 0 when otherwise (tender or very tender beef)

X_5 = "COOK.FTOUGH", total time (seconds) of the first fixation in the AOI related to the stimulus of tough beef;

X_6 = "COOK.FWEAKFLAVOUR", total time (seconds) of the first fixation in the AOI related to the stimulus of beef with weak flavor;

X_7 = "COOK.VPRICE580", number of times that the individual visits the product and fixates on the AOI related to the stimulus of beef price of USD 5.80;

X_8 = "COOK.VPRICE900", number of times that the individual visits the product and fixates on the AOI related to the stimulus of beef price of USD 9.00;

X_9 = "COOK.VINTENSEFLAVOUR", number of times that the individual visits the product and fixates on the AOI related to the stimulus of beef with intense flavor.

Appendix B

BREED

A group of animals that share similar physical and genetic characteristics.

Non-Breed	Nellore Breed	Angus Breed
Undefined origin	Originating from India	Originating from Scotland
Undefined hardiness	It is the hardiest breed	It is a less hardy breed
Undefined heat tolerance	Breed with high heat tolerance	Breed with low heat tolerance
Represents 10% of the total cattle herd in Brazil	Represents 80% of the total cattle herd in Brazil.	Represents 10% of the total cattle herd in Brazil.

COLOR

It can be observed on the surface of the meat.

Bright red

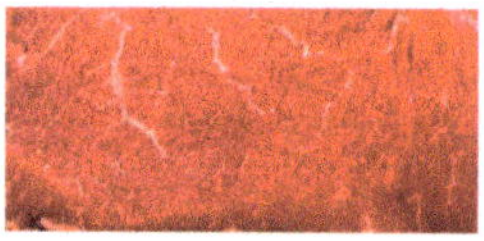

Brown

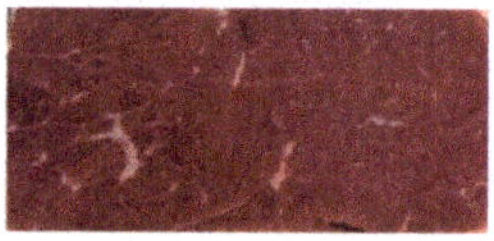

Dark red

MARBLING

It refers to the intramuscular fat within the meat.

Low marbling

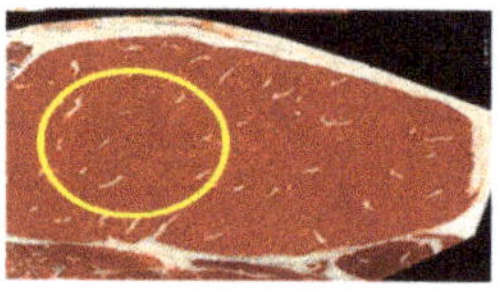

High marbling

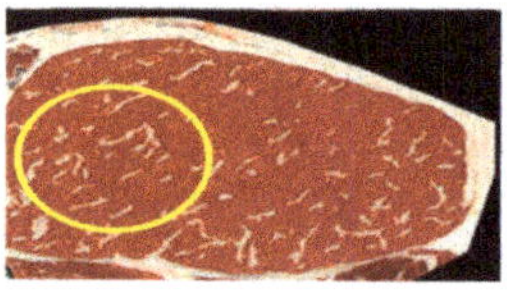

Moderate marbling

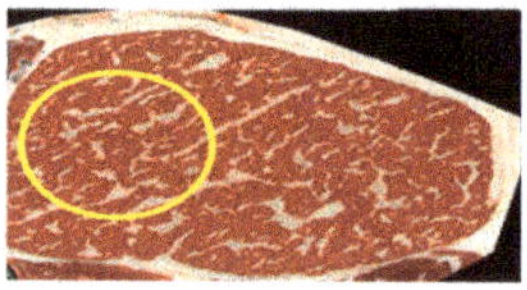

References

1. Barreto, A.M. Eye tracking como método de investigação aplicado às ciências da comunicação. *Rev. Comun.* **2012**, *1*, 168–186.
2. Chae, H.; Lee, Y.; Kim, S. The influence of visual attention on consumer choice in online grocery shopping: An eye-tracking study. *Sustainability* **2019**, *11*, 6347.
3. Font-I-Furnols, M.; Guerrero, L. Consumer preference, behavior and perception about meat and meat products: An overview. *Meat Sci.* **2014**, *98*, 361–371. [CrossRef]
4. Hoffman, D.E.; Solomon, M.B.; Lynch, M.P.; Nath, R. Physical and sensory characteristics of selected enhanced beef muscles. *Meat Sci.* **2005**, *71*, 105–110.
5. Troy, D.J.; Kerry, J.P. Consumer perception and the role of science in the meat industry. *Meat Sci.* **2010**, *86*, 214–226. [CrossRef]
6. Armel, C.; Beaumel, A.; Rangel, A. Biasing simple choices by manipulating relative visual attention. *Judgm. Decis. Mak.* **2008**, *3*, 396–403. [CrossRef]
7. Milosavljevic, M.; Navalpakkam, V.; Koch, C.; Rangel, A. Relative visual saliency differences induce sizable bias in consumer choice. *J. Consum. Psychol.* **2012**, *22*, 67–74. [CrossRef]

8. Malheiros, B.A.; Spers, E.E.; da Silva, H.M.R.; Contreras-Castillo, C.J. Southeast Brazilian Consumers' Involvement and Willingness to Pay for Quality Cues in Fresh and Cooked Beef. *J. Food Prod. Mark.* **2022**, *28*, 276–293. [CrossRef]
9. Rebollar, R.; Lidón, I.; Martín, J.; Puebla, M. The identification of viewing patterns of chocolate snack packages using eye-tracking techniques. *Food Qual. Prefer.* **2015**, *39*, 251–258. [CrossRef]
10. Fenko, A.; Schifferstein, H.N.J.; Hekkert, P. Shifts in sensory dominance between various stages of user–product interactions. *Appl. Ergon.* **2010**, *41*, 34–40. [CrossRef]
11. Siegrist, M.; Leins-Hess, R.; Keller, C. Which front-of-pack nutrition label is the most efficient one? The results of an eye-tracker study. *Food Qual. Prefer.* **2015**, *39*, 183–190. [CrossRef]
12. Duarte, R.B.; Silveira, D.S.; Brito, V.A.; Lopes, C.S. A systematic literature review on the usage of eye-tracking in understanding process models. *Bus. Process Manag. J.* **2021**, *27*, 346–367. [CrossRef]
13. Bialkova, S.; Grunert, K.G.; Trijp, H.V. From desktop to supermarket shelf: Eye-tracking exploration on consumer attention and choice. *Food Qual. Prefer.* **2020**, *81*, 103839. [CrossRef]
14. Hernandez, M.D.; Wang, Y.; Sheng, H.; Kalliny, M.; Minor, M. Escaping the corner of death? An eye-tracking study of reading direction influence on attention and memory. *J. Consum. Mark.* **2017**, *34*, 1–10. [CrossRef]
15. Duchowski, A.T. *Eye Tracking Methodology: Theory and Practice*; Springer: London, UK, 2007.
16. Mundel, J.; Huddleston, P.; Behe, B.; Sage, L.; Latona, C. An eye tracking study of minimally branded products: Hedonism and branding as predictors of purchase intentions. *J. Prod. Brand Manag.* **2018**, *27*, 146–157. [CrossRef]
17. Pentus, K.; Ploom, K.; Mehine, T.; Koiv, M.; Tempel, A.; Kuusik, R. Mobile and stationary eye tracking comparison—Package design and in-store results. *J. Consum. Mark.* **2020**, *37*, 259–269. [CrossRef]
18. Leon, F.A.D.; Spers, E.E.; de Lima, L.M. Self-esteem and visual attention in relation to congruent and non-congruent images: A study of the choice of organic and transgenic products using eye tracking. *Food Qual. Prefer.* **2020**, *84*, 103938. [CrossRef]
19. Boardman, R.; Mccormick, H. Attention and behaviour on fashion retail websites: An eye-tracking study. *Inf. Technol. People* **2022**, *35*, 2219–2240. [CrossRef]
20. Manippa, V.; Van der Laan, L.N.; Brancucci, A.; Smeets, P.A.M. Health body priming and food choice: An eye tracking study. *Food Qual. Prefer.* **2019**, *72*, 116–125. [CrossRef]
21. Bialkova, S.; Grunert, K.G.; Trijp, H.V. Standing out in the crowd: The effect of information clutter on consumer attention for front-of-pack nutrition labels. *Food Policy* **2013**, *41*, 65–74. [CrossRef]
22. Navalpakkam, V.; Kumar, R.; Li, L.; Sivakumar, D. Attention and selection in online choice tasks. In Proceedings of the User Modeling, Adaptation, and Personalization: 20th International Conference, UMAP 2012, Montreal, QC, Canada, 16–20 July 2012; pp. 200–211.
23. Peschel, A.O.; Orquin, J.L.; Loose, S.M. Increasing consumers' attention capture and food choice through bottom-up effects. *Appetite* **2019**, *132*, 1–7. [CrossRef] [PubMed]
24. Bialkova, S.; Trijp, H.V. An efficient methodology for assessing attention to and effect of nutrition information front of pack. *Food Qual. Prefer.* **2011**, *22*, 592–601. [CrossRef]
25. Scott, G.G.; Hand, C.J. Motivation determines Facebook viewing strategy: An eye movement analysis. *Comput. Hum. Behav.* **2016**, *56*, 267–280. [CrossRef]
26. Wedel, M.; Pieters, R. A review of eye-tracking research in marketing. *Rev. Mark. Res.* **2008**, *4*, 123–147.
27. Ares, G.; Giménez, A.; Bruzzone, F.; Vidal, L.; Antúnez, L.; Maiche, A. Consumer Visual Processing of Food Labels: Results from an Eye-Tracking Study. *J. Sens. Stud.* **2013**, *28*, 138–153. [CrossRef]
28. Banović, M.; Polymeros, C.; Grunert, K.G.; Rosa, P.J.; Gamito, P. The effect of fat content on visual attention and choice of red meat and differences across gender. *Food Qual. Prefer.* **2016**, *52*, 42–51. [CrossRef]
29. Van der Laan, L.N.; Papies, E.K.; Hooge, I.T.C.; Smeets, P.A.M. Goal-Directed Visual Attention Drives Health Goal Priming: An Eye-Tracking Experiment. *Health Psychol.* **2017**, *36*, 82–90. [CrossRef] [PubMed]
30. Rayner, K. Eye movements in reading and information processing: 20 years of research. *Psychol. Bull.* **1998**, *124*, 372–422. [CrossRef]
31. Rayner, K. Eye movements and attention in reading, scene perception, and visual search. *Q. J. Exp. Psychol.* **2009**, *62*, 1457–1506. [CrossRef]
32. Muñoz-Leiva, F.; Faísca, L.M.; Ramos, C.M.Q.; Correia, M.B.; Sousa, C.M.R.; Bouhachi, M. The influence of banner position and user experience on recall. The mediating role of visual attention. *Span. J. Mark. ESIC* **2021**, *25*, 85–114. [CrossRef]
33. Colourado, D.A.; Taba, M.C.; Parra, H.C. Las etiquetas nutricionales: Una mirada desde el consumidor. *Rev. En-Contexto* **2015**, *3*, 121–140.
34. Sousa, H.D. O estudo do neuromarketing como ferramenta de percepção da reação dos consumidores. *Rev. Tecnol. Fatec Am.* **2016**, *3*, 17.
35. Huang, Y.; Kuo, F. An eye-tracking investigation of internet consumers' decision deliberateness. *Internet Res.* **2011**, *21*, 541–561. [CrossRef]

36. Silva, A.R.C.S.; Braga, L.V.M.; Anastácio, L.R. Coexistence of high content of critical nutrients and claims in food products targeted at Brazilian children. *Rev. Paul. Pediatr.* **2023**, *41*, 2021355. [CrossRef]
37. Jacintho, C.L.A.B.; Jardim, P.C.B.V.; Sousa, A.L.L.; Jardim, T.S.V.; Souza, W.K.S.B. Brazilian food labeling: A new proposal and its impact on consumer understanding. *Food Sci. Technol.* **2020**, *40*, 222–229. [CrossRef]
38. Grunert, K.G. Future and consumer lifestyles with regard to meat consumption. *Meat Sci.* **2006**, *74*, 149–160. [CrossRef]
39. Wadhera, D.; Capaldi-Phillips, E.D. A review of visual cues associated with food on food acceptance and consumption. *Eat. Behav.* **2014**, *15*, 132–143. [CrossRef]
40. Imram, N. The role of visual cues in consumer perception and acceptance of a food product. *Nutr. Food Sci.* **1999**, *99*, 224–230. [CrossRef]
41. Aboah, J.; Lees, N. Consumers use of quality cues for meat purchase: Research trends and future pathways. *Meat Sci.* **2020**, *166*, 108142. [CrossRef]
42. Ballco, P.; Gracia, A. An extended approach combining sensory and real choice experiments to examine new product attributes. *Food Qual. Prefer.* **2020**, *80*, 103830. [CrossRef]
43. Morquecho-Campos, P.; Graaf, K.; Boesveldt, S. Smelling our appetite? The influence of food odors on congruent appetite, food preferences and intake. *Food Qual. Prefer.* **2020**, *85*, 103959. [CrossRef]
44. Strydom, P.; Burrow, H.; Polkinghorne, R.; Thompson, J. Do demographic and beef eating preferences impact on South African consumers' willingness to pay (WTP) for graded beef? *Meat Sci.* **2019**, *150*, 122–130. [CrossRef]
45. Wijk, R.A.; Kaneko, D.; Dijksterhuis, G.B.; Zoggel, M.V.; Schiona, I.; Visalli, M.; Zandastra, E.H. Food perception and emotion measured over time in-lab and in-home. *Food Qual. Prefer.* **2019**, *75*, 170–178. [CrossRef]
46. Fernández, J.; Melo, O.; Larraín, R.; Fernández, M. Valuation of observable attributes in differentiated beef products in Chile using the hedonic price method. *Meat Sci.* **2019**, *158*, 107881. [CrossRef]
47. Sonoda, Y.; Oishi, K.; Chomei, Y.; Hirooka, H. How do human values influence the beef preferences of consumer segments regarding animal welfare and environmentally friendly production? *Meat Sci.* **2018**, *146*, 75–86. [CrossRef] [PubMed]
48. Borgogno, M.; Favotto, S.; Corazzin, M.; Cardello, A.V.; Piasentier, E. The role of product familiarity and consumer involvement on liking and perceptions of fresh meat. *Meat Sci.* **2015**, *44*, 139–147. [CrossRef]
49. Meyerding, S.G.H.; Gentz, M.; Altmann, B.; Meier-Dinkel, L. Beef quality labels: A combination of sensory acceptance test, stated willingness to pay, and choice-based conjoint analysis. *Appetite* **2018**, *127*, 324–333. [CrossRef]
50. Holman, B.; Mao, Y.; Coombs, C.; Van de Ven, R.J.; Hopkins, D.J. Relationship between colourimetric (instrumental) evaluation and consumer-defined beef colour acceptability. *Meat Sci.* **2016**, *121*, 104–106. [CrossRef]
51. Jackman, P.; Sun, D.W.; Allen, P.; Bradon, K.; White, A.M. Correlation of consumer assessment of longissimus dorsi beef palatability with image colour, marbling and surface texture features. *Meat Sci.* **2010**, *84*, 564–568. [CrossRef]
52. Chonpracha, P.; Ardoin, R.; Gao, Y.; Waimaleongora-ek, P.; Tuuri, G.; Prinyawiwatkul, W. Effects of Intrinsic and Extrinsic Visual Cues on Consumer Emotion and Purchase Intent: A Case of Ready-to-Eat Salad. *Foods* **2020**, *9*, 396. [CrossRef]
53. Barcellos, M.D.; Aguiar, L.K.; Ferreira, G.C.; Vieira, L.M. Beef consumption in Brazil: Consumer behavior and environmental considerations. *Meat Sci.* **2019**, *156*, 77–84.
54. Brasil. Instrução normativa nº 22, de 24 de novembro de 2005. In *Aprova o Regulamento Técnico Para Rotulagem de Produto de Origem Animal Embalado*; Diário Oficial da União: Brasília, Brasil, 2005; seção 1.
55. Chow, I.H.S.; Lau, V.P.; Lo, T.W.C.; Sha, Z.; Yun, H. Service quality in restaurant operations in China: Decision- and experiential-oriented perspectives. *Int. J. Hosp. Manag.* **2007**, *26*, 698–710. [CrossRef]
56. Vermeir, I.; Roose, G. Visual design cues impacting food choice: A review and future research agenda. *Foods* **2020**, *9*, 1495. [CrossRef]
57. Hallez, L.; Qutteina, Y.; Raedschelders, M.; Boen, F.; Smits, T. That's My Cue to Eat: A Systematic Review of the Persuasiveness of Front-of-Pack Cues on Food Packages for Children vs. Adults. *Nutrients* **2020**, *12*, 1062. [CrossRef] [PubMed]
58. Coucke, N.; Vermeir, I.; Slabbinck, H.; Kerckhove, A.V. Show Me More! The Influence of Visibility on Sustainable Food Choices. *Foods* **2019**, *8*, 186. [CrossRef]
59. Kunz, S.; Haasova, S.; Florack, A. Fifty shades of food: The influence of package colour saturation on health and taste in consumer judgments. *Psychol. Mark.* **2019**, *37*, 900–912. [CrossRef]
60. Kuster, I.; Vila, N.; Sarabia, F. Food packaging cues as vehicles of healthy information: Visions of millennials (early adults and adolescents). *Food Res. Int.* **2019**, *119*, 170–176. [CrossRef]
61. Doolan, K.J.; Breslin, G.; Hanna, D.; Gallagher, A.M. Attentional bias to food-related visual cues: Is there a role in obesity? *Proc. Nutr. Soc.* **2014**, *74*, 37–45. [CrossRef]
62. Scheibehenne, B.; Todd, P.M.; Wansink, B. Dining in the dark. The importance of visual cues for food consumption and satiety. *Appetite* **2010**, *55*, 710–713. [CrossRef]
63. Clement, J. Visual influence on in-store buying decisions: An eye-track experiment on the visual influence of packaging design. *J. Mark. Manag.* **2007**, *23*, 917–928. [CrossRef]

64. Clement, J.; Kristensen, T.; Grønhaug, K. Understanding consumers' in-store visual perception: The influence of package design features on visual attention. *J. Retail. Consum. Serv.* **2013**, *20*, 234–239. [CrossRef]
65. Pieters, R.; Wedel, M. Attention capture and transfer in advertising: Brand, pictorial, and text-size effects. *J. Mark.* **2004**, *68*, 36–50. [CrossRef]
66. Mormann, M.; Fehr, E.; Hare, T.A. The impact of visual attention on value computation in goal-directed choice. *Front. Neurosci.* **2012**, *6*, 146.
67. Van Loo, E.J.; Caputo, V.; Nayga, R.M.; Meullenet, J.F.; Ricke, S.C. Consumers' willingness to pay for organic chicken breast: Evidence from choice experiment using eye tracking. *Food Qual. Prefer.* **2015**, *41*, 156–165.
68. CONEP—Conselho Nacional de Saúde. Resolução nº 510. 2016. Available online: http://conselho.saude.gov.br/resolucoes/2016/Reso510.pdf (accessed on 23 April 2025).
69. Kytö, E.; Järveläinen, A.; Mustonen, S. Hedonic and emotional responses after blind tasting are poor predictors of purchase behavior. *Food Qual. Prefer.* **2018**, *70*, 49–56. [CrossRef]
70. Murias, M.; Major, S.; Davlantis, K.; Franz, L.; Harris, A.; Rardin, B.; Sabatos-DeVito, M.; Dawson, G. Validation of eye-tracking measures of social attention as a potential biomarker for autism clinical trials. *Autism Res.* **2018**, *11*, 166–174. [CrossRef]
71. Stock, J.H.; Watson, M.W. Regressão com uma variável dependente binária. In *Econometria*; Addison Wesley: São Paulo, Brasil, 2004; pp. 202–218.
72. Hosmer, D.W.; Lemeshow, S.; Sturdivant, R.X. *Applied Logistic Regression*, 3rd ed.; John Wiley & Sons: Hoboken, NJ, USA, 2013.
73. Wooldridge, J. Limited Dependent Variable Models and Sample Selection Corrections. In *Introdutory Econometrics: A Modern Approach*, 5th ed.; South-Western; Cengage Learning: Mason, OH, USA, 2013; pp. 604–609.
74. Hair-Junior, J.F.; Black, W.C.; Babin, B.J.; Anderson, R.E.; Tatham, R.L. *Analise Multivariada de Dados*; Bookman: Porto Alegre, Brazil, 2009; 688p.
75. Pino, F. Modelos de decisão binários: Uma revisão. *Rev. Econ. Agric.* **2007**, *54*, 44–57.
76. Kubberød, E.; Ueland, Ø.; Rødbotten, M.; Westad, F.; Risvik, E. Gender specific preferences and attitudes towards meat. *Food Qual. Prefer.* **2002**, *13*, 285–294. [CrossRef]
77. Mottin, C.; Eiras, C.E.; Chefer, D.M.; Barcelos, V.D.C.; Ramos, T.R.; Prado, I.N.D. Influencing factors of consumer willingness to buy cattle meat: An analysis of survey data from three Brazilian cities. *Acta Sci. Anim. Sci.* **2019**, *41*, e43871. [CrossRef]
78. Wu, S.; Han, J.; Liang, R.; Dong, P.; Zhu, L.; Hopkins, D.L.; Zhang, Y.; Luo, X. Investigation of muscle-specific beef colour stability at different ultimate pHs. Asian-Australas. *J. Anim. Sci.* **2020**, *33*, 1999–2007.
79. Yang, X.; Zhang, Y.; Luo, X.; Zhang, Y.; Zhu, L.; Xu, B.; Hopkins, D.L.; Liang, R. Influence of oxygen concentration on the fresh and internal cooked colour of modified atmosphere packaged dark-cutting beef stored under chilled and superchilled conditions. *Meat Sci.* **2022**, *188*, 108773. [CrossRef]
80. Lombard, W.A.; Van Zyl, J.H.; Beeldersc, T.R. Eye-tracking consumers' awareness of beef brands. *Agrekon* **2020**, *59*, 156–168. [CrossRef]
81. Graham, D.J.; Orquin, J.L.; Visschers, V.H.M. Eye tracking and nutrition label use: A review of the literature and recommendations for label enhancement. *Food Policy* **2012**, *37*, 378–382. [CrossRef]
82. Nassu, R.T.; Uttaro, B.; Aalhus, J.L.; Zawadski, S.; Juárez, M.; Dugan, M.E.R. Type of packaging affects the colour stability of vitamin E enriched beef. *Food Chem.* **2012**, *135*, 1868–1872. [CrossRef]
83. Venturini, A.C.; Contreras-Castillo, C.J.; Faria, J.A.F.; Gallo, C.R.; Silva, T.Z.; Shirahigue, L.D. Microbiological, colour and sensory properties of fresh beef steaks in low carbon monoxide concentration. *Packag. Technol. Sci.* **2010**, *23*, 327–338. [CrossRef]
84. Velasco, C.; Michel, C.; Youssef, J.; Gamez, X.; Cheok, A.D.; Spence, C. Colour–taste correspondences: Designing food experiences to meet expectations or to surprise. *Int. J. Food Des.* **2016**, *1*, 83–102. [CrossRef] [PubMed]
85. Lavieri, S.; Williams, K. Effects of packaging systems and fat concentrations on microbiology, sensory and physical properties of ground beef stored at 4 ± 1 °C for 25 days. *Meat Sci.* **2014**, *97*, 534–541. [CrossRef]
86. Bialkova, S.; Sasse, L.; Fenko, A. The role of nutrition labels and advertising claims in altering consumers' evaluation and choice. *Appetite* **2016**, *96*, 38–46. [CrossRef]
87. Ferraz, J.B.S.; Felício, P.E. Production systems—An example from Brazil. *Meat Sci.* **2010**, *84*, 238–243. [CrossRef]
88. Ardeshiri, A.; Rose, M.J. How Australian consumers value intrinsic and extrinsic attributes of beef products. *Food Qual. Prefer.* **2018**, *65*, 146–163. [CrossRef]
89. Bressan, M.C.; Rodrigues, E.C.; Rossato, L.V.; Ramos, E.M.; Gama, L.T. Physicochemical properties of meat from Bos taurus and Bos indicus. *Rev. Bras. Zootec.* **2011**, *40*, 1250–1259. [CrossRef]
90. Shackelford, S.D.; Koohmaraie, M.; Miller, M.F.; Crouse, J.D.; Reagan, J.O. An evaluation of tenderness of the longissimus muscle of Angus by Hereford versus Brahman crossbred heifers. *J. Anim. Sci.* **1991**, *69*, 171–177. [CrossRef] [PubMed]
91. Wheeler, T.L.; Cundiff, L.V.; Koch, R.M. Effect of marbling degree on beef palatability in Bos taurus and Bos indicus cattle. *J. Anim. Sci.* **1994**, *72*, 3145–3151. [CrossRef] [PubMed]

92. Banović, M.; Fontes, A.; Barreira, M.M.; Grunert, K.G. Impact of product familiarity on beef quality perception. *Agribusiness* **2012**, *28*, 157–172. [CrossRef]
93. Hess, R.; Visschers, V.; Siegrist, M. The role of health-related, motivational and sociodemographic aspects in predicting food label use: A comprehensive study. *Public Health Nutr.* **2012**, *15*, 407–414. [CrossRef]
94. Grunert, K.G.; Will, J.M.; Fernández-Celemín, L. Nutrition knowledge and use and understanding of nutrition information on food labels among consumers in the UK. *Appetite* **2010**, *55*, 177–189. [CrossRef]

MDPI AG
Grosspeteranlage 5
4052 Basel
Switzerland
Tel.: +41 61 683 77 34

Applied Sciences Editorial Office
E-mail: applsci@mdpi.com
www.mdpi.com/journal/applsci

www.ingramcontent.com/pod-product-compliance
Lightning Source LLC
Chambersburg PA
CBHW061759210926
51850CB00006B/599